POWER TECHNOLOGY

POWER TECHNOLOGY

FOURTH EDITION

George E. Stephenson

Delmar Publishers Inc.®

Delmar Staff
Administrative Editor: Mark W. Huth
Developmental Editor: Marjorie A. Bruce
Production Editor: Carol Micheli

For information, address Delmar Publishers Inc.
2 Computer Drive West
Box 15015
Albany, New York 12212-5015

Printed in the United States of America
Published simultaneously in Canada
by Nelson Canada,
A division of International Thomson Limited

10 9 8 7 6 5 4 3

Library of Congress Cataloging in Publication Data
Stephenson, George E.
Power technology.

Includes index.
1. Power (Mechanics) 2. Power resources. I. Title.
TJ163.9.S73 1986 621.4 85-4367
ISBN 0-8273-2446-4
ISBN 0-8273-2447-2 (instructor's guide)

CONTENTS

SECTION 9: POWER TECHNOLOGY AND THE FUTURE

PREFACE

The second half of the twentieth century has brought to public awareness one topic formerly taken for granted—energy. The abundant energy that supported the Industrial Revolution and brought the United States to its position as one of the leading industrial nations in the world is now recognized as a fragile resource. The fossil fuels that spurred the American economy for so many years are not renewable and are being depleted. At the same time, our energy needs have increased and will continue to do so. In recent years aggressive research has been undertaken to identify new sources of energy and to develop the conversion technology to make this energy usable on a daily basis. The challenge for the future is to continue research into alternative energy sources—to develop new sources and to increase the efficiency of sources now being used. An important goal is to make their costs comparable to or lower than the costs of our dwindling fossil fuels. Another major goal is to develop these resources while keeping any effects on the environment within acceptable limits.

The fourth edition of POWER TECHNOLOGY addresses the issues of energy sources and applications, their effect on the environment, and future development. The text presents major concepts underlying each of the traditional and alternative sources of energy. The text is not designed to present each concept in great detail, but will help make the student aware of the following: (1) ways in which each concept has been developed, (2) the major conversion systems used to provide usable power, and (3) problems to be solved to increase usefulness and efficiency. The content has been updated to keep pace with development in the 1980s. As in previous editions, safety in the use of energy and power is stressed. Features of the fourth edition include:

- Addition of metric units to help students gain the ability to work in both inch and metric systems

- Revised section on small gasoline engines to reflect systems and components on late model engines
- Updated unit on automobile engines to reflect current technology
- Many new photographs and illustrations throughout the text
- Updated units on other internal combustion engines (diesel, jet, rocket and gas turbine)
- Additional information on earth orbiting satellites and the space shuttle
- Expanded section on alternative energy sources includes units on solar energy, wind energy, hydroelectric energy, energy from biomass sources, and geothermal, wood and tidal energy
- New section covering the fossil fuels (coal, petroleum and natural gas), including formation, characteristics, sources, mining or drilling, processing, uses, environmental problems, supply and future
- New unit describing the basic principles of computers, a short history of the evolution of computers, applications of computers in manufacturing and the use of computers in modern automobiles
- Addition of an extensive glossary
- Expanded appendix to include sections on troubleshooting, tune-up and reconditioning of engines; horsepower and buying considerations for engines; and various specifications, disassembly and reassembly procedures and diagnosis charts for selected engines

An Instructor's Guide is available, giving answers to the review questions for each unit of the text. A Student Manual provides activities for each unit to give students the opportunity to investigate energy and power in greater depth. Many laboratory-type exercises are provided to further enhance student understanding and to allow the student to develop valuable skills.

ACKNOWLEDGMENTS

A number of teachers evaluated the author's changes for this edition. Their criticisms and recommendations provided valuable guidance. Appreciation is expressed to the following teachers for their assistance:

Robert Belle
North Syracuse Junior High School
North Syracuse, NY 13212

Samual J. Lapp
Wissahickon High School
Ambler, PA 19002

John Toth
North Syracuse Junior High School
North Syracuse, NY 13212

Greg Stamford
Appleton West High School
Appleton, WI 54914

Dr. Stephen Fardo
Eastern Kentucky University
Richmond, KY 40475

Richard Janos
Appleton East High School
Appleton, WI 54915

Michael Pierno
Seaholm High School
Birmingham, MI 48009

Paul N. Eshoo
New Britain Senior High School
New Britain, CT 06050

Roger Anthony
Everett High School
Everett, WA 98201

Introduction: The Struggle to Harness Energy

OBJECTIVES

After completing this unit, the student should be able to:

- **Discuss how the sun affects other sources of energy.**
- **List sources of energy.**
- **Trace the history of the steam engine and how it affected society.**
- **Explain the difference between a steam engine and a steam turbine.**
- **Trace the development of the internal combustion engine and discuss its place in society.**

Prehistoric civilizations used energy. They observed its existence in the force of the wind, the heat of the sun, in flowing water, fire, and in the wild animals that preyed on them. These forces were sometimes a help, but very often, a real danger. Heat, cold, hurricanes, floods, and predatory animals could threaten existence.

When people began to think about how they could use energy to work for them, the mechanics of power were born. Slowly through the centuries, machines and devices were developed that harnessed this energy to do work.

Power technology is a study of the many energy-converting machines and devices which have been developed as technical knowledge has expanded. Nature provided the energy sources, and human ingenuity has provided the machines, a combination that is responsible for a high degree of civilization.

For the general purposes of this introduction, the term *power*, in its general sense, will be used interchangeably with *force* and *energy*. Later, these and other terms will be defined in their true, technical descriptions.

THE SUN — MAJOR SOURCE OF ENERGY

Without the sun, no plants or animals could have lived, so it is easy to understand that almost all sources of energy—with the exception of nuclear energy—can be traced to the sun. Coal was formed through the ages as plant life died and accumulated in the swampy areas of the earth. This plant life grew because of the sun's energy. When coal is burned, this energy, which has been stored for perhaps millions of years, is released.

The story of petroleum is similar. It is believed that petroleum was formed where large amounts of plants and animal life accumulated. When petroleum fuels are burned they release stored energy.

Wood was probably the first known source of energy—used for many years as the main form of heat, light, and cooking. It became the forerunner of charcoal which possesses energy and releases its energy when it is burned.

Wind and water energy are also traceable to the sun. Wind is the circulation of air caused

Fig. 1

by the sun's heat. Water energy is harnessed from a flowing stream, which is filled by rainfall. The rain is caused by the sun evaporating the earth's water. In a short time, the water vapor condenses and rain falls, filling the streams.

EARLY ATTEMPTS TO CONTROL ENERGY

One of the first attempts at controlling energy, as just mentioned, was to use fire for warmth and cooking. Another early step forward was the domestication of wild animals. Taming them and using them as beasts of burden provided new power.

Of course, the beast of burden's load was lightened with the evolution of the wheel, which enabled the animal to move a larger load with less effort. Again, it further increased the rate of doing work. Animals were used for land transportation for centuries; in some parts of the world animals are still the main form of transportation.

For most of the world it has only been within the last century that the beast has been taken out of harness and replaced by the en-

Fig. 2

gine. The truck, tractor, and automobile take the pace of the donkey, ox, and horse.

When the first person discovered the power of the wind, the horizons expanded even more. For centuries the exploration and conquest of the world was done with the aid of the sailing ship. Ancient traders and merchants sent their wares around the world in sailing ships. The first colonists sailed to America using wind energy.

The power of the wind was not limited to the water, of course. Vertical-shaft windmills

Fig. 3

Fig. 4

were used to grind grain in Persia as early as 200 BC. Later, windmills fitted with sails were used in the Mediterranean regions. Windmills of the Netherlands that pumped water and ground grain date back to the eleventh century AD.

Water wheels date back many centuries and have been used for grinding grain and powering the machines of early manufacturing. In the case of both windmills and water wheels the geography of any location determined whether or not the energy source could be used. Only certain locations were suitable for these energy-converting devices; if there were no river or if the flow was not correct the location could not be used. Early commerce developed around the prime locations where the wind was steady, or in the case of water wheels, where the flow was proper.

Fig. 5

BEGINNING OF THE MECHANICAL AGE

The Development of Steam Power

In 1765 James Watt of Scotland produced a successful steam engine. Although others had worked on steam engines earlier, Watt's engine was much more efficient. It consisted of three parts: boiler, cylinder and piston, and condenser. When steam was let into the bottom of the cylinder, the piston was forced upward. The steam was then shut off, and the condenser opened. The condenser turned the steam back into water. When this happened, a vacuum was created, which pulled the piston down. The repetition of this cycle produced energy. Watt's engine was first used to power water pumps in the coal mines of England.

Watt's steam engine became an extremely efficient and refined machine through the years and was the key to rapid industrial growth. Because steam engines could be built anywhere, the factory was no longer tied to the river and water power. The steam engine provided the power for the newly developing factories of the era. Workers and craftspeople went from small shops to the more efficient factories. Steam-powered machinery increased the output of workers. England attempted to monopolize the use of steam engines, but in 1789 Samuel Slater came to America with memorized plans. By 1807, fifteen steam-powered cotton mills were operating successfully in America.

On August 18, 1807, Robert Fulton boarded his new ship, the *Clermont* and made the first successful steamship voyage. Fulton

Fig. 6

traveled from New York to Albany in a record time of 32 hours.

Steam engines were soon used to power railroads in the United States. Although Peter Cooper's *Tom Thumb* lost to a horse-drawn train in a race of 1830 because of mechanical difficulties, the steam engine *DeWitt Clinton* followed by making a successful run from Albany, New York to nearby Schenectady.

A refinement of the steam engine, the turbine, was developed toward the end of the nineteenth century. In this engine, a jet of steam directed against the turbine blade rotated the blade, producing power in much the same way as a windmill. The steam turbine was used mainly for powering large ships and producing electrical power; it is still widely used.

The Development of the Internal Combustion Engine

Steam power was dominant as a power source until the development of the internal combustion engine. The internal combustion engine has taken over many steam engine jobs and has found countless new jobs on its own. The smaller internal combustion engine was ideally suited for land vehicles and led to the development of automobiles, trucks, and tractors.

One of the earliest attempts to produce an internal combustion engine was made by Christian Huygens late in the seventeenth century. In this engine, a piston was forced down a cylinder by the explosion of gunpowder, but gunpowder presented problems and was not the answer. From the late 1700s

Fig. 7

Fig. 8

Fig. 9

to the 1800s others worked on internal combustion engines and experienced some mechanical success. However, their engines were large and not commercially successful.

In 1860, J. J. E. Lenoir developed a three horsepower (2.2 kilowatt) engine that burned coal gas. In several years, over fourteen hundred of his engines were in operation in France and England powering water pumps and printing presses.

In 1877 Nikolaus A. Otto patented his silent gas engine. This engine combined Otto's inventive theories and the work of others into a precision piece of machinery. This successful engine firmly established the four-stroke cycle principle on which the majority of today's internal combustion engines operate. Although the original invention was successful, it was large and awkward by today's standards, weighing about 1110 pounds per horsepower (670 kilograms per kilowatt).

By 1886 Gottleib Daimler had developed a four-stroke cycle engine that operated at a much higher speed, 800 revolutions per minute, and weighed only 88 pounds per horsepower (50 kilograms per kilowatt). During the late 1800s and early 1900s, continued improvements in fuels and ignition systems led to the successful application of engines to automobiles. Today, engines are manufactured that weigh as little as one pound per horsepower (0.6 kilograms per kilowatt).

Small horsepower internal combustion engines have also earned their place in advancing civilization. Although they have been helping us do our work for more than sixty-five years, the tremendous popularity of small engines around the home is a rather recent phenomenon. Many people can remember when they first saw such now common products as chain saws, rotary lawn mowers, snowmobiles, and even weed eaters. In a few years, the small engine and its applications have become part of our daily life. It is estimated that there are more than 70 million small engines in use today, including approximately 37 million power mowers, 7 million outboard engines, and 1 million snowmobiles.

Most of us use engines every day. Sometimes they are directly controlled by us, for example the automobile, chain saw, or outboard motor. At other times they are used indirectly. Television may be viewed using electricity produced by a steam turbine. The television program may come via a satellite which was put into orbit with a rocket engine.

Our lives are very much involved with engines, which are prime movers, and the many energy sources that surround us. Energy sources are of special importance today because of fuel shortages and high prices. It is to everyone's advantage to learn about the broad field of power technology.

REVIEW QUESTIONS

1. What is the original source of nearly all energy? Explain.
2. What sources of energy were used before the steam engine?
3. In what ways did the development of the steam engine affect society?
4. Who is credited with producing the first successful steam engine?
5. When did steam power come to the United States?
6. What is the basic difference between a steam engine and a steam turbine?
7. How did the internal combustion engine affect the use of steam engines?
8. When did automobiles first appear?
9. List as many applications of small gasoline engines as possible.

SECTION 1
INTRODUCTION

Unit 1 Work, Energy, and Power

OBJECTIVES

After completing this unit, the student should be able to:

- **Define the terms work, energy, power and foot-pounds.**
- **State and use the formulas for work, efficiency, power and horsepower.**
- **State the difference between kinetic and potential energy.**
- **State the law of the conservation of energy.**
- **Name the forms that energy can take.**
- **Explain how friction affects the efficiency of machines.**

Internal combustion engines are the workhorses of our industrial society. These engines are often taken for granted even though they have been in existence for less than 200 years. Before the era of engines, energy was derived mainly from the wind, water power, and draft animals.

James Watt developed the first practical steam engine in 1765. The engine burned fuel outside the engine in order to produce the steam needed to run the engine. These first steam engines were called external combustion engines. They were large and heavy and suited for use at factory sites, in coal mines, on ships and locomotives. The development of external combustion engines paved the way for the modern internal combustion engine which burns fuel inside the engine itself.

Early pioneers in internal combustion engines, such as Christian Huygens and J.J.E. Lenoir, developed workable engines. By 1877, Nikolas A. Otto had built a successful internal combustion engine. It was large and heavy by today's standards but the principles on which it operated are still used today. Other pioneers such as Gottleib Daimler soon developed lighter engines suitable for use on trucks and automobiles.

Continued improvements in engine design have made the lightweight small gasoline engine a reality for almost every industry and household in the country. More than 70 million small engines are in use today powering lawn mowers, chain saws, outboard motors and many other machines that are used for work and recreation.

A basic understanding of small gasoline engines can benefit almost every person. Before starting a detailed study of the elements of small gasoline engines, it is necessary to review some of the terms which are basic to an understanding of mechanical power. What is work? What is energy? What is power? Are these terms all the same and if not how do they differ? It is the purpose of this unit to define and explain the meaning of these terms.

WEIGHT, MASS, AND FORCE

Throughout this textbook, metric measurements are used along with the U.S. Customary system. To eliminate any problems when the metric system is referred to, the proper terminology must be used. In everyday language, the word *weight* can mean either the mass of an object or the force of gravity acting on it. However, in technical and scientific usage, a clear distinction is made between the mass, which is measured in grams; and the force of gravity, measured in newtons. For example:

> An object may have a weight of 2.2 pounds; in the metric system it has a mass of 1 kilogram. On earth, the force of gravity on this 1 kilogram mass is about 9.8 newtons. But on the moon, the force of gravity on this 1 kilogram mass is only 1.6 newtons.

WORK

Work is a scientific term as well as an everyday term. To many people, work means engaging in an occupation. One person may work hard in an office while another person may work hard carrying bricks on a construction job. Both individuals may come home tired but, scientifically, only the one carrying bricks has done work.

Work, in the scientific sense, involves the moving of things by applying force. *Work* is defined as applying a force through a distance. In other words, work is motion caused by applying a force.

Measurement of Work

The common unit for measuring work is the foot-pound (joule). Raising one pound one foot is one foot-pound of work. This can be translated; the force of one newton moving one meter is one joule. A person carrying 50 pounds (225 newtons) of bricks up a 10-foot (3-meter) ladder does 500 foot-pounds of work (675 joules). Stated as a formula:

U.S. CUSTOMARY

Work = Force × Distance
Foot-pounds = Pound × Feet

METRIC

Work = Force × Distance
Joules = Newtons × Meters

As a simple example, calculate the work involved in lifting a 20-pound (90 newton) weight 5 feet (15 meters), as shown in Figure 1-1.

U.S. CUSTOMARY

Work = Force × Distance
Work = 20 pounds × 5 feet = 100 foot-pounds

METRIC

Work = Force × Distance
Work = 90 newtons × 1.5 meters = 135 joules

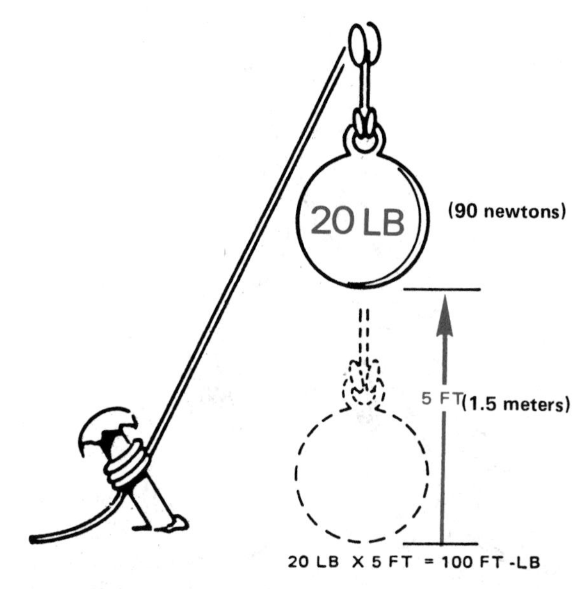

Fig. 1-1 A twenty pound weight lifted five feet in the air (90 newtons lifted 1.5 meters = 135 joules)

By scientific definition, the person who struggles to move a boulder but fails to budge it is not doing work because there is no "distance moved." Engines can do work because they are capable of applying force to move machines, loads, implements, and so forth through distances.

ENERGY

Energy is the ability to do work. The energy stored in the human body, for instance, supplies the body with a potential to do work. A person can work for a long period of time on stored energy before needing to refuel with food. The human body is a good example of potential energy.

Potential Energy

Potential energy is the energy a body has due to its position, its condition, or its chemical state. The following are some examples of potential energy: water at the top of a waterfall (position); a tightly wound watch spring (condition); fuels such as gasoline or coal, and foodstuffs to be burned by the body (chemical state).

Stored or potential energy is measured in the same way as work, in foot-pounds or joules. How much potential energy is there when a 20-pound (90 newton) weight is lifted 5 feet (1.5 meters)?

U.S. CUSTOMARY

Potential Energy = Force × Distance
Potential Energy = 20 pounds × 5 feet
= 100 foot-pounds

METRIC

Potential Energy = Force × Distance
Potential Energy = 90 newtons × 1.5 meters
= 135 joules

With the weight at the 5-foot (1.5-meter) height, there are 100 foot-pounds (135 joules) of potential energy.

Kinetic energy is the energy of motion. Some examples are the energy of a thrown ball, the energy of water falling over a dam, and the energy of a speeding automobile. Kinetic energy is, in effect, released potential energy.

Consider a thrown ball, the work of throwing it, and what becomes of it. If a ball is thrown by applying a force of 8 pounds (36 newtons) through 6 feet (1.8 meters), how much work is done?

U.S. CUSTOMARY

Work = Force × Distance
Work = 8 pounds × 6 feet = 48 foot-pounds

METRIC

Work = Force X Distance
Work = 36 newtons X 1.8 meters = 65 joules

What becomes of this energy? It exists as kinetic energy, the *energy of motion* of the ball. The ball in flight has 48 foot-pounds (65 joules) of energy which it will deliver to whatever it hits.

Refer back to the example of potential energy. When the 20-pound (90 newton) weight which rests 5 feet (1.5 meters) above the floor is allowed to fall, its potential energy of 100 foot-pounds (135 joules) is converted to an equal amount of kinetic energy (energy of motion) as seen in figure 1–2.

Energy can change form but it cannot be destroyed. This is the *law of the conservation of energy*. Energy can take the form of light, sound, heat, motion, and electricity. Consider the potential chemical energy of gasoline in an engine. Upon ignition, its potential energy is converted mainly into heat energy. The heat energy is used to push the pistons down and now can be seen as the kinetic energy of the rotating crankshaft.

Refer back again to the 20-pound (90-newton) weight poised at a 5-foot (1.5-meter) height. When the weight falls to the floor its energy is not lost. Rather, it is transformed into other types of energy. The floor would become slightly warmer at the point of impact because of the creation of heat energy. There would be noise at the moment of impact, an indication of the presence of sound energy. The floor would move slightly, thus absorbing some energy.

The efficiency of machines that transform energy is not 100 percent. Some energy is wasted in moving the machine parts to overcome friction which is inherent in all machines. The *efficiency* of a machine is the ratio of the work done by the machine to the work put into it.

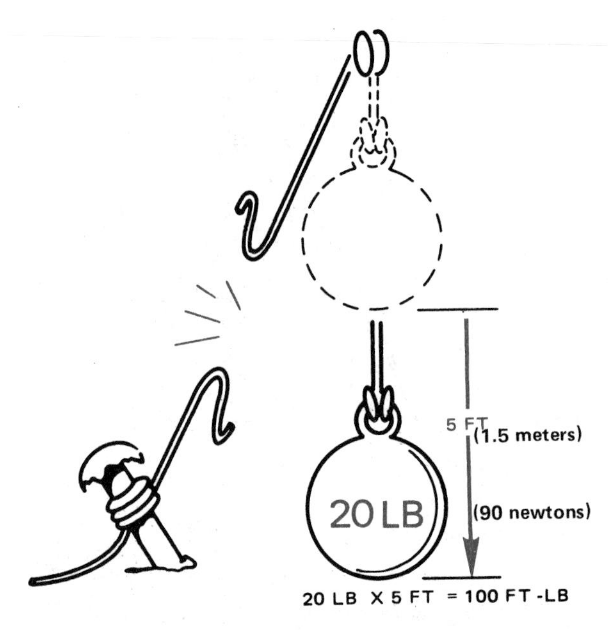

Fig. 1–2 A twenty pound weight dropped to the floor from a height of five feet delivers 100 foot-pounds of kinetic energy. (In the metric system, a 90 newton weight dropped to the floor from a height of 1.5 meters delivers 135 joules of kinetic energy.)

Stated as a formula:

$$\text{Efficiency} = \frac{\text{Output}}{\text{Input}} \times 100$$

or

$$\text{Efficiency} = \frac{\text{Input} - \text{Losses}}{\text{Input}} \times 100$$

or

$$\text{Efficiency} = \frac{\text{Output}}{\text{Output} + \text{Losses}} \times 100$$

The figure 100 in these formulas is a way of converting the fraction to a percentage.

POWER

In everyday language the word *power* can mean different things; for example, political power is different than physical power. However, in the language of scientists and engineers, power refers to how fast work is done or how fast energy is transferred. *Power* is the rate of doing work or the rate of energy conversion.

How fast a machine can work is an important consideration for engineers and consumers. For example, buying gasoline is a purchase of potential energy. The size and efficiency of the engine determine how fast this energy can be converted into power. A 10-horsepower (7.5-kilowatt) engine can work only half as fast as a 20-horsepower (15-kilowatt) engine.

Measurement of Power

Power is measured in foot-pounds (joules) per second or in foot-pounds (joules) per minute. Stated as a formula:

$$\text{Power} = \frac{\text{Work}}{\text{Time}}$$

Refer again to the 20-pound (90-newton) weight that was lifted 5 feet (1.5 meters). How much power is required to lift the weight in 5 seconds?

U.S. CUSTOMARY

Work = Force × Distance
Work = 20 pounds × 5 feet = 100 foot-pounds

$$\text{Power} = \frac{\text{Work}}{\text{Time}}$$

$$\text{Power} = \frac{\text{100 foot-pounds}}{\text{5 seconds}}$$

$$= \text{20 foot-pounds per second}$$

METRIC

Work = Force × Distance
Work = 90 newtons × 1.5 meters = 135 joules

$$\text{Power} = \frac{\text{Work}}{\text{Time}}$$

$$\text{Power} = \frac{\text{135 joules}}{\text{5 seconds}} = \text{27 watts}$$

As another example: if an elevator lifts 3500 pounds (16 000 newtons) a distance of 40 feet (12 meters) and it takes 25 seconds to do it, what is the rate of doing work?

U.S. CUSTOMARY

Work = Force × Distance
Work = 3500 pounds × 40 feet
= 140,000 foot-pounds

$$\text{Power} = \frac{\text{Work}}{\text{Time}}$$

$$\text{Power} = \frac{\text{140,000 foot-pounds}}{\text{25 seconds}}$$

$$= \text{5600 foot-pounds per second}$$

METRIC

Work = Force × Distance
Work = 16 000 newtons × 12 meters
= 192 000 joules

$$\text{Power} = \frac{\text{Work}}{\text{Time}}$$

$$\text{Power} = \frac{\text{192 000 joules}}{\text{25 seconds}} = \text{7 700 watts}$$

Power can also be expressed in foot-pounds (or joules) per minute. If a pump needs 10 minutes to lift 5500 pounds (22 000 newtons) of water 60 feet (18 meters), it is doing 300,000 foot-pounds (396 000 joules) of work in 10 minutes. This is a rate of 30,000 foot-pounds (39 600 joules) per minute.

Horsepower

The power developed by most machinery is measured in horsepower. The unit originated many years ago when James Watt, attempting to sell his new steam engines, found it necessary to rate his engines in comparison with the horses they were to replace. He found that an average horse, working at a steady rate, could do 550 foot-pounds of work per second, as illustrated by figure 1-3. This rate is the definition of one *horsepower*. A more modern definition can be seen in figure 1-4. The term horsepower is not used in the metric system; instead, all power is measured in watts.

The formula for horsepower is:

$$\text{Horsepower} = \frac{\text{Work}}{\text{Time (in seconds)} \times 550}$$

If time in the formula above is expressed in minutes, it is multiplied by 550 × 60 (seconds) or 33,000. The formula, then, may be expressed as:

$$\text{Horsepower} = \frac{\text{Work}}{\text{Time (in minutes)} \times 33{,}000}$$

What horsepower motor would the elevator previously referred to have?

$$\text{Horsepower} = \frac{\text{Work}}{\text{Time (in seconds)} \times 550}$$

$$\text{Horsepower} = \frac{140{,}000 \text{ foot-pounds}}{25 \text{ seconds} \times 550}$$

$$\text{Horsepower} = 10+$$

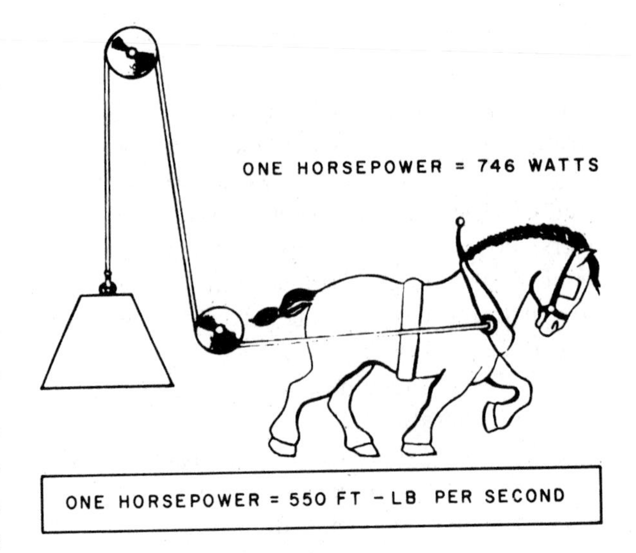

Fig. 1-3 Horsepower

SUMMARY

Terms which are basic to an understanding of the applications of mechanical power and which will follow in later units have now been discussed. The measurements which these terms involve have been shown as formulas. To review, the following formulas are a means of expressing each of the terms covered:

a. $\text{Work} = \text{Force} \times \text{Distance}$

b. $\text{Efficiency} = \frac{\text{Output}}{\text{Input}} \times 100$

c. $\text{Power} = \frac{\text{Work}}{\text{Time}} = \frac{\text{Force} \times \text{Distance}}{\text{Time}}$

d. $\text{Horsepower} = \frac{\text{Work}}{\text{Time (in seconds)} \times 550}$

$$\frac{\text{Work}}{\text{Time (in minutes)} \times 33{,}000}$$

Fig. 1-4 Both outboard motor engines are entirely suitable for their job requirements. However, the larger horsepower engine can convert energy to work at a faster rate. (Courtesy Johnson Motors Co.)

REVIEW QUESTIONS

1. Define work.
2. What is the unit of measurement for work?
3. Define energy.
4. Explain the difference between kinetic and potential energy.
5. Explain the law of conservation of energy. What are some common forms of energy?
6. Explain why a machine cannot be 100% efficient.
7. Define power.
8. What is the unit of measurement for power in the U.S. Customary system?
9. How much power is needed to lift a 100-pound (445-newton) bag of cement onto a 3-foot (0.9-meter) high truck bed in 2 seconds?

10. What is the formula for calculating horsepower?
11. How much horsepower does an engine on a grain elevator deliver if it can load 705 pounds (3137 newtons) of corn into a 25-foot (7.6-meter) storage bin in 55 seconds?

Unit 2 Safety and Power Technology

OBJECTIVES After completing this unit, the student should be able to:

- Explain how personal attitudes toward safety affect other people.
- List rules for storing gasoline.
- Name safety precautions for gasoline use.
- List safety rules for operating power equipment.
- Identify the proper fire extinguisher to be used with each class of fire.
- State how carbon monoxide is produced.
- List the effect that carbon monoxide has on the body.
- State rules for the safe use of hand tools.

The maintenance, operation, repair and testing which are involved in a study of power technology requires that careful and constant emphasis be placed on safety. It is beyond the scope of this unit to deal with the particular safety considerations of each different prime mover and its vast number of applications. Instead, this unit deals with safety considerations in several common areas: safety attitudes, safety with gasoline, fires, carbon monoxide safety, safety with basic hand tools, safety with basic machines and applicances, and safety with small gasoline engines.

Specific safety information is included in later units as it applies. However, it is useful to review this unit before beginning laboratory experiences.

SAFETY ATTITUDES

Safety is important to every student, both as an individual and as a member of a community. The machine age brought many safety hazards. In transportation alone millions of power-driven vehicles presented a number of hazards. Industrial, construction, and scientific employees work with the tools of power technology every day. There are also many hazards in the home – power appliances, household machines, fire, boiling liquids, electricity, and so forth. Considering these hazards, it is necessary that an emphasis on safety begin early in childhood and continue throughout life.

Accidents can be compiled into statistics. Accidents can be measured in terms of frequency, probability, medical costs, and time lost from work. This information can serve a useful purpose in identifying hazards and in developing safety programs.

On an individual basis, the statistical approach loses all of its importance. How can a number or a dollar sign be put on the personal cost and human suffering that result when accidents occur? People do not like to think of the unpleasantness of an accident happening to them. Fortunately, most try to protect themselves and those close to them from hazardous situations and accidents.

People must consider both their own safety, and the safety of others. Each person's actions and safety attitudes do have an effect on other individuals. Sometimes the effect is slow and indirect, such as parents' carelessness becoming part of their children's safety standards. At other times, the effect on others is direct, such as a head-on collision on the highway caused by an irresponsible driver. The point is that in power safety, individuals are responsible for themselves as well as for the safety of others.

Students should consider the safety aspect of every activity in which they engage. Most basic safety approaches to prime movers and machines are similar and the knowledge gained from one situation can be applied to another. Safety is never sacrificed for the sake of speed or expediency. It should have priority over everything else. How a person operates a machine can reveal that person's entire safety code. It can also indicate personal qualities or deficiencies.

Learning safety by trial and error alone is not enough. The safety considerations of each situation must be studied. A defensive attitude of, "if this happens, what might the result be?" must be a part of a person's overall judgment. Some safety is common sense. However, many safety requirements must be learned from machinery instruction booklets and repair manuals. To ensure personal safety as well as safety for others, the operator must know specific information that applies to the particular prime mover and its uses.

SAFETY WITH GASOLINE

Understanding the nature of gasoline and its safe use is very important. Gasoline is used as a common item in many households. The presence of this liquid, which is more powerful than TNT, is readily accepted. Many people have become too casual about this volatile and explosive liquid.

Gasoline is around the home (in the garage, toolshed, or barn) largely because the use of small gasoline engines has become so widespread. Lawn mowers, garden tractors, snow throwers, go-carts, utility vehicles, motorbikes, outboard motors, and chain saws need gasoline for operation.

Unfortunately, accidents do happen. It is estimated that every year several hundred persons lose their lives due to accidents involving gasoline and other flammable liquids. In addition, several million dollars of property damage is caused by such accidents.

The volatility of gasoline is the characteristic that makes it such an ideal fuel. This same characteristic is also the prime danger. Gasoline can vaporize and explode in a closed or nonventilated room just as it can in the engine. A concentration of one to seven cubic feet of gasoline vapor per 100 cubic feet of air represents a flammable condition. Any spark or flame can touch off gasoline as a fire or explosion.

Gasoline should not be stored inside the home – it is too dangerous. It should be kept in a garage, toolhouse, or some other outside building. Storage in the basement is dangerous as a gas can could be tipped over. If the seal was imperfect, gasoline would leak out and begin to vaporize. Given certain circumstances, the whole basement could blow up – triggered by the flame from a furnace or water heater or from the arc of an electric switch.

Many states have recognized the hazards of gasoline around the home and are developing safety regulations that help to protect the public. The proper use of storage containers is one important area. Ideal storage is in a red metal can clearly labeled GASOLINE. Red is a universal color indicating danger. A metal can, such as shown in figure 2-1, will not break if it is accidentally dropped or struck. Follow these rules for the storage of gasoline:

- Store gasoline in an approved metal container.
- Store gasoline in an outbuilding – not inside the home.
- Do not store gasoline during the off season. It is too dangerous to have around.
- Do not allow small children to have access to the gasoline supply.
- Store gasoline away from flames, excessive heat, sparks caused by static electricity, sparks caused by electrical contacts, or sparks caused by mechanical contact.
- Have the proper portable fire extinguisher available for use if needed.

Gasoline should be regarded as dangerous; using it safely involves decisions and judgments. Furthermore, gasoline should be regarded as fuel for gasoline engines and not as a handy all-purpose solvent, cleaner, or fire starter. Again, the volatility and explosive nature of gasoline make it unsuitable for any type of general household use. Gasoline used as a cleaning agent for paint brushes or to cut grease creates real safety hazards. Accidental ignition can occur while the gasoline is being used. Also, disposing of the dirty gasoline may be hazardous.

Never use gasoline to start a fire. The result may be a fire or explosion much larger than the intended size resulting in injury to the person with the gasoline.

Fig. 2-1 Gasoline properly labeled and stored in proper container

Observe these precautions when using gasoline:

- Always regard gasoline as dangerous.
- Use gasoline only as a fuel for engines or devices where its use is clearly intended.
- Never smoke cigarettes, pipes, or cigars around gasoline.
- Never fill the tank of a hot engine or a running engine if the tank and engine are at all close to each other. Gasoline splashed on a hot engine can ignite or explode.

- When pouring gasoline from a container to the tank, reduce the possibility of a static electricity spark by having metal-to-metal contact. A metal spout held against the tank opening is good to use. A funnel that contacts both the container and the tank is also safe. Pouring gasoline through a chamois may present a static electricity hazard.

Fig. 2-2 Using a fire extinguisher

FIRES

An understanding of fires and fire extinguishers is very useful, figure 2-2. Most fires fall into one of three categories:

Class A – Wood, cloth, paper, rubbish

Class B – Oil, gasoline, grease, paint

Class C – Electrical equipment

Class A Fires

Class A fires are the most common type of fire. Extinguishing Class A fires consists of quenching the burning material and reducing the temperature below that of combustion. Some extinguishing methods also have a smothering effect on the fire.

The Class A fire can be effectively fought with water from a fire hose, garden hose, or a water-base extinguisher. The water stream should be directed at the base of the fire first, then back and forth following the flames upward.

The dry chemical extinguisher is the most common extinguisher used in the home, office, or shop. Dry chemical extinguishers are effective in fighting almost all types of fires (Class A, B, or C). These extinguishers expel a chemical that interrupts and smothers the fire.

Class B Fires

Class B fires involve burning liquids such as gasoline, oil, paint, thinners, solvents, etc. Careless or unknowing action against Class B fires can place the fire fighter in great danger. Burning liquids such as gasoline are unpredictable; they can explode into a ball of fire as well as burn. The basic extinguishing technique is to cut off the oxygen supply which feeds the burning liquid and interrupt the flame without splashing the burning liquid around. Dry chemical extinguishers are suitable for fighting Class B fires.

Carbon dioxide (CO_2) extinguishers contain carbon dioxide under high pressure; they also are used to extinguish Class B fires. Carbon dioxide does not support combustion and, therefore, has the effect of smothering the flame. The extinguisher should be used with a slow sweeping action that travels from side to side, working to the back of the flame area. The discharge horn of the extinguisher becomes very cold during discharge and so should not be touched. Another hazard is that using the extinguisher in a small room may decrease the oxygen supply which is dangerous to the fire fighter.

Foam fire extinguishers discharge a water-base foam that can smother the fire. When using a foam extinguisher on Class B fires, direct the foam at the back of the fire allowing the foam to spread onto the flame area. This minimizes the possibility of splashing the flaming liquid of an open container fire.

Class C Fires

Class C fires involve electrical equipment. Water-base extinguishing materials are not suitable for putting out Class C fires. Water on or around an electrical fire creates a shock hazard. If the electrical energy can be completely and positively shut off, the fire is no longer considered a Class C fire. The method used to extinguish the fire then depends on the material that is burning. For example, shorted wires in an electrical outlet may be arcing and causing insulation to smolder. If the fire has spread to the wooden walls and if the electric current is turned off, the burning walls are considered a Class A fire. A dry chemical extinguisher would be suitable for both the electrical fire and the wood fire. Using water would be hazardous unless the electricity had been shut off. If the electrical energy cannot be completely shut off, the fire must be fought with carbon dioxide extinguishers, dry chemical extinguishers, or vaporizing liquid extinguishers.

CARBON MONOXIDE SAFETY

The hazard of carbon monoxide is very real. Carbon monoxide (CO) is a poison that is colorless, odorless and tasteless. It is a killer that is responsible for more poisoning deaths than any other deadly poison.

Carbon monoxide is the result of incomplete combustion of solid, liquid, or gaseous fuels of a carbonaceous nature. At home the gas may be present due to an improperly adjusted hot water heater. In industry carbon monoxide can be found with kilns, oven stoves, foundries, smelters, mines, forges, and in the distillation of coal and wood, to list a few. However, it also exists in a much more common circumstance – the exhaust gases of all internal combustion engines from lawn mowers to automobiles. When an engine is in operation, carbon monoxide is produced. Knowledge about this poison and how to eliminate or minimize the danger is an important topic for everyone.

The action of this poison on the human body can be quite rapid. The hemoglobin of the blood has a great attraction for CO, three hundred times greater than that for oxygen. When the CO is combined with the hemoglobin, it has the effect of reducing the amount of hemoglobin available to carry oxygen to the body tissues. If large amounts of CO combine with the hemoglobin, the body becomes starved for oxygen and literally suffocates.

Ventilation is of prime importance in preventing carbon monoxide poisoning, figure 2-3. Unburned gases and exhaust must be carried away as effectively as possible by using chimneys, ventilation systems, and exhaust systems, figure 2-4. These systems must be efficient because even small amounts of the gas can cause a dulling of the senses, which indicates danger.

Carbon monoxide in exhaust gas can get inside an automobile in several ways: through a defective exhaust system – tail pipe, muffler, or manifold; or through rusted-out or defective floor panels. If there is excessive CO around the automobile due to faulty exhausting or a poorly tuned engine, it can come in through an open window.

Knowing the symptoms of carbon monoxide poisoning and heeding their warnings could be a life-and-death matter. These symptoms include a tightness across the forehead followed by throbbing temples, weariness, weakness, headache, dizziness, nausea, decrease in muscle control, increased pulse rate, and increased rate of respiration. If anyone has these symptoms while riding in a car, the car should be stopped and the person should get some fresh air.

Fig. 2-3 Ventilation is of prime importance in preventing carbon monoxide poisoning

Fig. 2-4 Using an exhaust system to carry away carbon monoxide

If a person is discovered who has passed out from carbon monoxide poisoning, remove the victim to the fresh air immediately. If breathing has stopped or if the victim is only gasping occasionally, begin artificial respiration at once. Also, have someone call a doctor and/or emergency squad.

To prevent carbon monoxide poisoning, follow these rules:

- Do not drive an auto with all the windows closed if a carbon monoxide leak is suspected.
- Do not operate internal combustion engines in closed spaces such as garages or small rooms, unless the space is equipped with an exhaust system.
- Keep carbon monoxide producing engines and devices tuned or adjusted properly in order to reduce the output of CO gas.

- **Do not sit in parked cars with the engine running.**
- **Remember the importance of properly working ventilation and exhaust systems.**

SAFETY WITH BASIC HAND TOOLS

Tools should be treated with respect and should be well cared for. They should be stored in a way that will protect their vital surfaces, cutting edges, and true surfaces. A wall tool panel with individual hangers or holding devices for each tool is ideal since the tools are easily seen, ready for use, and well protected. Tools carelessly thrown in a drawer can scratch, dull, or damage each other and are a hazard to the person who is looking for the right tool.

When tools must be carried in a toolbox, they should have individual compartments, if possible. If the box lacks individual compartments, the more sensitive tools and those with cutting edges should be wrapped in cloth.

Moisture and the rusting that results is a problem. If tools get wet, wipe them dry before storing. Also, the possibility of rusting can be lessened by wiping the tools with a slightly oily rag. If a tool picks up dirt during use, wipe it clean before putting it away.

The basic rules for the safe use of hand tools are:

- **Use tools that are in good condition and well sharpened. A dull tool is inefficient and is more likely to slip during use.**
- **Use the correct tool for the job and use it as it should be used. The incorrect use of tools causes many accidents.**

In some operations, hand tools can present the hazard of flying metal chips. A misstruck nail may fly across the room. Even a hammer can be dangerous. Especially dangerous are chipping operations with a cold chisel. All struck tools such as chisels and punches can mushroom out at the end over a period of use, figure 2-5. The ends should be dressed on a grinder to eliminate the possibility of tool chips breaking and flying from the tool when it is hit. Hammers can chip if they are struck side blows or glancing blows; hammers should strike the work with their full face. figure 2-6. Safety glasses should be worn whenever there is a possibility of flying chips or breakage. Grinding operations are especially hazardous and safety glasses should always be worn when grinding.

When properly used, a wrench is a safe tool. However, if used incorrectly, the wrench can be dangerous. Be certain to use the exact size wrench for the nut or bolt to prevent slipping or damage to the flats of the nut or bolt. If an adjustable wrench is used, place it tightly against the flats of the nut or bolt and be sure that the direction of force places the major strain on the fixed jaw and not on the movable jaw, figure 2-7. Be sure

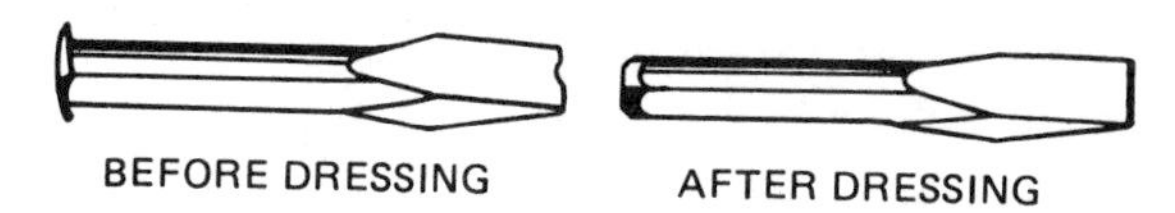

Fig. 2-5 All struck tools should be properly dressed on the ends

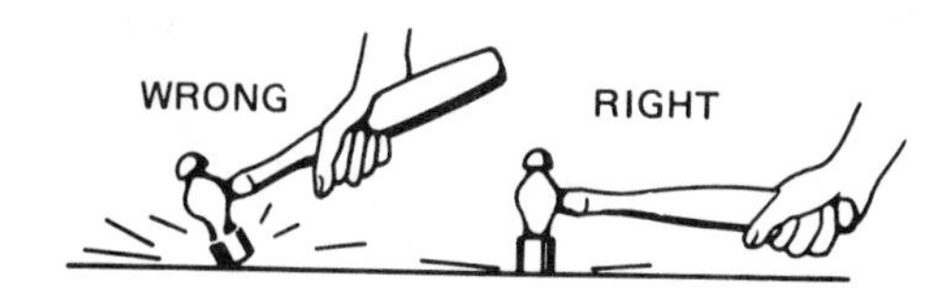

Fig. 2-6 The hammer face should strike the work with its full, flat face

WRONG

RIGHT

Fig. 2-7 Use an adjustable wrench so the strain is on the fixed jaw

to pull the wrench toward you; never push it. Pushing can be dangerous; the nut or bolt may suddenly loosen, or the tool may slip causing the user to be injured.

Screwdrivers are probably the most misused of all common tools. The list of abuses ranges from using it to open paint cans to using it as a wrenching bar. The tool is intended to be used for tightening or loosening various types of screws; its tip must be preserved for this purpose. Use a screwdriver that is the correct size, figure 2-8. It should fit the screw slot snugly in width; the blade should be as wide as the slot is long. Have several sizes of screwdrivers in a tool assortment. It is a poor practice to hold the work in one hand in such a manner that a slip would send the screwdriver into the palm of the hand. All small or irregularly shaped pieces should be held securely in a vise. Screwdrivers, pliers, or other tools that are used for electrical or electronic work should have insulated handles, figure 2.9.

Files are rarely sold with handles attached but they should be so equipped. The tang is

TOO SMALL

CORRECT SIZE

Fig 2-8 Use the correct size screwdriver for the job

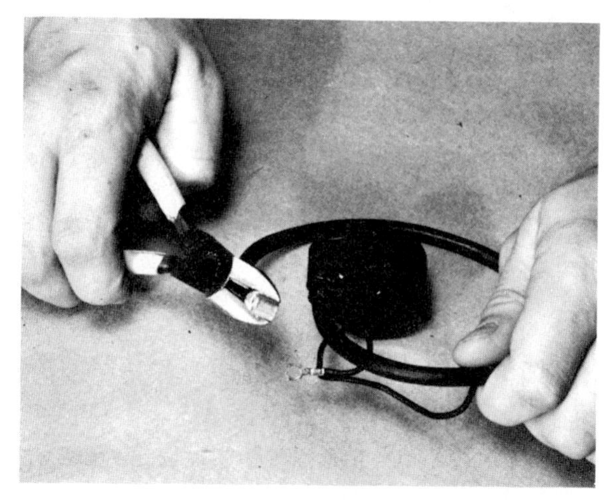

Fig. 2-9 Tools used in electrical and electronic work should have insulated handles

fairly sharp and can cut or puncture the hand if the file suddenly hits an obstruction. The brittle nature of files makes them a poor choice and dangerous as a pry bar; do not use them for this purpose.

SAFETY WITH BASIC MACHINES AND EQUIPMENT

Safe use of basic household machines, equipment, and appliances during normal use, adjustment, and repair is important to all, figures 2-10 and 2-11. Reference is made to power hand tools; power tools; appliances such as stoves, refrigerators, dishwashers, clothes driers, washing machines, air conditioners, electronic equipment such as television sets, record players, and radios; powered hobby or sports equipment; small home appliances for kitchen or personal use; and so forth. Several common rules should be followed when using these machines:

- Before attempting to adjust or repair a machine, always unplug or disconnect the machine or device from the electrical power source. Be certain that someone else does not accidentally reconnect or plug in the machine while you are working on it. If the power source is in a remote location, tag the source "Do not connect. Equipment under repair."

Fig. 2-10 This man is practicing proper safety techniques while using a filament trimmer. Note the heavy work clothes, rugged shoes, safety glasses and hardhat. (Courtesy of Toro Company)

- Before attempting to adjust or repair machines driven by engines, stop the engine, disengage the clutch, and remove the spark plug lead to eliminate the possibility of accidentally restarting the engine.
- Be sure to ground electrical equipment that requires grounding. Use the three-prong plug on portable tools or devices and properly ground the convenience outlet.

- Keep loose clothing away from rotating parts that could grab the cloth and thereby pull the operator into the machine.
- Protect eyes with safety glasses if there is any chance of flying chips or breakage.
- Thoroughly guard all belts, pulleys, chains, gears, etc. Do not remove these guards or allow them to be removed by others.
- Study the instructions that come with the machine. Be certain that the operating principles and the safety precautions are understood before operating or repairing the machine.
- Use the machine only in the manner and for the purpose for which it was designed.
- During adjustment and repair, retighten all nuts, bolts, and screws securely.
- Unless qualified to be there, keep away from high voltage areas in electronic equipment.
- Do not pull a lawn mower – push it. If a person falls while pulling the mower, it may be pulled over the body.
- Inspect the lawn for rocks, sticks, wire and other foreign objects that can be converted into projectiles by the mower blade.
- Allow only a responsible person to operate a power mower.
- Small children should be kept away from the mowing area.

Fig. 2-11 Keep feet away from the blade when startin the mower

- Do not use a riding lawn mower as a play vehicle.
- Keep self-propelled lawn mowers under full control.
- Never leave a mower running unattended.
- Wear safety glasses while mowing.

REVIEW QUESTIONS

1. Name two factors that contribute to the safe operation of an engine.
2. What characteristics of gasoline makes it such a dangerous liquid?
3. List four rules for storing gasoline.
4. List three precautions to follow when using gasoline.
5. What type of extinguishers are best for putting out gasoline fires?
6. How is carbon monoxide produced? What is its effect on the body?
7. State two rules for the safe use of hand tools.
8. List six safety rules for operating power lawn mowers.

SECTION 2 THE SMALL GASOLINE ENGINE

Unit 3 Construction of the Small Gasoline Engine

OBJECTIVES

After completing this unit, the student should be able to:

- **Identify the basic parts of an engine and explain how they work together.**
- **Explain the principles of the four-stroke cycle.**
- **Explain the principles of the two-stroke cycle.**
- **Explain the differences in the construction of four-stroke cycle and two-stroke cycle engines.**

The internal combustion engine is classified as a *heat* engine; its power is produced by burning a fuel. The energy stored in the fuel is released when it is burned. *Internal combustion* means that the fuel is burned inside the engine itself. The most common fuel is gasoline. If gasoline is to burn inside the engine, there must be oxygen present to support the combustion. Therefore, the fuel needs to be a mixture of gasoline and air. When ignited, a fuel mixture of gasoline and air burns rapidly; it almost explodes. The engine is designed to harness this energy.

The engine contains a cylindrical area commonly called the *cylinder* that is open at both ends. The top of the cylinder is covered with a tightly bolted-down plate called the *cylinder head.* The cylinder contains a *piston,* which is a cylindrical part that fits the cylinder with little clearance. The piston is free to slide up and down within the cylinder. The air/fuel mixture is brought into the cylinder, then the piston moves up and compresses it into a small space called the combustion chamber. The *combustion chamber* is the area where the fuel is burned; it usually consists of a cavity in the cylinder head and perhaps the uppermost part of the cylinder. When the fuel is ignited and burns, tremendous pressure builds up. This pressure forces the piston back down the cylinder; thus the untamed energy of combustion is harnessed to become useful mechanical energy. The basic motion within the engine is that of the piston sliding up and down the cylinder, a *reciprocating* motion.

There are still many problems, however. How can the up-and-down motion of the piston be converted into useful rotary motion? How can exhaust gases be removed? How can new fuel mixture be brought into the combustion chamber? Studying the engine's basic parts can help answer these questions.

BASIC ENGINE PARTS

The internal combustion engine parts discussed here are those of the four-stroke cycle gasoline engine, the most common type. The most essential of these parts are briefly listed.

- *Cylinder:* hollow; stationary; piston moves up and down within the cylinder.
- *Piston:* fits snugly into the cylinder but still can be moved up and down.
- *Connecting rod:* connects the piston and the crankshaft.
- *Crankshaft:* converts reciprocating motion into more useful rotary motion.
- *Crankcase:* the body of the engine; it contains most of the engine's moving parts.
- *Valves:* "doors" for admitting fuel mixture and releasing exhaust.

Figure 3-1 shows the basic parts of two-cycle and four-cycle engines. Cross sections of two small gas engines are shown in figures 3-2 and 3-3.

Cylinder

The cylinder is a finely machined part which can be thought of as the upper part

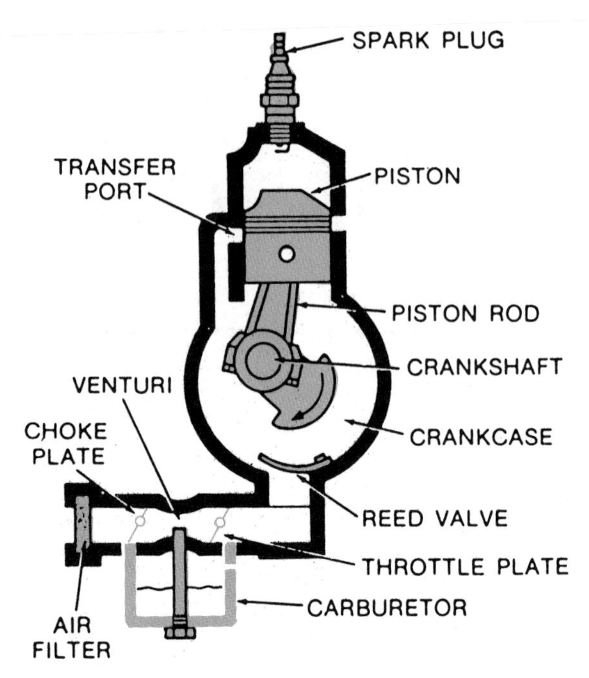

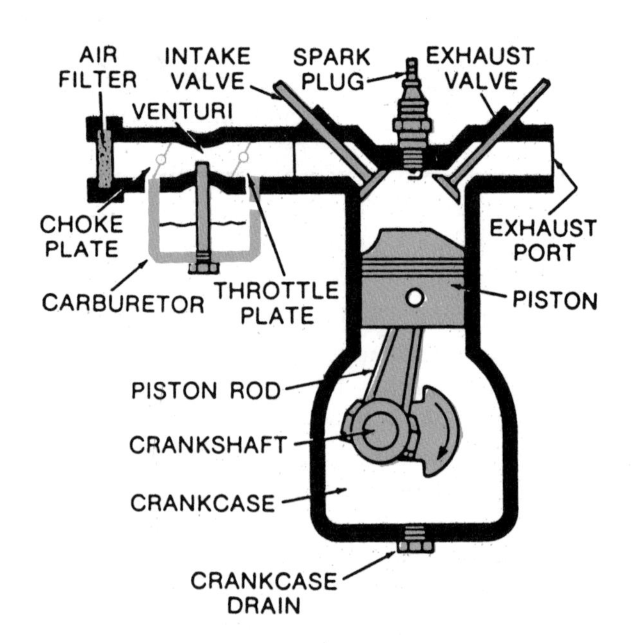

Fig. 3-1 Basic parts of two-cycle and four-cycle engines

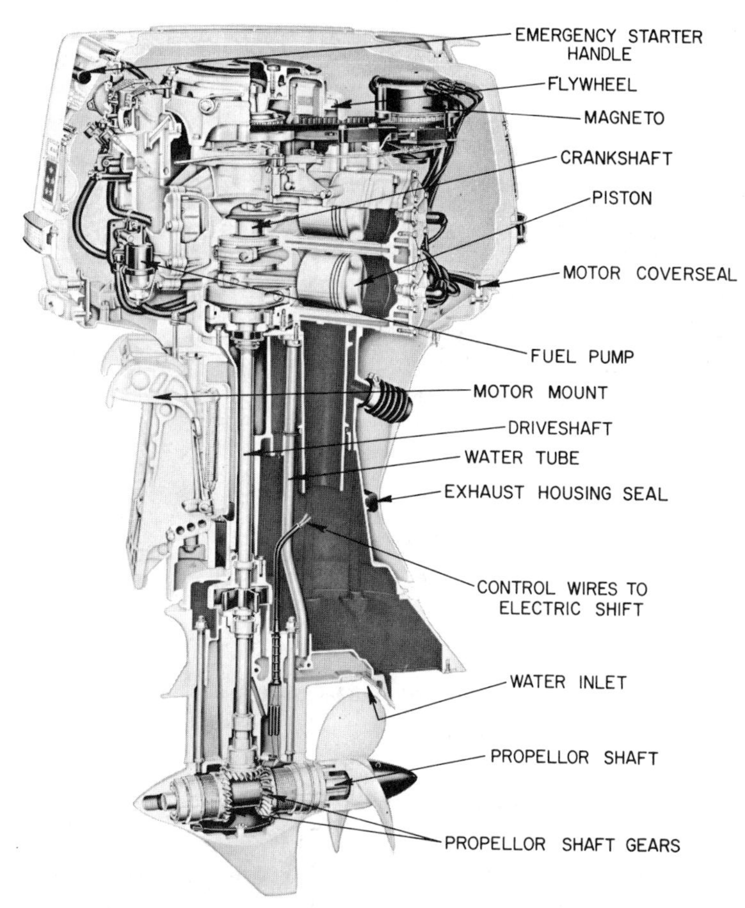

Fig. 3-2 Cross section of an outboard motor (Courtesy Johnson Company)

of the crankcase. The piston slides up and down in the cylinder. The fit is close; the piston is only a few thousandths of an inch smaller in diameter than the cylinder.

Light duty aluminum engines often use the aluminum engine block for the cylinder. However, some aluminum engines have a cast iron cylinder sleeve cast into the engine and the two are inseparable. A cast iron engine has a cast iron block for the cylinder. Some engines have a chrome-plated cylinder.

In engine specifications, the diameter or width across the top of the cylinder is referred to as the *bore* of the engine; this is an important measurement. Figure 3-4 shows four different cylinder arrangements.

Cylinder Head

The cylinder head forms the top of the combustion chamber and is exposed to great heat and pressure. This is where the action is – the few cubic inches made up by the top of the piston, the upper cylinder walls and the cylinder head.

The cylinder head must be tightly bolted to the cylinder. It is important to tighten all cyl-

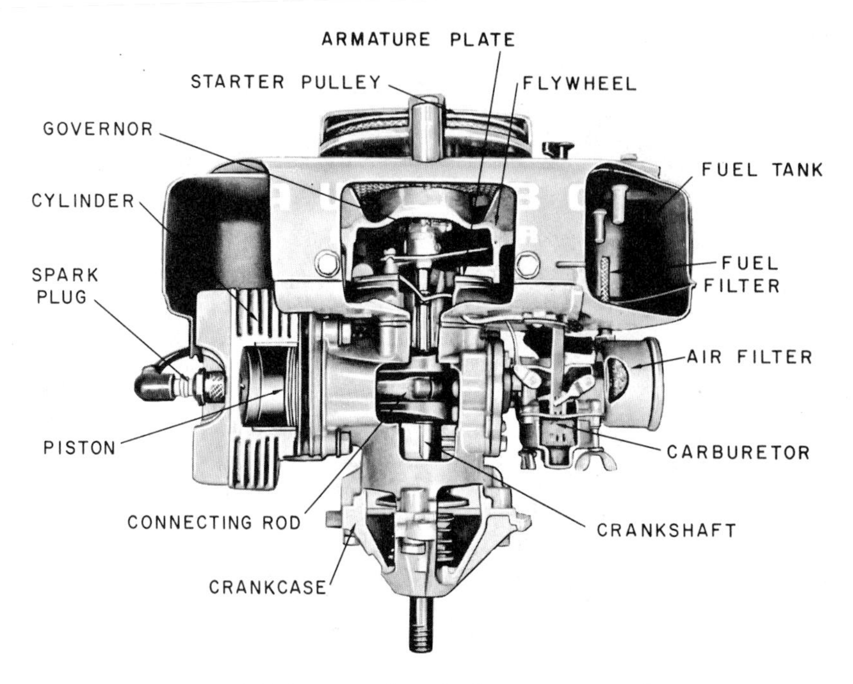

Fig. 3-3 Cross section of a lawn mower engine

inder head bolts with an even pressure and in their correct order so that uneven stress is not set up in the cylinder walls, figure 3-5. Such stress can distort the cylinder walls or warp the cylinder head. The cylinder head also contains the threaded hole for the spark plug.

Cylinder Head Gasket

A gasket between the two metal surfaces makes an airtight seal, figure 3-6. Gaskets are made of relatively soft material such as fiber, soft metal, cork, asbestos, and rubber. Cylinder head gaskets must be able to hold the high pressure of combustion without blowing out and they must be heat resistant.

There are many engines that have the cylinder and the cylinder head cast as one piece, therefore, there are no gaskets or head bolts. A threaded spark plug hole is still located at the top of the cylinder head. Many small two-cycle engines use this design.

Piston

The piston slides up and down in the cylinder. It is the only part of the combustion chamber that is free to move when the pressure of the rapidly burning fuel mixture is applied. The piston can be made of cast iron, steel, or aluminum alloy; aluminum is commonly used for small engines because of its light weight and its ability to conduct heat away rapidly.

There must be clearance between the cylinder wall and the piston to prevent excessive wear. The piston is 0.003 to 0.004 of an inch smaller in diameter on the thrust sides. To check this clearance, a special flat feeler gauge is necessary.

The piston top, or crown, may be flat, convex, concave, or many other shapes. Manufacturers select the shape that causes turbulence of the fuel mixture in the combustion chamber and promotes smooth burning of the

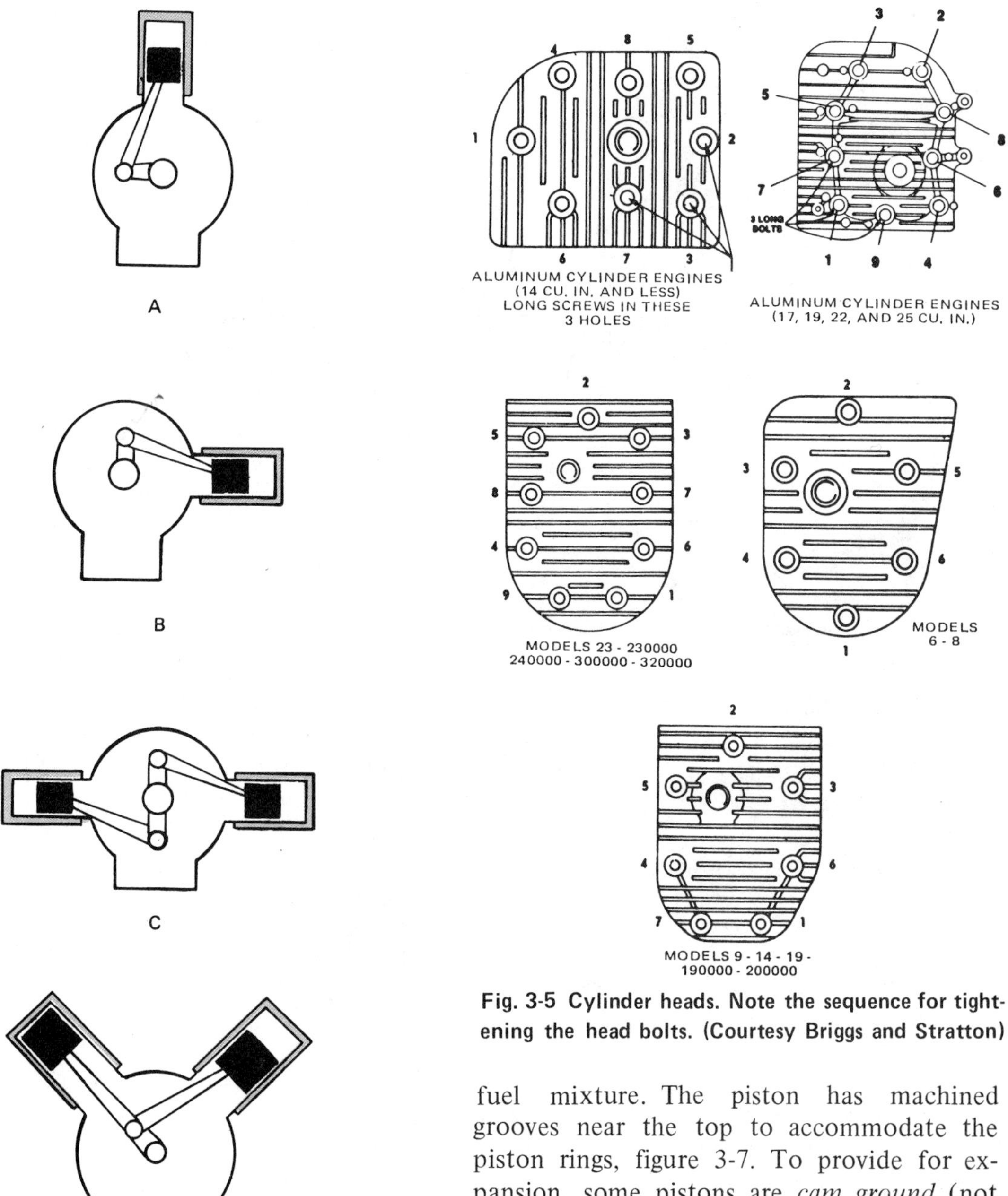

Fig. 3-5 Cylinder heads. Note the sequence for tightening the head bolts. (Courtesy Briggs and Stratton)

Fig. 3-4 Cylinder arrangement: (A) vertical, (B) horizontal, (C) opposed, and (D) V-type

fuel mixture. The piston has machined grooves near the top to accommodate the piston rings, figure 3-7. To provide for expansion, some pistons are *cam ground* (not perfectly round), or have a vertical slot cut up the side.

The distance the piston moves up and down is called the *stroke.* The uppermost point of the piston's travel is called its *top*

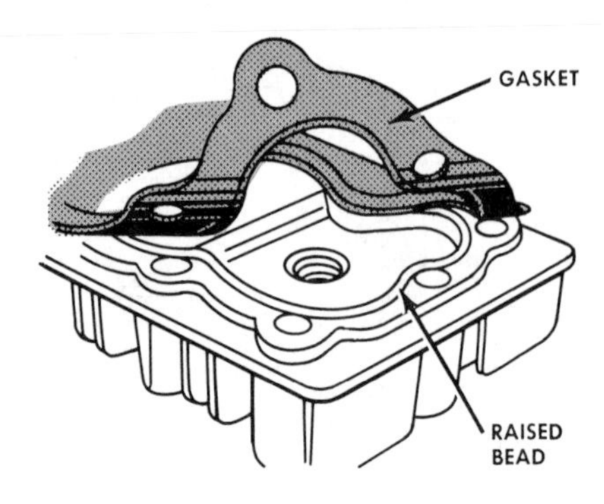

Fig. 3-6 The cylinder head gasket provides an airtight seal.

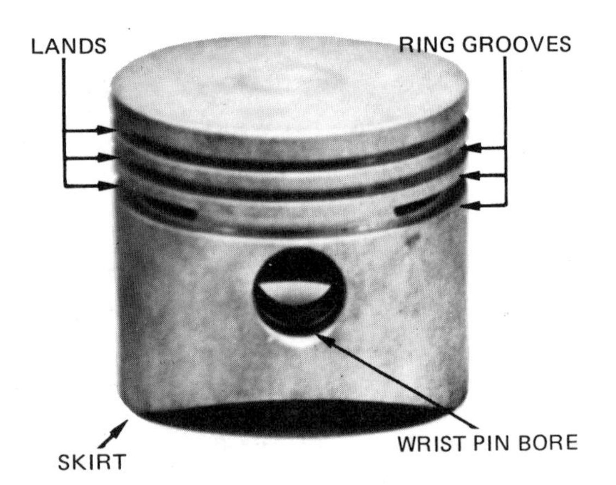

Fig. 3-7 The parts of a piston

dead center (TDC) position. The lowest point of the piston's travel is called its *bottom dead center* (BDC). Therefore, stroke is the distance between TDC and BDC.

Piston Rings

Piston rings provide a tight seal between the piston and the cylinder wall. Without the piston rings much of the force of combustion would escape between the piston and cylinder walls into the crankcase. The piston rings mainly act as a power seal, ensuring that a maximum amount of the combustion pressure is used in forcing the piston down the cylinder. Since the piston rings are between the piston and cylinder wall, there is a small area of metal sliding against the cylinder wall. Therefore, the piston rings reduce friction and the accompanying heat and wear that are caused by friction. Another function of the piston ring is to control the lubrication of the cylinder wall.

The job of the piston rings might appear simple. However, the piston rings may have to work under harsh conditions such as (1) distorted cylinder walls due to improper tightening of head bolts, (2) cylinder out of round, (3) worn or scored cylinder walls, (4) worn piston, and (5) conditions of expansion due to intense heat. The important job of providing a power seal can become difficult if the engine is worn or has been abused.

Piston rings are made of cast iron or steel and are finely machined. Sometimes the surfaces of piston rings are plated with other metals to improve their action. There are many designs of piston rings. Some rings, especially oil rings, may consist of more than one piece.

Blow-by and oil pumping are two serious problems that can be caused by bad piston rings, a bad piston, a warped cylinder, distortion, and/or scoring. In *blow-by,* the pressure of combustion is great enough to break the oil seal provided by the piston rings. When this happens, the gases of combustion force their way into the crankcase. The result is a loss in engine power and contamination of the crankcase oil.

Piston rings that are not fitted properly may start pumping oil. This pumping action leads to fouling in the combustion chamber, excessive oil consumption, and poor combustion characteristics. The rings have a side clearance which can cause a pumping action as the piston moves up and down if the piston and rings become worn.

A piston usually has three or four piston rings. Some have only one ring. The top piston ring is a compression ring, figure 3-8. This pounds on the cylinder wall. The ring has a clearance in its groove: it can move slightly up and down and expand slightly in and out. The second ring from the top is also a compression ring. These two rings form the power seal. The second ring serves two purposes. The upper half helps with compression control and the lower outside edge acts as a scraper for oil control. In small engines, the second ring is commonly referred to as a scraper ring.

The third ring down (and fourth is one is present), is an oil control ring, figure 3-9. The job of this piston ring is to control the lubrication of the cylinder wall. These rings spread the correct amount of oil on the cylinder wall, scraping the excess from the wall and returning it to the crankcase. The rings

Fig. 3-8 Compression ring

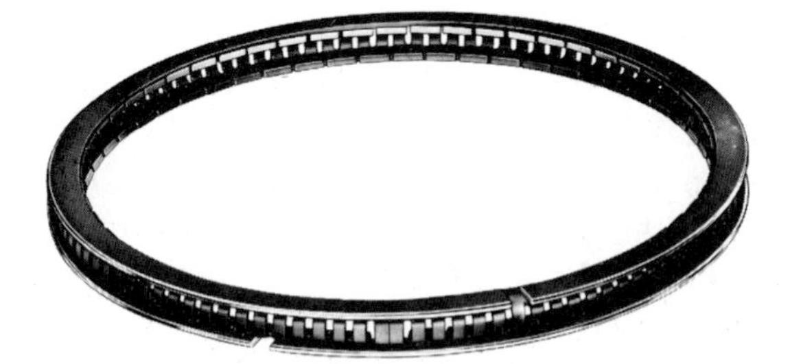

Fig. 3-9 Oil control ring

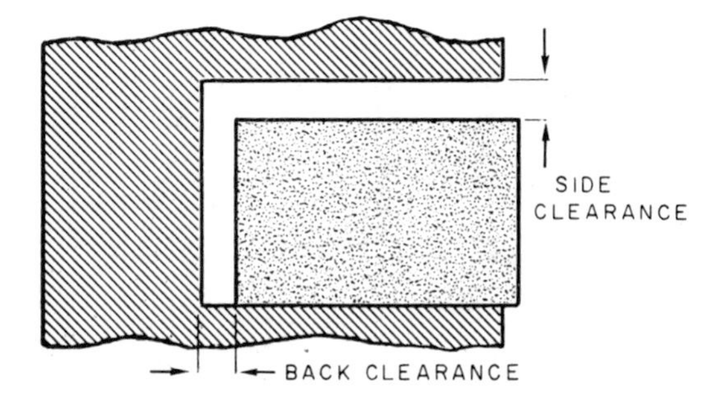

Fig. 3-10 Cross section showing a piston ring in its groove. Notice back clearance and side clearance.

are slotted and grooves are cut into the piston behind the ring. This enables much of the excess oil to be scraped through the piston and it then drips back into the crankcase. It should be noted that compression rings have a secondary function of oil control.

Two clearances (or tolerances) that are important in piston rings are end-gap and piston ring side groove clearance. The end gap is the space between the ends of the piston ring, measured with a flat feeler gauge when the ring is in the cylinder. The gap must be large enough to allow for expansion due to heat but not so large that power loss due to blow-by will result. Often 0.004 inch is allowed for each inch of piston diameter (top ring), and 0.003 inch is allowed for rings under the top ring. Since the second and third rings are exposed to less heat their allowance can be smaller.

Side clearance or ring groove clearance is also provided for heat expansion, figure 3-10. Often 0.0025 inch is allowed for the top ring and 0.003 inch is allowed on the second and third rings.

Piston Pin or Wrist Pin

The piston pin is a precision ground steel pin that connects the piston and the connecting rod, figure 3-11. This pin can be solid or

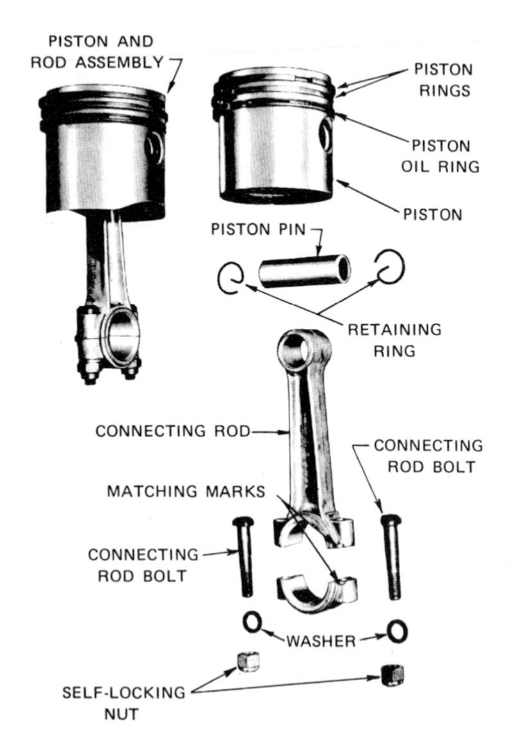

Fig. 3-11 Piston, piston pin, connecting rod, and associated parts for free-floating piston assembly (Courtesy Tecumseh Products Company)

hollow; the hollow piston pin has the advantage of being lighter in weight. Because piston pins are subjected to heavy shocks when combustion takes place, high-tensile strength steel is used. To keep noise at a minimum, the piston pin fits to a very close tolerance in the connecting rod. The piston pin does not rotate. It has a rocking motion similar to the action of the wrist when the wrist is moved back and forth, hence the name *wrist pin.* Heavy shocks, close tolerances, and the rocking motion make the lubrication of this part difficult. In most engines, lubricating oil is squirted or splashed on it.

Connecting Rod

The connecting rod connects the piston (with the aid of the piston pin) and the crankshaft, figure 3-12. Many small engines have cast aluminum connecting rods; larger engines can have steel connecting rods. Generally, the cross section of the connecting rod is an I-beam shape for strength. The lower end of the connecting rod is fitted with a cap. Both the lower end and the cap are accurately machined to form a perfect circle. This cap is bolted to the rod, encircling the connecting rod bearing surface on the crankshaft. Many connecting rods are fitted with replaceable bearings surfaces called rod bearings or inserts. These are used especially on heavy-duty or more expensive engines.

Crankshaft

The crankshaft is the vital part that converts the reciprocating motion of the piston into rotary motion. One end of the crankshaft has a provision for power takeoff (connecting accessory to use engine's power) and the other is machined to accept the flywheel.

Crankshafts are made of high-quality steel; they are carefully forged and then machined to close tolerances. Their bearing surfaces are large because they must accept a great amount of force. Since they revolve at high speeds, they must be well balanced. Figure 3-13 is an example of a single-throw crankshaft.

Figure 3-14 shows two types of small engine designs. Engines designed with *vertical* crankshafts are shown in figure 3-15. Vertical crankshafts are excellent for applications such as rotary power lawn mowers where the blade is bolted directly to the end of the crankshaft. Most wheeled applications such as motorbikes and go-carts use *horizontal* crankshaft engines which enable the power

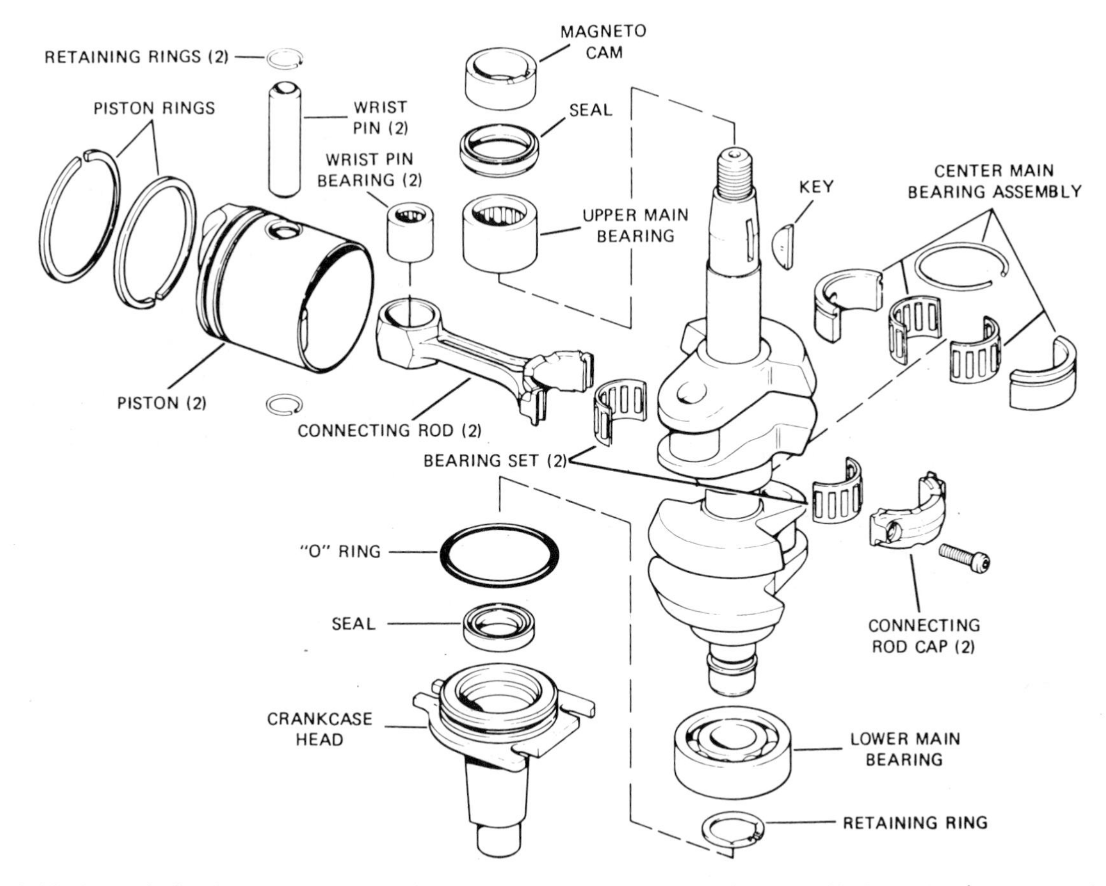

Fig. 3-12 Crankshaft, pistons, and connecting rod components – two-cycle, two-cylinder engine (note crankshaft)

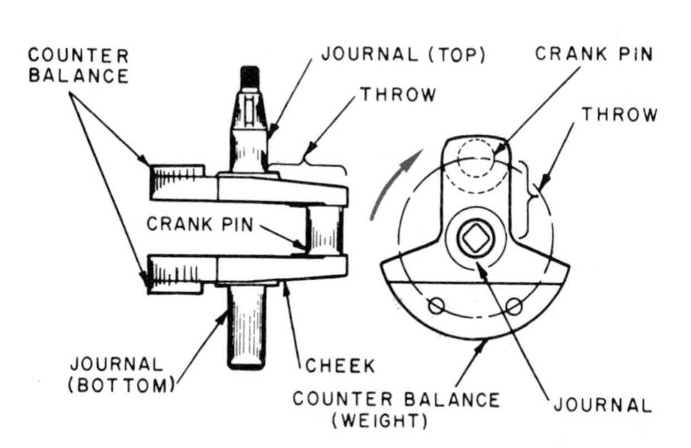

Fig. 3-13 Single-throw crankshaft: one crank pin for one cylinder

to be delivered to an axle through belts or chains. Multiposition engines such as chain saws are another variation. The crankshafts themselves are much the same but other changes in engine design are necessary to meet the requirements of crankshaft position. An engine can have clockwise or counterclockwise crankshaft rotation, depending upon its application and design.

Crankcase

The crankcase is the body of the engine. It houses the crankshaft and has main bearing

Fig. 3-14 Horizontal and vertical shaft engines

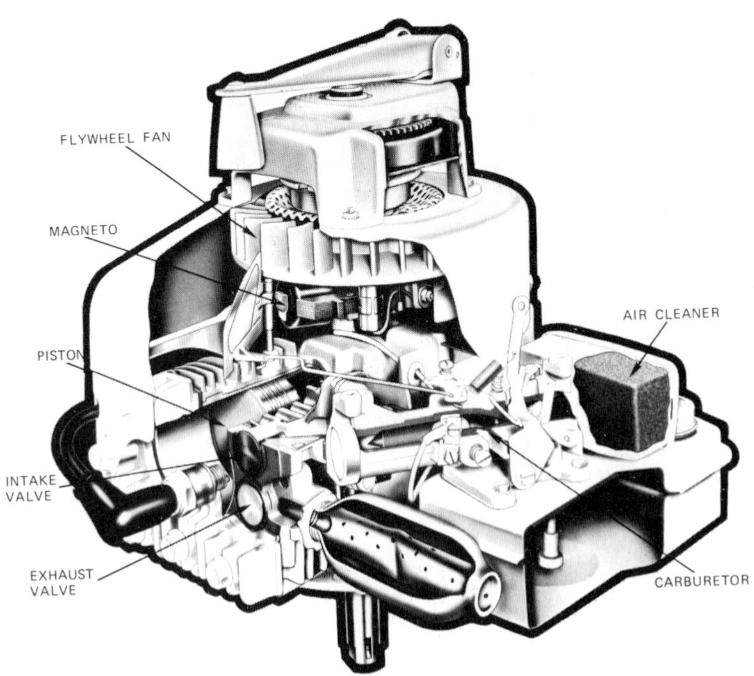

Fig. 3-15 Four-cycle engine with vertical shaft

surfaces on which the crankshaft revolves. The connecting rod, cam gear, camshaft, and lubrication mechanism are also located within the crankcase. The cylinder and bottom of the piston are exposed to the crankcase. On four-cycle engines a reservoir of oil is found within the crankcase. This lubricating oil is splashed, pumped, or squirted onto all the moving parts located in the crankcase. Figure 3-16 shows the components of a short block assembly (crankcase and its associated parts). A single-cylinder engine is illustrated in figure 3-17.

Generally the crankcase is made of cast aluminum for engines that are for light duty and especially where a lightweight engine is needed. Medium duty engines are die cast aluminum with a cast iron sleeve pressed into the aluminum. Crankcases for heavy duty engines are usually made of cast iron.

Flywheel

The flywheel is mounted on the tapered end of the crankshaft. A *key* keeps the two parts solidly together and they revolve as one part, figure 3-18. The key is also called a *shear pin* which acts as a safety device to protect the crankshaft in case of a sudden impact. (The slot that the key fits into is called the *keyway.*) The flywheel is found on all recipro-

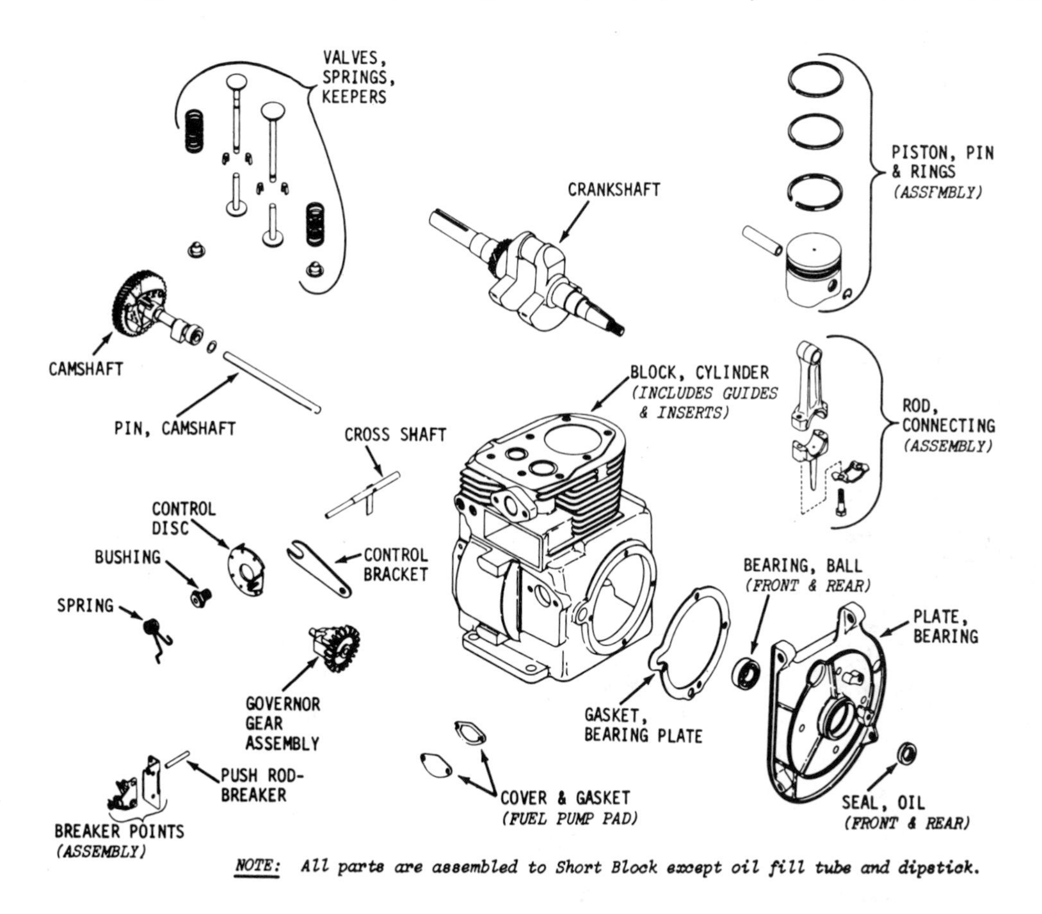

Fig. 3-16 Components of a short block assembly

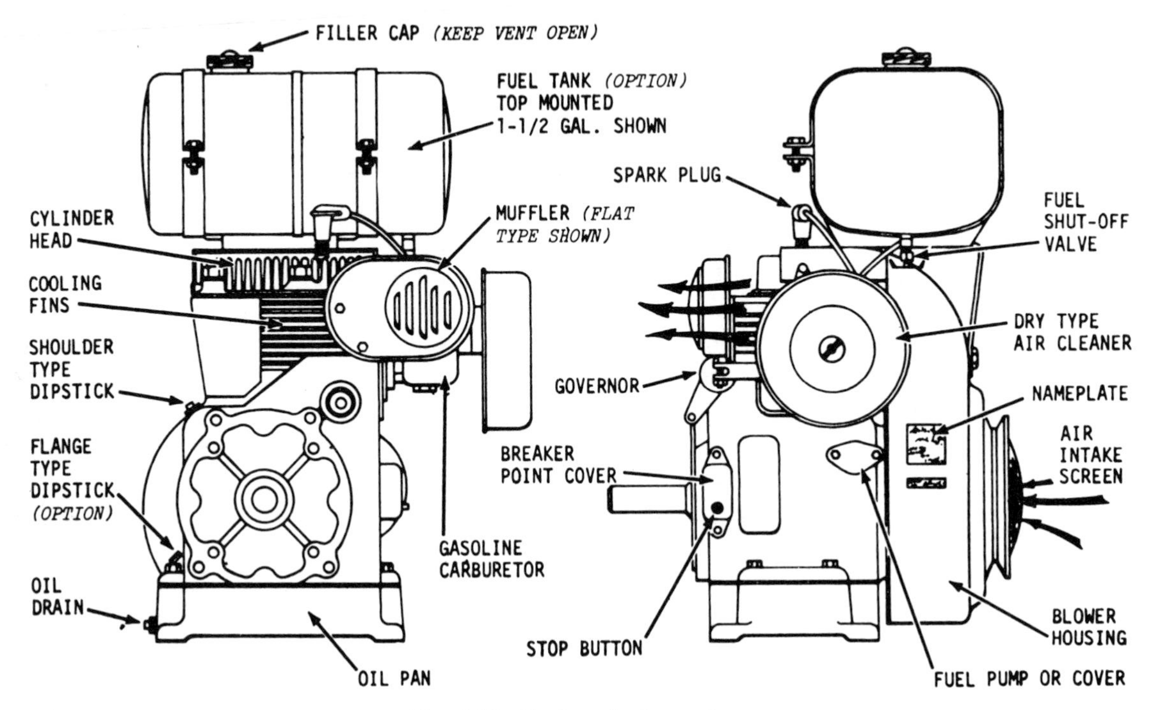

Fig. 3-17 Single-cylinder engine

cating piston engines. It is relatively heavy and helps to smooth out the operation of the engine. That is, after the force of combustion has pushed the piston down, the momentum of the flywheel helps to move the piston back up the cylinder. This tends to minimize or eliminate sudden jolts of power during combustion. The more cylinders the engine has, the less important this smoothing out action becomes.

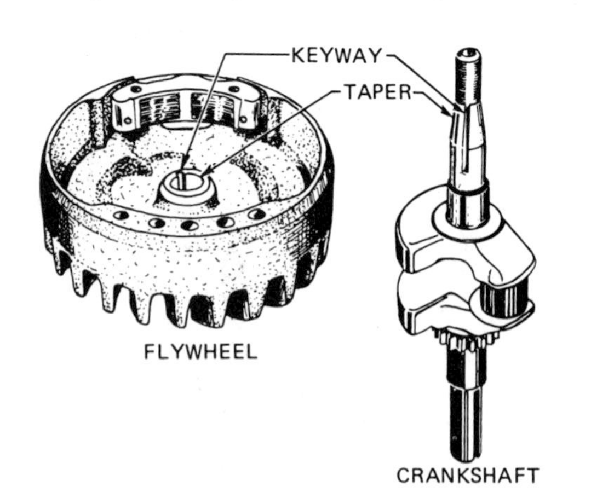

Fig. 3-18 A key locks the flywheel and crankshaft together

The flywheel can also be a part of the engine's cooling system by having air vanes to scoop air as the flywheel revolves. This air is channeled across the hot engine by the cover or shroud, to carry away heat.

Most flywheels on small gasoline engines are part of the ignition systems; they have permanent magnets mounted in them. These permanent magnets are an essential part of the magneto ignition system.

Valves

Figure 3-19 shows the parts of a value. The most common valves are called *poppet valves*. With these valves, exhaust gases can be

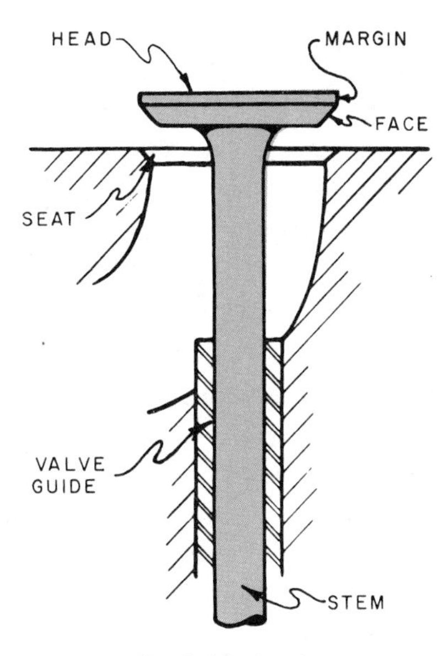

Fig. 3-19 A valve

removed from the combustion chamber and new fuel mixture can be brought in. To accomplish this, the valve actually pops open and snaps closed. In the normal four-cycle combustion chamber there are two valves – one to let the exhaust out (exhaust valve) and one to allow the air/fuel mixture to come in (intake valve).

The two valves may look almost identical, but they are different. The exhaust valve must be designed to take an extreme amount of punishment and still function perfectly. Not only is it subjected to the normal heat of combustion (about 4500°F), but when it opens, the hot exhaust gases rush by it. When the valve returns to its seat, a small area touches, making it difficult for the valve's heat to be conducted away. This rugged valve opens and closes 1800 times a minute if the engine is operating at 3600 rpm. Exhaust valves are made from special heat-resisting alloys.

Valves can be made to rotate a bit each time they open and close. This action tends to prevent deposits from building up at the valve seat and margin. In addition, rotation provides for even wear and exposure to temperatures which greatly increases the life of the valves, figure 3-20.

Intake values operate in the same manner as exhaust values except they are not subjected to the extreme heat conditions. Each time they open new fuel mixture rushes by, helping to cool the intake valve.

Operation of the valve requires the help of associated parts – the valve train and camshaft, figure 3-21. The valve train consists of the valve (intake or exhaust), valve spring, tappet or valve lifter, and cam. The valve spring closes the valve, holding it tightly against its seat. The valve lifter rides on the cam; there is a small clearance between it and the valve stem. Some valve lifter clearances are adjustable especially on larger and more expensive engines. The cam or lobe pushes up on the valve lifter; it, in turn, pushes up on the valve stem. The valve opens when the high spot on the cam lobe is reached. The opening and closing of the valves must be carefully timed in order to get the exhaust out at the correct time and the new fuel mixture in at the correct time.

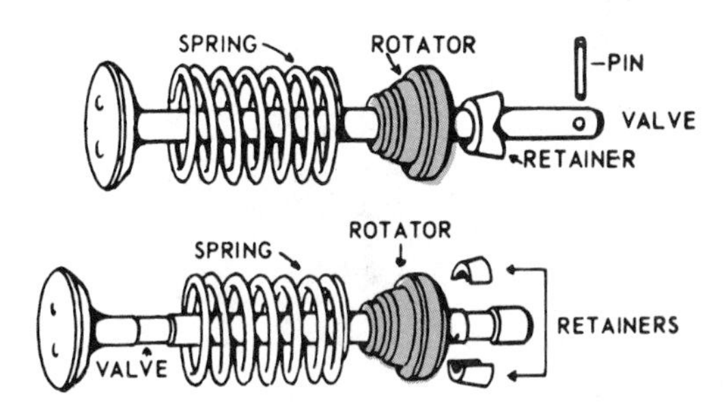

Fig. 3-20 "Rotocap" provides for even wear and long valve life

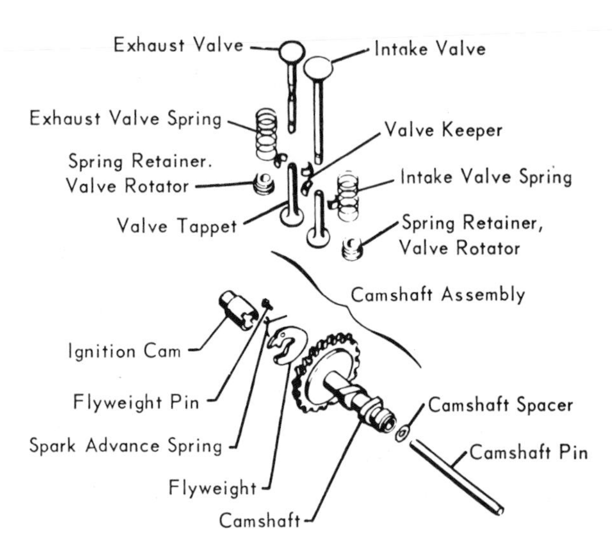

Fig. 3-21 Complete valve train and camshaft assembly

Fig. 3-22 L-head design used for most four-stroke cycle small engines

Valve Arrangement

Engines may be designed and built with one of several valve arrangements, as follows:

- *The L-head:* both valves open up on one side of the cylinder. This is the most common arrangement in small gasoline engines, figures 3-22 and 3-23.
- *The I-head:* both valves open down over the cylinder, figure 3-24.
- *The F-head:* one valve opens up and one valve opens down; both are on the same side of the cylinder, figure 3-25.
- *The T-head:* one valve is on one side of the cylinder and the other valve is on the opposite side; both valves open up, figure 3-26.

Camshaft

The camshaft and its associated parts control the opening and closing of the valves. The cams on this shaft are egg-shaped. As they revolve, their noses cause the valves to open and close. A tappet rides on each cam. When the nose of the cam comes under the tappet, the valve is pushed open admitting new fuel mixture or allowing exhaust to escape, depending on which valve is opened.

The camshaft is driven by the crankshaft. On four-cycle engines, the camshaft operates

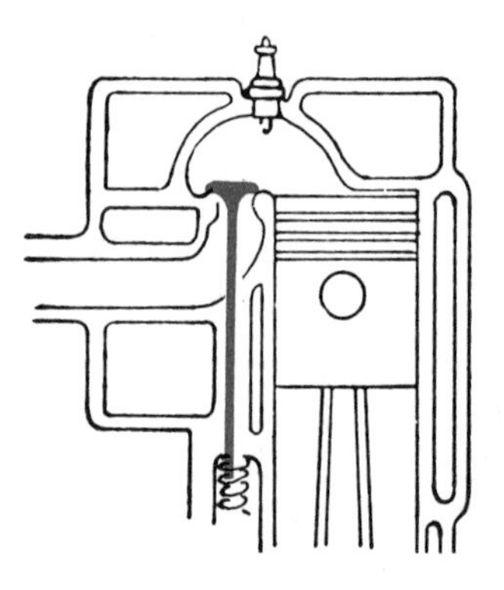

Fig. 3-23 L-head valve arrangement

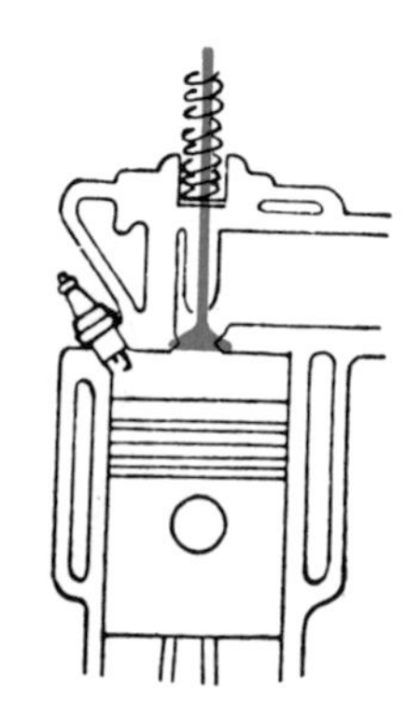

Fig. 3-24 I-head valve arrangement

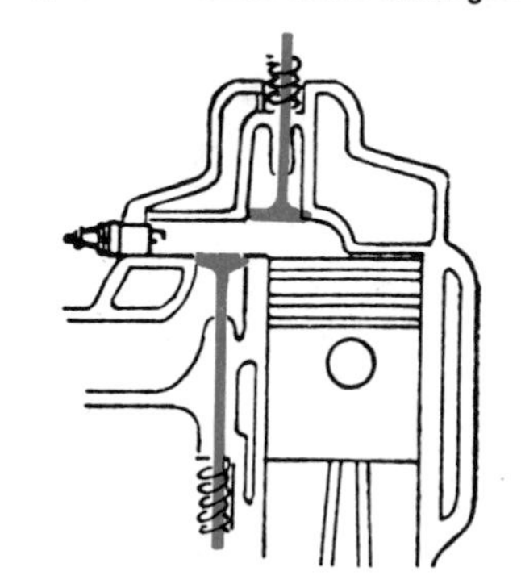

Fig. 3-25 F-head valve arrangement

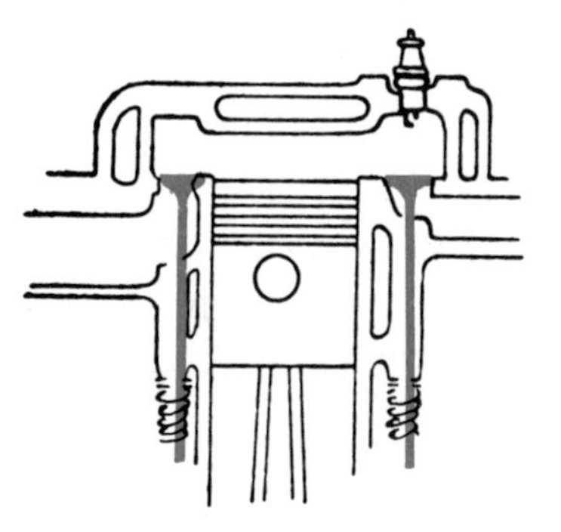

Fig. 3-26 T-head valve arrangement

at one-half the crankshaft speed. (If the engine is operating at 3400 rpm, the camshaft revolves at 1700 rpm.) The crankshaft and camshaft must be in perfect synchronization. Therefore, most engines have timing marks stamped or cut into the gears to ensure the correct reassembly of the gears, figure 3-27.

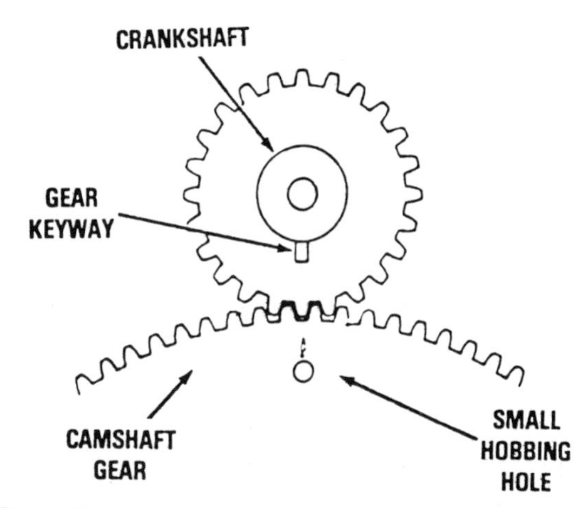

Fig. 3-27 Aligning timing marks on the crankshaft gear and camshaft gear. The larger camshaft gear revolves at one-half crankshaft speed.

FOUR-STROKE CYCLE OPERATION

All of these parts and many others must be assembled correctly if the gasoline engine is to operate properly. The most common method of engine operation is the four-stroke cycle. Most small gasoline engines as well as automobile engines operate on this principle. The four-stroke cycle engine is a very successful producer of power.

Four-stroke cycle means that it takes four strokes of the piston to complete the operating cycle of the engine, figure 3-28. The piston goes down the cylinder, up, down, and back up again to complete the cycle. This takes two revolutions of the crankshaft; each stroke is one-half revolution.

Intake is when the piston travels down the cylinder and fuel mixture enters the combustion chamber. Reaching the bottom of its stroke, the piston comes up on compression and the fuel mixture is compressed into the combustion chamber. Power comes when the fuel mixture is ignited, pushing the piston back down. The piston moves up once more

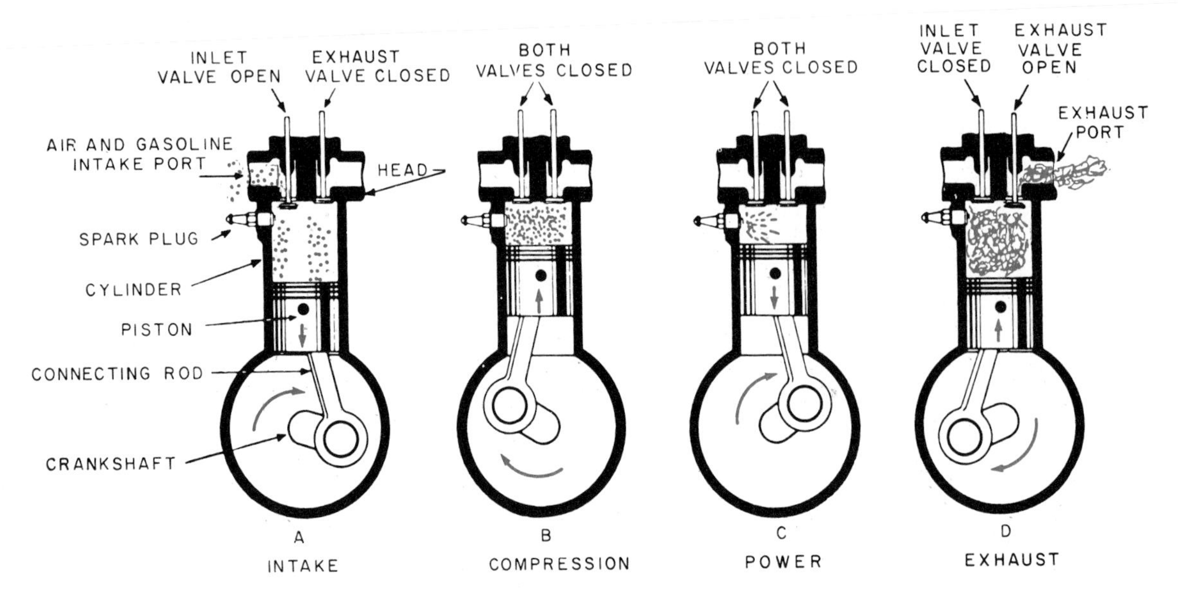

Fig. 3-28 The four-stroke cycle

pushing the exhaust gases out of the combustion chamber. This is exhaust. These four strokes complete the engine's cycle and require two revolutions of the crankshaft. As soon as one cycle is complete, another begins. If an engine is operating at a speed of 3600 rpm there are 1800 cycles each minute.

Intake Stroke

If the fuel mixture is to enter the combustion chamber, the intake valve must be open and the exhaust valve must be closed. With the intake valve open and the piston traveling down the cylinder, the fuel mixture rushes in easily. This is the *intake stroke.*

Pushing down on everything is an atmospheric pressure of 14.7 pounds per square inch (psi) at sea level. When the piston is at the top of its stroke there is normal atmospheric pressure in the combustion chamber. When the piston travels down the cylinder there is more space for the same amount of air and the air pressure is reduced; a partial vacuum is created. The intake valve opens and the normal atmospheric pressure rushes in to equalize this lower air pressure in the combustion chamber. The air/fuel mixture is actually pushed into the combustion chamber, even though it seems to be sucked in by the partial vacuum.

Compression Stroke

When the piston reaches the bottom of its stroke and the cylinder is filled with air/fuel mixture, the intake valve closes and the exhaust valve remains closed. The piston now travels up the cylinder compressing the fuel mixture into a smaller and smaller space. When the fuel mixture is compressed into this small space it can be ignited more easily and it expands very rapidly.

The term *compression ratio* applies to this stroke. *Compression ratio* is the ratio of the total cylinder volume when the piston is at the bottom of its stroke to the volume remaining when the piston is at the top of its

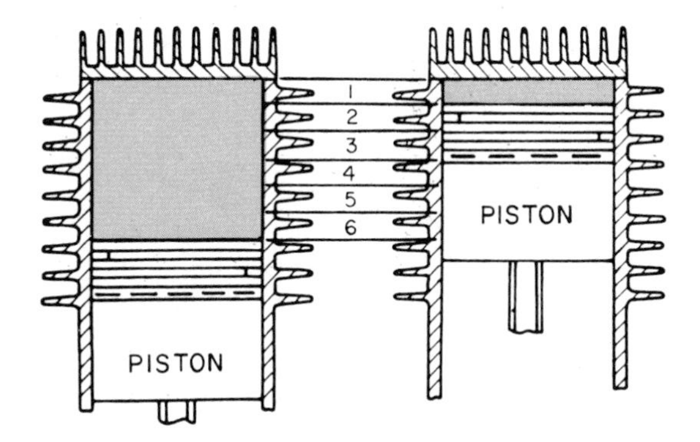

Fig. 3-29 An engine with a six-to-one compression ratio

stroke. If the piston compresses the fuel mixture into one-sixth of the original space, the compression ratio is 6:1, figure 3-29. An engine whose piston compresses the fuel mixture into one-eighth the original space has a compression ratio of 8:1. The higher the compression, the more powerful the force of combustion.

Power Stroke

When the piston reaches the top of its stroke and the fuel mixture is compressed, a spark jumps across the spark plug igniting the fuel mixture. The fuel mixture burns very rapidly. The burning, expanding gases exert a great pressure in the combustion chamber. The combustion pressure is felt in all directions, but only the piston is free to react. The piston reacts by being pushed rapidly down the cylinder. This is the power stroke. *Power stroke* occurs when the rapidly burning fuel creates a pressure that pushes the piston down thus turning the crankshaft and producing usable rotary motion.

Both intake and exhaust valves must be closed for the power stroke. The force of combustion must not leak out the valves or blow-by the piston rings. Piston rings that fit properly prevent this. The cylinder head gasket and the spark plug gasket should be airtight. The power must be transmitted to the top of the piston and not lost elsewhere.

Exhaust Stroke

When the piston reaches the bottom of its power stroke, the momentum of the flywheel and crankshaft bring the piston back up the cylinder. This is the *exhaust stroke.* The exhaust valve opens, the intake valve remains closed, and the piston pushes the burned exhaust gases out of the cylinder and combustion chamber.

TWO-STROKE CYCLE OPERATION

Another common operating principle for gasoline engines is the two-stroke cycle engine, figure 3-30A. This engine is also very successful and in common use today, especially for outboard motors, chain saws, and lawn mowers, figure 3-30B. The two-stroke cycle engine is constructed somewhat differently from the four-stroke cycle engine. See figure 3-31, or a comparison of moving parts.

Two-stroke cycle means that it takes two strokes of the piston to complete the operating cycle of the engine.

The basic two-stroke cycle consists of a pumping action in the crankcase – high and low pressure caused by the movement of the piston up and down the cylinder. On a four-stroke cycle engine the crankcase plays no part in fuel induction but on a two-stroke cycle engine the crankcase is a vital part.

The two-stroke cycle starts with the piston moving up the cylinder as seen in figure 3-32. As the piston moves up the cylinder, a low pressure or partial vacuum is created in the crankcase; there is a larger space for the

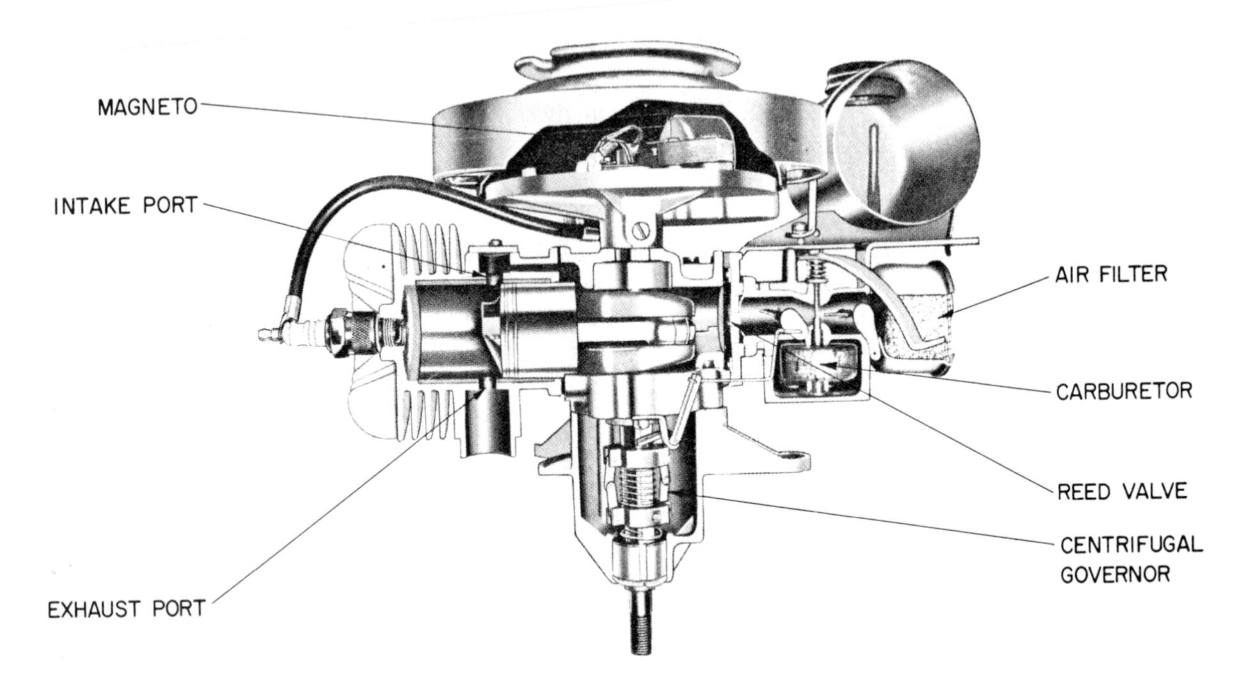

Fig. 3-30A Two-cycle engine

Fig. 3-30B The lightweight and constant lubrication features of a two-cycle engine make it ideal for chain saws. (Photo courtesy of John Deere and Company)

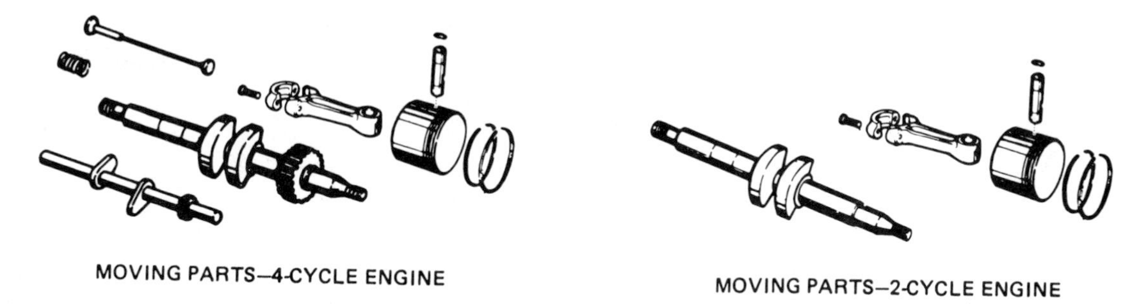

Fig. 3-31 Comparison of four-cycle and two-cycle engines

same amount of air. The greater atmospheric pressure outside rushes through the carburetor pushing open the springy reed valve, filling the crankcase with fuel mixture. When the piston is at the top of its stroke, the pressure in the crankcase and the atmospheric pressure are just about equal and the leaf valve springs shut.

As the piston moves back down the cylinder, the fuel mixture trapped in the crankcase is put under pressure. The reed valve can only open in the opposite direction. Near the bottom of the piston's stroke the intake transfer ports are uncovered by the piston. The pressurized fuel mixture in the crankcase now has a path to escape into the cylinder and it rushes into the cylinder where it can be compressed and burned.

The fuel charge, now in the cylinder, is compressed as the piston moves up the cylinder. A spark ignites the fuel mixture and the great pressure of combustion pushes the piston rapidly down the cylinder. As the piston nears the bottom of its stroke the exhaust ports begin to uncover and the exhaust gases, which are still hot and under pressure, start to rush through the exhaust ports and out of the engine.

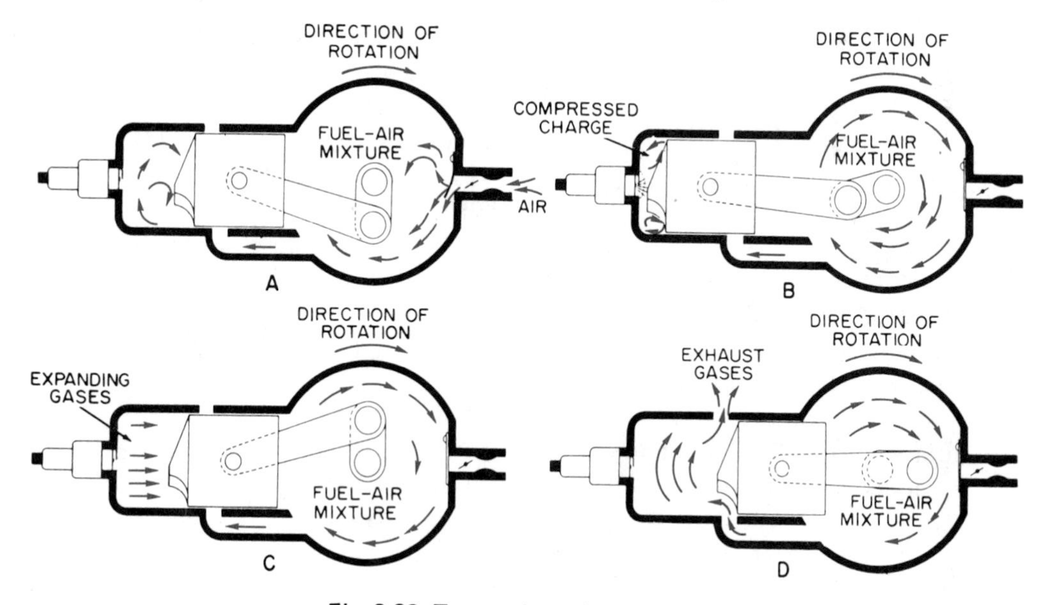

Fig. 3-32 Two-cycle engine operation

Just after the exhaust ports begin to uncover, the downward motion of the piston uncovers the intake transfer ports and compressed fuel mixture from the crankcase begins to rush in. Exhaust is leaving the cylinder at the same time new fuel mixture is entering. The incoming fuel mixture is directed toward the top of the cylinder and its action helps to scavenge out the exhaust gases. With the transfer of fuel complete, the piston moves up covering the intake transfer ports and the exhaust ports thus trapping the fuel mixture for compression.

The cycle is then repeated. If the engine operates at 4000 rpm, the cycle is completed 4000 times each minute. There is a power stroke for each revolution of the crankshaft.

To summarize the two-stroke cycle operation:

- Air/fuel mixture in the crankcase is put under pressure when the piston is going down on the power stroke.
- At the bottom of the power stroke, exhaust gases leave and air/fuel mixture is transferred from the crankcase to the cylinder.
- During the compression stroke, air/fuel mixture is brought into the crankcase.

Crankcase

The crankcase of a two-stroke cycle engine is designed to be as small in volume as possible. The smaller the volume, the greater the pressure created as the piston comes back down the cylinder. Crankcase pressure moves the fuel mixture through the engine. The greater the crankcase pressure, the more efficient the transfer of fuel from the crankcase through the transfer ports into the cylinder. The crankcase must be air tight. The only openings are the reed valves and the intake transfer ports which open only at specific times.

Moving parts in the crankcase must be lubricated; however, there is no reservoir of oil in the crankcase. Instead, the lubricating oil is mixed with the gasoline. A small amount of oil enters the crankcase each time the gasoline and air-fuel mixture enters the crankcase.

Intake Transfer Ports and Exhaust Ports

Intake transfer ports and exhaust ports are holes drilled in the cylinder wall which allow exhaust gases to escape and new fuel mixture to enter from the crankcase. The ports are located near the bottom of the cylinder and are covered and uncovered by the piston. By using ports, many parts seen in the four-cycle engine are eliminated: the valves, tappets, valve springs, cam gears, and camshaft. The remaining major moving parts are the piston, connecting rod, and crankshaft.

Piston (Cross Scavenged)

The piston still serves the same function in the cylinder, but for two-cycle engines the top is usually designed differently, figure 3-33. The two-stroke cycle pistons have a contoured top. On the intake side there is a sharp deflection that sends the incoming air/fuel mixture to the top of the cylinder. On the exhaust side there is a gentle slope so that exhaust gases have a clear path to escape. The loop scavenged two-cycle engine does, however, have a flat top piston.

Loop Scavenging

The loop scavenged two-cycle engine is basically the same as any other two-cycle engine. However, it does not need the contoured piston of the cross scavenged design because it does not have to deflect incoming

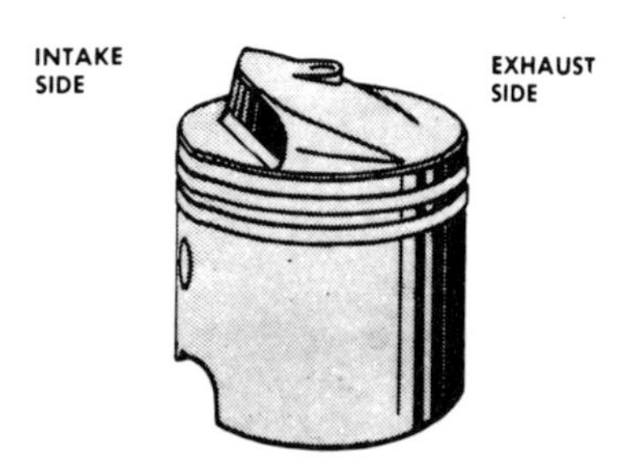

Fig. 3-33 A piston for a two-cycle engine showing the exhaust side and the intake side

gases. The difference is that the intake ports are drilled into the cylinder walls at an angle which aims the incoming fuel mixture toward the top of the cylinder. Usually there are two pairs of intake ports angling in from the sides directing the fuel mixture to the top of the cylinder around and back through the center of the cylinder to scavenge out the last of the exhaust through the exhaust ports. Loop scavenging provides a more complete removal of exhaust gases and produces somewhat more horsepower per unit weight.

Reed or Leaf Valves

Reed, or leaf valves, are located between the carburetor and crankcase. The valve itself is a thin sheet of springy alloy steel. The valve is pushed open by atmospheric pressure which allows fuel mixture to enter the crankcase; it springs closed to seal the crankcase. This occurs when a low pressure is created in the crankcase by the action of the piston moving up the cylinder. There may be only one reed valve or there may be several reed valves working together. The majority of small gasoline engines operating on the two-cycle principle use reed valves even though it is possible to use a poppet valve in the crankcase. Several kinds of reed plates are shown in figure 3-34.

Not all two-cycle engines use reed or leaf valves. Some engines have the compression and power strokes with intake and exhaust taking place between the two but use different valves to achieve the result.

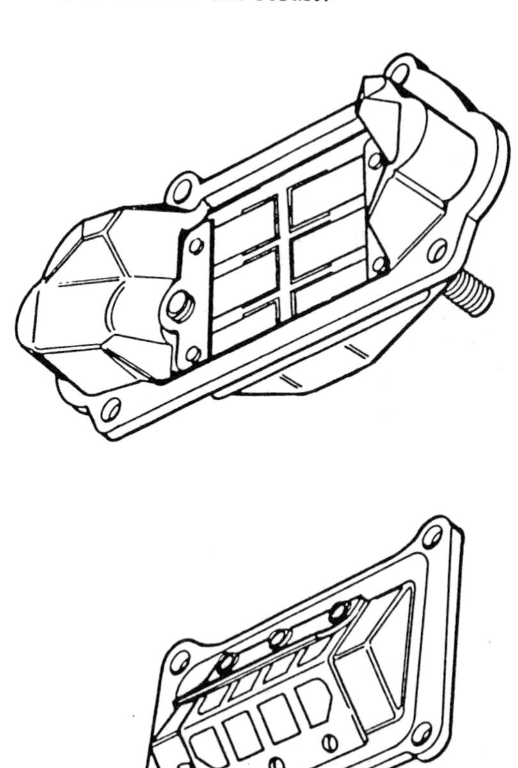

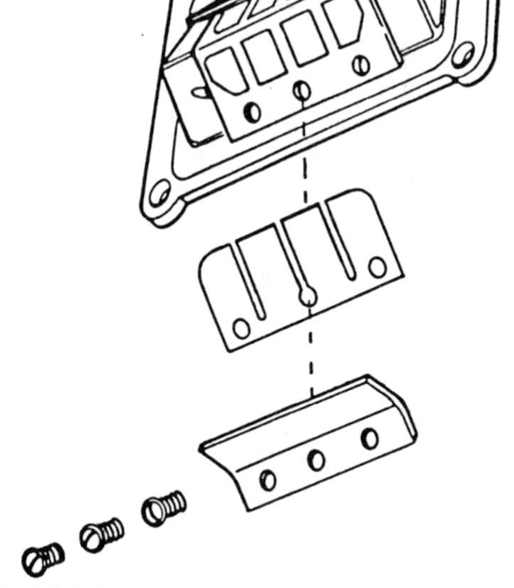

Fig. 3-34 Reed plates (Courtesy Tecumseh Products Company)

Rotary Valve

Some two-cycle engines use a rotary value for admitting fuel mixture, figure 3-35. The rotary valve is a flat disc with a section removed; it is fastened to the crankshaft. The rotary valve normally seals the crankcase. However, as the piston nears the top of its stroke and there is a slight vacuum in the crankcase, the valve is rotated to its open position. This allows fuel mixture to travel from the carburetor through the open valve and into the crankcase.

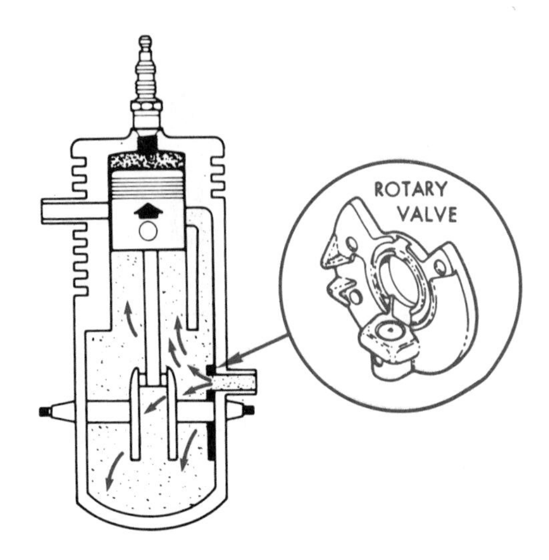

Fig. 3-35 Rotary valve

Third Port Design

Third port design has the regular exhaust ports and intake ports on the cylinder wall but in addition it has a third port in the cylinder wall, figure 3-36. This third port is for admitting fuel mixture to the crankcase. The bottom of the piston skirt uncovers the third port as the piston nears the top of its stroke. The piston's upward travel creates a low pressure in the crankcase. When the port is uncovered, air/fuel mixture rushes into the crankcase. As the piston comes down on the power stroke, the port is sealed off and the trapped mixture is pressurized. It is now ready for transfer through the intake ports into the cylinder.

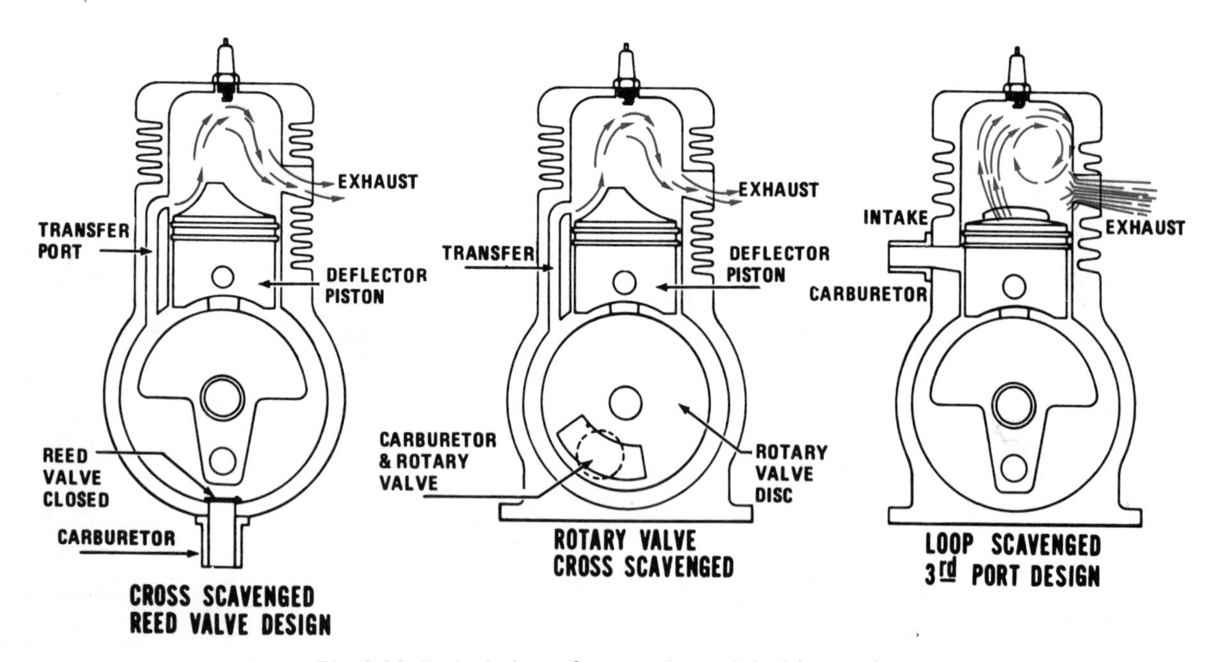

Fig. 3-36 Basic designs of two-cycle spark ignition engines

Poppet Valves

There are some two-cycle engines that use poppet valves to admit new fuel mixture into the crankcase. These poppet valves may be spring loaded and operated by differences in crankcase pressure. They open when the partial vacuum in the crankcase overcomes a slight spring tension. This happens when the piston is on its upward stroke. With the poppet valve open, fuel mixture rushes into the crankcase. The poppet valve may also be operated by cam action. In this case a crankshaft cam opens the valve for the piston's upward stroke.

Fig. 3-37 An all-terrain-vehicle has a powerful small gas engine.

REVIEW QUESTIONS

1. Explain the term *internal combustion.*
2. List the most essential moving parts of the four-stroke gasoline engine.
3. List the most essential stationary parts of the four-stroke gasoline engine.
4. Define the terms: bore, stroke, TDC, and BDC.
5. What purpose does a gasket serve?
6. What is a combustion chamber?
7. What two purposes do piston rings serve?
8. What is the function of the crankshaft?
9. What parts do the connecting rod and wrist pin work with?
10. What three things does the flywheel accomplish?
11. What does a key and keyway do?
12. Explain how the valves operate.
13. Explain what a valve lifter does.
14. Prepare a sketch of a cam.

15. On a four-cycle engine, at what speed does the camshaft revolve relative to the crankshaft?

Four-Stroke Cycle Review Questions

1. Explain what causes the fuel mixture to rush into the cylinder during the intake stroke.
2. What is accomplished on the compression stroke?
3. Explain the power stroke.
4. How many revolutions are required for the complete cycle?

Two-Stroke Cycle Review Questions

1. What parts are replaced by the intake and exhaust ports?
2. What are ports?
3. How are some two-stroke cycle pistons different from four-stroke cycle pistons?
4. Explain the action of the reed or leaf valves.
5. Explain how intake and exhaust take place on a two-cycle engine.
6. What part does the crankcase play in regard to the fuel mixture?
7. How many revolutions are required for the complete cycle?
8. What is the advantage of loop scavenging?

Unit 4 Fuel Systems, Carburetion, and Governors

OBJECTIVES

After completing this unit, the student should be able to:

- Identify the basic parts of a fuel system and explain the function of each part.
- Identify the basic parts of the carburetor and explain the function of each.
- Discuss the theory of carburetor operation.
- Discuss several common types of carburetors.
- Discuss how gasoline vaporizes and burns and common problems such as vapor lock, detonation and preignition.
- Identify the basic parts of air vane and mechanical governors and explain how they work.
- Adjust the carburetor for maximum power and efficiency.
- Adjust the governed speed and discuss the action of the governor on the throttle.

The fuel system must maintain a constant supply of gasoline for the engine under all operating speeds and conditions. The carburetor must correctly mix the gasoline and air together to form a combustible mixture which burns rapidly when ignited in the combustion chamber.

A typical fuel system contains a *gasoline tank* (reservoir for gasoline); a *carburetor* (mixing device for gasoline and air); the *fuel line* (tubes made of rubber, copper, or steel through which the gasoline passes from the gas tank to the carburetor); and an *air cleaner* (device for filtering air brought into the carburetor), figure 4–1. In addition, the system may have a *shutoff valve* (valve at the gas tank that can cut off the gasoline supply when the engine is not in use), figure 4–2, a *fuel pump* (pump that supplies the carburetor with a constant supply of gasoline); a *sediment bowl* (small glass bowl attached to the fuel line where dirt and other foreign matter can settle out), figure 4–3, and a *strainer* (fine screen in the gas tank to prevent leaves and dirt from entering the fuel line).

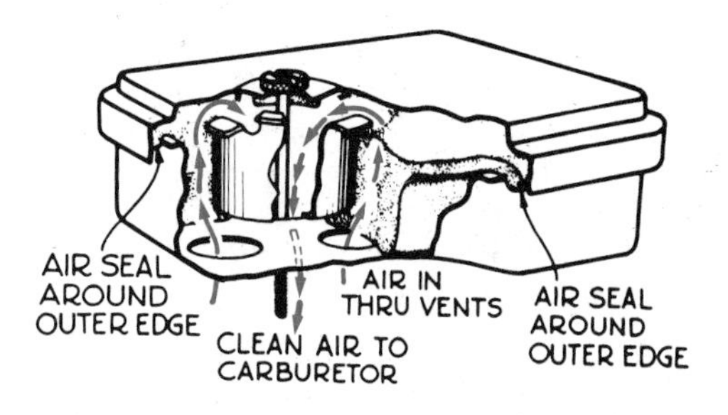

Fig. 4-1 Air cleaner

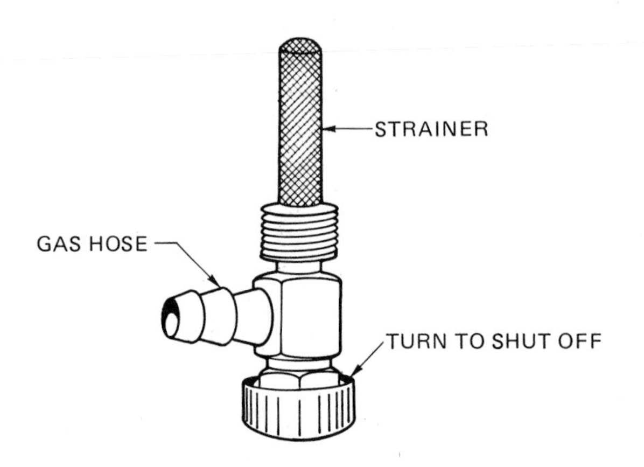

Fig. 4-2 Fuel shutoff valve

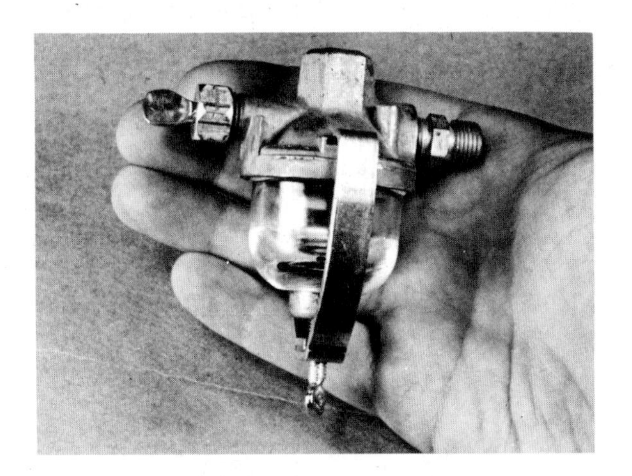

Fig. 4-3 Sediment bowl with built-in fuel shut-off valve

It should be noted that not all fuel systems contain all eight of the basic parts. Shut-off valves, sediment bowls, fuel pumps, and air cleaners are not common to all engines.

A constant supply of gasoline must be available at the carburetor. To provide this supply, four systems are in common use today. These systems are the suction system, gravity system, fuel pump system, and pressurized tank system.

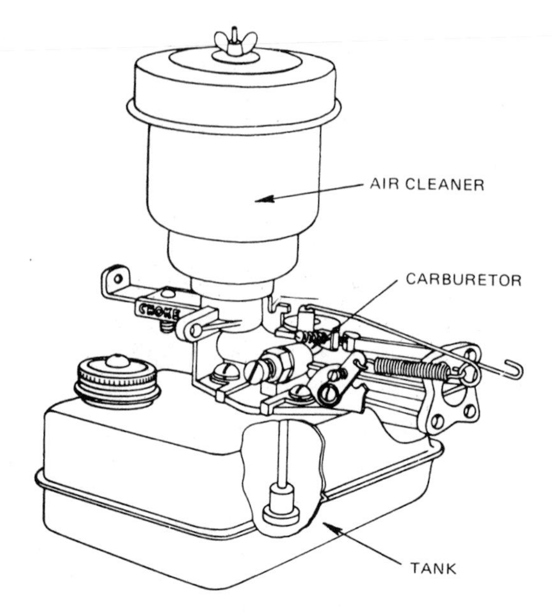

Fig. 4-4 Suction feed fuel system

SUCTION SYSTEM

The suction system is the simplest fuel system, figure 4–4. With this system, the gas tank is located below the carburetor and the gasoline is pushed up into the carburetor. However, the gas tank cannot be very far away from the carburetor or the carburetor action will not be strong enough to allow the gasoline to be pushed from the tank.

GRAVITY SYSTEM

With the gravity system the gas tank is located above the carburetor and the gasoline runs downhill to the carburetor, figure 4–5. To prevent gasoline from continuously pouring through it, the carburetor has incorporated a float and float chamber. The *float bowl* provides a constant level of gasoline without flooding the carburetor. When gasoline is used, the float goes down, opening a valve to admit more gasoline to the float bowl. The float rises

Fig. 4-5 Gravity feed fuel system

and shuts off the gasoline when it reaches its correct level. In actual practice the float and float valve do not rapidly open and close but assume a position allowing the correct amount of fuel to constantly enter the float chamber, figure 4–6. If the engine is speeded up, a new position is assumed by the float and float valve supplying an increased flow of gasoline.

FUEL PUMP SYSTEM

On many engines it is necessary to place the gas tank some distance from the carburetor. Therefore, the fuel must be brought to the carburetor by some means other than gravity or suction. One method is by using a

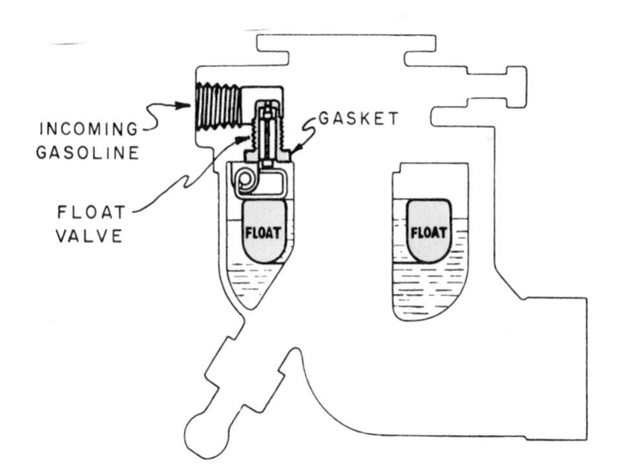

Fig. 4-6 Cutaway of a carburetor showing float and float valve

fuel pump. The automobile engine uses a fuel pump. The outboard motor with a remote fuel tank is a good example of a smaller engine that commonly uses a fuel pump.

The fuel pump used on many two-stroke cycle outboard motors is quite simple. The fuel pump consists of a chamber, inlet and discharge valve, a rubber diaphragm, and a spring. It is operated by the crankcase pressure. As the piston goes up, low pressure in

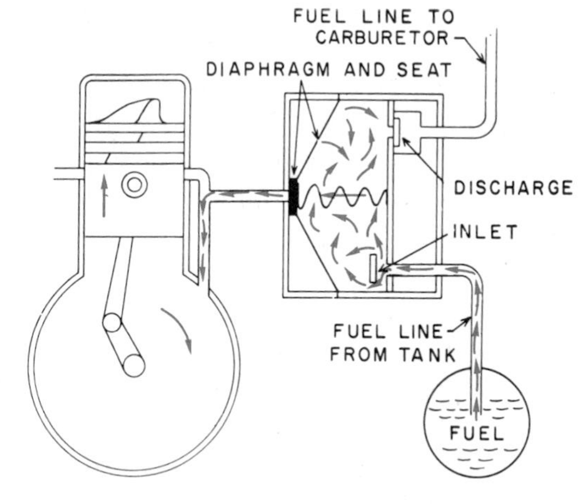

Fig. 4-7 Low pressure in the crankcase allows the fuel chamber to be filled.

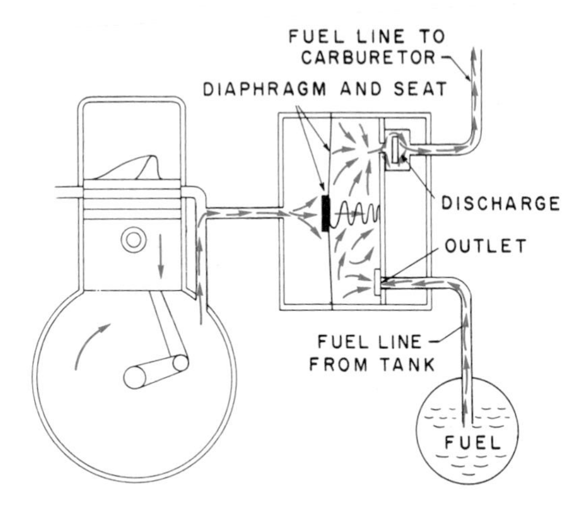

Fig. 4-8 High pressure in the crankcase forces the trapped fuel onto the carburetor.

the crankcase pulls the diaphragm toward the crankcase bringing gasoline through the inlet valve into the fuel chamber, figure 4–7. As the piston comes down, pressure in the crankcase pushes the diaphragm away from the crankcase. When this happens, the intake valve closes and the discharge valve opens allowing the trapped fuel to be forced to the carburetor, figure 4–8.

PRESSURIZED FUEL SYSTEM

Another method for forcing fuel to travel long distances to the carburetor is the pressurized fuel system, figure 4–9. This method is sometimes used with outboard motors. Engines using this system have two hoses or lines between the gas tank and the engine. One line brings gasoline to the carburetor; the other line brings air under pressure from the crankcase to the gas tank. If this method is used, the gas tank must be airtight so that enough air pressure can build up to force the gasoline to flow to the carburetor.

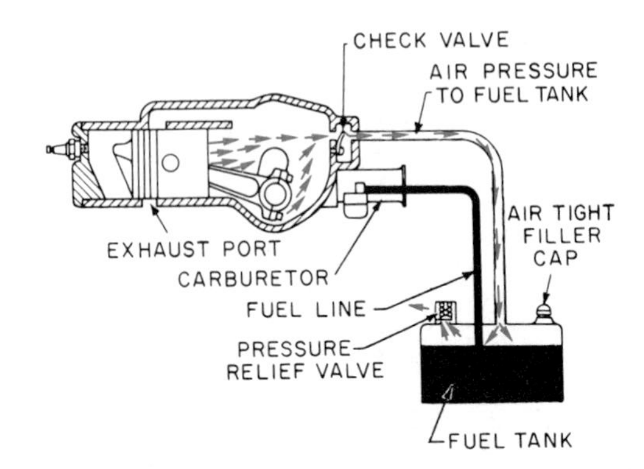

Fig. 4-9 Pressurized fuel systems can be found on some outboard engines.

THE CARBURETOR

The carburetor must prepare a mixture of gasoline and air in the correct proportions for burning in the combustion chamber, figures 4–10 and 4–11. The carburetor must function correctly under all engine speeds, under varying engine loads, in all weather conditions, and at all engine temperatures. To meet all these requirements, carburetors have many built-in parts and systems. Before studying the carburetor's operation in detail, an examination of the basic carburetor parts and their functions is helpful.

Throttle. The throttle controls the speed of the engine by controlling the amount of fuel mixture that enters the combustion chamber by regulating the air flow. The more fuel mixture admitted, the faster the engine speed.

Choke. The choke controls the air flow into the carburetor. It is used only for starting the engine. In starting the engine, the operator closes the choke, cutting off most of the engine's air supply. This produces a *rich mixture* (one containing a higher percentage of gasoline) which ignites and burns readily in a cold engine. As soon as the engine starts, the choke is opened.

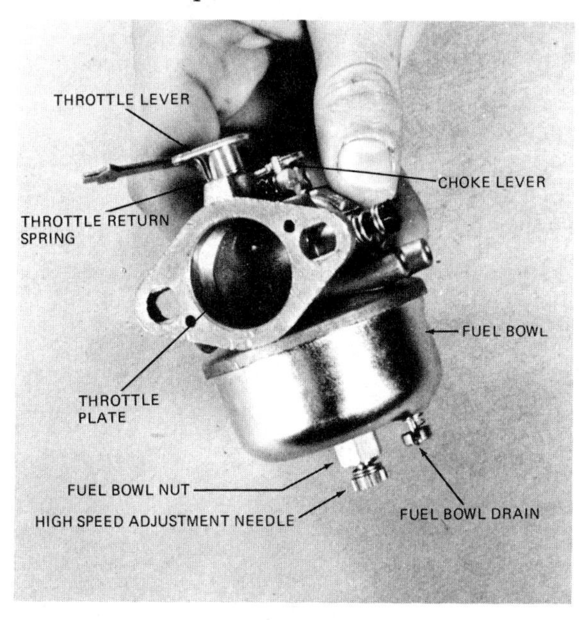

Fig. 4-10 Typical carburetor

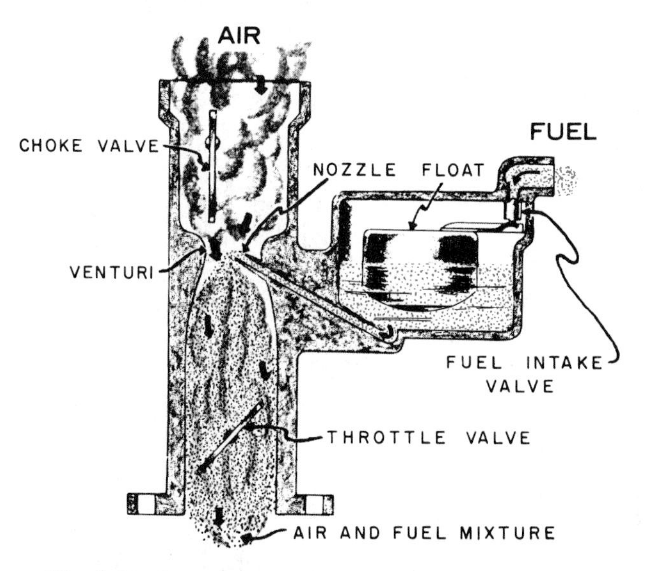

Fig. 4-11 Cross section of a simplified carburetor

Needle Valve. The needle valve controls the amount of gasoline that is allowed to pass out of the carburetor into the engine, figure 4–12. It controls the richness or leanness of the fuel mixture.

Idle Valve. The idle valve also controls the amount of gasoline that is available to the carburetor but it functions only at low speeds or idling. Some carburetors have what is called a *slow-speed needle valve* which performs basically the same job as the idle valve.

Float and Float Bowl. The float and float bowl are found on all carburetors except the suction-fed carburetor and diaphragm carburetors. The float and float bowl maintain a constant gasoline level in the carburetor.

Venturi. The venturi is a narrow section of the carburetor that has a smaller cross-sectional area for the air to flow through. In the venturi, the gasoline and air are brought together and begin to mix.

Jets. Carburetor jets are small openings through which gasoline passes within the carburetor.

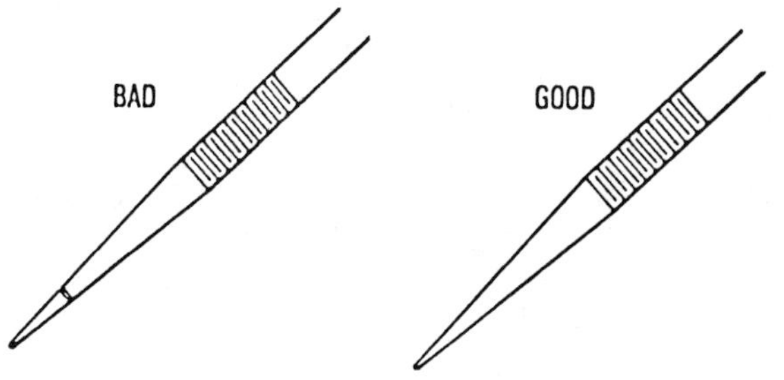

Fig. 4-12 Needle valves (Courtesy Tecumseh Products Company)

CARBURETOR OPERATION

In the beginning, air flows through the carburetor because there is a partial vacuum or low pressure in the combustion chamber as the piston travels down the cylinder. One usually thinks of the air being sucked into the engine by the piston action. Atmospheric pressure outside the engine actually pushes the air through the carburetor to equalize the lower pressure that is in the combustion chamber.

Air flows rapidly through the carburetor as the piston moves down. In the carburetor the air must pass through a constriction called the venturi, figure 4-13. For the same amount of incoming air to pass through this smaller opening, the air must travel faster and this it does. Here a principle of physics comes into use: the greater the velocity of air passing through an opening, the lower the static air pressure exerted on the walls of the opening. The venturi thus creates a low-pressure area within the carburetor, figure 4-14.

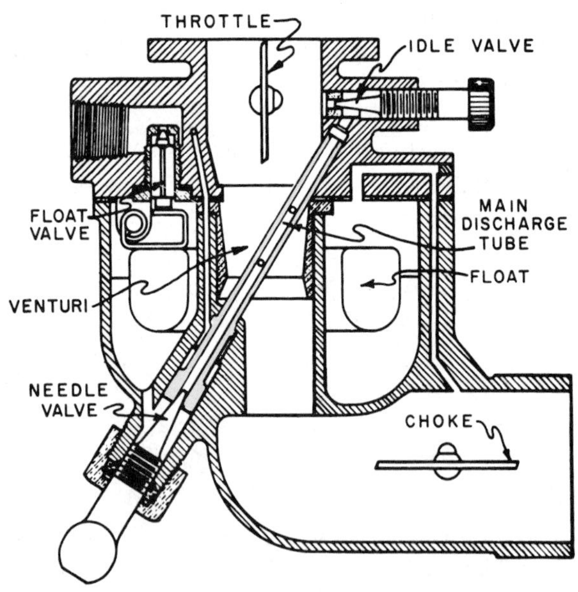

Fig. 4-13 A gravity fed carburetor

The principle of the airfoil is also used to gain lower pressure conditions in the venturi. A fuel supply tube or jet is placed in the venturi section. The action of the incoming air causes a high pressure on the front of the jet but a very low pressure on the back of the jet. Gasoline is available in this jet and streams out of the jet because a low-pressure area has been created by the action of the venturi and the airfoil, and because the gasoline is under atmospheric pressure which is greater. Greater pressure pushes the gasoline out of the discharge jet into the airstream. See figure 4-15 for a simplified illustration of carburetor operation.

As gasoline streams into the airflow, it is mixed thoroughly with the air. The best mixture of gasoline and air is fourteen or fifteen parts of air to one part of gasoline, by weight. This air-to-gasoline ratio can be changed for different operating conditions; heavy load and fast acceleration require more gasoline (richer mixture). The needle valve is used to change this ratio; it controls the

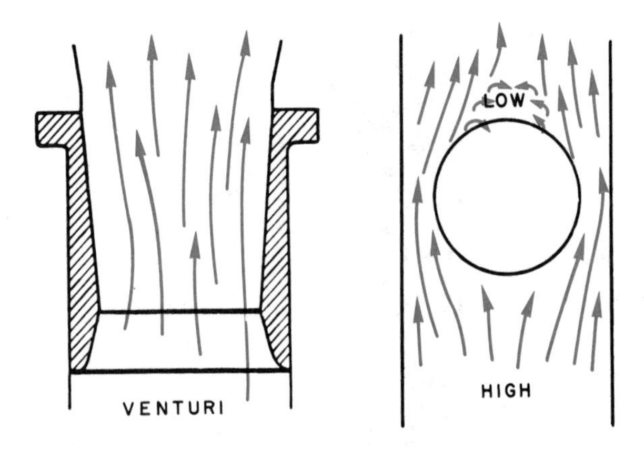

Fig. 4-14 (A) The venturi creates a low pressure in the carburetor. (B) The principle of the air foil also helps create a low pressure.

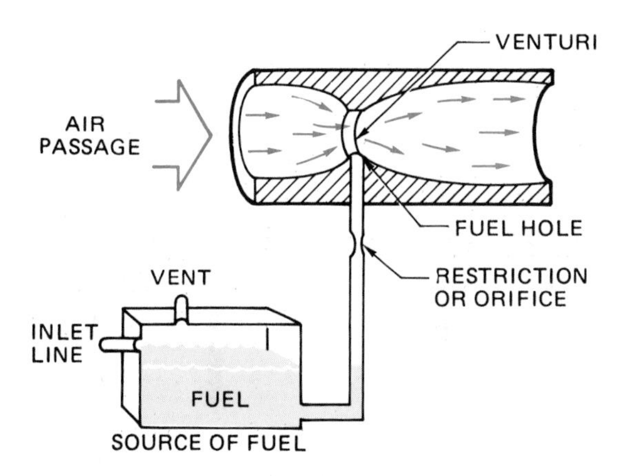

Fig. 4-15 Simplified carburetor operation

amount of gasoline that is available to be drawn from the discharge jet.

The throttle, or "butterfly" valve, is mounted on a shaft beyond the venturi section. The operator of the engine controls its setting to regulate the engine speed. When the throttle valve is wide open it does not restrict the flow of air; air flows easily through the carburetor and the engine is operating at its top speed. As the operator closes the throttle, the flow of air is restricted. A smaller amount of air can rush through the carburetor, therefore, the air pressures at the venturi section are not as low and less gasoline streams from the discharge holes. With less fuel mixture in the combustion chamber, the piston is pushed down with less force during combustion; power and speed are reduced.

The ratio of air to fuel remains about the same through the different throttle settings. However, when the throttle is closed and the engine begins to idle, very little air is drawn through the carburetor and the difference between atmospheric pressure and venturi air is slight. Little gasoline streams from the discharge jet. In fact, the mixture of gasoline is so lean that a special idling device must be built into the carburetor to provide a richer mixture for idling.

In some carburetors, the main discharge jet is continued past the venturi section to the area of the throttle. It discharges fuel into a small well and jet that are behind the throttle when it is closed. The air pressure behind the throttle is very low. Therefore, gasoline streams from the idle jet readily and mixes with the small amount of air that is coming through the carburetor and a rich fuel mixture is provided for idling. A threaded needle valve called the *idle valve* controls the amount of gasoline that can flow from the idle jet. Many carburetors use the air bleed principle to help atomize the gasoline and to make the needle valve easier to adjust. See figure 4-16 for an illustration of these conditions.

When a cold engine is to be started, an extremely rich mixture of gasoline and air must be provided if the engine is to start easily. The choke provides this rich mixture, figure 4-17. The choke is a plate, similar to the throttle, placed in the air horn before the venturi section. For starting, the choke is closed, shutting off most of the carburetor's air supply. When the engine is turned over slowly, usually by hand, little air is drawn through the carburetor. The air pressure within the carburetor is very low and gasoline streams from the main discharge jet, mixing with the air that does get by the choke. As soon as the engine starts, the choke is opened. The fuel mixture provided when the engine is choked is so rich that all the gasoline may not vaporize with the air and raw liquid gasoline may be drawn into the combustion chamber. Continued operation with the choke closed can cause crankcase dilution (the raw gasoline seeps into the crankcase diluting the lubricating oil).

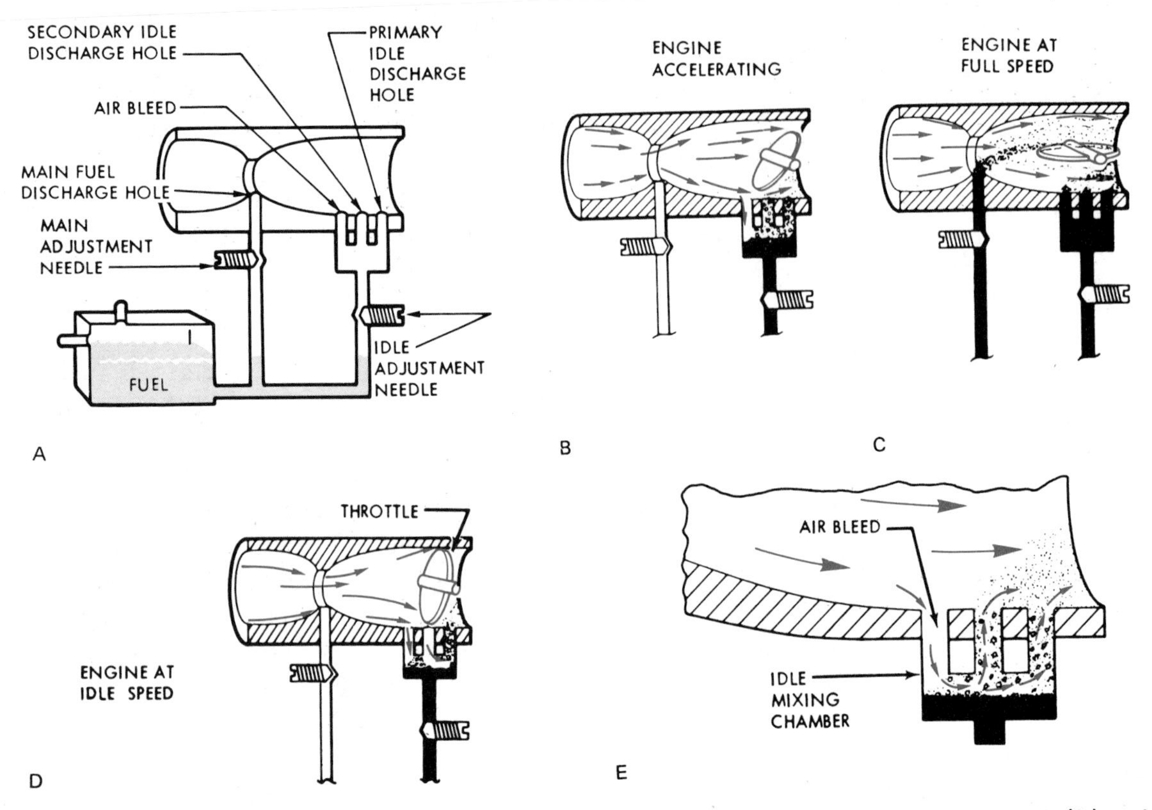

Fig. 4-16 (A) Schematic of basic carburetor parts, (B) engine accelerating, (C) engine at full speed, (D) engine at idle speed, and (E) air bleed principle

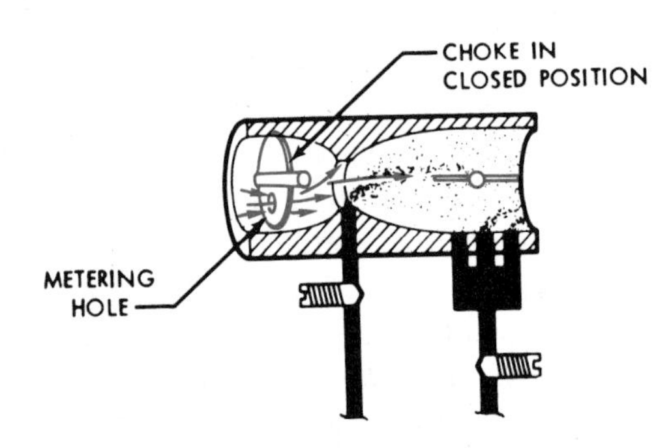

Fig. 4-17 The choke provides a rich mixture for starting.

The complete carburetor operation includes:

- the float and float bowl maintaining a constant reservoir of gasoline in the carburetor
- the venturi section producing a low-pressure area
- the needle valve controlling the richness or leanness of fuel mixture
- the main discharge jet spraying gasoline into the airstream
- the idle valve providing a rich mixture for idling conditions
- the choke producing an extremely rich mixture for easy starting.

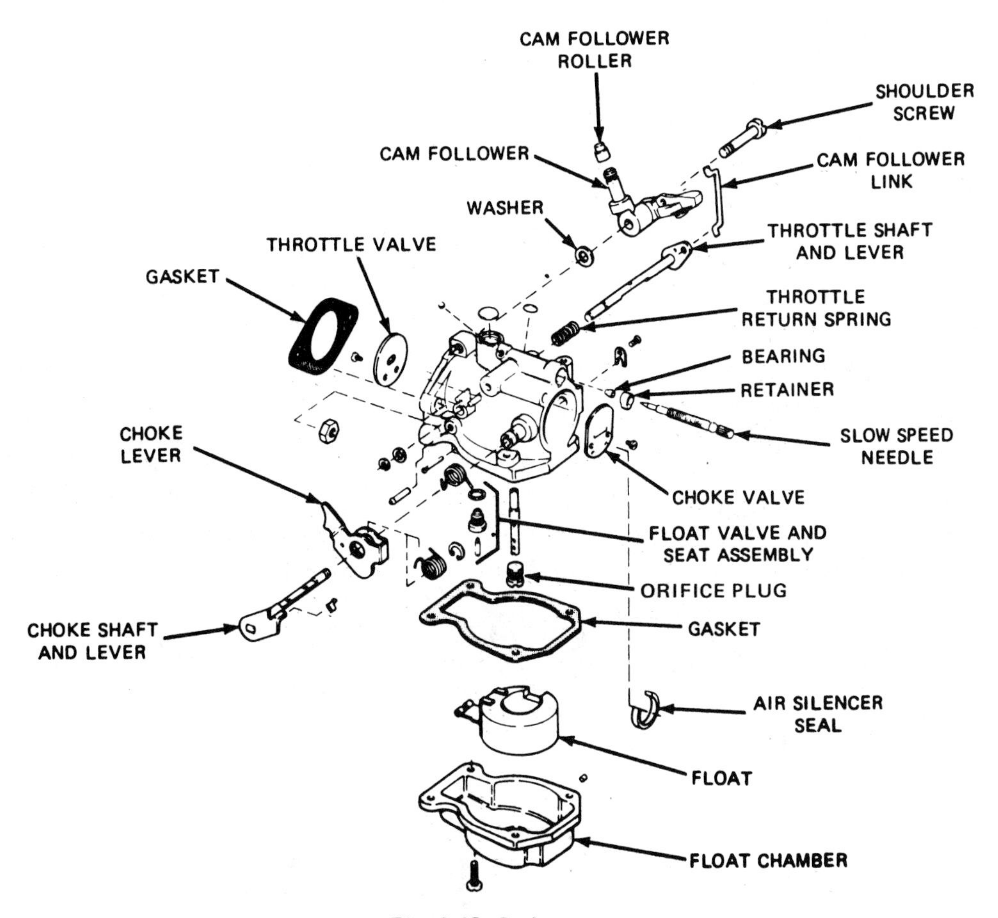

Fig. 4-18 Carburetor components

See figure 4–18 for an exploded view of a carburetor and 4–19 for an illustration of where it fits into the fuel system.

Primer

It should be noted that some carburetors have a *primer* instead of a choke for providing the extra gasoline and rich mixture for starting. Pushing the primer causes liquid or raw gasoline to be squirted into the venturi section of the carburetor. One or two pushes on the primer provides enough gasoline for starting the engine.

DIAPHRAGM CARBURETOR

The diaphragm carburetor has come into wide use, especially on chain saws. It is also found on other applications where the engine may be tipped at extreme angles. The diaphragm supplies the carburetor with a constant supply of gasoline.

The diaphragm carburetor may be gravity fed. Crankcase pressure moves the diaphragm and associated linkages to allow gasoline to enter the fuel chamber. The gasoline is

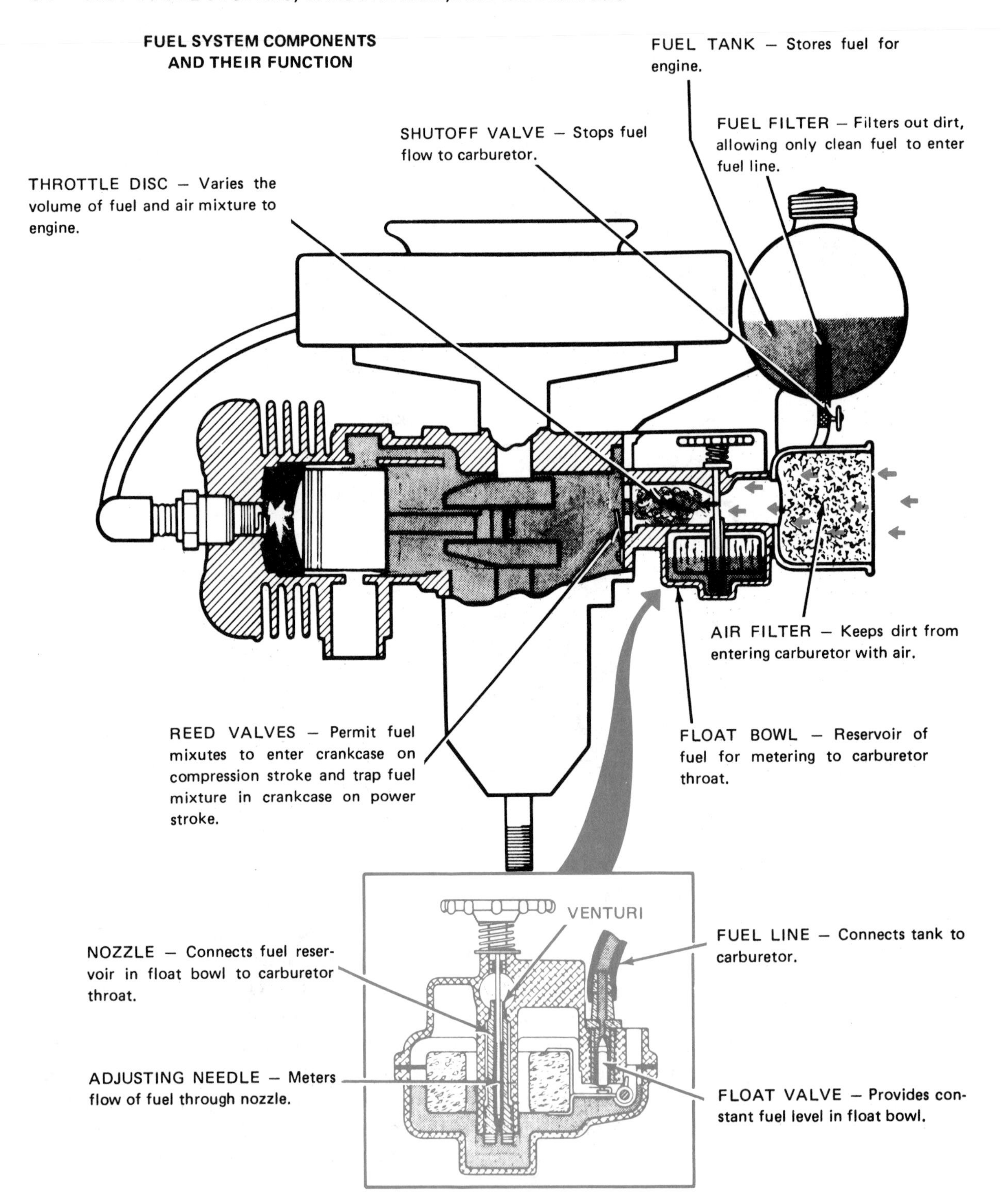

Fig. 4-19 Two-stoke cycle carburetor and fuel system

Fig. 4-20 Chain saws commonly use diaphragm carburetors (Courtesy Beaird Poulan Weed Eater, Emerson Electric Company)

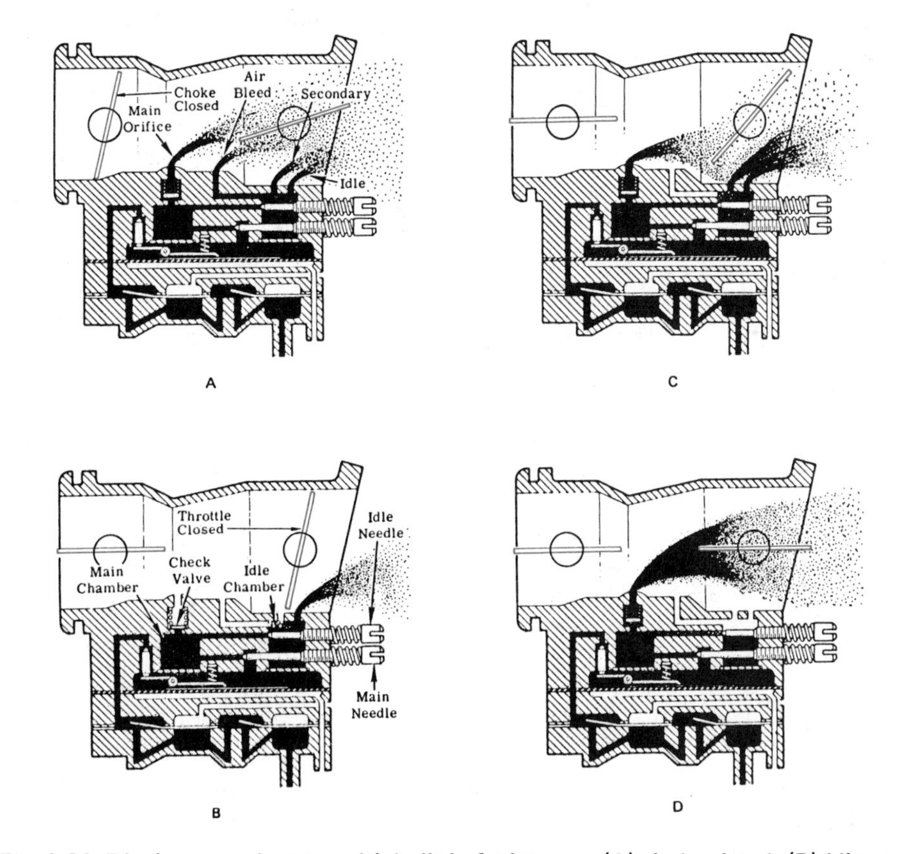

Fig. 4-21 Diaphragm carburetor with built-in fuel pump. (A) choke closed, (B) idle, (C) part throttle, (D) full throttle

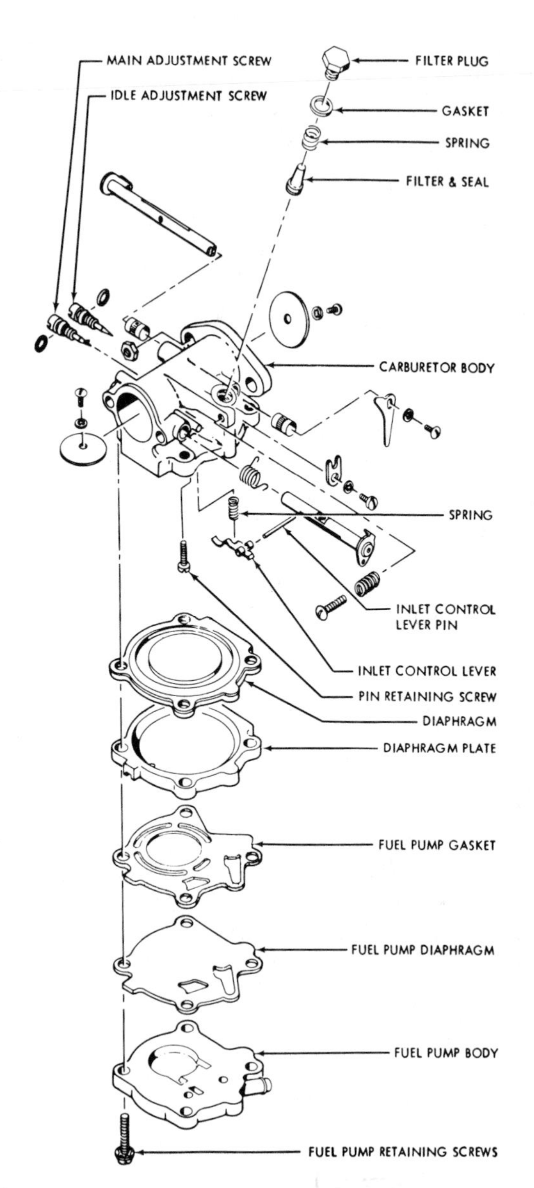

Fig. 4-22 An exploded view of a diaphragm carburetor with a built-in fuel pump

available by gravity and the in-and-out motion of the diaphragm meters out fuel in the correct amount.

Diaphragm carburetors are also made with a built-in fuel pump, figures 4–21 and 4–22. The fuel pump operates on differences in crankcase pressure; its action is much the same as that of the outboard fuel pump previously discussed. Diaphragm carburetors are widely used on small engines.

Other Carburetors

A common carburetor variation uses both the fuel pump and suction feed principles. The Briggs and Stratton Pulsa-Jet® is such a carburetor. At first this carburetor may appear to be an ordinary suction type carburetor but closer examination shows there are two fuel pipes into the carburetor – one long and one short.

The low pressure created by the intake stroke activates the fuel pump drawing gasoline through the long pipe and discharging it into a smaller constant level fuel chamber at the top of the tank. This provides a constant level of gasoline regardless of the amount of gasoline in the tank. It also reduces the amount of lift required to deliver gasoline into the venturi. This carburetor permits a larger venturi and improves the engine's horsepower rating. The Briggs and Stratton Pulsa-Jet® is shown in figures 4-23 and 4-24.

AIR BLEEDING CARBURETORS

There is a tendency for carburetors to supply too rich a fuel mixture at high speeds; the ratio of gasoline to air increases as the velocity of the air passing through the carburetor increases. One method that is commonly used to correct this condition is *air bleeding* (see figure 4–25, view E). Air bleeding involves introducing a small amount of air into the main discharge well vent to restrict the flow of gasoline from the main discharge jet. As engine speeds are increased, greater amounts of air are brought into the main discharge well vent placing a greater restriction on the gasoline flow. The additional air

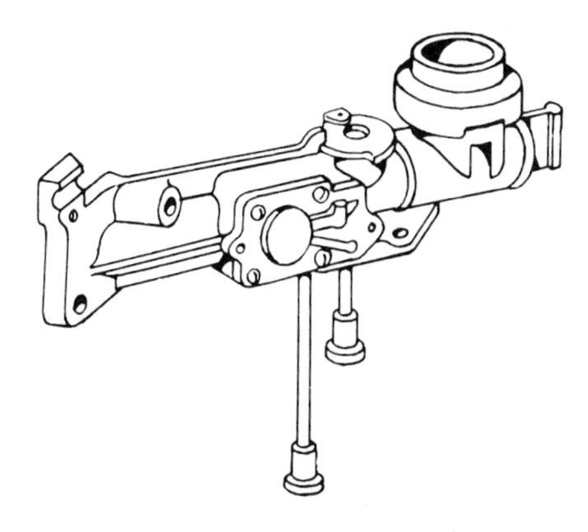

Fig. 4-23 Pulsa-Jet® carburetor

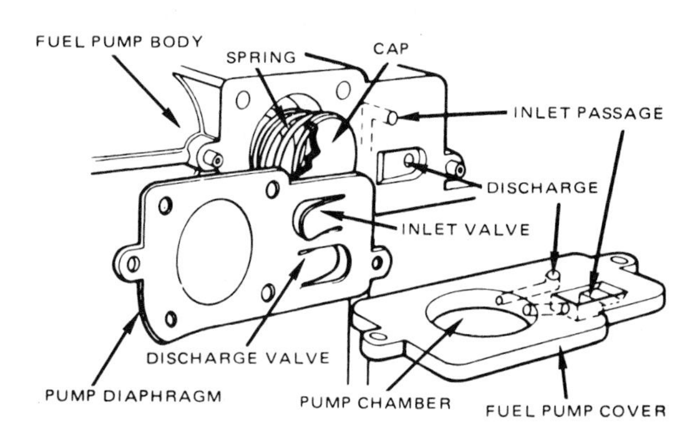

Fig. 4-24 Pulsa-Jet® carburetor pump

overcomes the carburetor's natural tendency to provide too rich a mixture at high speeds. This action maintains the proper ratio of fuel and air between a throttle setting of one-fourth to wide open. The air that enters the discharge well vent mixes with the gasoline and is drawn through the main discharge jet into the main airstream. The Zenith® carburetor shown in figure 4-25, uses the principle of air bleeding.

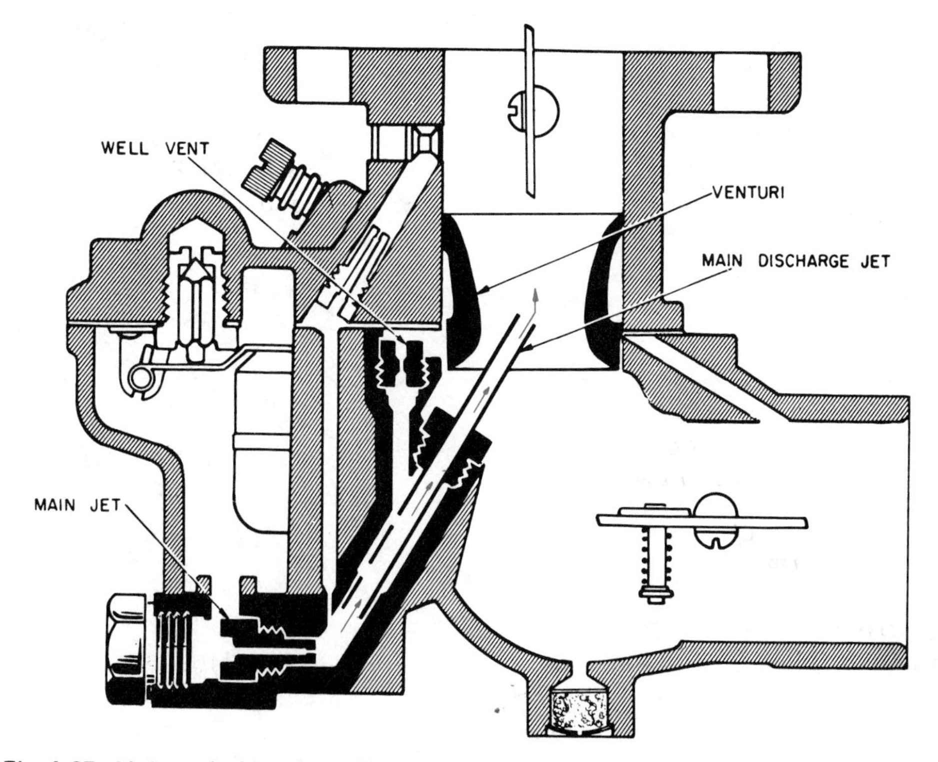

Fig. 4-25 Air brought into the well vent bleeds into the main discharge jet, maintaining the correct air-fuel ratio throughout throttle range.

CARBURETOR ADJUSTMENTS

The engine manufacturer sets the carburetor adjustments at the factory. These settings cover normal operation. However, after a long period of usage, or under special operating conditions, it may be necessary to adjust the carburetor. In adjusting the carburetor, the main needle valve and the idle valve are both reset to give the desired richness or leanness of fuel mixture. Too lean a mixture can be detected by the engine missing and backfiring. Too rich a mixture can be detected by heavy exhaust and sluggish operation.

Procedure

This is a typical procedure for adjusting a carburetor for maximum power and efficiency.

1. Close the main needle valve and idle valve finger tight. Excessive force can damage the needle valve. Turn clockwise to close.
2. Open the main needle valve one turn. Open the idle valve 3/4 turn. Turn counterclockwise to open.
3. Start the engine, open the choke, and allow the engine to reach operating temperature.
4. Run the engine at operating speed (2/3 to 3/4 of full throttle). Turn the main needle valve clockwise slowly, 1/8 turn at a time, until the engine slows down indicating too lean a mixture. Note the position of the valve. Turn the needle valve counterclockwise until the engine speeds up and then slows down indicating too rich a mixture. Note the position of the valve. Reposition the valve halfway between the rich and lean settings.
5. Close the throttle so the engine runs slightly faster than normal idle speed. Turn the idle mixture valve clockwise until the engine slows down, then turn the idle valve counterclockwise slowly, 1/8 turn at a time, until the engine speeds up and idles smoothly. Adjust the idle-speed regulating screw to the desired idle speed, figure 4–26:

   NOTE: Idle speed is not the slowest speed at which the engine can operate; rather, it is a slow speed that maintains good airflow for cooling and a good take-off spot for even acceleration. A tachometer and the manufacturer's specifications regarding proper idle speed are necessary for the best adjustment.
6. Test the acceleration of the engine by opening the throttle rapidly. If acceleration is sluggish, a slightly richer fuel mixture is usually needed.

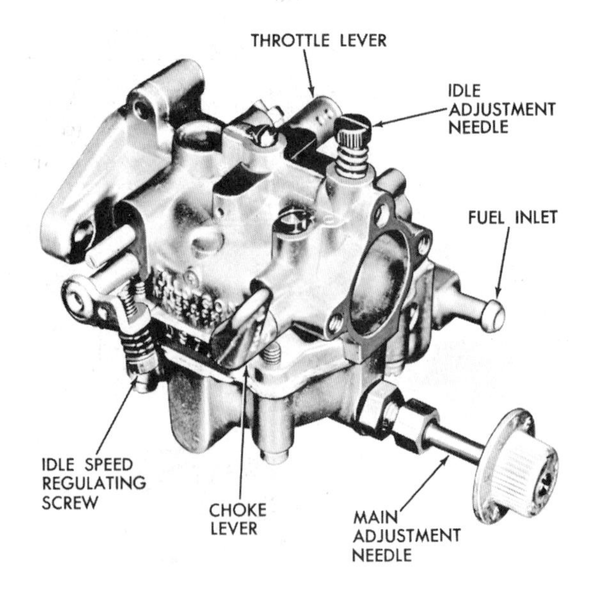

Fig. 4–26 Float-feed type carburetor

CARBURETOR ICING

Carburetor icing can occur when the engine is cold and certain atmospheric conditions are present, figure 4–27. If the temperature is between 28°F and 58°F and the relative humidity is above 70 percent, carburetor icing can take place.

Fuel mixture of gasoline and air rushes through the carburetor and the rapid action of evaporating gasoline chills the throttle plate to about 0°F. Moisture in the air condenses and freezes on the throttle plate when the relative humidity is high, figure 4–28. This formation of ice restricts the airflow through the carburetor; at low or idle settings the ice can completely block off the airflow stalling the engine.

When this condition is present, the engine can be restarted but stalls again at low or idle speeds. As soon as the carburetor is warm enough to prevent ice formation, normal operation can take place.

VAPOR LOCK

Vapor lock can occur anywhere along the fuel line, fuel pump, or in the carburetor when temperatures are high enough to vaporize the gasoline. Gasoline vapor in these places cuts off the liquid fuel supply stalling the engine. If vapor lock occurs, the operator must wait until the carburetor, gas line, and fuel pump cool off and the gasoline vapor returns to liquid before the engine will restart. Vapor lock usually occurs on unseasonably hot days and is more troublesome at high altitudes.

GASOLINE

Most internal combustion engines burn gasoline as their fuel. Gasoline comes from petroleum, also called crude oil. Crude oil is actually a mixture of different hydrocarbons such as gasoline, kerosene, heating oil, lubricating oil, and asphalt. These chemicals are all hydrocarbons but they have characteristics that are quite different. Hydrogen is a light, colorless, odorless gas; carbon is black and solid. Different combinations of carbon and hydrogen give the different characteristics of hydrocarbon products.

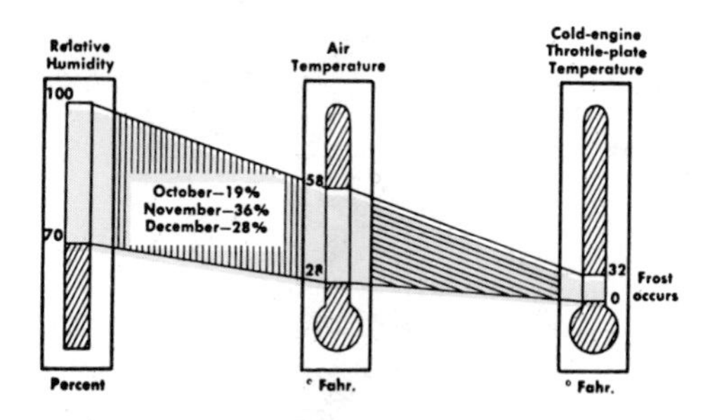

Fig. 4–27 Humidity and temperature conditions which may lead to carburetor icing

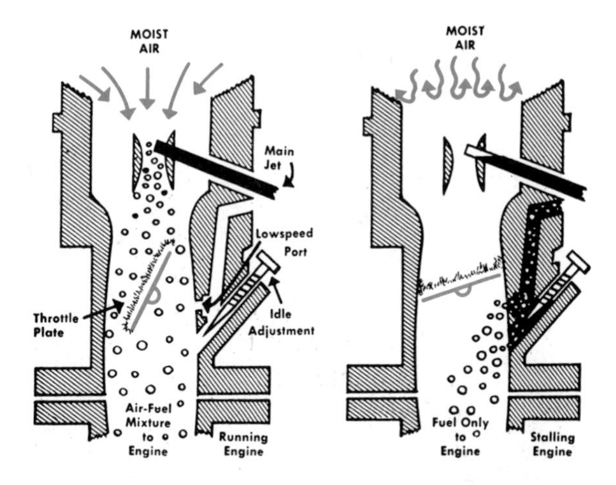

Fig. 4–28 Under carburetor icing conditions, ice forms on the throttle plate cutting off air to the engine when the throttle closes.

The various hydrocarbons are separated by distillation of the crude oil. Crude oil is first heated to a temperature of 700° to 800°F and then released into a fractionating or bubble tower, figure 4–29. The tower contains 20 to 30 trays through which hydrocarbon vapors can rise from below. When the heated crude oil is released into the tower, most of it flashes into vapor. The vapors cool as they rise. Each type of hydrocarbon condenses at a different tray level. Heavier hydrocarbons condense first at relatively high temperatures. The lighter hydrocarbons, such as gasoline, condense high in the tower at relatively low temperatures. Gasoline is then further processed to improve its qualities.

Good gasoline must have several characteristics.

- It must vaporize at low temperatures for good starting.
- It must be low in gum and sulfur content.
- It must not deteriorate during storage.
- It must not knock in the engine.
- It must have proper vaporizing characteristics for the climate and altitude.
- It must burn cleanly to reduce air pollution.
- It must remain a liquid until it mixes with air in the carburetor venturi.

The vaporizing ability of gasoline is the key to its success as a fuel. In order to burn inside the engine, the hydrocarbon molecules of the gasoline must be mixed with air since oxygen is also necessary for combustion. To mix the gasoline and air, there must be rapid motion and turbulence to keep the molecules suspended in the air, figure 4–30. There is rapid motion and turbulence in the carburetor;

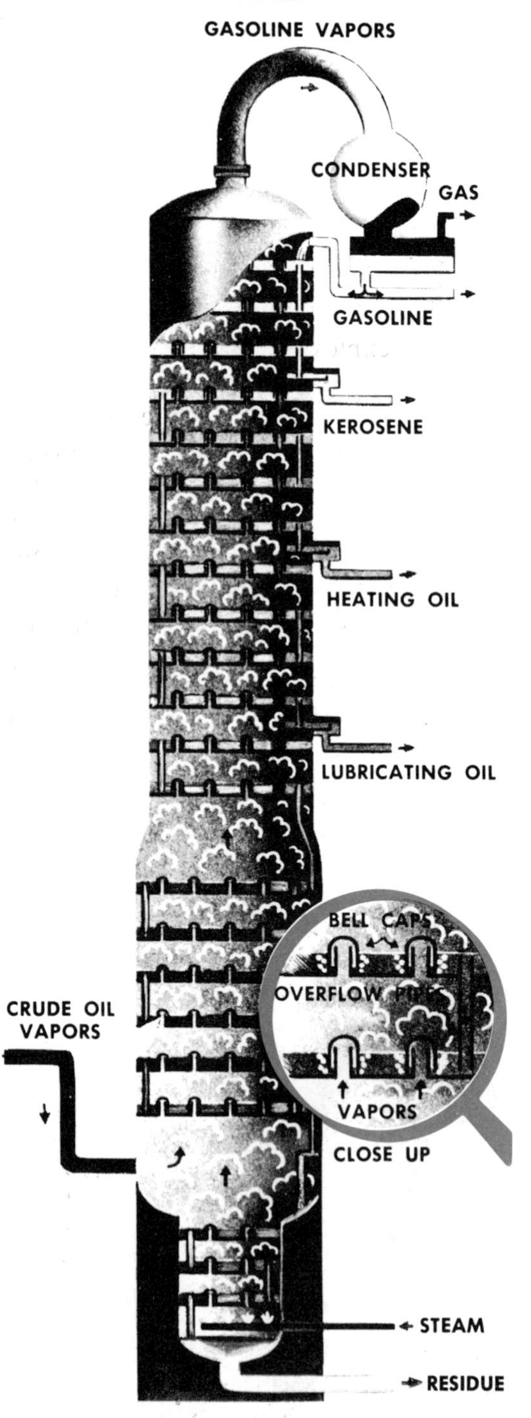

Fig. 4-29 Distillation of petroleum in the bubble tower

this turbulence continues into the combustion chamber. Combustion chambers are designed to create the maximum turbulence so that gasoline molecules will stay suspended in the air until they are ignited. A thorough mixture ensures smooth and complete burning of gasoline which delivers maximum power.

Although the vaporizing ability of gasoline makes it an excellent fuel, it also presents a great fire and explosion hazard. When vaporized, a gallon of gasoline produces 21 cubic feet of vapor. If this vapor were combined with air in a mixture of 1.4 to 7.6 percent gasoline by volume, it would burn rapidly when ignited. Therefore, only one gallon of gasoline properly vaporized would completely fill an average living room with explosive vapor. The safety rules given in unit 2 for the use and storage of gasoline should always be followed.

In engines, the compression of the fuel mixture of gasoline and air results in high combustion pressures and the force necessary to move the piston. This compression may also cause the gasoline to explode (*detonate*) in the engine instead of burning smoothly.

Detonation is also called knocking, fuel knock, spark knock, carbon knock, and ping. To understand it, one has to think in slow motion. Refer to figure 4-31. The spark plug ignites the fuel mixture, and a flame front moves out from this starting point. As the flame front sweeps across the combustion chamber, heat and pressure build. The unburned portion of the fuel mixture ahead of the flame front is exposed to this heat and pressure. If it self-detonates, two flame fronts are created which race toward each other. The last unburned portion of fuel caught between the two fronts explodes with hammerlike force. Detonation causes a knocking sound in the engine and power loss. Re-

Petroleum fuel – a mixture of many hydrocarbon compounds of different weights.

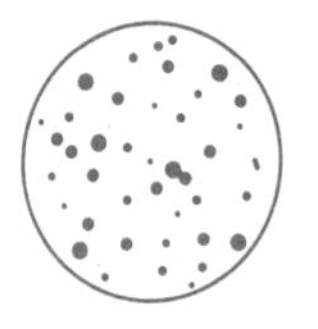

Like a stack of gravel, consisting of many pebbles basically alike, but of many different sizes—from larger, heavier ones to particles of fine dust.

Extreme turbulence is required to maintain suspension of all particles of fuel vapor – rapid movement through the carburetor and manifold, turbulence in the crankcase and combustion chamber.

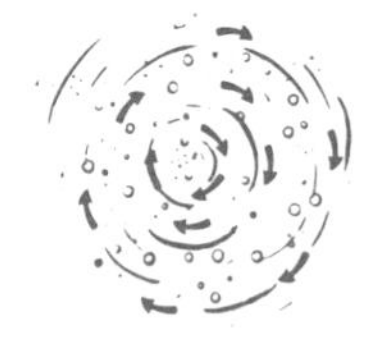

Like a whirlwind of gravel on a country road, all particles remain suspended as long as agitation and motion continue.

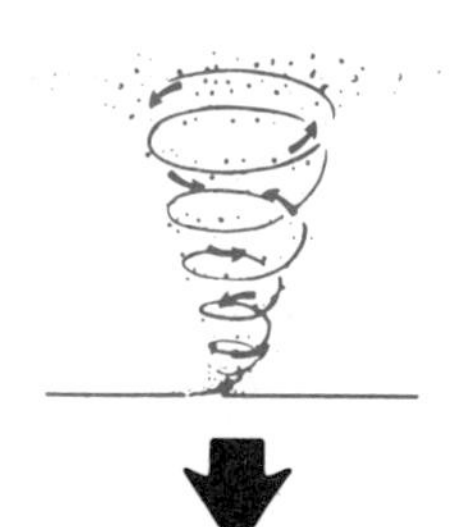

Heavier particles of the fuel settle or rain out of the fuel vapor with lessening turbulence, to a puddle below.

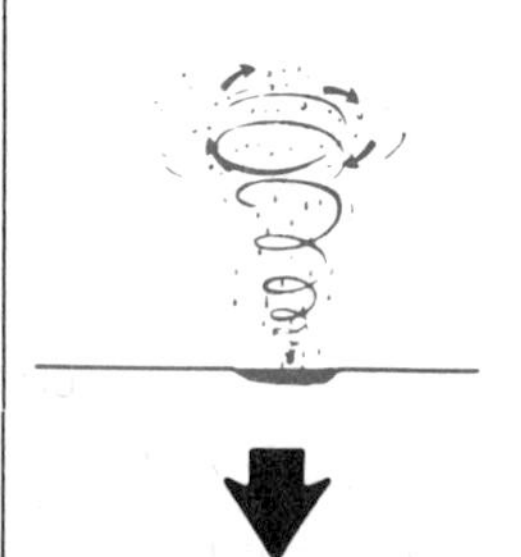

The very lightest particles (hydrocarbons) remain suspended like the heavier pebbles falling out of the whirlpool and returning to the roadway with lessening of the whirling wind. The very lightest particles remain suspended as dust.

Fig. 4-30 Turbulence is necessary to keep the gasoline molecules suspended in the air.

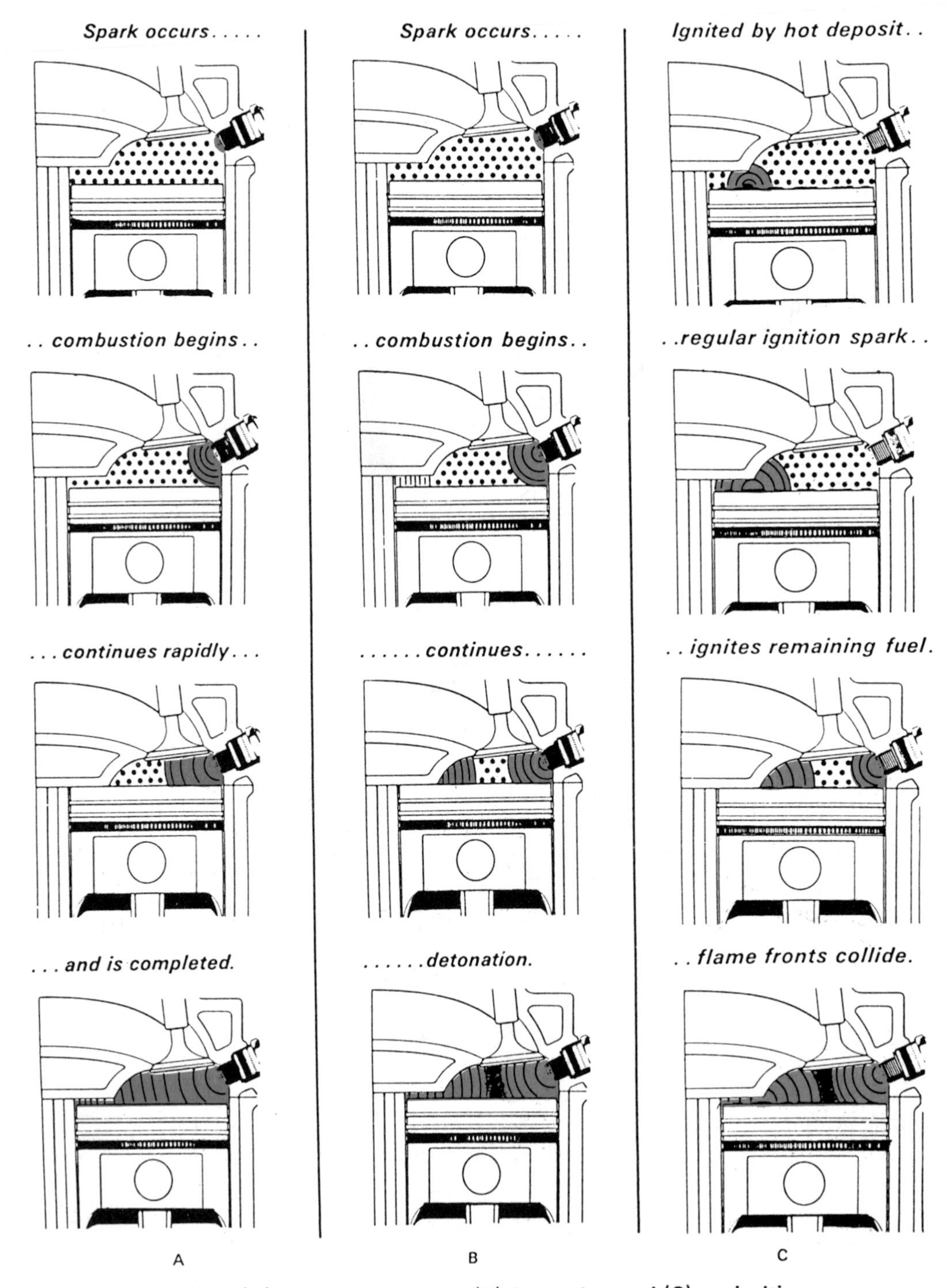

Fig. 4-31 (A) Normal combustion, (B) detonation, and (C) pre-ignition

Fig. 4-32 Detonation damage

Fig. 4-33 Pre-ignition damage

peated detonation can damage the piston, figure 4-32.

Pre-ignition causes the same undesirable effects as detonation, including engine damage, figure 4-33. The cause of pre-ignition, however, is somewhat different. Hot spots, or red-hot carbon deposits, actually ignite the fuel mixture and begin combustion before the spark plug fires.

The antiknock quality of a gasoline, or its ability to burn without knocking, is called its *octane rating.* Gasoline that has no knocking characteristics at all is rated at 100. Usually, tetraethyl lead is added to gasoline to give it the proper antiknock quality. If lead is not used as in unleaded gasoline, special aromatic compounds are used to obtain smooth burning.

The higher the compression ratio of the engine, the higher the octane requirements for the engine's fuel. Low-grade gasoline with an octane rating of 70–85 is suitable for compression ratios of 5–7:1. Regular grade gasoline with an octane rating of 88–94 is suitable for compression ratios of 7–8.5:1. Most small engine manufacturers recommend the use of regular gasoline. High-octane premium gasoline, leaded or unleaded, does not improve the performance of small engines. Premium gasoline with an octane rating of about 100 is suitable for compression ratios of 9–10:1. Super premium has an octane rating of over 100 and is good for engines with compression ratios of 9.5–10.5:1.

ENGINE SPEED GOVERNORS

Speed governors are used to keep engine speed at a constant rate regardless of the load. For instance, a lawn mower is required to cut tall as well as short grass. A governor ensures that the engine operates at the same speed in spite of the varying load conditions. Engines on many other applications also need this constant speed feature.

Speed governors are also used to keep the engine operating speed below a given, preset rate, so that the engine speed will not surpass this rate. This maximum rate is established and the governor set accordingly by the engineers at the factory. This type of speed governor protects both the engine and the operator from speeds that are dangerously high.

In the case of lawn mowers the governor also limits the top speed to conform to federal safety standards which limit the blade tip speed to 19,000 feet per minute.

There are two main governor systems used on small gasoline engines: (1) mechanical or centrifugal type and (2) pneumatic or air vane type. Although there are other types of governors, most of the governors used on small engines are included in one of these categories.

MECHANICAL GOVERNORS

The mechanical governor operates on centrifugal force, figures 4–34 to 4–36. With this method, counterweights mounted on a geared shaft, a governor spring, and the associated governor linkages keep the engine speed at the desired rpm. For constant speed operation the action follows this pattern: when the engine is stopped, the mechanical governor's spring pulls the throttle to an open position. The governor spring tends to keep

Fig. 4–34 A mechanical governor which operates on centrifugal force

Fig. 4-35 Mechanical governor (centrifugal)

the throttle open but as the engine speed increases centrifugal force throws the hinged counterweights further and further from their shaft. This action puts tension on the spring in the other direction to close the throttle. At governed speed, the spring tension is overcome by the counterweights and the throttle opens no further. A balanced position is maintained and the engine assumes its maximum speed.

If, however, a greater load is placed on the engine, it slows down; the hinged counterweights swing inward due to lessened centrifugal force; the governor spring becomes dominant, opening the throttle wider. With a wider throttle opening, the engine speeds up until governed speed is reached again. The action described is fast and smooth; little time is needed for the governor to meet revised load conditions.

The governed speed can be changed somewhat by varying the tension on the governor spring. The more tension there is on the governor spring, the higher the governed speed.

Another commonly used centrifugal or

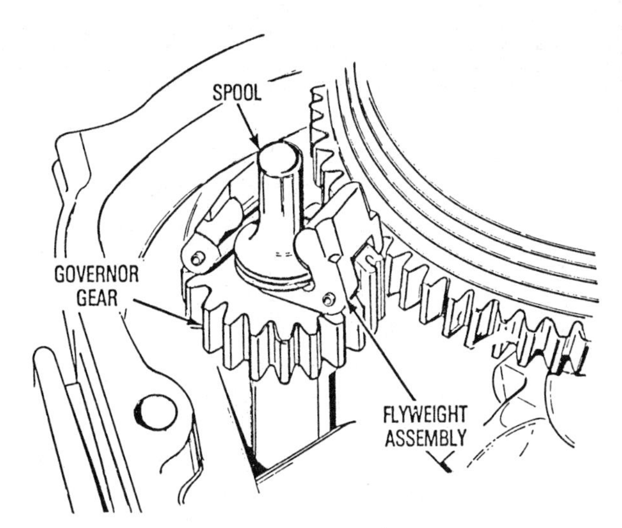

Fig. 4-36 Mechanical governor (Courtesy Tecumseh Products Company)

mechanical governor is the flyball type. With this governor, round steel balls in a spring loaded raceway move outward as the engine speed increases. The centrifugal force applied by these balls increases until it balances the spring tension holding the throttle open. At this point, the governed speed is reached.

PNEUMATIC OR AIR VANE GOVERNORS

Many engines have a pneumatic or air vane governor, figures 4-37 through 4-39. Again, the governor holds the throttle at the governed speed position which prevents it from opening further. An air vane, which is located near the flywheel blower, controls the speed. As engine speed increases, the flywheel blower pushes more air against the air vane causing it to change position. At governed speed the air vane position overcomes the governor spring tension and a balance is assumed.

Engines are often designed so that they can

be accelerated freely from idle speed up to governed speed. With these engines, the governor spring takes on very little tension at low speeds. The operator has full control. As the throttle is opened further and further, there is more tension on the spring, both from the operator and from the strengthening governor action. At governed speed, the force to close the throttle balances the spring tension to open the throttle.

It should be noted that not all engines have engine speed governors. Engines that are always operated under a near constant load do not need the governor. The load of an outboard engine is a good example of a near constant load. Although different loads are placed on the outboard (number of people in boat, position of people in boat, use of trolling mechanism to decrease boat's speed, etc.), the load is constant at full throttle for a given period of use.

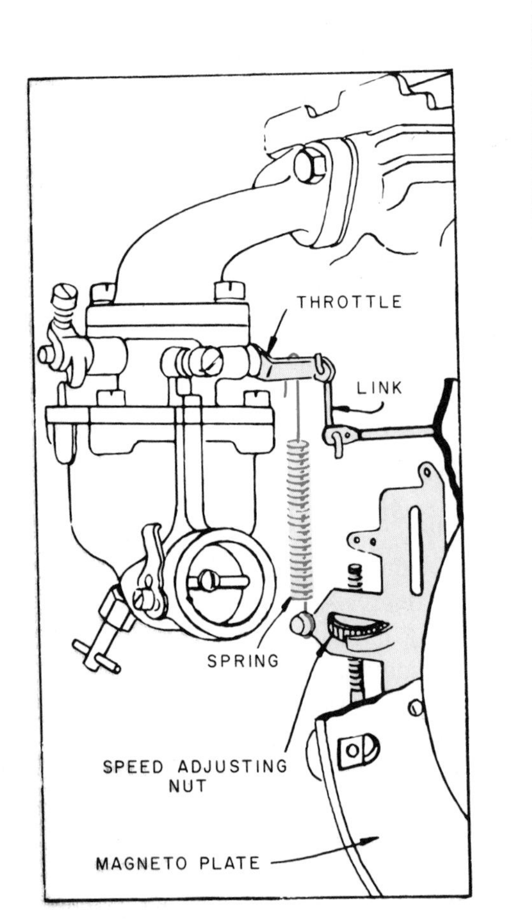

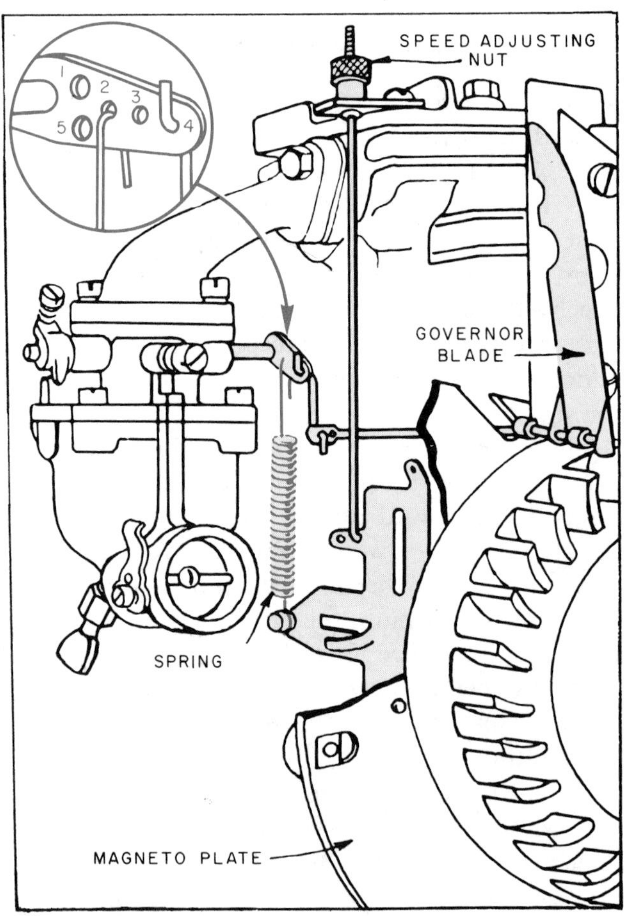

Fig. 4-37 A pneumatic or air vane governor

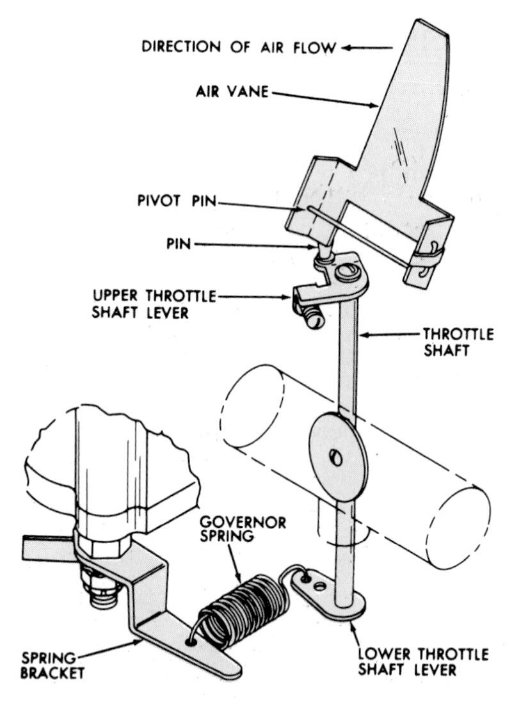

Fig. 4-38 A pneumatic or air vane governor

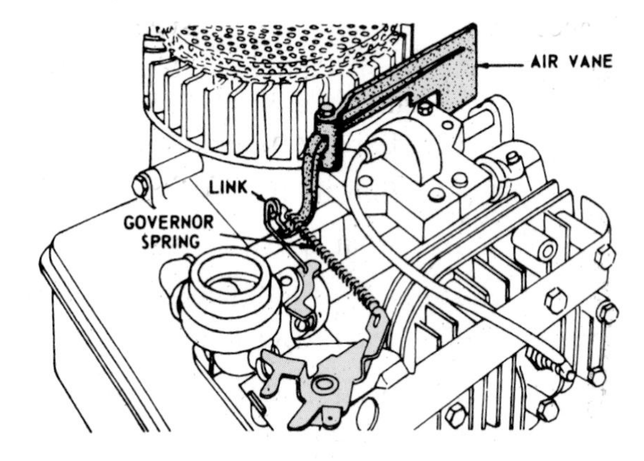

Fig. 4-39 Air vane governor

REVIEW QUESTIONS

1. Name the basic parts of a typical fuel system.
2. What force is used to operate most outboard fuel pumps?
3. How can a pressurized fuel system be recognized?
4. Briefly describe the task of the carburetor.
5. Name the main parts that make up the carburetor.
6. What causes air to flow through the carburetor?
7. What causes low air pressure in the venturi section of the carburetor?
8. What is the function of the throttle?
9. What is the function of the choke?
10. What is the function of the needle valve?

11. Explain the action of the float in a float-type carburetor.
12. What is the advantage of a diaphragm carburetor?
13. Explain the principle of air bleeding.
14. Explain the operation of the accelerating pump.
15. What is a rich mixture?
16. Why is turbulence important inside the combustion chamber?
17. What two main purposes do engine speed governors serve?
18. What are the two main governor systems used on small gasoline engines?
19. Why are some engines made that do not have a speed governor?
20. Can the governed speed be changed? How?

Unit 5 Lubrication

OBJECTIVES **After completing this unit, the student should be able to:**

- **Discuss the several functions of engine lubricating oil.**
- **Discuss friction bearings and antifriction bearings.**
- **Discuss lubricating oil quality designations and SAE numbers.**
- **Discuss how a two-stroke cycle engine is lubricated.**
- **Correctly drain and refill the crankcase with oil.**
- **Explain how a two-stroke cycle engine is lubricated.**

Whenever surfaces move against one another they cause friction which results in heat and wear. Lubricating oils have one main job to perform in the engine – to reduce friction. The lubricating oil provides a film that separates the moving metal surfaces and keeps the contact of metal against metal to an absolute minimum, figure 5-1.

Without a lubricating oil or with insufficient lubrication, the heat of friction builds up rapidly. Engine parts become so hot that they fail. That is when the metal begins to melt, bearing surfaces *seize* (adhere to other parts), parts warp out of shape, or parts actually break. The common expression is to say the engine burns up.

As an example of friction, lay a book on the table and then push the book slowly. Notice the resistance. Friction makes the book difficult to slide. Now place three round pencils between the book and the table top and push the book. Notice how easily it moves. Friction has been greatly reduced. Oil molecules correspond to the pencils by forming a coating between two moving surfaces. With oil, the metal surfaces roll along on the oil molecules and friction is greatly decreased.

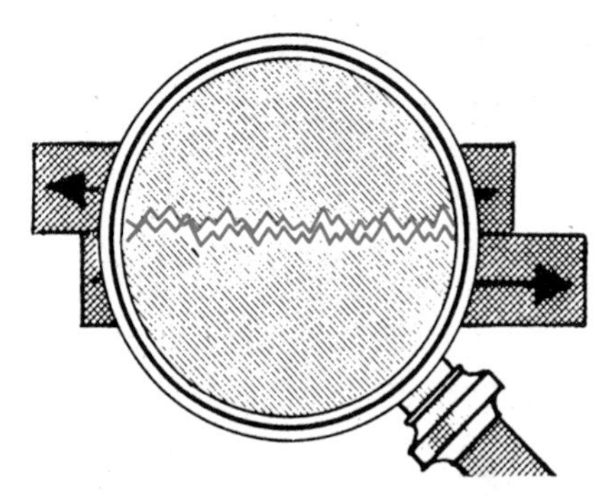

Fig. 5-1 Exaggerated view of metal surfaces in contact

Besides reducing friction and the wear and heat it causes, the lubricating oil serves several other important functions.

Oil Seals Power. The oil film seals power, particularly between the piston, piston rings, and cylinder walls. The great pressures in the combustion chamber cannot pass by the airtight seal which the oil film provides. If this oil film fails, a condition called blow-by exists. Combustion gases push by the film and enter the crankcase. Blow-by not only reduces engine power but also has a harmful effect on the oil's lubricating quality.

Oil Helps to Dissipate Heat. Oil helps to dissipate heat by providing a good path for heat transfer. Heat conducts readily from inside metal parts through an oil film to outside metal parts that are cooled by air or water. Also, heat is carried away as new oil arrives from the crankcase and the hot oil is washed back to the crankcase.

Oil Keeps the Engine Clean. Oil keeps the engine clean by washing away microscopic pieces of metal that have been worn off moving parts. These tiny pieces settle out in the crankcase or they are trapped in the oil filter, if one is used.

Oil Cushions Bearing Loads. The oil film has a cushioning effect since it is squeezed from between the bearing surfaces relatively slowly. It has a shock absorber action. For example, when the power stroke starts, the hard shock of combustion is transferred to the bearing surfaces; the oil film helps to cushion this shock. Figure 5-2 illustrates the oil clearance between the journal and the bearing.

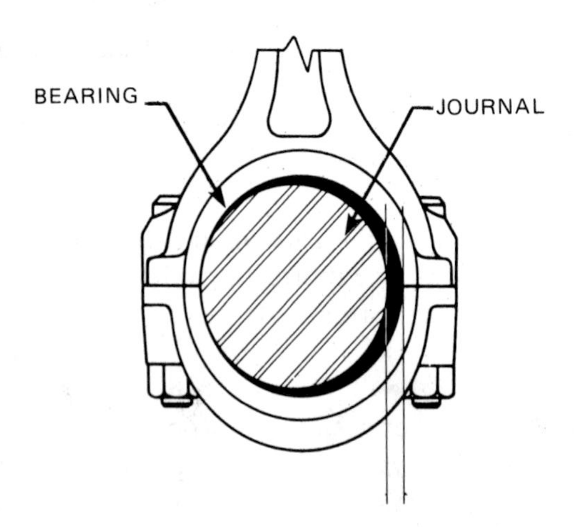

Fig. 5-2 Oil clearance between journal and bearing (exaggerated)

Oil Protects Against Rusting. The oil film protects steel parts from rusting. Air, moisture, and corrosive substances cannot reach the metal to oxidize or corrode the surface.

FRICTION BEARINGS

There are three types of friction bearings used in a small gasoline engine: journal, guide, and thrust, figure 5-3. The *journal bearing* is the most familiar. This bearing supports a revolving or oscillating shaft. The connecting rod around the crankshaft and the main bearings are examples. The *guide bearing* reduces the friction of surfaces sliding longitudinally against each other, such as the piston in the cylinder. The *thrust bearing* supports or limits the longitudinal motion of a rotating shaft.

Bearing inserts are commonly used in larger engines or heavy-duty small engines, particularly on the connecting rod and cap and the main bearings. The inserts are precision-made of layers of various metals and alloys. Alloys such as babbit, copper-lead, bronze, aluminum, cadmium, and silver are commonly used. Crankshaft surfaces are made of steel. Most small engines, however, do not have bearing inserts. They use only the aluminum connecting rod around the steel journal of the crankshaft.

ANTIFRICTION BEARINGS

Antifriction bearings are also commonly used in engines, figure 5-4. They substitute rolling friction for sliding friction. Ball bearings, roller bearings and needle bearings

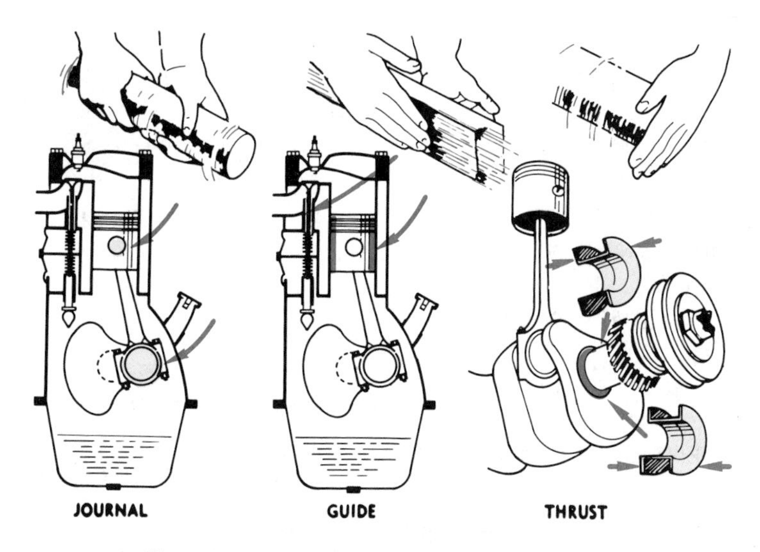

Fig. 5-3 Three major types of friction bearings

are of this type, figure 5-5. On many small engines, the main bearings are of the antifriction type.

QUALITY DESIGNATION OF OIL AND SAE NUMBER

Selecting engine lubricating oils can be confusing. One must be aware of different viscosities, different qualities, different refining companies, different additives, and the meaning of many advertising phrases. Generally, lubricating oils used for various small gasoline engines are the same ones that are used for automobile engines. However, two-stroke cycle engines almost always use special two-cycle oil.

Quality Designation by the American Petroleum Institute (API)

Figure 5-6 is an example of lubricating

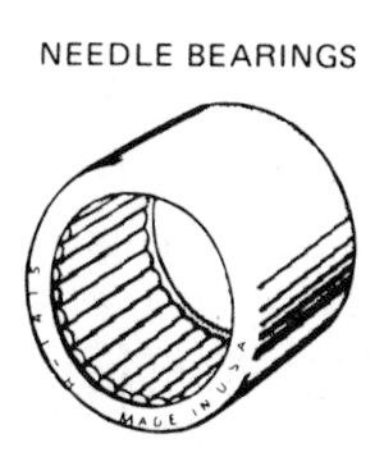

Fig. 5-4 Types of antifriction bearings

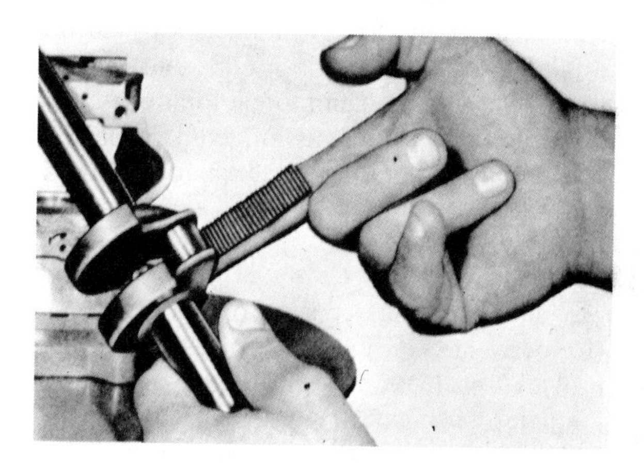

Fig. 5-5 Needle bearings

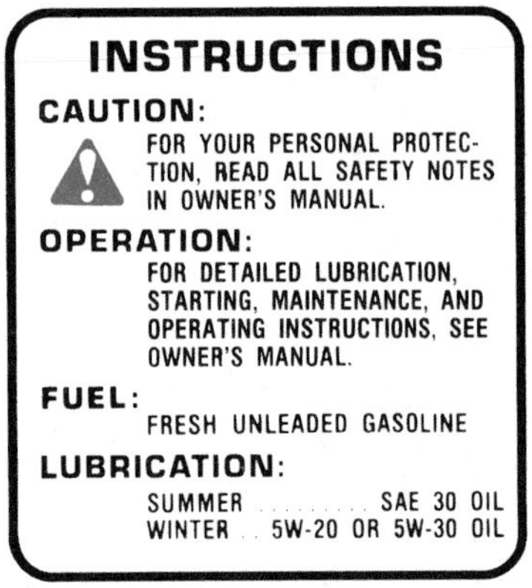

INSTRUCTIONS

CAUTION:
FOR YOUR PERSONAL PROTECTION, READ ALL SAFETY NOTES IN OWNER'S MANUAL.

OPERATION:
FOR DETAILED LUBRICATION, STARTING, MAINTENANCE, AND OPERATING INSTRUCTIONS, SEE OWNER'S MANUAL.

FUEL:
FRESH UNLEADED GASOLINE

LUBRICATION:
SUMMER .... SAE 30 OIL
WINTER .. 5W-20 OR 5W-30 OIL

Fig. 5-6 Lubricating instructions on decal (Courtesy Tecumseh Products Co.)

instructions given on the decal. The API classifications are usually found on the top of the oil can and refer to the quality of the oil.

Oils for Service SF. This oil was developed for automotive engines manufactured in 1980 or later. It is the highest-quality oil and meets the manfacturers' maintenance requirements.

Oils for SE (Service Extreme). SE oils are suitable for the most severe type of operation beginning with 1972 models and some 1971 automobiles. Extreme conditions of start-stop driving; short trip, cold weather driving; and high speed, long distance, hot weather driving can be handled by this oil. The oil meets the requirements of automobiles that are equipped with emission control devices and engines operating under manufacturers' warranties.

Oils for SD (Service Deluxe). Most small engine manufacturers approve of the use of SD (formerly MS) oil in their engines. These oils provide protection against high and low temperature engine deposits, rust, corrosion, and wear.

Oils for Service SC. This oil was also formerly classified as suitable for MS. It has much the same characteristics as SD but is not quite as effective as SD oil. SC oil was developed for auto engines of 1964-67, while SD oil was developed for engines of 1968-70 manufacture.

Oils for Service SB (Formerly MM). This oil is recommended for moderate operating conditions such as moderate speeds in warm weather; short-distance, high-speed driving; and alternate long and short trips in cool weather. It is satisfactory for certain older autos but not for new autos under warranty. It is seldom recommended for small engine use.

Oils for Service SA (Formerly ML). No performance requirements are set for this oil. It is straight mineral oil and may be suitable for some light service requirements. SA oil is not recommended by small engine manufacturers.

The new classifications for oils suitable for diesel service are CA, CB, CC, and CD. These replace the old classifications of DG, AM (Supp-1), DM (MIL-L2104B), and DS respectively.

It is wise to buy the best quality oil for an engine. Sacrificing on quality for a few cents savings may be more expensive in the long run due to engine wear.

Viscosity Classification by the Society of Automotive Engineers

The SAE (Society of Automotive Engineers) number of an oil indicates its *viscosity*, or thickness. Oils may be very thin (light), or they may be quite thick (heavy). The range is from SAE 5W to SAE 50; the higher the number, the thicker the oil. The owner should consult the engine manufacturer's instruction book for the correct oil. SAE 10W, SAE 20W, SAE 20, and SAE 30 are the most commonly used oil weights. Usually, manufacturers recommend SAE 30 for summer use and either SAE 10W or SAE 20W for cold, subfreezing weather. In winter, thinner oil should be added since oil thickens in cold weather. The *W*, as in 5W, 10W, and 20W, indicates that the oil is designed for service in subfreezing weather.

Multigrade Oils. Multigrade oils may also be used in small gasoline engines, figure 5-7. These oils span several SAE classifications because they have a very high viscosity index. Typical classifications are SAE 5W-20, or 5W-10W-20; 5W-30 or 5W-10W-20W-30; SAE 10W-30 or 10W-20W-30; 10W-40, and 20W-40. The first number represents the low temperature viscosity of the oil; the last number represents its high temperature viscosity. For example, 10W-30 passes the viscosity test of SAE 10W at low temperatures and the viscosity test of SAE 30 at high temperatures. These oils are also referred to as all-season, all-weather oils, multiviscosity, or multiviscosity grade oils.

Motor Oil Additives

The best oil cannot do its job properly in a modern engine if additives are not blended into the base oil.

Pour-Point Depressants. Pour-Point depressants keep the oil liquid even at very low temperatures when the wax in oil would otherwise thicken into a buttery consistency making the oil ineffective.

Oxidation and Bearing Corrosion Inhibitors. These additives prevent the rapid oxidation of the oil by excessive heat. Viscous, gummy materials are formed without these inhibitors. Some of these oxidation products attack metals such as lead, cadmium, and silver which are often used in bearings. The inhibiting compounds are gradually used up and, thus, regular oil changes are needed.

Rust and Corrosion Inhibitors. These inhibitors protect against the damage that might be caused by acids and water which are by-products of combustion. Basically, acids are neutralized by alkaline materials much the same as vinegar can be neutralized with baking soda. Special chemicals surround or capture molecules of water preventing their contact with the metal. Other chemicals which have a great attraction for metal form an unbroken film on the metal parts. These inhibitors are also used up in time.

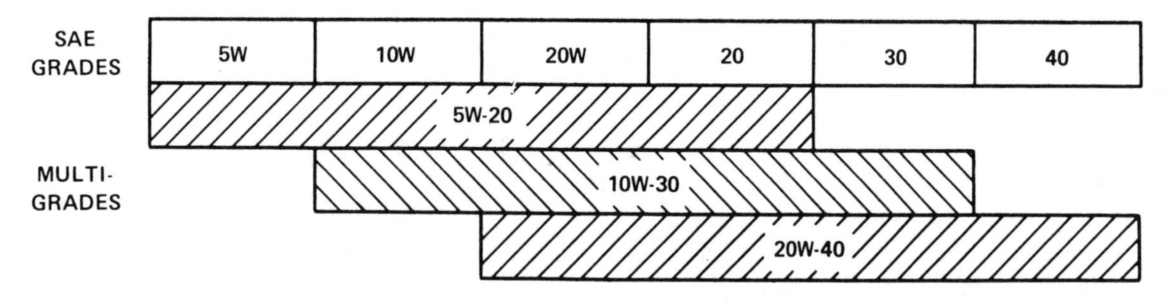

Fig. 5-7 Multigrade motor oils span several SAE (single) grades

Detergent/Dispersant Additives. This type of additive prevents the formation of sludge and varnish. Detergents work much the same as household detergents in that they have the ability to disperse and suspend combustion contaminants in the oil but do not affect the lubricating quality of the oil. When the oil is changed, all the contaminants are discarded with the oil so that the engine is kept clean. Larger particles of foreign matter in the oil either settle out in the engine base or are trapped in the oil filter if one is used.

Foam Inhibitors. Foam inhibitors are present in all high-quality motor oils to prevent the oil from being whipped into a froth or foam. The action in the crankcase tends to bring air into the oil; foaming oil is not an effective lubricant. Foam inhibitors called silicones have the ability to break down the tiny air bubbles and cause the foam to collapse.

LUBRICATION OF FOUR-CYCLE ENGINES

Since all moving parts of the engine must be lubricated to avoid engine failure, a constant supply of oil must be provided. Each engine, therefore, carries its own reservoir of oil in its crankcase where the main engine parts are located.

Basically, the oil is either pumped to or splashed on the parts and bearing surfaces that need lubrication. There are several lubrication systems used on small engines. The following systems are among the most common.

- Simple Splash
- Constant-level Splash
- Ejection Pump
- Barrel-type Pump
- Full-pressure Lubrication

Simple Splash System

The *simple splash system* is perhaps the simplest system for lubrication. It consists of a splasher or dipper that is fastened to the connecting rod cap. There are several different sizes and shapes of dippers, figure 5-8. The dipper may be bolted on with the connecting rod cap or may be cast as a part of the connecting rod cap. Each time the piston nears the bottom of its stroke the dipper splashes into the oil reservoir in the crankcase, splashing oil onto all parts inside the crankcase. Since the engine is operating at 2000 to 3000 rpm, the parts are literally drenched by millions of oil droplets. Vertical crankshaft engines use an oil slinger that is driven by the camshaft, figure 5-9. The slinger performs a similar function to the dipper.

Constant-level Splash System

The *constant-level splash system* has three refinements over the simple splash system: a

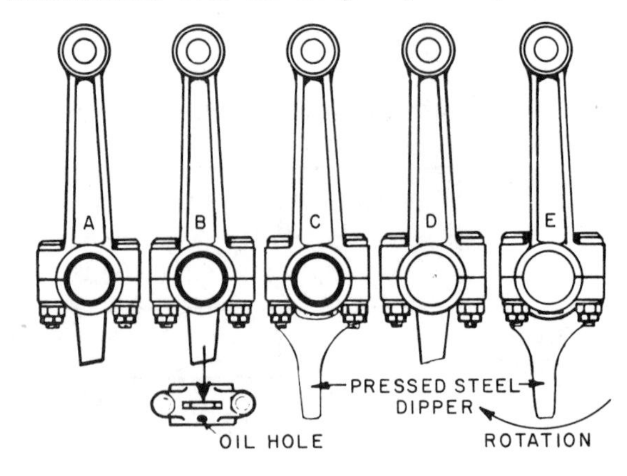

Fig. 5-8 Dippers used on different connecting rods

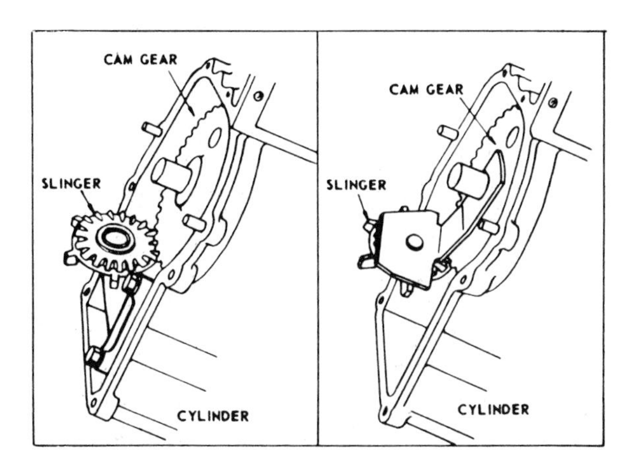

Fig. 5-9 Oil slinger

pump, a splash trough, and a strainer, figure 5-10. With this system a cam-operated pump brings oil from the bottom of the crankcase into a splash trough. Again the splasher on the connecting rod dips into the oil splashing it on all parts inside the crankcase. The strainer prevents any large pieces of foreign matter from recirculating through the system. The pump maintains a constant oil supply in the trough regardless of the oil level in the crankcase.

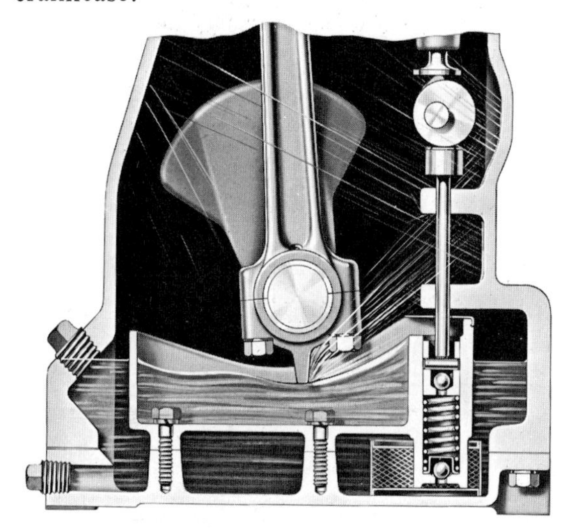

Fig. 5-10 Constant-level splash system used on some engines

Ejection Pump System

Ejection pumps of various types are found on many small engines, figure 5-11. With this method, a cam-operated or electric-operated pump draws oil from the bottom of the crankcase and sprays or squirts it onto the connecting rod. Some of the oil enters the connecting rod bearing through small holes; the remainder of the oil is deflected onto the other parts within the crankcase.

The parts of the ejection pump are the base, screen, check valve, spring, and plunger. The spring tends to push the plunger up but the plunger is driven down every revolution by the cam. A check valve allows the chamber to be filled with oil when the plunger goes up but when the cam pushes the plunger down, the check valve closes, and the trapped oil is squirted or sprayed from the pump.

Barrel-type Pump System

The *barrel-type pump* is also driven by an eccentric on the camshaft, figures 5-12 and

Fig. 5-11 Ejection pump as it is used on some engines

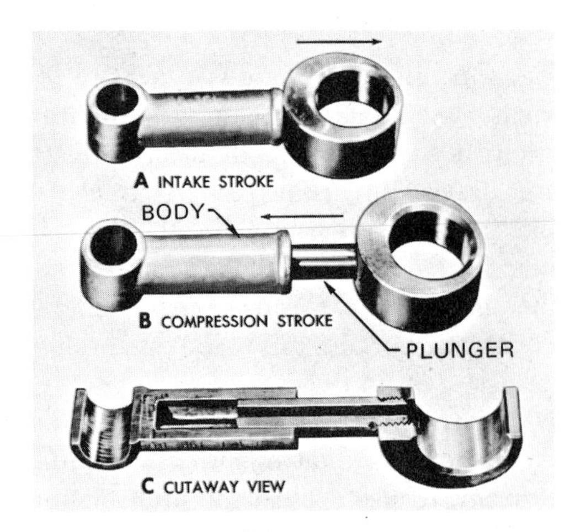

Fig. 5-12 Barrel-type pump

5-13. The camshaft is hollow and extends to the pump of the vertical crankshaft. As the pump plunger is pulled out on intake, an intake port in the camshaft lines up allowing the pump body to fill. When the plunger is forced into the pump body on discharge, the discharge ports in the camshaft line up allowing the oil to be forced to the main bearing and to the crankshaft connecting rod journal. Small drilled passages are used to channel the oil. Oil is also splashed onto other crankcase parts.

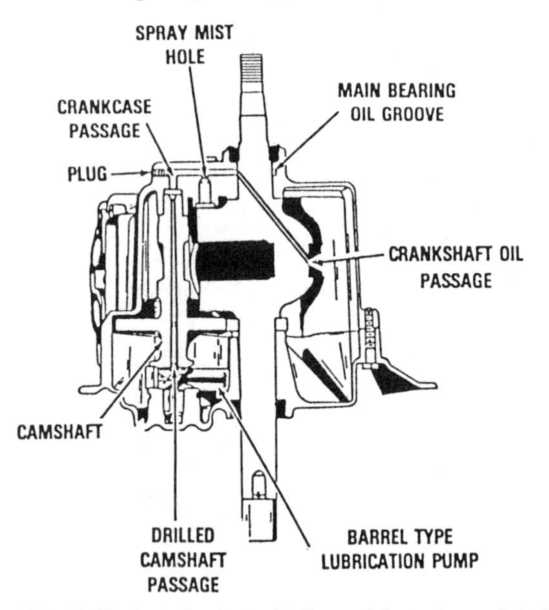

Fig. 5-13 Lubrication oil flow of barrel-type lubrication system (Courtesy Tecumseh Products Company)

Full-pressure Lubrication System

Full-pressure lubrication is found on many engines especially the larger of the small engines and on automobile engines. Lubricating oil is pumped to all main, connecting, and camshaft bearings through small passages drilled in these engine parts, figure 5-14. Oil is also delivered to tappets, timing gears, etcetera under pressure. The pump used is usually a positive displacement gear type. It is also common to use a splash system in conjunction with full-pressure systems.

LUBRICATING CYLINDER WALLS

Oil is splashed or sprayed onto cylinder walls and the piston rings spread the oil evenly for proper lubrication. Piston rings

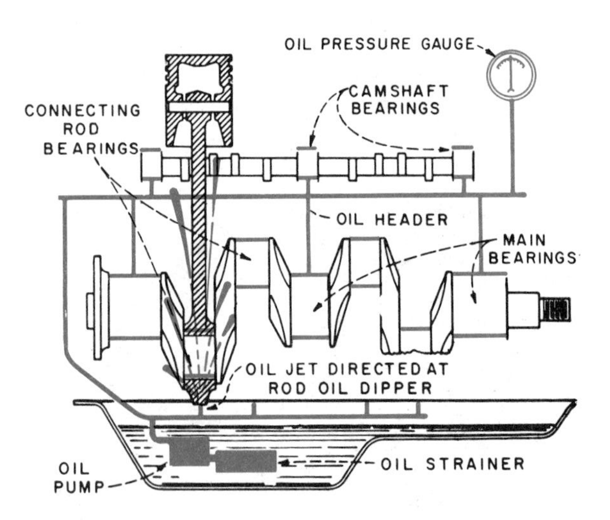

Fig. 5-14 Full-pressure lubrication

must function properly to avoid excessive oil consumption. The rings must put an even pressure on the cylinder walls and give a good seal. If piston ring grooves are too large, the piston rings begin a pumping action as the piston moves up and down. As a result, excess oil is brought into the combustion chamber. Worn cylinder walls, worn pistons, and worn rings can all add to high oil consumption as well as loss of compression and eventual loss of power.

BLOW-BY

Blow-by, the escape of combustion gases from the combustion chamber to the crankcase, occurs when piston rings are worn or are too loose in their grooves. Carbon and soot from burned fuel are forced into the crankcase by the rings. Much of the carbon is deposited around the rings which hinders their operation further.

Another damaging result of worn rings can be crankcase dilution. If raw, unburned gasoline collects in the combustion chamber, it can leak by the piston rings and into the crankcase. This dilutes the crankcase oil and reduces the oil's lubricating properties.

CRANKCASE BREATHERS

Four-stroke cycle engines do not have completely airtight, oiltight crankcases. Engine crankcases must breathe. Without the ability to breathe, pressures build up in the crankcase and cause oil seals to rupture or allow contaminants to remain in the crankcase. Pressure buildup may be caused by the expansion of the air as the engine heats up, by the action of the piston coming down the cylinder, and by the blow-by of combustion gases along cylinder walls.

Most single-cylinder engines use breathers, figure 5-15. These breathers allow air to leave but not reenter a reed-type check valve or a ball-type check valve. The breathers place the crankcase under a slight vacuum and are called closed breathers.

Open breathers allow the engine to breathe in and out freely and are usually equipped with air filters. They are located where the splashing of oil is not a problem. Open breathers are often incorporated with the valve access cover. If an engine with this type of breather is tipped on its side, oil may run

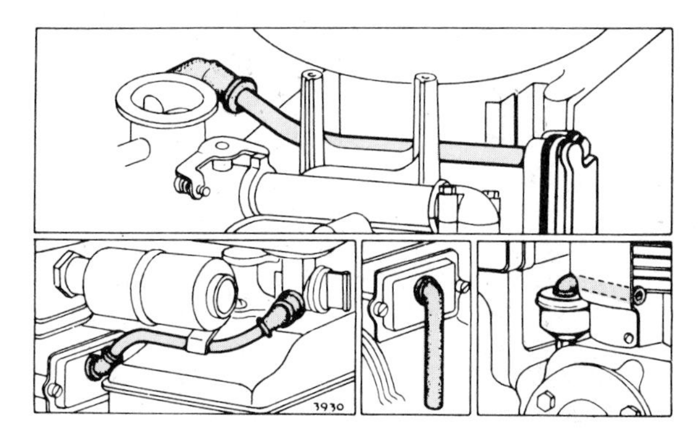

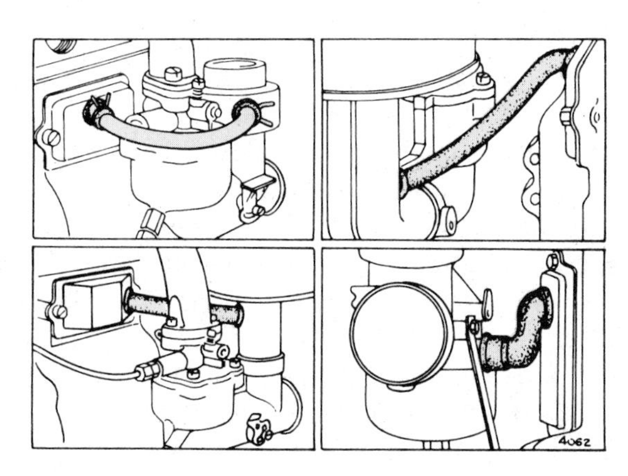

Fig. 5-15 Breather assemblies

out through it. Some breathers are vented to the atmosphere while others are vented back through the carburetor.

OIL CHANGES

The crankcase oil should be changed periodically. The exact number of engine operating hours between oil changes varies depending on the manufacturer. It may be as short a time as every 20 hours or as long a time as every 100 hours. (Every 25 hours is most common.) To ensure minimum engine wear and maximum engine life, follow the manufacturer's lubrication suggestions.

It is best to drain the oil when the engine is hot since more dirt and slightly more oil can be removed. Dirty oil should be replaced because it cannot give proper, high-quality lubrication.

LUBRICATION OF TWO-CYCLE ENGINES

Lubrication of the two-cycle engine is quite different from the four-cycle engine. Since the fuel mixture must travel through the crankcase, a reservoir of oil cannot be stored there. The lubricating oil is mixed with gasoline and then put into the gas tank. The lubricating oil for all crankcase parts enters the crankcase as a part of the fuel mixture. Millions of tiny oil droplets suspended in the mixture of gasoline and air settle onto the moving parts in the crankcase providing lubrication. Oil droplets are relatively large and heavy and quickly drop out of suspension. Much of the oil is carried on into the combustion chamber where it is burned along with the gasoline and air.

On a two-cycle engine, oil must be mixed with the gasoline, figure 5-16. The engine is not operated on gasoline alone. If it were, the heat of friction would burn up the engine in a short time. Some two-cycle engine manufacturers are developing and marketing engines with oil metering devices that eliminate the need to premix the oil and gasoline. Mixing is done automatically in the correct proportions.

Too much oil mixed with the gasoline results in incomplete combustion and a very heavy exhaust. Deposits build up rapidly in the engine, possibly fouling the tip of the spark plug and collecting around the exhaust ports. Clogged exhaust ports reduce the engine's power output considerably and can cause overheating.

Too little oil mixed with the gasoline can cause the engine to overheat. It may even cause scuffing and scoring of the cylinder

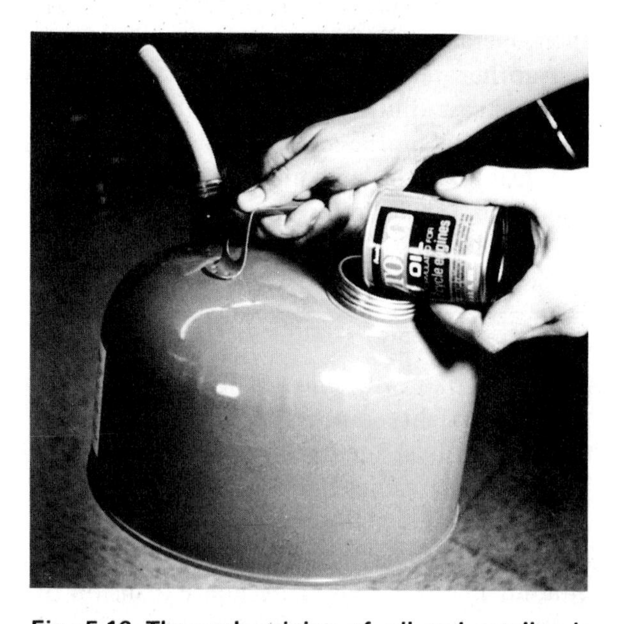

Fig. 5-16 Thorough mixing of oil and gasoline is necessary for two-cycle engines.

walls, bearing damage, or the piston to seize in the cylinder.

In preparing the mixture of gasoline and oil for most two-cycle engines, observe the following rules:

- Mix a good grade of regular gasoline (leaded or unleaded) and oil in a separate container. Do not mix in the gas tank unless it is a remote tank such as is found on many outboard motors.
- Pour the oil into the gasoline to ensure good mixing, and shake the container well. If poorly mixed, the oil settles to the bottom of the tank causing difficult starting.
- Strain the fuel with a fine mesh strainer when pouring it into the tank.
- Use the oil that is specified for the particular two-cycle engine, figures 5-17 and 5-18. Most manufacturers specify their own brand of two-cycle oil. In an emergency other oils may be used, usually SAE 30SB (formerly MM) or SAE 30SD (formerly MS) nondetergent oil. The manufacturer's brand, however, provides the best lubrication with a minimum of deposit formation.
- Mix gasoline and oil in the proportions recommended by the engine manufacturer. One common proportion is three-fourths pint of oil to one gallon of gasoline when breaking in a new engine, and one-half pint of oil to one gallon of gasoline for normal use.

Fig. 5-17 Proper mixing of oil and gasoline for this two-cycle portable generator is essential. (Courtesy of Johnson Motor Co.)

Fig. 5-18 Most outboard motors use two-cycle engines which require mixing of gasoline and oil. (Courtesy Johnson Motor Co.)

NOTE: Mixing proportions vary widely; check the engine's instruction book.

MIXING RATIOS

OUNCES TO BE MIXED WITH GASOLINE

For a Dilution Ratio of	Gallons of gasoline					
	1	2	3	4	5	6
	Ounces of Two-Cycle Oil					
24:1	5	11	16	21	27	32
20:1	6	13	19	26	32	38
16:1	8	16	24	32	40	48

Add oil to gasoline in proportions recommended by manufacturer.

ADDITIONAL LUBRICATION POINTS

Whether the engine has two-cycle or four-cycle lubrication, there may be other lubrication to consider besides the crankcase area. On an outboard engine do not neglect the lower unit which needs a special gear lubricant at several points, figure 5-19. If an engine is powering an implement or other machinery, there may be transmissions, gear boxes, chains, axles, wheels, shafts, and linkages that need periodic lubrication. Lubricate these additional parts with the oil or lubricant recommended by the manufacturer.

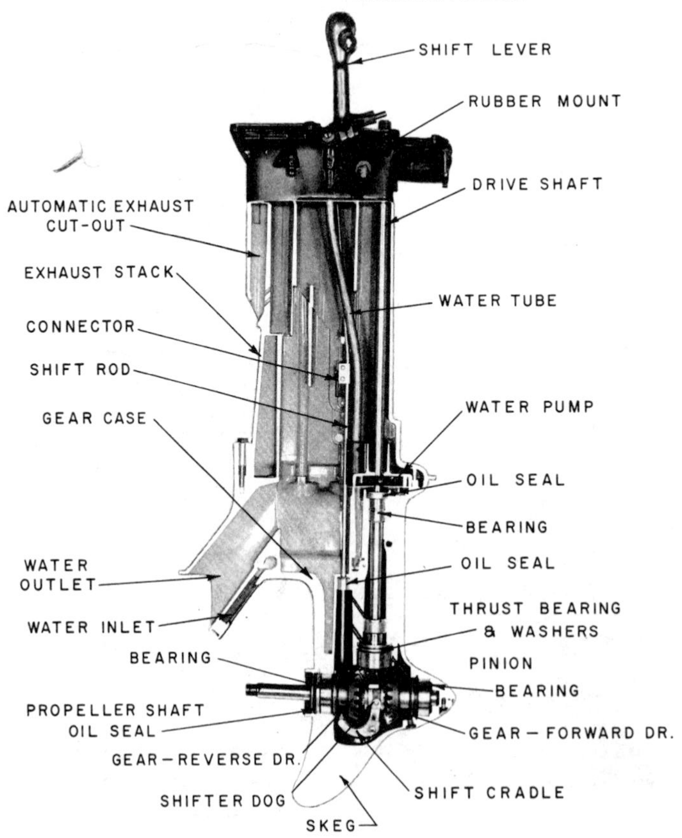

Fig. 5-19 Gear case – lower unit assembly

REVIEW QUESTIONS

1. What is the main job of a lubricant?
2. Explain how the lubricating oil also:
   a. Seals power
   b. Helps to dissipate heat
   c. Keeps the engine clean
   d. Cushions bearing loads
   e. Protects against rusting
3. What are the three types of friction bearings?
4. What is an antifriction bearing?
5. What does an oil's SAE number refer to?
6. What is a multigrade oil?
7. Explain how a detergent oil works.
8. What are the quality designations of oil?
9. List five common four-cycle engine lubrication systems.
10. How are cylinder walls lubricated?
11. Explain blow-by.
12. Briefly explain how two-cycle lubrication is accomplished.
13. What is one common proportion of gasoline to oil for two-cycle engines?
14. What kind of oil should be used for two-cycle engines?

Unit 6 Cooling Systems

OBJECTIVES After completing this unit, the student should be able to:

- Discuss the construction of the water jacket and trace the path of cooling water through the engine.
- Explain the construction and operation of the water pump and trace the path of water through the pump.
- List the basic parts of the air-cooling system and trace the path of cooling air through the engine.

In internal combustion engines, the temperature of combustion often reaches over 4000° Fahrenheit, a temperature well beyond the melting point of the engine parts. This intense heat cannot be allowed to build up. A carefully engineered cooling system is, therefore, a vital part of every engine. A cooling system must maintain a good engine operating temperature without allowing destructive heat to build up and cause engine part failure, figure 6-1.

The cooling system does not have to dispose of all the heat produced by combustion. A good portion of the heat energy is converted into mechanical energy by the engine, benefiting engine efficiency. Some heat is lost in the form of hot exhaust gases. The cooling system must dissipate about one-third of the heat energy caused by combustion.

Engines are either air-cooled or water-cooled; both systems are in common use. Generally, air-cooled engines are used to power machinery, lawn mowers, garden tractors, chain saws, etc. The air-cooled system is usually lighter in weight and simpler; hence, its popularity for portable equipment, figure 6-2.

The water-cooled engine is often used for permanent installation or stationary power plants. Most automobile engines and outboard motors are water-cooled.

AIR-COOLING SYSTEM

The air-cooling system consists of heat radiating fins, flywheel blower, and shrouds for channeling the air. The path of airflow can be seen in figure 6-3.

Heat radiating fins are located on the cylinder head and cylinder because the greatest concentration of heat is in this area. The fins increase the heat radiating surface of these parts allowing the heat to be carried away to the atmosphere more quickly, figure 6-4.

The flywheel blower consists of air vanes cast as a part of the flywheel. As the flywheel revolves, these vanes blow cool air across the fins carrying away the heated air and replacing it with cool air.

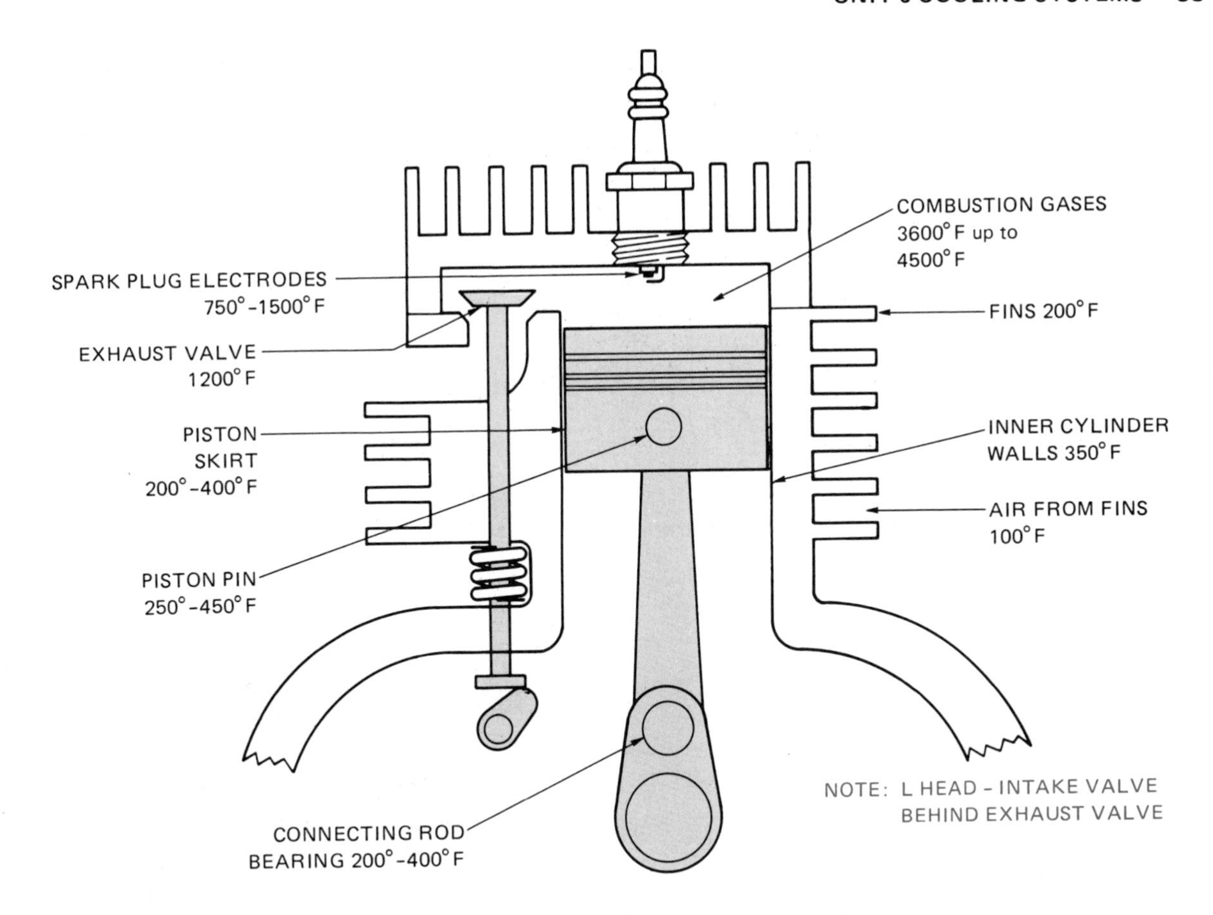

Fig. 6-1 Approximate engine operating temperatures

The shrouds direct the path of the cool air to the areas that demand cooling, figure 6-5. Shrouds must be in place if the cooling system is to operate at its maximum efficiency.

Care of the Air-cooling System

The air-cooling system is almost trouble free. Several points, however, should be considered. Heat radiating fins are thin and often fragile, especially on aluminum engines. If through carelessness they are broken off, a part of the cooling system is gone. Besides losing some cooling capacity, hot spots can develop which may warp the damaged area. Also, it is easy for dust, dirt, grass, and oil

Fig. 6-2 Dirt bikes use air-cooled engines as they are lighter in weight than water-cooled engines. (Courtesy Kawasaki Motor Corp USA)

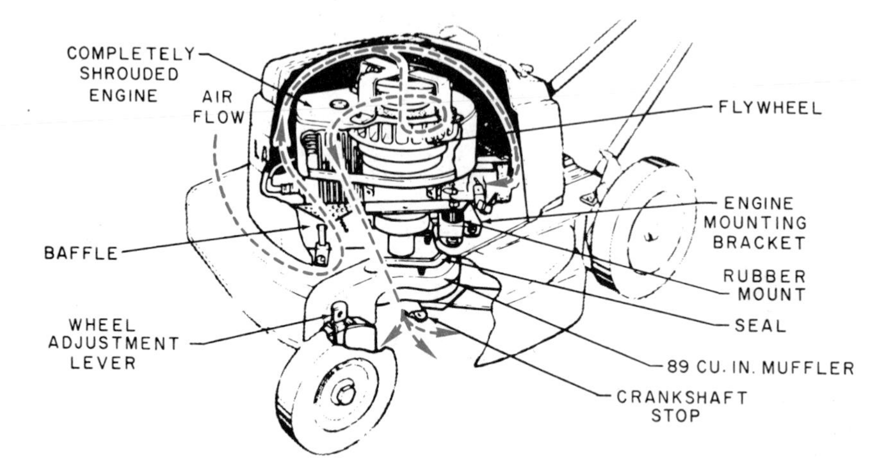

Fig. 6-3 The path of airflow on certain lawn mower engines

STRAIGHT FIN HEADS

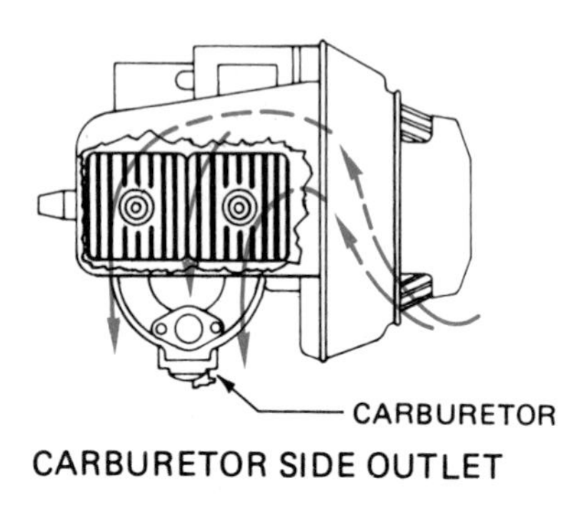

OBLIQUE FIN HEADS

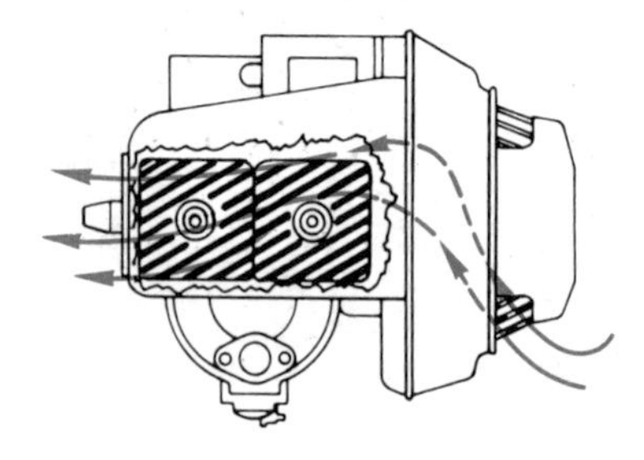

TOP OR END OUTLET

Fig. 6-4 Airflow across the cylinder head fins

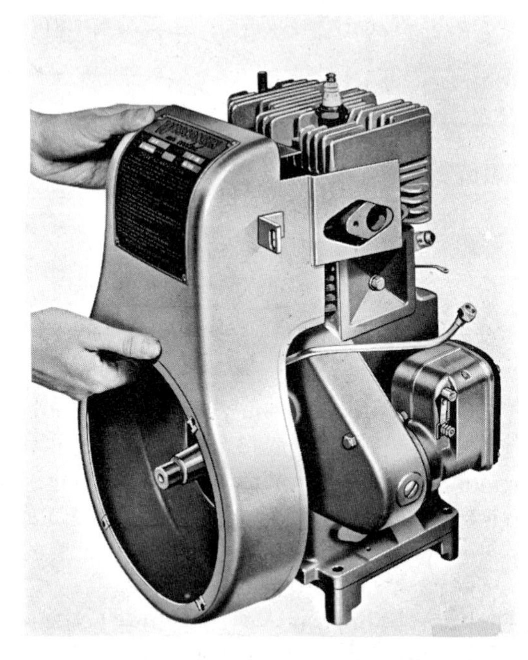

Fig. 6-5 The air shroud is an important part of the air-cooled engine. (Courtesy Wisconsin Motor Corp.)

to accumulate between the fins. Any buildup of foreign matter reduces the cooling system's efficiency. All parts of the system should be kept clean, especially the radiating fins.

The flywheel vanes should not be chipped or broken. Besides reducing the cooling capacity of the engine, such damage can destroy the balance of the flywheel. An unbalanced flywheel causes vibration and an excessive amount of wear on engine parts.

WATER-COOLING SYSTEM

Some small engines are water-cooled as are most larger engines. Such engines have an enclosed water jacket around the cylinder walls and cylinder head. Water jackets are relatively trouble free. However, damaging deposits can build up over a period of time. Salt corrosion, scale, lime, and silt can restrict the flow of water and heat transfer. Water jackets are found on most outboard motors and on the majority of automobile engines. Cool water is circulated through this jacket picking up the heat and carrying it away.

Figure 6-6 shows a cross section of water passage in the head and block. Water cooling systems often found on stationary small gas engines or automobile engines include the following basic parts: radiator, fan, thermostat, water pump, hoses, and water jacket.

The water pump circulates the water throughout the entire cooling system. The hot water from the combustion area is carried from the engine block and head to the radiator. In the radiator, many small tubes and radiating fins dissipate the heat into the atmosphere. A fan blows cooling air over the radiating fins. From the bottom of the radiator the cool water is returned to the engine.

Engines are designed to operate with a water temperature of between 160 degrees and 180 degrees Fahrenheit. To maintain the correct water temperature a thermostat is used in the cooling system. When the temperature is below the thermostat setting, the thermostat remains closed and the cooling water circulates only through the engine, figure 6-7A. However, as the heat builds up to the thermostat setting, the thermostat opens and the cooling water moves throughout the entire system, figure 6-7B.

Fig. 6-6 Cross section showing water passages in head and block

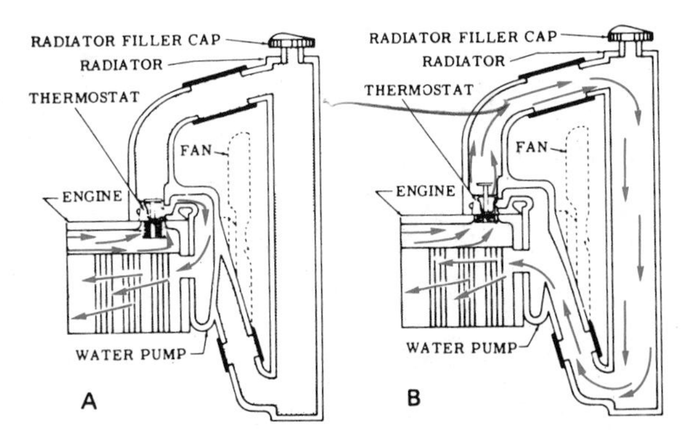

Fig. 6-7 (A) Thermostat closed: water recirculated through engine only. (B) Thermostat open: water circulated through both engine and radiator.

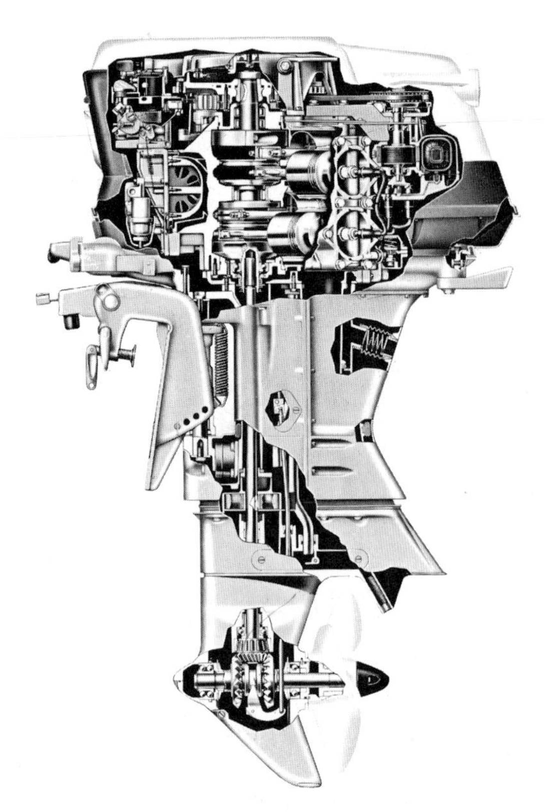

Fig. 6-8 Cutaway of an outboard motor (Courtesy Johnson Motors)

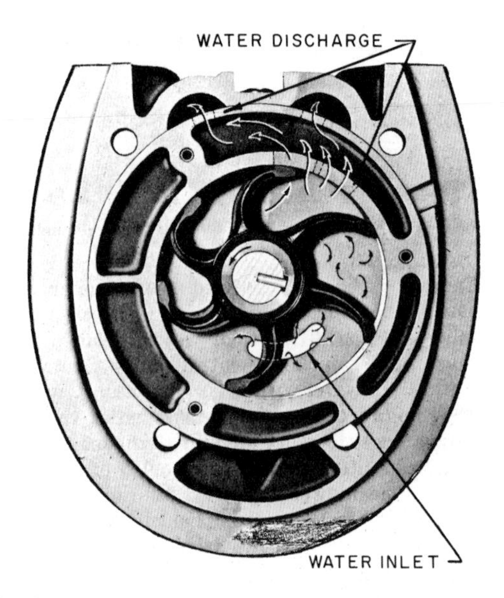

Fig. 6-10 An impeller water pump with the cover removed to show the impeller

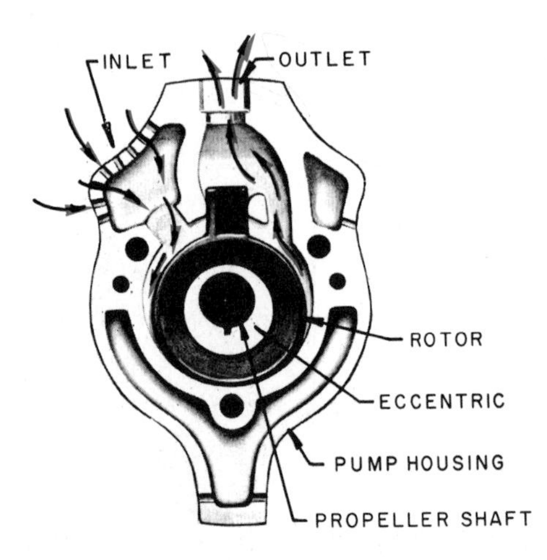

Fig. 6-9 An eccentric rotor water pump with the cover removed to show the eccentric and rotor

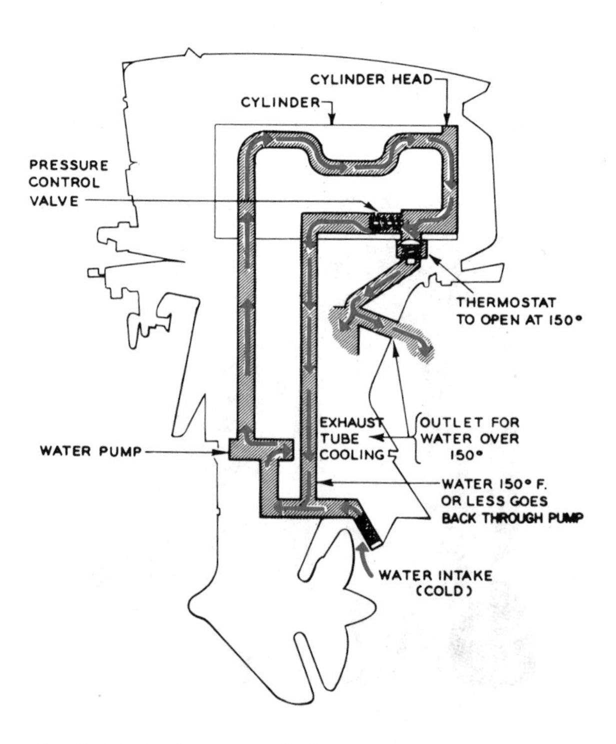

Fig. 6-11 A thermostatically controlled cooling system used on outboard engines

WATER COOLING THE OUTBOARD ENGINE

Cooling the outboard engine with water is a simpler process because there is an inexhaustible supply of cool water present where the engine operates. Outboards pump water from the source through the engine's water jacket and then discharge the water back into the source. A cutaway of an outboard motor is shown in figure 6-8.

The water pump on outboards is located in the lower unit. It is driven by the main driveshaft or the propeller shaft. The cool water is pumped up copper-tube passages to the water jacket. After the cooling water picks up heat, it is discharged into the exhaust area of the lower unit and out of the engine. Several types of water pumps are used on outboard motors. Plunger type pumps; eccentric rotor pumps, figure 6-9; and impeller pumps, figure 6-10, are some examples.

Many outboard motors, especially large horsepower models, are equipped with a thermostatically controlled cooling system, figure 6-11. The temperature of the water circulating through the water jacket is maintained at about 150° Fahrenheit.

REVIEW QUESTIONS

1. How high can the temperature of combustion reach?
2. Does the cooling system remove all the heat of combustion? Explain.
3. Why are air-cooling systems often used for portable equipment?
4. What are the main parts of the air-cooling system?
5. Why are cylinders and cylinder heads equipped with fins instead of being cast with a smooth surface?
6. What are the basic parts of the water-cooling system found on stationary small gas engines?
7. What is the function of the thermostat?
8. What are several types of water pumps used on outboard motors?

Unit 7 Ignition Systems

OBJECTIVES After completing this unit, the student should be able to:

- Discuss electron theory and magnetism.
- Discuss how the basic parts of the magneto are constructed and how they are mounted on the engine.
- Explain the complete magneto cycle.
- Discuss spark advance.
- Discuss solid state ignition.
- Discuss battery ignition systems.

Small gasoline engines normally use a magneto for supplying the ignition spark. A *magneto* is a self-contained unit that produces the spark for ignition, figure 7-1. No outside source of electricity is needed to produce the spark for ignition. The magneto is a simple and very reliable ignition system. On small gas engines that do not have electric starters, lighting systems, radios, and other electrical accessories, a storage battery is not necessary. The magneto is, therefore, ideally suited for these small gasoline engines.

The basic parts of the magneto ignition system are the: permanent magnets, high-voltage coil (primary and secondary windings), laminated iron core, breaker points, breaker cam, condenser, spark plug cable, and spark plug. Before discussing how these parts work together, a review is given of some essentials of electricity and magnetism which explains the construction and function of each individual part.

ELECTRON THEORY

All matter is composed of tiny particles called atoms. The atom is composed of electrons, protons, and neutrons, figure 7-2. The number and arrangement of these particles determines the type of atom: hydrogen, oxygen, carbon, iron, lead, copper, or any other element. Weight, color, density, and all other characteristics of an element are determined by the structure of the atom. Electrons from an atom of copper are the same as electrons from any other element.

The electron is a very light particle that spins around the center of the atom. Electrons move in an orbit. The number of electrons orbiting around the center or nucleus of the atom varies from element to element. The electron has a negative (–) electrical charge.

The proton is a very large and heavy particle in relation to the electron. One or more pro-

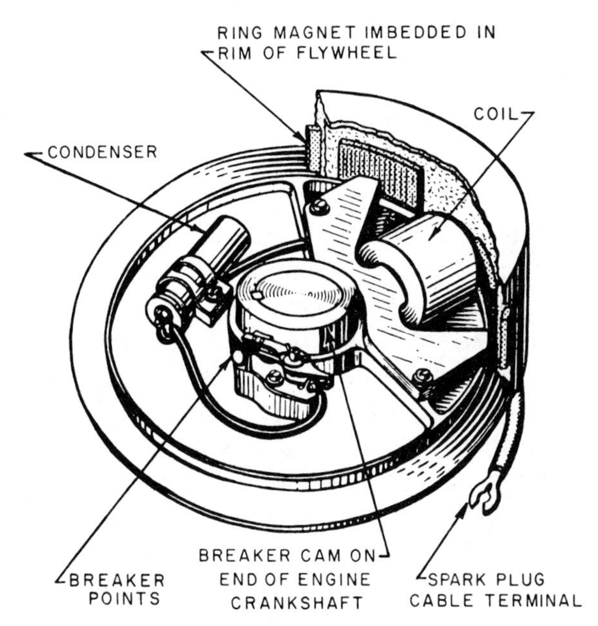

Fig. 7-1 Flywheel magneto

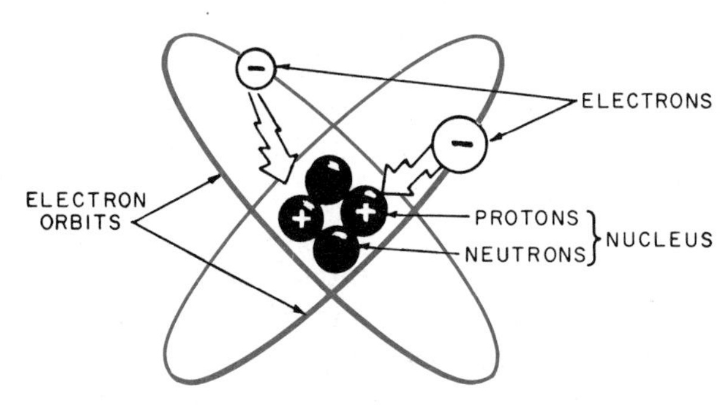

Fig. 7-2 Atomic structures: electron, proton, and neutron

tons form the center or nucleus of the atom. The proton has a positive (+) electrical charge.

The neutron consists of an electron and proton bound tightly together. Neutrons are located near the center of the atom. The neutron is electrically neutral; it has no electrical charge.

Atoms are normally electrically neutral; that is, the number of electrons and protons are the same, cancelling out each other's electrical force. Atoms stay together because, unlike electrical charges, they attract each other. The electrical force of the protons holds the electrons in their orbits. Like electrical charges repel each other, so negatively charged electrons do not collide with each other.

In most materials, it is very difficult, if not impossible, for electrons to leave their orbit around the atom. Materials of this type are called nonconductors of electricity or *insulators.* Some typical insulating materials are glass, mica, rubber and paper. Electricity cannot flow through these materials easily.

In order to have electric current, electrons must move from atom to atom. Insulators do not allow this electron movement. However, in many substances, an electron can jump out of its orbit and begin to orbit in an adjoining or nearby atom. Substances which permit this movement of electrons are called *conductors* of electricity. Some typical examples are copper, aluminum, and silver.

Electron flow in a conductor takes place when there is a difference in electrical potential and there is a complete circuit or path for electron flow, figure 7-3. In other words, when the source of electricity is short of electrons, it is positively charged. Since unlike charges attract each other, electrons (being negatively charged) move toward the positive source.

A source of electricity can be produced or seen in several basic forms: mechanical, chemical, static, and thermal. Electricity is produced mechanically in the electrical generator which is commonly connected to water power or steam turbines. The electricity used in homes and factories is produced mechanically. In the magneto, mechanical energy is

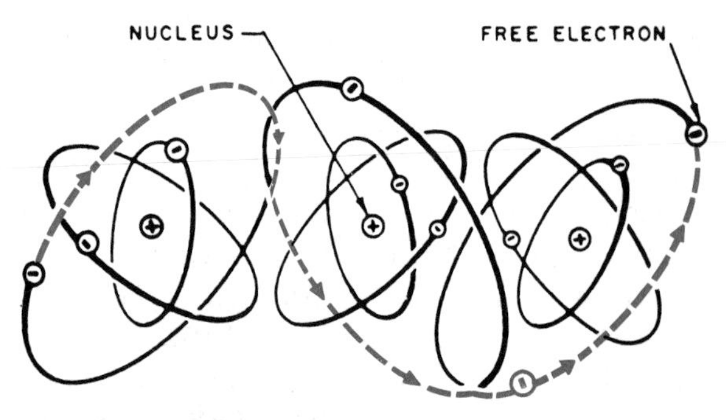

Fig. 7-3 Current: flow of electrons within a conductor

used to rotate the permanent magnet. Electricity produced by chemical action is seen in the storage battery and dry cell. Static electricity can be seen in nature when lightning strikes. The lightning occurs when the air insulation breaks down and electrons are in a positive area. The lightning may be between clouds, from cloud to earth, or from earth to cloud.

UNITS OF ELECTRICAL MEASUREMENT

There are three basic units of electrical measurement: *ampere* (rate of electron flow), *volt* (force or pressure causing electron flow), and *ohm* (resistance to electron flow).

Ampere

The *ampere* is the measurement of electrical current – the number of electrons flowing past a given point in a given length of time. If a person stands at a point on a wire and could count the electrons passing by in one second, 6,280,000,000,000,000,000 (6.28×10^{18}) electrons would be counted, which equals one ampere of current. To help visualize amperage, think of water flowing in a pipe. A small pipe might deliver two gallons of water a minute. A larger pipe might deliver five gallons of water a minute. Electric wires are generally the same; larger wires can handle more amperage or electron flow than smaller wires.

Volt

The *volt* is the measurement of electrical pressure or the difference in electrical potential that causes electron flow in an electrical circuit. The energy source is short of electrons and the electrons in the circuit want to go to the source. The pressure to satisfy the source is called *voltage.* Voltage might be compared to the pressure that water in a high tank places on the pipe located at the street level. The higher the water pressure, the faster the water flow from a pipe below. Likewise, a higher voltage tends to cause greater flow of electrons.

Ohm

The *ohm* is the unit of electrical resistance. Every substance puts up some resistance to the movement of electrons. Insulators such as porcelain, oils, mica, and glass put up a great resistance to electron flow. Conductors such as copper, aluminum and silver put up very little resistance to electron flow. Even though conductors readily permit the flow of electric current they do tend to put up some resistance. In the water pipe example, this resistance can be seen as the surface drag by the sides of the pipe, or scale and rust in the pipe. Using a larger pipe or, electrically, a larger wire is one way of reducing resistance.

OHM'S LAW

In every example of electricity flowing in an electrical circuit, amperes, volts, and ohms each play their part; they are related to each other. This relationship is stated in *Ohm's Law.*

$$\text{Amperes (rate)} = \frac{\text{volts (potential)}}{\text{ohms (resistance)}}$$

The formula is usually abbreviated to:

$$I = \frac{E}{R}$$

For example, if the voltage is 6 and the resistance is 12 ohms, the current is calculated as follows.

$$I = \frac{E}{R} \quad I = \frac{6}{12} \quad I = 0.5 \text{ amperes}$$

The formula can be written to find the resistance, or the voltage.

$$R = \frac{E}{I} \qquad E = IR$$

MAGNETISM

Most people have experimented with a magnet at one time, watching it pick up steel objects and attracting or repelling another magnet. These effects are not entirely explained but scientists generally agree on the molecular theory of magnetism. Molecules are the smallest divisions of substance that are still recognizable as that substance. Several different atoms may make up one molecule. For example, a molecule of iron oxide contains atoms of iron and oxygen. In many substances the atoms in the molecules are more positive at one spot and more negative at another spot. This is termed a *north pole* and a *south pole.* Usually the poles of adjoining molecules are arranged in a random pattern and there is no magnetic force since their effects cancel one another, figure 7-4A. However, in some substances such as iron, nickle, and cobalt the molecules are able to align

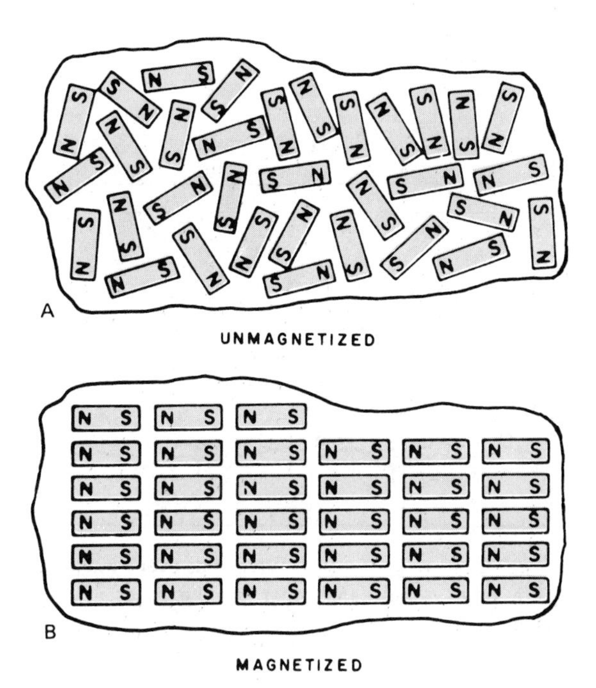

Fig. 7-4 In an unmagnetized bar the molecules are in a random pattern. In a magnetized bar the molecules align their atomic poles.

themselves so that all north poles point in one direction and all south poles point in the opposite direction, figure 7-4B. The small magnetic forces of many tiny molecules combine to make a noticeable magnetic force. In magnets, like poles repel each other and unlike poles attract each other just as like and unlike electrical charges react, figure 7-5. Electricity and magnetism are very closely tied together.

Some substances keep their molecular alignment permanently and are classed as *permanent magnets.* Hard steel has this ability. Other materials such as a piece of soft iron (a nail), can attain the molecular alignment of a magnet only when it is in a magnetic field. As soon as soft iron is removed from the magnetic field, its molecules dis-

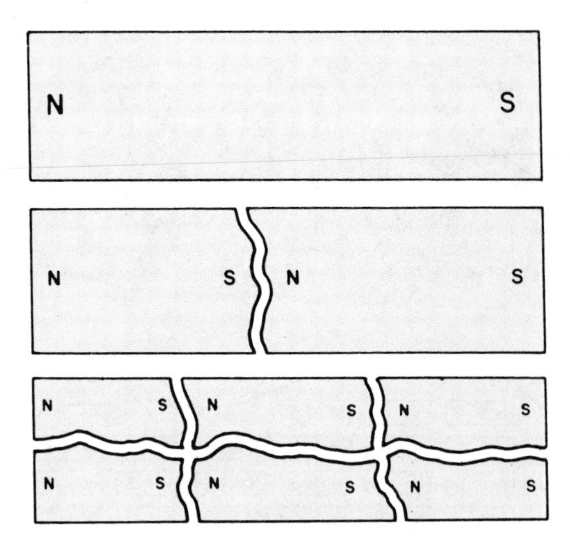

Fig. 7-5 A magnetic field surrounds every magnet. Like poles repel each other; unlike poles attract each other.

arrange themselves and the magnetism is lost. Such substances are termed *temporary magnets.*

More than 100 years ago Michael Faraday discovered that magnetism could produce electricity. Magnetos used on gasoline engines apply this discovery. Faraday found that if a magnet is moved past a wire, electrical current starts through the wire. If the magnet is stopped near the wire, the current stops. Electricity flows only when the magnetic

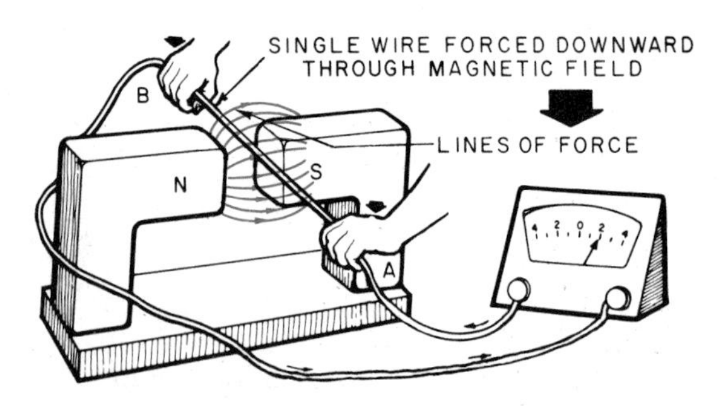

Fig. 7-6 Current flows in the wire as it moves down through the magnetic field.

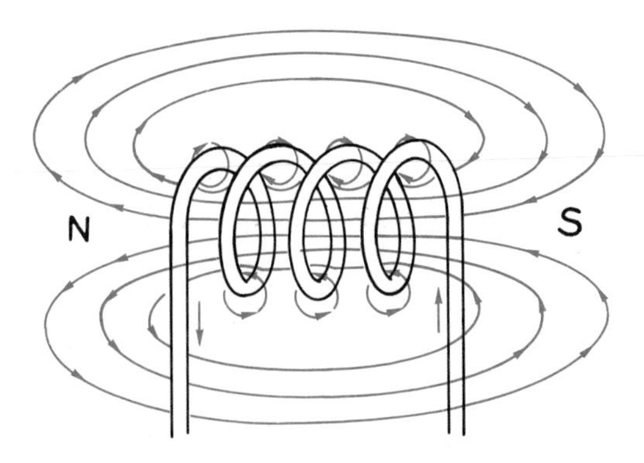

Fig. 7-7 When electricity flows through a coil of wire, a magnetic field is set up around the coil.

field or magnetic lines of force are being cut by the wire, figure 7-6.

Another principle that the magneto uses is that when electrons flow through a coil of wire, a magnetic field is set up around the coil, figure 7-7. The coil itself becomes a magnet. Therefore, when electrons flow through the coils in a magneto, a magnetic field is set up.

TRANSFORMERS

The principle of the transformer and induced voltage is also used in the magneto. In a transformer there is a primary coil and a secondary coil wound on top of the primary; the two are insulated from each other, figure 7-8. These coils are wound on a soft iron core. When alternating current passes through the primary coil there is an alternating magnetic field set up in the iron core. The magnetic lines of force cut the secondary coil and induce an alternating voltage within the coil. The voltage produced depends on the ratio of windings in the primary coil and secondary coil. If there are more windings in the secondary than in the primary, the secondary voltage is higher; a *step-up transformer.* If there are more wind-

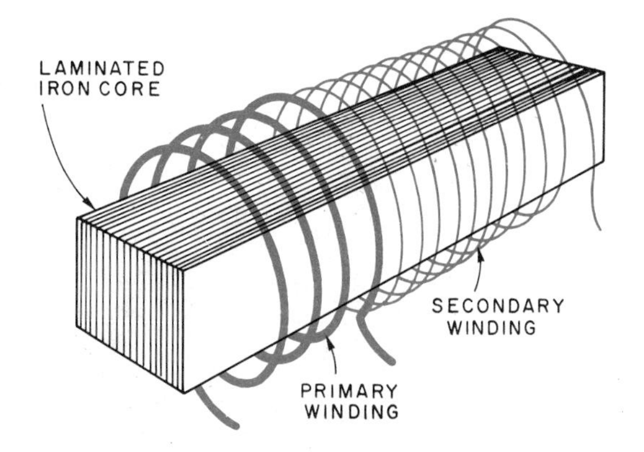

Fig. 7-8 A simple transformer

ings in the primary than in the secondary, the secondary voltage is smaller; a *step-down transformer.* Although the magneto does not operate on alternating current, it does use the principle of the step-up transformer.

In a small gasoline engine magneto, there is a magnetic field which induces current in the primary coil, thus setting up a magnetic field around both the primary and secondary coils. At the point of maximum current, the circuit is broken in the primary. Electrons can no longer flow; therefore, the magnetic field collapses. This rapidly collapsing magnetic field induces a very high voltage in the secondary coil igniting the fuel mixture.

Before studying in detail how the magneto operates, it is useful to understand the function of each part. The seven basic parts which work together to produce the ignition spark are the permanent magnets, high-voltage coil, laminated iron core, breaker points, condenser, spark plug lead, and spark plug. An example of a magneto ignition system is shown in figure 7-9.

PERMANENT MAGNETS

Permanent magnets are usually made of an alloy called *alnico,* a combination of aluminum, nickle, and cobalt. This magnet is quite strong and keeps its magnetism for a very long time. On a flywheel magneto the magnet is cast into the flywheel and cannot be removed. The other magneto parts are often mounted on a fixed plate underneath the flywheel. The magnet revolves around the other parts of the magneto. Sometimes the other magneto parts are mounted near the outside rim of the flywheel and the permanent magnets pass by each revolution.

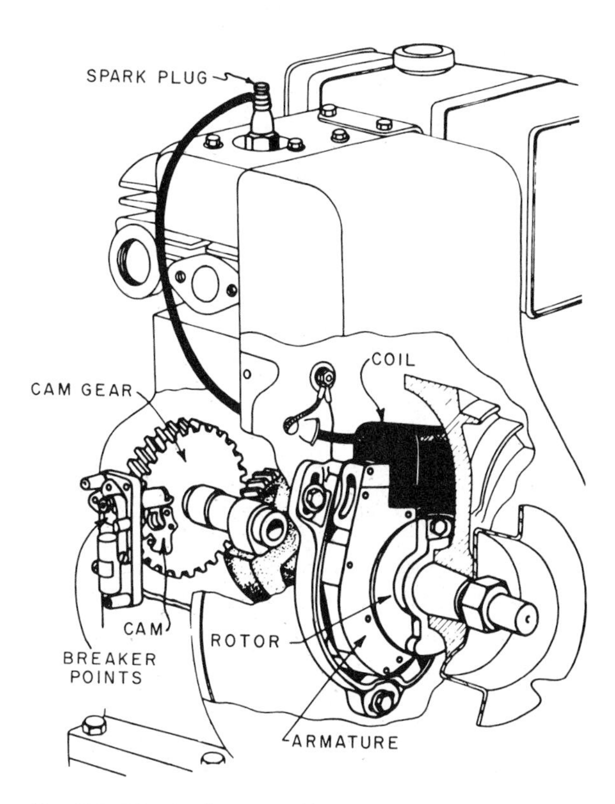

Fig. 7-9 The basic parts of a magneto ignition system

Rotor-type permanent magnets are also in use, figure 7-10. With this type of construction the permanent magnet rotor may be mounted on the end of the crankshaft or the rotor may be geared to the crankshaft. The rotor lies within the other magneto parts.

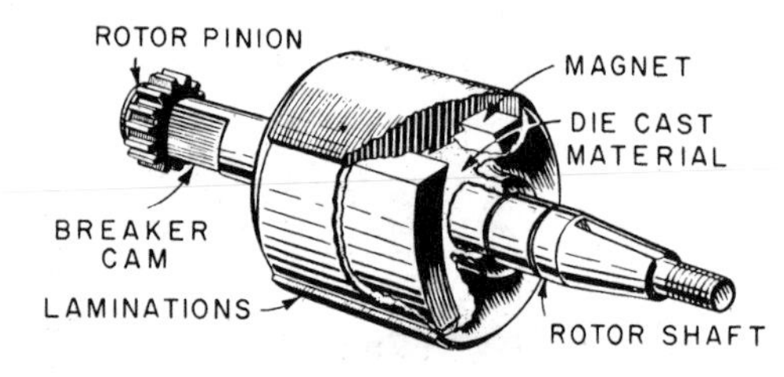

Fig. 7-10 Rotor-type permanent magneto

Care should be taken not to drop or pound the magnet as this causes it to lose some of its magnetism.

HIGH-VOLTAGE COIL (PRIMARY AND SECONDARY WINDINGS)

Refer to figure 7-11. The primary winding of wire consists of about 200 turns of a heavy wire (about 18 gauge) wrapped around a laminated iron core. The primary coil is in the electrical circuit containing the breaker points and condenser, figure 7-12. When a magnet is brought near this coil and iron core, magnetic lines of force cut the coil and electrical current is produced. Usually this current flows from the coil through the closed breaker points and is grounded. This makes a complete circuit.

The secondary winding of wire consists of about 20,000 turns of a very fine wire wrapped around the primary coil. The secondary coil is in the electrical circuit containing the spark plug. When current flows in the primary, a magnetic field gradually expands around both coils. However, voltages produced in the secondary are quite small because of the high resistance of the small diameter secondary windings. When the breaker points open, the circuit is broken, electricity stops flowing and the magnetic field suddenly collapses. The suddenly collapsing magnetic field induces

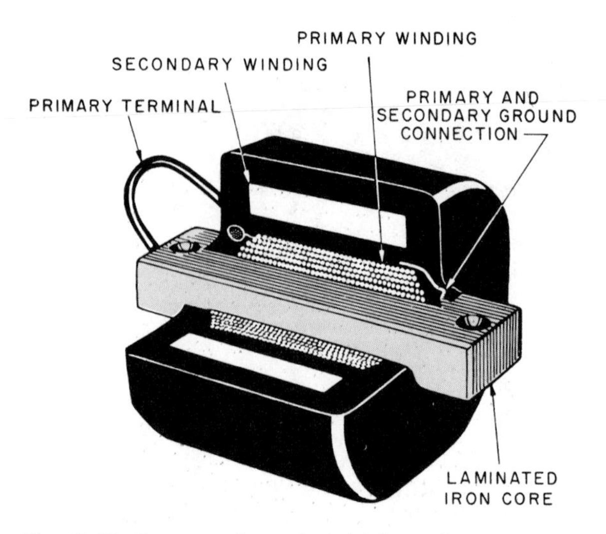

Fig. 7-11 Construction of a high-tension magneto ignition system

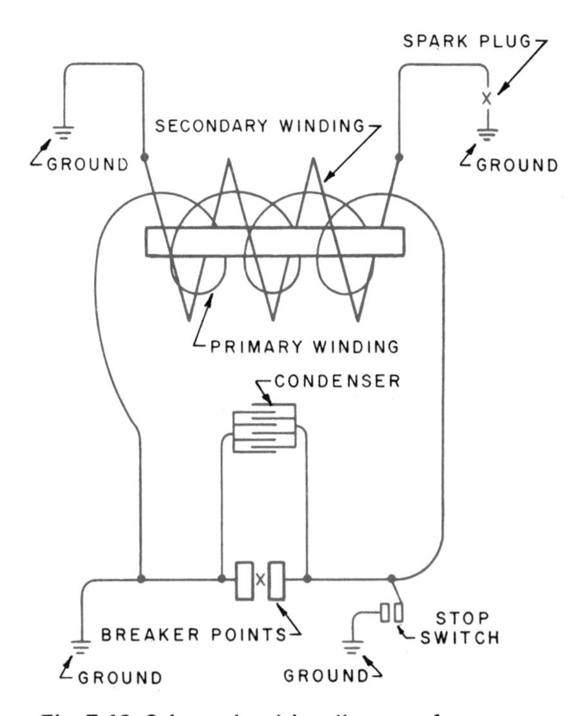

Fig. 7-12 Schematic wiring diagram of a magneto

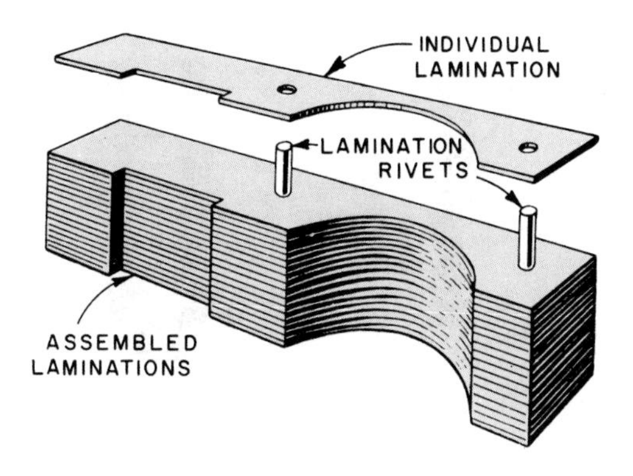

Fig. 7-13 Laminated iron core

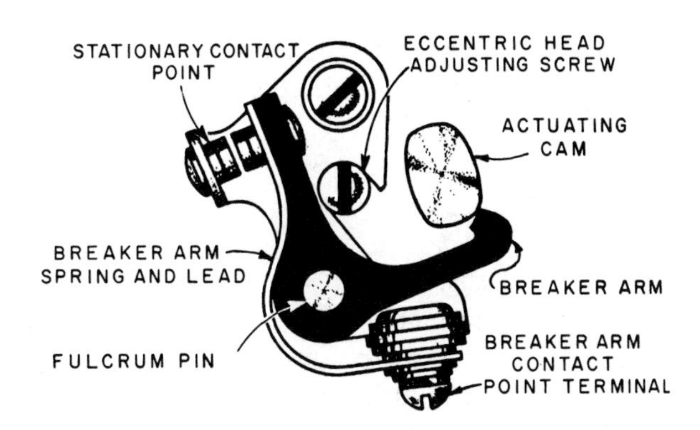

Fig. 7-14 Breaker point assembly

very high voltage in the secondary coil which is enough to jump the spark gap in the spark plug.

LAMINATED IRON CORE

The laminated iron core is made of many strips of soft iron fastened tightly together, figure 7-13. The soft iron core helps to strengthen the magnetic field around the primary and secondary coils, but the core cannot retain the magnetism and become permanently magnetized. The purpose of using many strips instead of a solid core is to reduce eddy currents (stray currents that try to flow in the opposite direction) which create heat in the core. The general shape of the laminated core may vary from magneto to magneto but its function remains the same.

BREAKER POINTS

The breaker points are made of tungsten and mounted on brackets, figure 7-14. The two points are usually closed or touching each other providing a path for electron flow. However, just before the spark is desired, the breaker points are opened by cam action breaking the electrical circuit. Breaker points open and close anywhere from 800 times per minute to 4500 or more times per minute depending on the engine speed.

When the breaker points are open they are usually separated by 0.020 of an inch; the exact opening varies from magneto to magneto. This separation is critical to the function of the magneto. Therefore, the breaker point gap must be adjusted correctly.

BREAKER CAM

The breaker cam actuates the breaker points. One bracket of the breaker assembly rides on this cam. As the cam rotates, it opens and closes the breaker points.

On most two-cycle engines, this cam is mounted on the crankshaft and opens the normally closed breaker points once each revolution. On most four-cycle engines, the breaker cam is mounted to operate from the camshaft which is turning at one-half crankshaft speed. By mounting the breaker points here, they open once every two revolutions of the crankshaft. This provides a spark for the power stroke but none for the exhaust stroke. The shape of the cam depends on the number of cylinders and the design of the magneto, figure 7-15.

CONDENSER

The condenser acts as an electrical storage tank in the primary circuit, figures 7-16 and 7-17. When the breaker points open quickly, the electrons tend to keep flowing. If no condenser was present, a spark could jump across the breaker points. If this happened, the breaker points would soon burn up resulting in a weakening effect on the voltage produced by the magneto secondary coil. The condenser provides an electrical storage tank for this last surge of electron flow. The action of the condenser provides for a very abrupt interruption of the primary circuit. The condenser's discharge helps to induce high voltages in the secondary coil as the magnetic fields reverse. The condenser can be easily located because it usually looks like a miniature tin can.

Fig. 7-15 Common breaker cam shapes

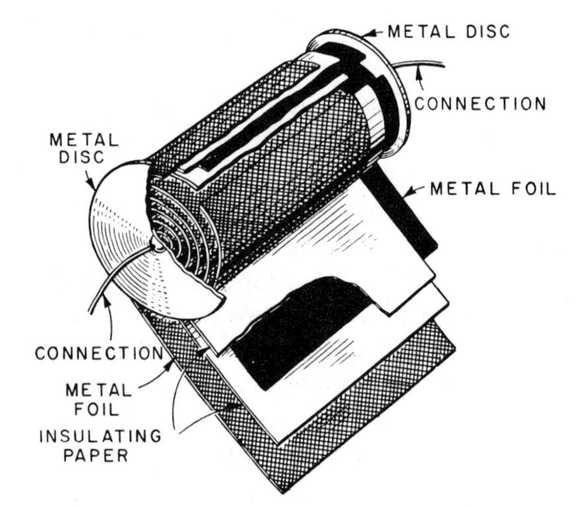

Fig. 7-16 Cutaway showing the construction of a condenser

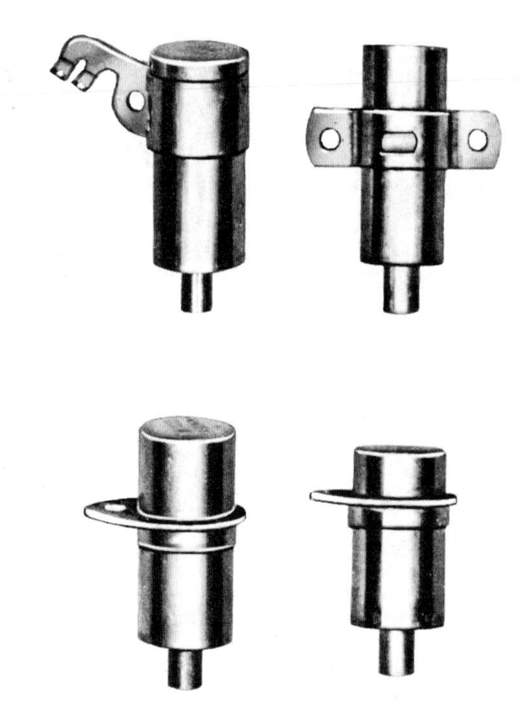

Fig. 7-17 Typical condensers

SPARK PLUG CABLE

The spark plug cable connects the secondary coil and the spark plug providing a path for the high-tension voltage. This part is often referred to as the high-tension lead.

SPARK PLUG

The spark plug is a vital part of the ignition system. In this part the resulting work of the magneto parts is seen. Basically, the spark plug consists of a shell, ceramic insulator, center electrode, and ground electrode. The two electrodes are separated by a gap of about 0.035 of an inch. The path of electricity is down the center electrode and across the air gap to the ground electrode. The voltage must jump the air gap. When a high-tension voltage of about 20,000 volts is reached, a spark jumps between the electrodes. This spark ignites the fuel mixture within the combustion chamber.

THE COMPLETE MAGNETO CYCLE

Refer to figure 7-18 for illustrations of the steps in the magneto cycle. When the permanent magnet is far away from the high-tension coil, it has no effect on the coil. As the permanent magnet comes closer and closer, the primary coil feels the increasing magnetic field; the coil is being cut by magnetic lines of force. Therefore, electrons flow in the primary coil. This current passes through the breaker points and into the ground. When the permanent magnet is nearly opposite the high-voltage coil the magnetic field around both coils is reaching its peak. Also, the piston is reaching the top of its stroke compressing the fuel mixture.

At this time, the position of the permanent magnet causes the polarity of the laminated iron core and the coils to reverse direction. This change of direction is momentarily choked or held back by the coil. The breaker points then open interrupting the current flow and allowing the reversal and rapid collapse of the intense magnetic field that had been built up around the primary and secondary coils. Magnetic lines of force are cutting coils very rapidly and high voltages are induced in the coils. In the secondary coil, the voltage may reach 18,000 to 20,000 volts which is enough to jump across the spark gap in the spark plug. This spark ignites the fuel mixture and the piston is forced down the cylinder.

SPARK ADVANCE

When fuel burns in the combustion chamber, it does not exert all of its power immediately. A very short period of time is required for the fuel to ignite and reach its full power. Even though this time lag is very short, it is very important in an engine operating at high speeds. For example, if the mixture was not ignited until the piston reached top dead center, the piston would already be starting back down the cylinder before the full force of the burning fuel is reached. This results in loss of power. Therefore, it is necessary to ignite the fuel slightly before the piston reaches the top of its stroke to realize the full force of combustion. Causing the spark to occur earlier in the engine cycle is called *spark advance.*

The spark is advanced more and more as the engine speed increases because there is less time for combustion to take place. For example, in a four-cycle engine operating at 2000 rpm, each power stroke takes about 1/64 of a second. If the speed is increased to 4000 rpm, each power stroke takes only 1/128 of a second. If the speed is doubled, there is only one-half as much time for combustion to take place. Also, high speeds give the engine higher compression and a more forceful mixture. At high speeds the spark jumps the spark gap before the piston reaches the top of its stroke.

At slow speeds the spark is delayed and occurs later in the cycle, slightly before the

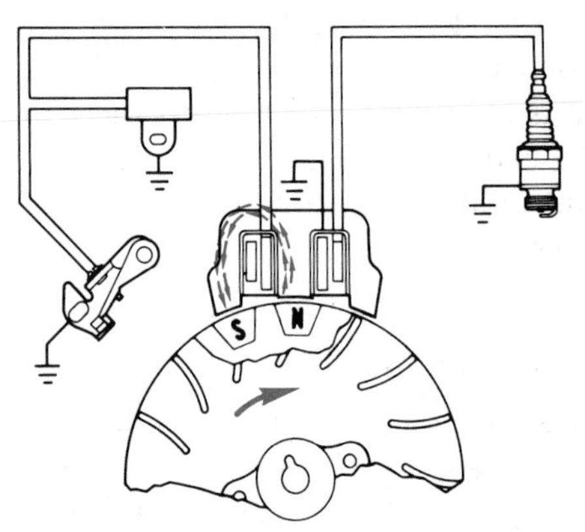

(A) Magnetic field as magnets approach the coil and laminated core

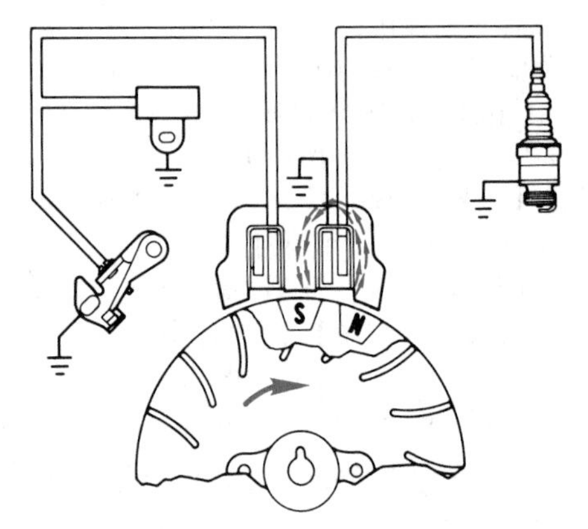

(B) Polarity of magnetic fields reverses rapidly

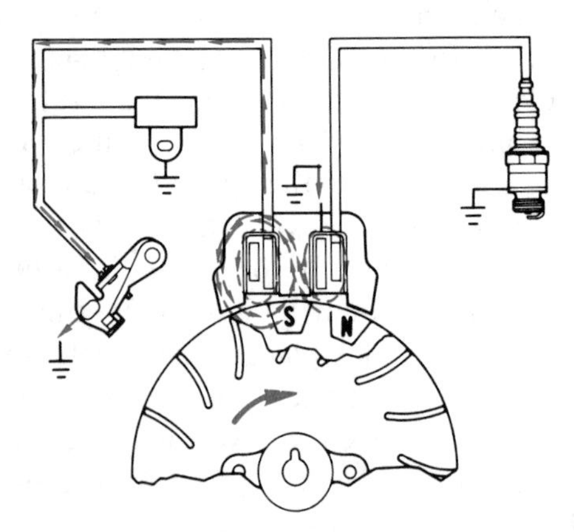

(C) Current flows in the primary circuit as magnetic fields reverse

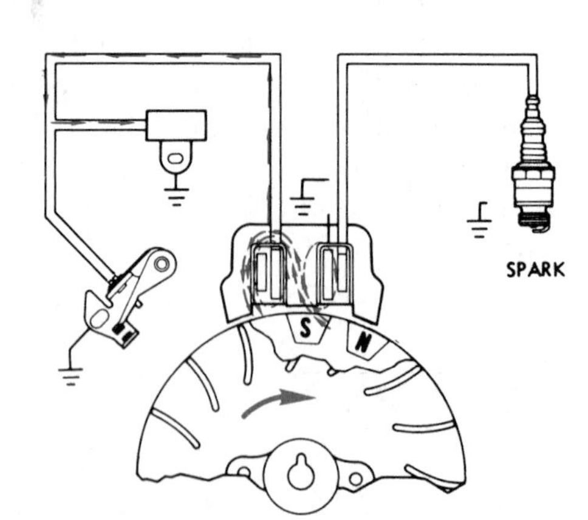

(D) Current stops as breaker points open

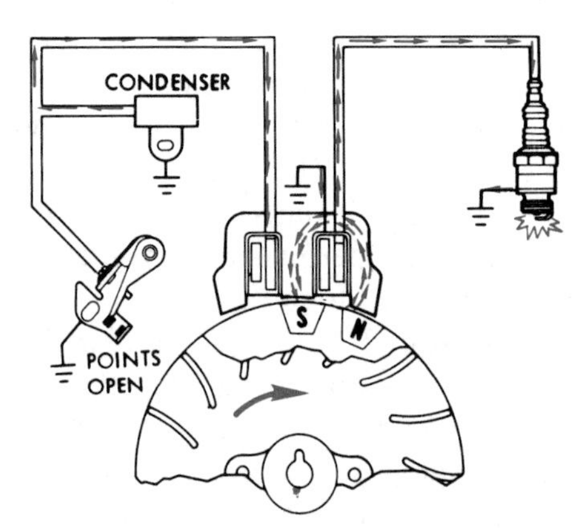

(E) Collapsing magnetic fields and condenser surge induces high voltage to jump the gap at the spark plug

Fig. 7-18 The magneto cycle

piston reaches top dead center. At slow speeds there is more time for combustion to take place. Also, compression is lower and not as much air/fuel mixture is drawn into the combustion chamber.

Regulating spark advance is done by controlling the time that the breaker points open. Advancing the spark can be done automatically or manually. Both methods are used on some small gasoline engines.

MANUAL SPARK ADVANCE

Manual spark advance is usually accomplished by loosening the breaker point assembly and rotating it slightly to an advanced position. It is then locked in its new position.

Many outboard motors have a type of manual spark advance although it is generally referred to by manufacturers as a spark-gas synchronization system. The magneto plate on which the breaker points are mounted is located underneath the flywheel. This plate can be rotated through about 25 degrees. The breaker cam is securely mounted on the crankshaft; its relative position cannot be changed. Spark advance is obtained by moving the magneto plate so the breaker points are opened earlier in the cycle. The throttle and magneto plates are linked together so that opening the throttle also advances the spark. At full throttle the spark is fully advanced. At idling speeds the throttle is closed and the magneto plate is positioned for minimum spark advance.

AUTOMATIC SPARK ADVANCE

Automatic spark advance is usually accomplished by a centrifugal mechanism which is capable of changing the relative position of the breaker cam and the breaker points, figures 7-19 and 7-20. Here the breaker cam can be rotated through a small distance on its shaft, about 25 degrees. A spring holds the cam in the retarded position at idling and slow speeds. As the engine speed increases, centrifugal force throws the mechanism's hinged weights outward. This outward motion overcomes the spring's tension and this motion is used to rotate the cam to a more advanced position. At full speed the cam has been rotated its full limit for maximum spark advance.

Some engines, especially automotive types, also have a vacuum advance mechanism working with a centrifugal mechanism to provide more accurate spark advance, especially at slow speeds. Automatic spark advance can

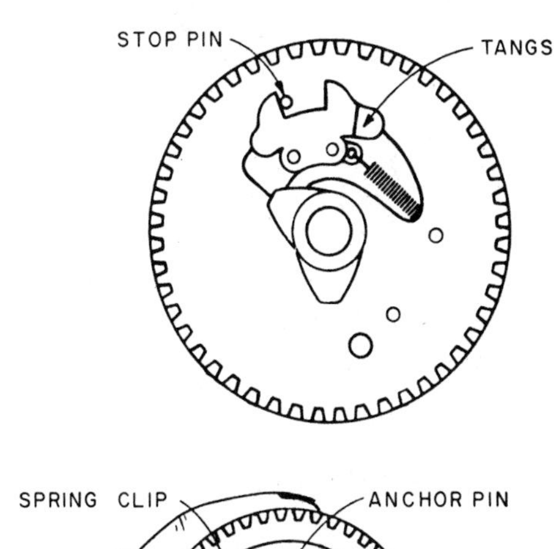

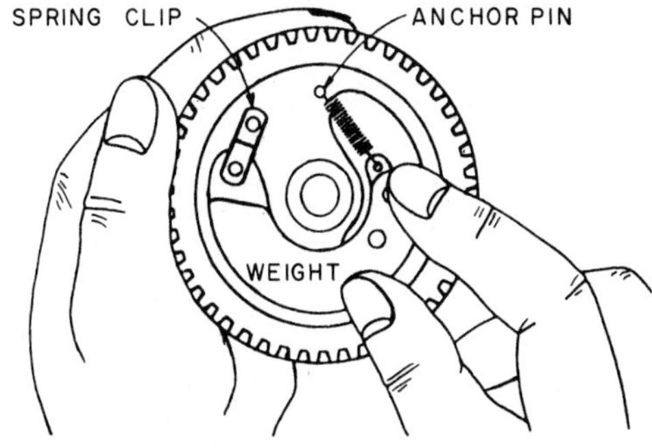

Fig. 7-19 Spark advance mechanism with breaker cam and weight mounted on the camshaft gear

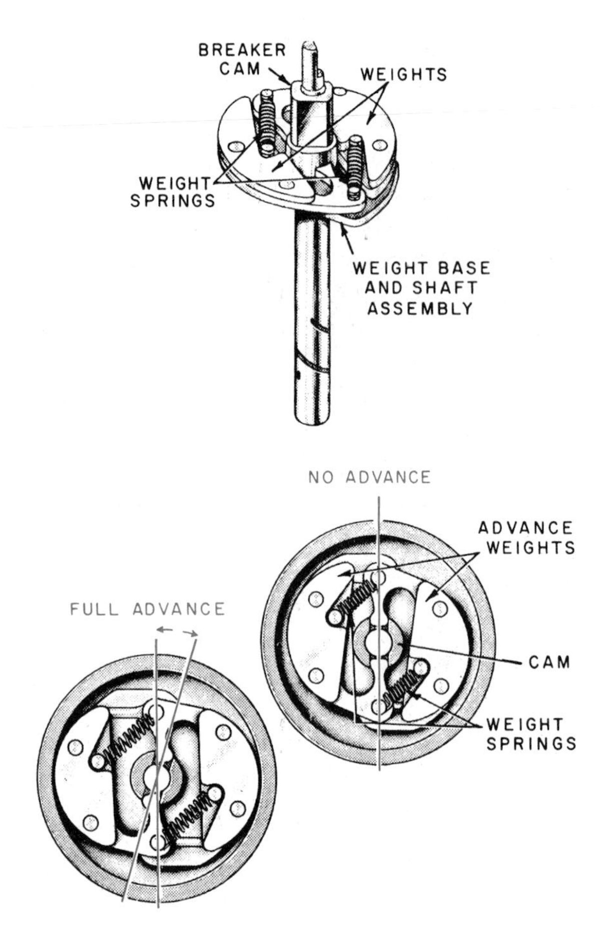

Fig. 7-20 **Automatic spark advance mechanism changes the breaker cam's position in relation to the points**

also be used to move the breaker point assembly to secure the proper advance.

IMPULSE COUPLING

When an engine is started by hand, it is turned over slowly. The voltages produced by the magneto depend upon the speed at which magnetic lines of force cut the primary coil. Therefore, the voltages produced for starting can be quite low, resulting in a weak spark.

Some engines are equipped with an impulse

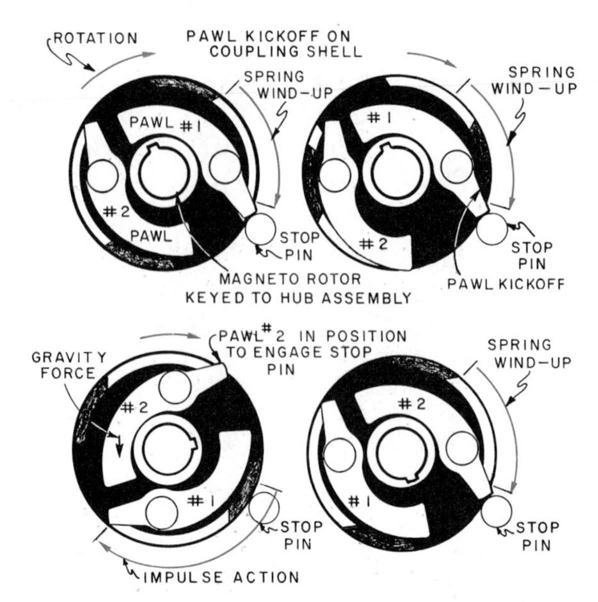

Fig. 7-21 Sequence of operation for impulse coupling for 180 degrees spark magneto

coupling device to supply higher voltages and a hotter spark for slow starting speeds. Using a tight spring and retractable pawls the rotation of the magneto's magnetic rotor can be stopped for all but the last few degrees of its revolution, figure 7-21. During the last few degrees of the revolution, the pawl retracts allowing the spring to snap the magnetic rotor past the firing position at a very high speed. A high voltage can be produced because magnetic lines of force are cutting the primary coil rapidly.

MAGNETO IGNITION FOR MULTICYLINDER ENGINES

Magnetos are often used for multicylinder engines. However, having more than one cylinder to supply with a spark does present a problem. Two solutions to this problem are in common use today.

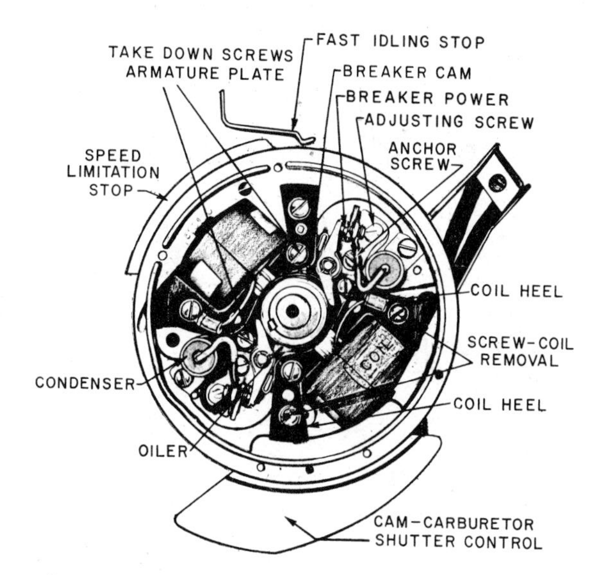

Fig. 7-22 Two complete magnetos are installed on this armature plate.

Two-, three- and four-cylinder engines can have magneto ignition by simply installing a separate magneto for each cylinder. This method is commonly used with outboard engines using a flywheel magneto. Instead of mounting just one magneto on the armature plate under the flywheel, a magneto is installed for each cylinder, figure 7-22. The breaker points of each magneto ride on the common cam. In two-cylinder engines the magnetos are 180 degrees apart; three-cylinder engines 120 degrees apart; and four-cylinder engines 90 degrees apart. Each magneto functions separately to supply its cylinder with a spark.

Another solution is to use a distributor and a two-pole or four-pole magnetic rotor, figure 7-23. Only one complete magneto is used even though the engine may have two or four cylinders. In the case of a two-pole rotor used on a two-cylinder engine, the magneto can produce two sparks every revolution of the magnetic rotor, one every 180 degrees rotation. A two-lobed cam is used to open the breaker points twice each rotor revolution. The distributor rotor channels the high-tension voltage to the correct spark plug.

ENGINE TIMING

The spark must jump the spark gap at exactly the right time. This is just before the piston reaches top dead center. In many engines, the breaker cam is driven by the camshaft, therefore, the gear on the crankshaft and the camshaft gear must be assembled correctly. These two gears are usually marked with punch marks so the repairer can easily make the correct assembly, figure 7-24. Incorrect alignment of the gears can cause poor operation or no operation at all.

SOLID STATE IGNITION

Solid state ignition refers to the fact that solid state electronic parts (namely transistors), replace the breaker points. The breaker points in a conventional ignition system can be a source of trouble. The breaker cam is also eliminated so there are no moving parts in the system. Sometimes the systems are called *capacitor discharge systems, breakerless ignition, transistorized ignition* or *electronic ignition.* These systems can be used with just a conventional flywheel magnet as the source of the magnetic force, or they may be used with a generator or an alternator battery system.

The main components of the solid state ignition system are a generator or alternator coil, trigger module, ignition coil assembly, and flywheel magnets. The system has a conventional spark plug and spark plug lead. A

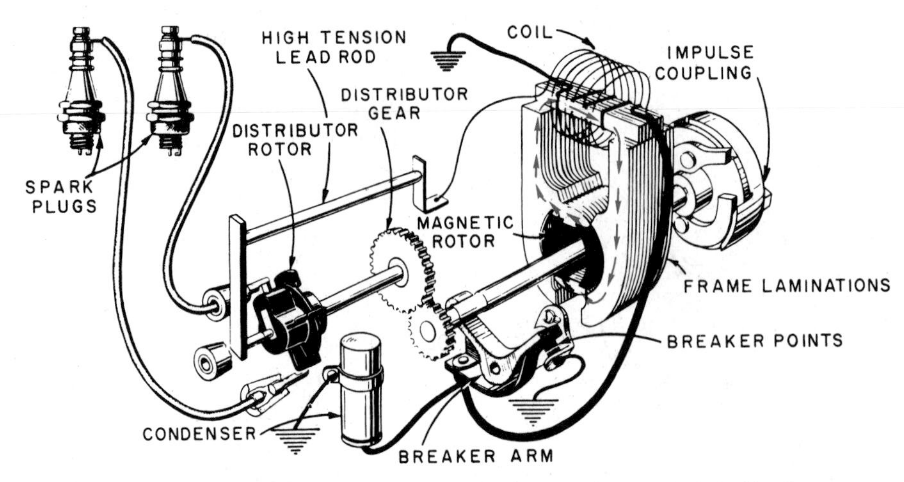

Fig. 7-23 Diagram of rotating-magnet magneto with jump-spark distributor

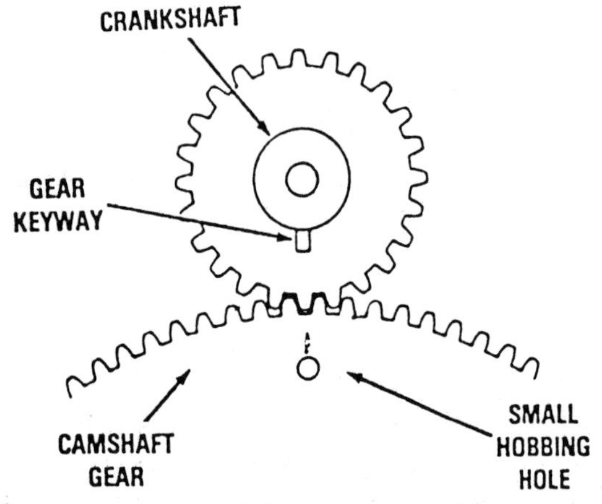

Fig. 7-24 Aligning timing marks on the crankshaft gear and camshaft gear. Notice the larger camshaft gear revolves at one-half crankshaft speed.

schematic drawing of a solid state ignition system is shown in figure 7-25.

The trigger module contains transistor diodes which rectify the alternating current, changing it into direct current. Transistors control the flow of electric current by acting somewhat like a valve. Their resistance to flow can be changed by a small current to the transistors. In addition to the diode rectifiers, the trigger module contains a resistor, a sensing coil and magnet, and a silicon controlled rectifier (SCR). The silicon controlled rectifier acts as a switch.

The ignition coil contains primary and secondary windings similar to those of a conventional magneto coil plus a condenser or capacitor. Capacitors and condensers are basically the same electrically. The operation of the solid state ignition system follows this cycle:

1. The rotating magnet on the flywheel sets up an alternating current in the charging coil. This alternating current is rectified into direct current and stored in the capacitor. The diode rectifiers permit flow in one direction only so the capacitor cannot discharge back through the diodes.
2. The magnet group passes the trigger coil, setting up a small current which triggers or gates the SCR. This gating makes the SCR conductive so that the stored voltage of the capacitor surges from the capacitor through the SCR.

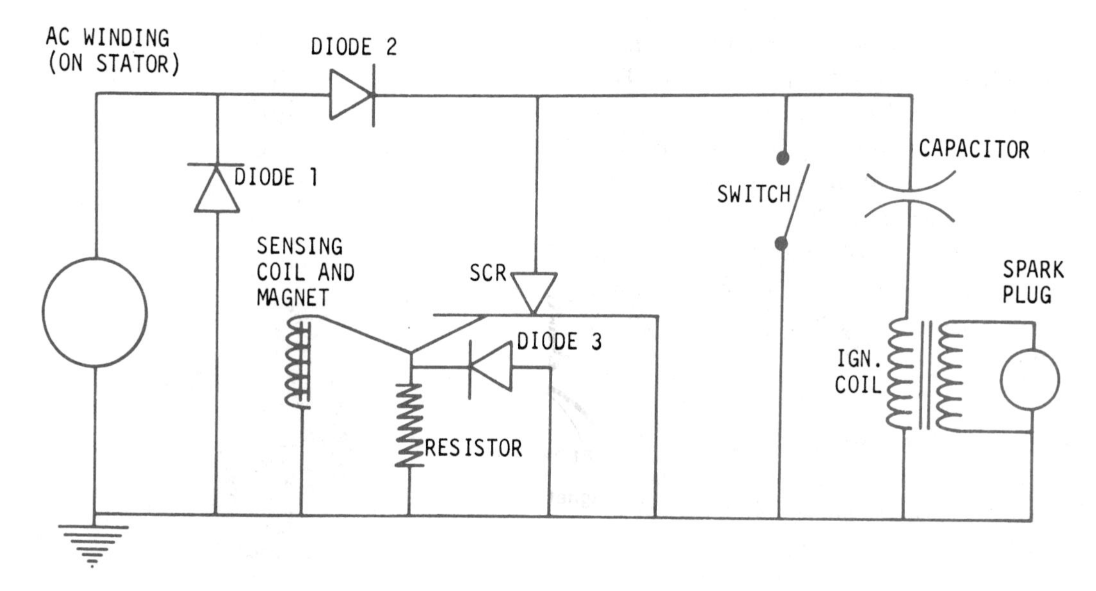

Fig. 7-25 Schematic of a solid-state ignition system

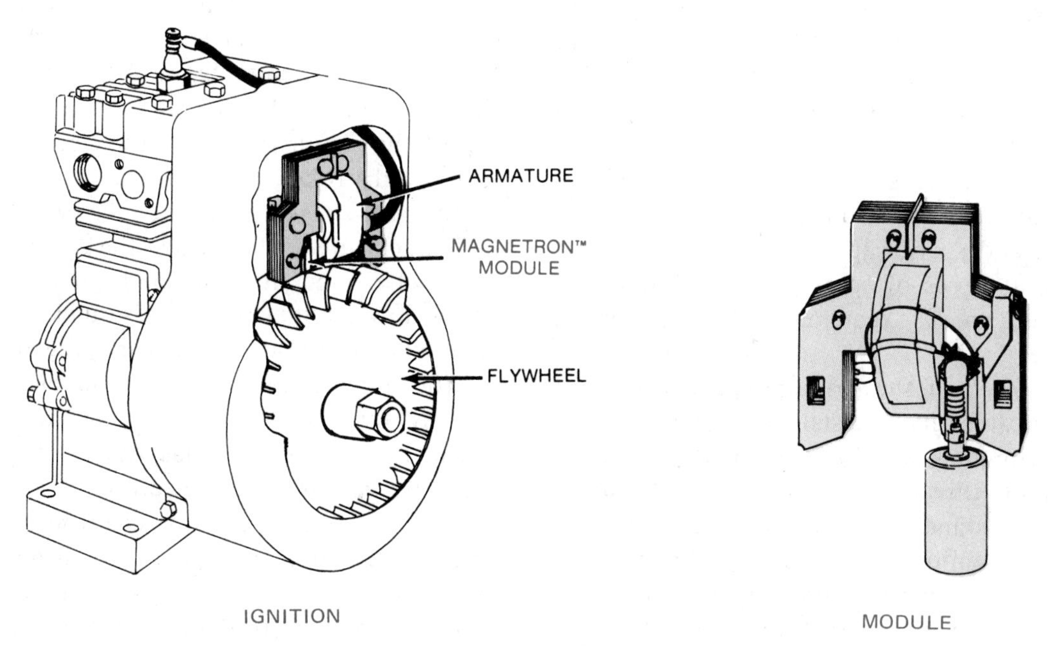

Fig. 7-26 Solid-state ignition "Magnetron™" used by Briggs and Stratton

It then is applied across the primary winding of the ignition coil. This surge of energy sets up a magnetic field in the primary winding of the ignition coil. A magnetic field is also induced around the secondary coil, and voltages of about 30,000 are produced which is sufficient voltage to jump the spark gap at the spark plug.

Solid state ignition offers several advantages such as automatic retarding of the spark at starting speeds, longer spark plug life, faster voltage rise, and high-energy spark which makes the condition of the plug and its gap less critical.

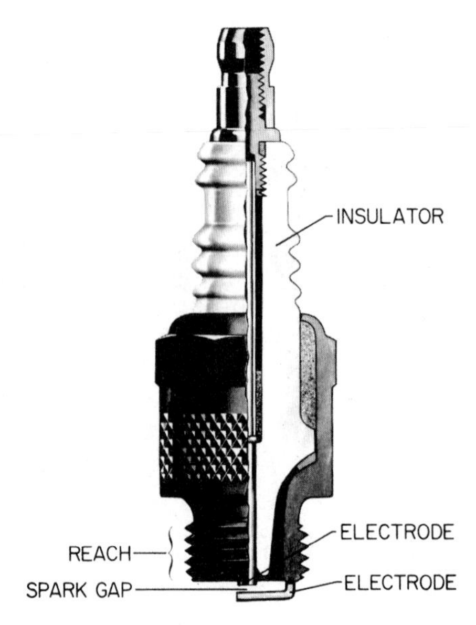

Fig. 7-27 Cutaway of a spark plug

SPARK PLUG

The spark plug is the part of the ignition system that ignites the fuel mixture, figure 7-27. It operates under severe and varying temperature conditions and is a critical part in engine operation.

All spark plugs are basically the same but they do differ in thread size, reach, heat range, and spark gap. There are hundreds of different types of spark plugs, each designed for the special requirements of a certain engine.

The *shell* of the spark plug is threaded so that it can easily be installed or removed from the cylinder head. Various spark plugs have different thread sizes. Some of the more common standard thread sizes are 10 millimeters, 14 millimeters, and 18 millimeters.

The *reach* of the spark plug is the distance between the gasket seat and the bottom of the spark plug shell, or about the length of the cut threads. The reach ranges from about 1/4 inch to 3/4 inch. Each engine must be equipped with a spark plug of the correct reach since reach determines how far the electrodes protrude into the combustion chamber. If the reach is too small, the electrodes have difficulty igniting the fuel when the spark jumps. If the reach is too great, the top of the piston may strike the electrodes on its upward stroke.

The *heat range* of a spark plug is the range of temperature within which the spark plug is designed to operate. If a spark plug operates at too low a temperature, it is quickly fouled with oil and carbon. If the spark plug operates at too high a temperature, the electrodes quickly burn up. The heat range for spark plugs depends on how fast the heat can be carried away from the electrodes. The path of heat transfer is from the electrodes to the ceramic insulator through the shell and into the cylinder head. The length of the ceramic insulator exposed to the combustion chamber determines the heat range of a spark plug. The longer this insulator, the longer the heat path and the hotter the spark plug's operating

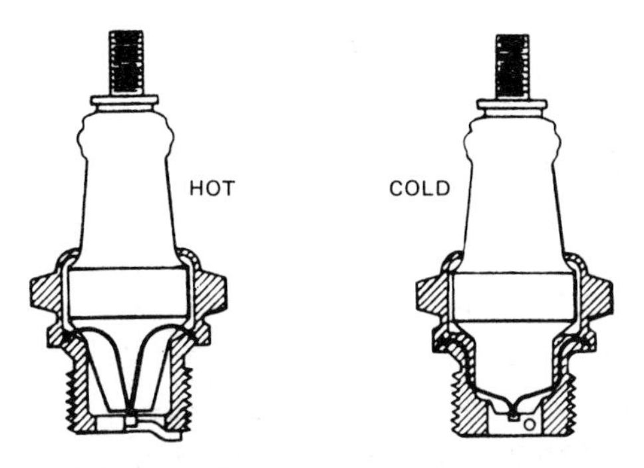

Fig. 7-28 Hot and cold spark plugs

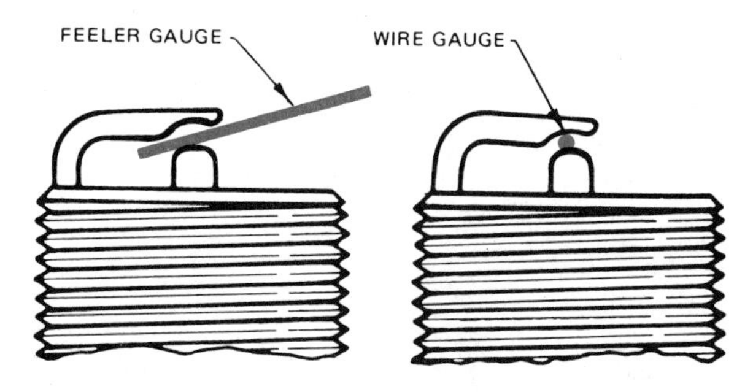

Fig. 7-29 A plain flat feeler gauge cannot accurately measure the true width of a spark gap.

temperature. Likewise, the shorter this insulator, the shorter the heat path and the colder the spark plug's operating temperature, figure 7-28.

In general, cold spark plugs are used for high-speed engine operation where excessive heat could burn the plug. Hot spark plugs would be suitable for slow speed operations where excessive carbon buildup may otherwise occur.

The *spark gap* (distance between the electrodes) must be correctly set. It is within this small space that the spark jumps and combustion begins. The gap must be large enough for sufficient fuel mixture to get between the electrodes but not so large as to prevent the spark from jumping across. The spark gap for various plugs ranges from about 0.020 inch to 0.040 inch.

If the spark plug is removed for cleaning, the gap should be checked and reset according to the manufacturer's specifications. A wire gauge should be used, figure 7-29.

BATTERY IGNITION

Battery ignition systems are often used on the larger types of small gasoline engines (4 to 15 horsepower), figure 7-30. With the battery a starter motor can be used to crank the en-

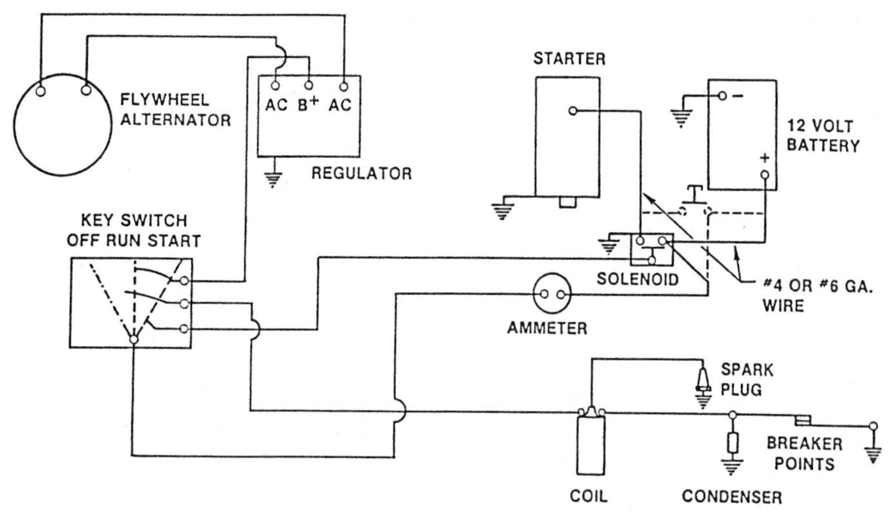

Fig. 7-30 Battery ignition system (Courtesy of Kohler Company)

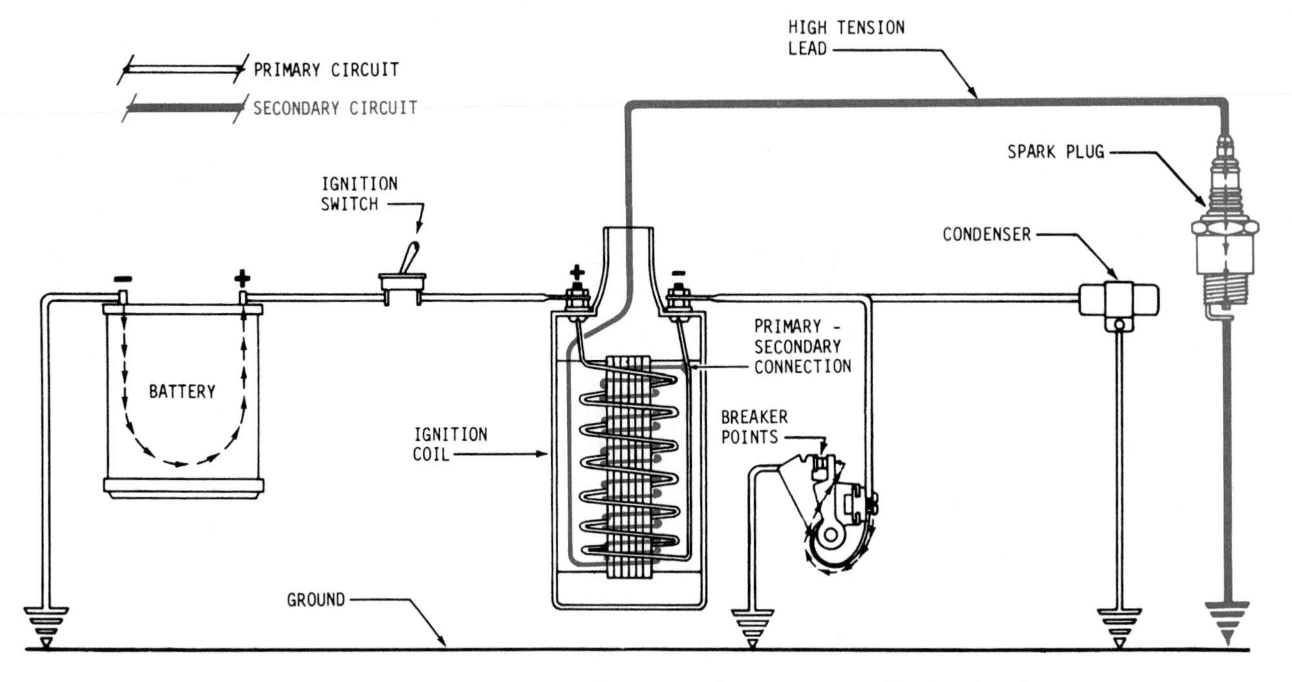

Fig. 7-31 Battery ignition system, primary (low voltage) and secondary (high voltage) circuits

gine. This eliminates the need for various types of hand-pulled or windup starters. The battery can also be used for lights or other accessories.

A typical battery ignition system includes virtually all of the magneto system parts that have been discussed earlier: the high-tension coil, breaker points, condenser, breaker cam, and spark plug. The one part that is replaced is the permanent magnet. The permanent magnet which supplies the energy to excite the coil is replaced by the battery. The most common systems for a small engine include a starter motor, battery, starter switch, voltage regulator, and the ignition system parts themselves, figure 7-31.

The starter motor uses electrical energy from the battery as a motor to crank the engine for starting, figure 7-32. After starting,

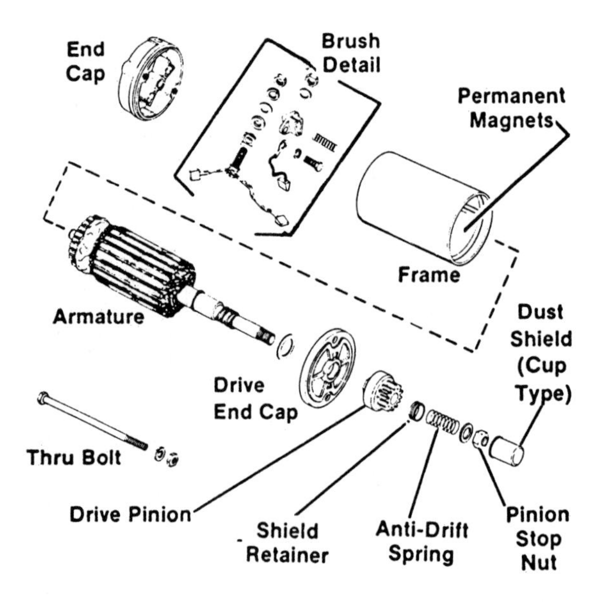

Fig. 7-32 Exploded view of a motor generator (Courtesy of Kohler Company)

the unit acts as a generator to keep the battery charged. Electrical energy is changed to mechanical energy for starting and mechanical energy is changed to electrical energy for charging the battery. With the starter switch closed, the motor turns the crankshaft through a V-belt drive. Upon starting, the starter switch is released but the motor does not disengage. The V-belt drive continues but now the motor acts as a generator for recharging the storage battery. It should be noted that a typical auto system has a separate starter which disengages upon starting and a separate generator for charging the battery.

The storage battery may be either the 6-volt or 12-volt type, figure 7-33. The battery case contains alternate plates of lead peroxide and spongy lead, both active but chemically quite different. Separators prevent the plates from touching each other, figure 7-34. The battery is filled with a solution of sulfuric acid and water. Electrically speaking, the spongy lead plates are cathodes or the negative section, lead peroxide plates are the anodes or the positive section, and the sulfuric acid and water are the electrolyte.

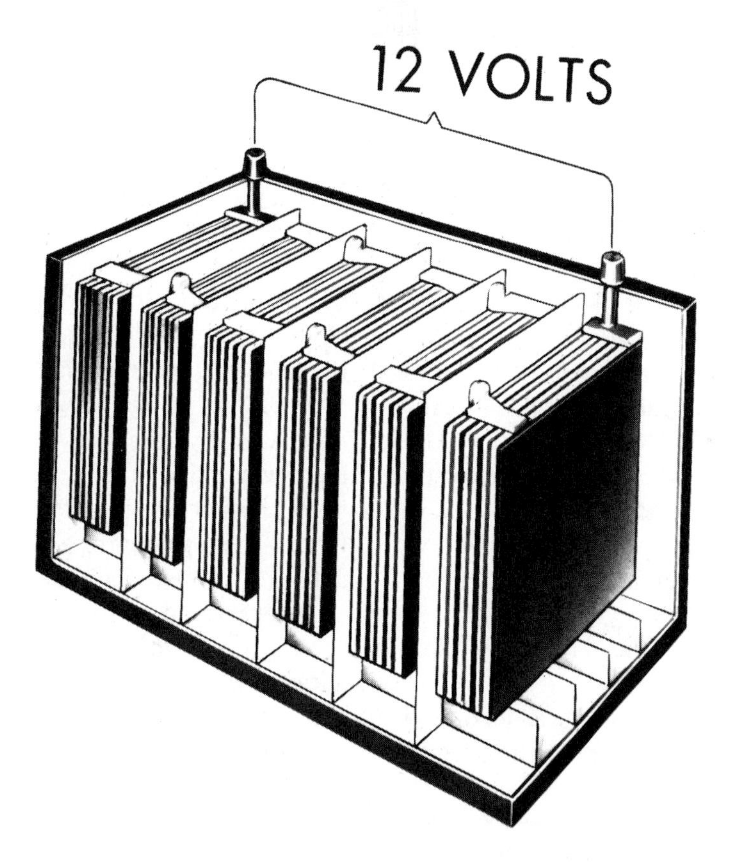

Fig 7-33 Twelve-volt battery with six cells

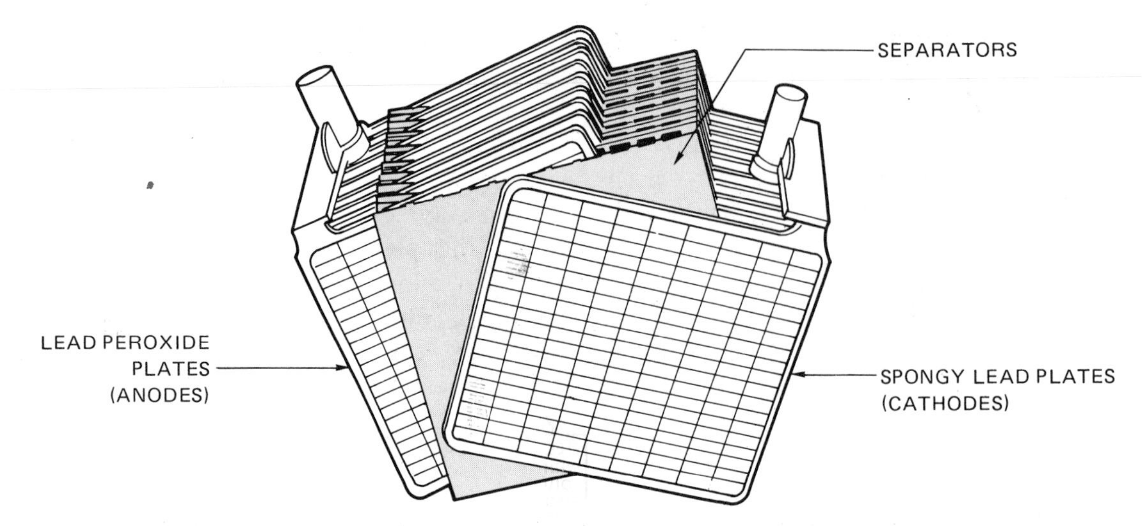

Fig 7-34 Separators prevent the lead peroxide plates and spongy plates from touching each other.

There is a voltage difference between the two groups of plates. When a complete circuit is made, current discharges from the battery. The circuit is closed and the electrolyte attacks the lead plates. The current is in motion and the battery is discharging. The lead peroxide plates start to become coated with lead sulfide and begin to look like the negative plates. The sulfuric acid breaks down forming the lead sulfide and water. The water further dilutes the electrolyte. When the battery is discharged, the negative and positive plates are very similar in chemical structure. Voltage depends on the chemical difference between the two active materials. Figure 7-35 illustrates the electrochemical action in a battery.

Since the solution of sulphuric acid and water gets weaker or less dense as the battery discharges, the condition of the battery can be determined by checking the specific gravity of the solution with a battery hydrometer. The solution in a fully charged battery will have a specific gravity of 1.300. In a discharged battery, the solution has a specific gravity of 1.120. Remember, the specific gravity of water is 1.000.

Fortunately, the action of the battery can be reversed. The starter generator recharges the battery without damage to the battery. The battery is recharged when the voltage produced by the generator is greater than battery voltage and the current being drawn from the battery is not greater than the input from the generator. Thus, the chemical action is reversed, and the battery is returned to its charged condition.

A current-voltage regulator comes into action when the engine has started and the starter switch is off and the engine is operating at 800 to 1000 rpm. The regulator prevents over charging the battery and also keeps the generator from producing voltages and current flow that could be damaging to the system.

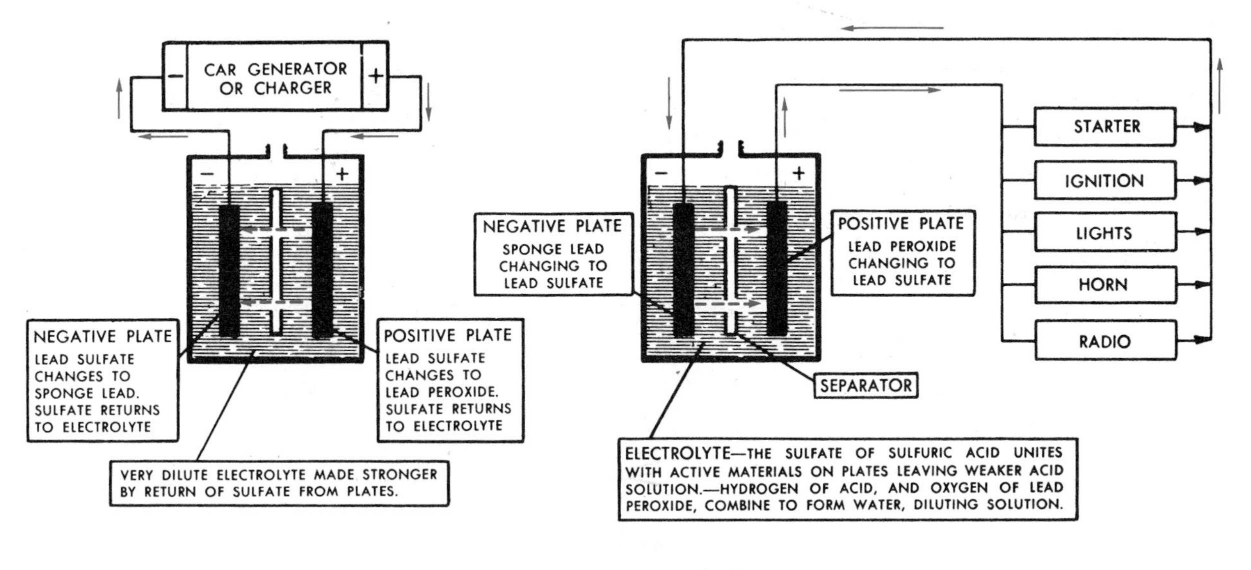

Fig. 7-35 Electrochemical action in a battery

REVIEW QUESTIONS

1. What three particles make up the atom?
2. Explain how electrons can move in a conductor.
3. What are the three ways that electricity can be produced or seen?
4. Explain the following electrical measurements: amperes, volts, and ohms.
5. State Ohm's Law.
6. Explain how permanent and temporary magnets differ.
7. What happens when a wire is cut by magnetic lines of force?
8. What happens to a coil of wire when electric current flows through it?
9. Explain how a transformer works.
10. In a magneto, what is the source of the magnetic field?
11. What are the two parts of the high-voltage coil?
12. What purpose does the laminated iron core serve?

13. What is the function of the breaker points? How are they opened?
14. What is the function of the condenser? The spark plug?
15. What are the two electrical circuits in the magneto?
16. Why must the spark be advanced at high speeds?
17. What is the advantage of impulse coupling?
18. Can magneto ignition be used for multicylinder engines?
19. What parts of a conventional magneto are replaced on a solid state ignition system?
20. Explain the function of the transistor diodes in a solid state ignition system.
21. Explain how a silicon controlled rectifier acts as a switch or gate.
22. In what four ways do spark plugs differ?
23. What is the main advantage of having a battery in the ignition system?
24. a. When does the motor generator work as a motor?
    b. When does it work as a generator?
25. Storage battery plates are made of what material?
26. What is an electrolyte?

Unit 8 Routine Care, Maintenance, and Winter Storage

OBJECTIVES After completing this unit, the student should be able to:

- Discuss the important areas of routine care and maintenance.
- Identify types of spark plug fouling.
- Clean and regap spark plugs.
- Discuss the various types of air cleaners and their care.
- Correctly prepare an engine for winter storage.
- Locate routine care and maintenance information in an owner's instruction book.

A gasoline engine represents an investment. To safeguard this investment the engine operator must perform certain routine steps in care and maintenance. Routine care and maintenance help to ensure the longest life possible for engine parts and can save on repair bills.

The best source of information on engine care is the owner's manual or operator's instruction book that comes with every new engine. The book contains the information that the manufacturer considers necessary for the owner to know. The information included allows the owner to obtain top performance and long engine life. Some typical topics discussed are lubrication procedures, carburetor adjustment, ignition system data, air cleaner cleaning instructions, winter storage, and care of the machine that the engine is used on. The information in the book is important. A person with a new engine should read and thoroughly understand the instruction book before touching the engine. A plastic cover helps to keep the book in good condition.

ROUTINE CARE AND MAINTENANCE

Routine care and maintenance is given to an engine during its normal use. It is provided to keep the engine operating at its peak efficiency and to prevent undue wear of engine parts. There are four basic points to consider: oil supply, cooling system, spark plugs, and air cleaner, figure 8-1.

Oil Supply

On a four-cycle engine, the oil supply in the crankcase should be checked daily or each time the engine is used. If the oil level has dropped below the *add* mark on the dipstick, fill the crankcase to the proper level. If the engine does not have a dipstick, fill the crank-

WARNING: *Before servicing engine or equipment, always remove the spark plug leads to prevent engine from starting accidentally. Ground the leads to prevent sparks that could cause fires.*

✓ If engine is operated under dirty, dusty conditions, maintenance procedures should be performed more frequently than stated.

MAINTENANCE PROCEDURES	FREQUENCY
Clean Air Intake Screen ✓	DAILY
Check Oil Level	DAILY
Replenish Fuel Supply	DAILY
Service Precleaner ✓	25 HRS.
Change Oil	50 HRS.
Clean Cooling Fins and External Surfaces ✓	50 HRS.
Check Air Filter ✓	50 HRS.
Change Oil Filter	100 HRS.
Check Spark Plugs	100 HRS.
Have Breaker Points Checked*	500 HRS.
Have Ignition Timing Checked*	500 HRS.
Have Valve & Tappet Clearance Checked*	500 HRS.
Have Cylinder Heads Serviced	500 HRS.•

*Have these services done only by a Kohler engine specialist.
•250 hours when leaded gasoline is used

Fig. 8-1 All engines should have a service schedule similar to this one for a Kohler Model K582 engine. (Courtesy of Kohler Company)

case until the oil can be seen in the filler hole. A new engine or one that is being used for the first time must be carefully checked to be sure that oil is in the crankcase.

Most manufacturers recommend that the oil be changed after a short period of time when breaking in a new engine. Depending on the manufacturer, this first oil change should be after two to twenty hours of operation.

After the oil has been changed once, the engine can go for longer periods of time between oil changes. Lauson specifies changing engine oil after every 10 hours of operation; Briggs and Stratton after every 20 hours; Kohler after every 25 hours; Gravely and Wisconsin after every 50 hours, Onan after every 100 hours. Each manufacturer's instruction book gives the length of time between oil changes and the exact SAE number and quality of oil to be used.

Two-cycle engines have their lubricating oil mixed with the gasoline. The owner's manual specifies the proportions of oil to gasoline and the type of oil to be used.

Cooling System

The most important consideration regarding the cooling system is to keep it clean. On air-cooled engines, clean the radiating fins, flywheel vanes, and shrouds whenever they begin to accumulate grass, dirt, etc., figure 8-2. Do not allow deposits to build up which reduce the engine's cooling capacity. How often the cooling system is cleaned depends on the dust conditions under which the engine operates.

On a water-cooled engine, check the water level in the radiator before operating the engine. If the water level is low, fill it to the correct level.

Outboard motors need little routine care for the cooling system. However, each time the engine is started the operator should

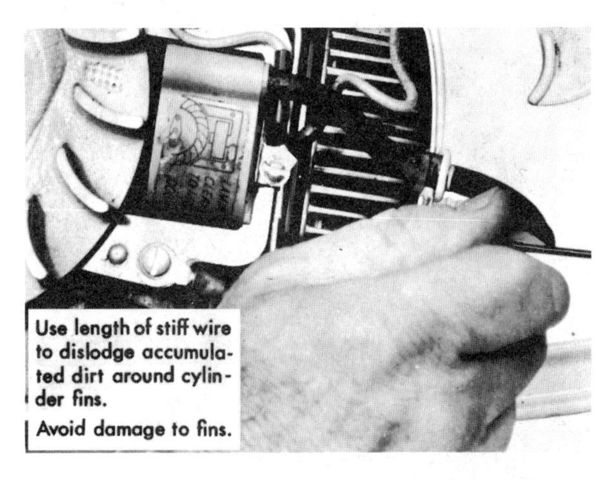

Fig. 8-2 Keep heat radiating fins clean

check to see that the engine is pumping water. This can be detected on most engines by a small amount of water coming from the telltale holes. If the engine is not pumping water, it should be stopped immediately and the source of the trouble corrected. Do not put hands by the underwater discharge to determine if the engine is pumping water as the propeller may be dangerously close. If the engine's cooling system is thermostatically controlled, it may take a few moments for the thermostat to open to allow a flow of water from the engine.

Spark Plugs

Spark plugs need to be cleaned and regapped periodically. Spark plugs are usually cleaned after 100 hours of operation; some manufacturers recommend cleaning after as little as 50 hours, while a few manufacturers do not recommend cleaning at all, but advise regular replacement. Since spark plugs operate under severe conditions, they become fouled easily. Knowing the signs of fouling can often tell about the engine's overall condition. Common problems that develop with spark plugs are carbon fouling, burned electrodes, chipped insulator, and splash fouling, figure 8-3.

Normal spark plugs have light tan or gray deposits but show no more than 0.005 inch increase in the original spark gap. These can be cleaned and reinstalled in the engine.

Worn out plugs have tan or gray colored deposits and show electrode wear of 0.008 inch to 0.010 inch greater than the original gap. Throw away such plugs and install new ones.

Oil fouling is indicated by wet, oily deposits. This condition is caused when oil is pumped (by the piston rings) to the combustion chamber. A hotter spark plug can help but the condition may have to be corrected by engine overhaul.

Gas fouling or fuel fouling is indicated by a sooty, black deposit on the insulator tips, electrodes, and shell surfaces. The cause may be a too rich fuel mixture, light loads, or long periods at idle speed.

Carbon fouling is indicated if the plug has dry, fluffy, black deposits. This condition may be caused by a too rich fuel mixture, improper carburetor adjustment, choke partly closed, or clogged air cleaner. Also, slow speeds, light loads, long periods of idling, and the resulting cool operating temperature can cause deposits not to be burned away. A hotter spark plug may correct carbon fouling.

Lead fouling is indicated by a soft, tan, powdery deposit on the plug. These deposits of lead salts build up during low speeds and light loads. They cause no problem at low speed but at high speeds, when the plug heats up, the fouling often causes the plug to misfire, thus limiting the engine's top performance.

NORMAL SPARK PLUG

WORN OUT SPARK PLUG

CARBON--FOULED SPARK PLUG

CHIPPED INSULATOR ON SPARK PLUG

OVERHEATED SPARK PLUG

OIL-FOULED SPARK PLUG

SPLASH-FOULED SPARK PLUG

BENT SIDE ELECTRODE

Fig. 8-3 Normal and damaged spark plugs

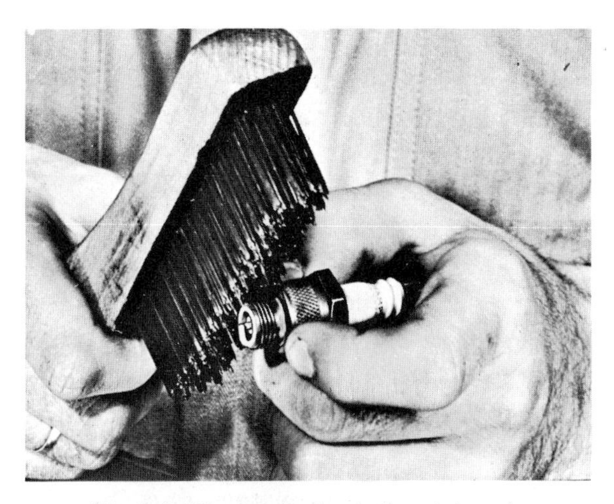

Fig. 8-4 Wire brush the shell and threads

Fig. 8-5 File electrodes

Burned electrodes are indicated by thin, worn away electrodes. This condition is caused by the spark plug overheating. This overheating can be caused by lean fuel mixture, low octane fuel, cooling system failure, or long periods of high-speed heavy load. A colder plug may correct this trouble.

Splash fouling can occur if accumulated cylinder deposits are thrown against the spark plugs. This material can be cleaned from the plug and the plug reinstalled. If spark gap tools or pliers are not properly used, the electrode can be misshaped into a curve.

Cleaning Spark Plugs. If inspection indicates that the spark plug needs cleaning, it can be cleaned within a few minutes. First, wire brush the shell and threads not the insulator and electrodes, figure 8-4. Then wipe the plug with a rag that has been saturated with solvent. This removes any oil film on the plug. Next, file the sparking surfaces of the electrodes with a point file, figure 8-5. Finally, regap the electrodes setting them according to specifications, figure 8-6. Check the gap with a wire spark gap gauge, figure 8-7. Abrasive blast cleaning machines found

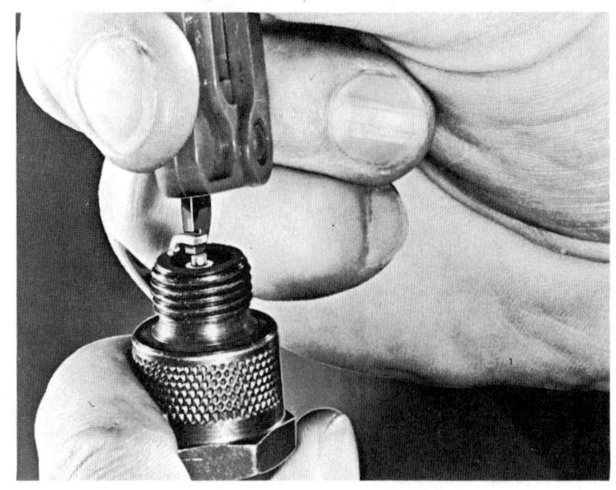

Fig. 8-6 Regap the plug

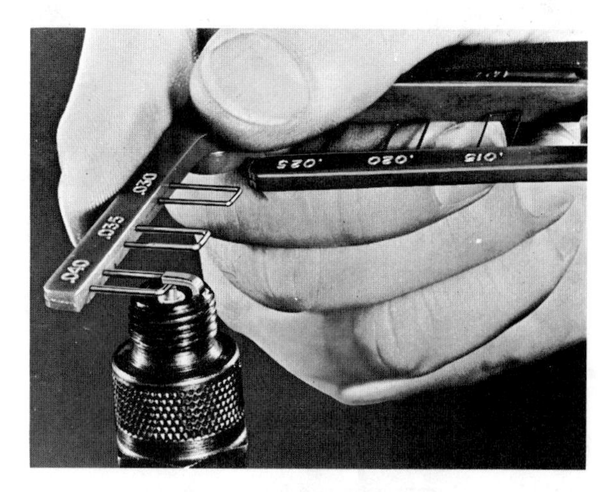

Fig. 8-7 Check the spark gap

in many service stations are not recommended by several engine manufacturers. Do not allow any foreign material to fall into the cylinder while the spark plug is out. Also, do not forget the spark plug gasket when reinstalling the spark plug. This copper ring provides a perfect seal and the proper cooling for the spark plug.

Cleaning cannot repair a broken, cracked, or otherwise severely damaged spark plug. Cleaning is designed to maintain good operation and prolong spark plug life. Many small engine manufacturers recommend the installation of a new spark plug at the beginning of each season. On a single-cylinder engine, the spark plug is more critical than on a six- or eight-cylinder engine.

Air Cleaner

The air cleaner serves the important function of cleaning the air before it is drawn through the carburetor and into the engine. Small abrasive particles of dust and dirt are trapped in the air cleaner, figure 8-8. The air cleaner should be periodically cleaned; how

Fig. 8-8 About this much abrasive material would enter a six-cylinder engine every hour if the air cleaner was not used.

often depends mainly on the atmosphere in which the engine is operating, figure 8-9. In a dusty atmosphere, (a garden tractor used in a dry garden for example), the air cleaner should be cleaned every few hours. Under normal conditions, the air cleaner should be cleaned every 25 hours or sooner.

If an engine is always operated in a clean atmosphere it may not be equipped with an air cleaner. Examples are snowmobiles, snow throwers, and outboard motors.

Air cleaners can be classified as either oil

Fig. 8-9 Engines operating in different environments have different air cleaner requirements. (Courtesy of Kawasaki Motor Corporation and Evinrude Motor Division)

type or dry type. Oil type includes oil-bath air cleaners and oil-wetted polyurethane (foam) filter element air cleaners. Dry type includes foil, moss, or hair element, felt or fiber hollow element, and metal cartridge air cleaners.

Fig. 8-10 Oil-bath air cleaner

Oil-type Air Cleaners

Oil-bath Air Cleaner. The oil-bath air cleaner carries a small amount of oil in its bowl, figure 8-10. The level of this oil should be checked before each use of the engine. If the oil level is low, fill it to the mark but not above the mark. Use the same type of oil as is used in the crankcase. As this cleaner works, both the oil and filter element become dirty. To clean an oil-bath air cleaner, disassemble the unit and then pour out the dirty oil. Wash the bowl, cover, and filter element in solvent. Dry all parts. Refill the bowl to the correct level and then reassemble the unit, figure 8-11.

Oil-wetted Polyurethane Filter Element Air Cleaner. Shown in figure 8-12, this type of air cleaner is perhaps the most common. Clean it by washing the element in kerosene, liquid detergent and water, or an approved solvent. Dry the element by squeezing it in a cloth or towel. Apply considerable oil to the foam and work it throughout the element.

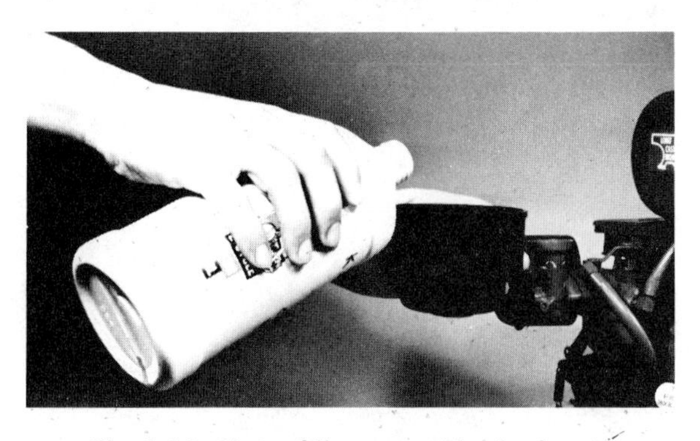

Fig. 8-11 Clean, fill, reassemble air cleaner

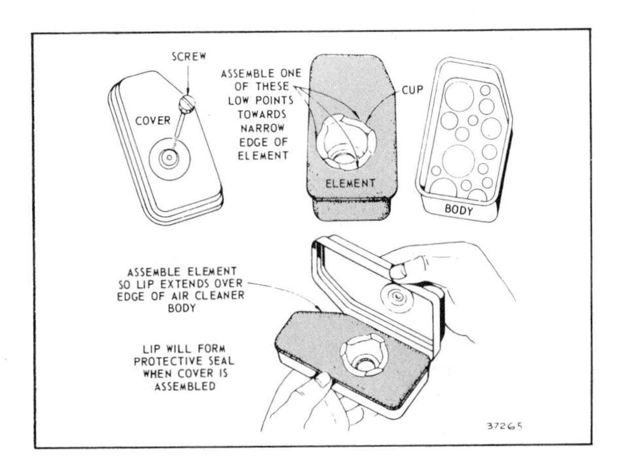

Fig. 8-12 Oil foam air cleaner

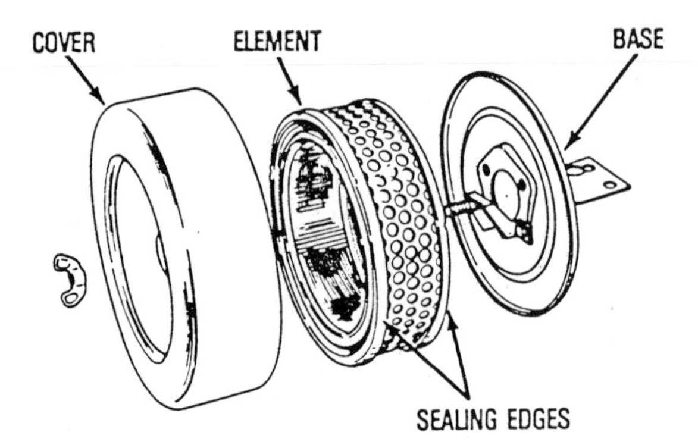

Fig 8-13 Paper element-type air cleaner

Squeeze out any excess oil and then reassemble the air cleaner. Make sure it overhangs the edges. Do *not* neatly tuck it in.

Dry-type Air Cleaners

Foil, Moss, or Hair Element Air Cleaner. This type of element is also washed in solvent. After washing, allow the element to dry, then return it to the air cleaner body.

Paper Element Air Cleaners. These air cleaners should be cleaned by blowing compressed air through them in the opposite direction from normal flow, figure 8-13. Dirt and dust particles will be blown out of the element.

Fig. 8-14 Metal-cartridge type air cleaner

Metal-cartridge Air Cleaners. These air cleaners are cleaned by tapping or shaking to dislodge dirt accumulations, figure 8-14.

Air cleaner elements are replaceable parts. If they wear out or become very clogged, they should be thrown away and replaced with a new filter element. As a simple test to determine if the air cleaner is badly clogged: (1) run the engine with the air cleaner removed; (2) then, with the engine still running, replace the air cleaner; and (3) notice if the engine speed remains constant or drops down. Any noticeable drop in speed probably indicates that the filter element is clogged.

Periodically inspect the engine for loose parts. Do not neglect to lubricate the engine's associated linkages, gear reduction units, chains, shafts, wheels, or the machinery the engine powers. These requirements vary greatly from engine to engine. On an outboard engine remember that the lower unit requires special lubrication.

EXHAUST PORTS (TWO-CYCLE)

Cleaning the exhaust ports of a two-cycle engine can also be considered a part of routine care and maintenance, figure 8-15. Over a period of time these ports can become clogged with carbon deposits which greatly reduce power and performance. After removing the muffler to expose the ports, the carbon can be cleaned out with a soft object such as a dowel rod. Be sure to move the piston so that it covers the ports. This prevents loose deposits from going into the cylinder.

WINTER STORAGE AND CARE

Preparing an engine for winter storage at the end of the season should be done carefully. Several procedures should be done:

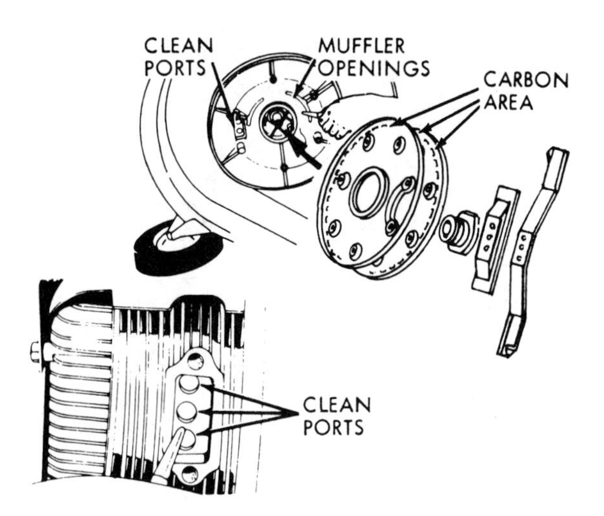

Fig. 8-15 Cleaning the exhaust ports

- Drain the fuel system.
- Inject oil into the cylinder (upper cylinder).
- Drain the crankcase (four-cycle).
- Clean the engine.
- Wrap the engine in a canvas or blanket and store in a dry place.

The entire fuel system should be drained: tank, fuel lines, sediment bowl, and carburetor. Gasoline that is stored over a long period of time has a tendency to form gummy deposits or varnish. These deposits could damage and block the fuel system if gasoline stands in the system for a long time. Do not plan to hold gasoline over from season to season. Purchase a new supply when the engine is returned to use.

Many manufacturers recommend that a small amount of oil be poured into the cylinder prior to storage. Remove the spark plug and pour in the oil (about a tablespoon in most cases). Turn the engine over by hand a few times to spread the oil evenly over the cylinder walls. Replace the spark plug.

Some manufacturers suggest that the oil be introduced into the cylinder while the engine is running at a slow speed and just before the engine is stopped. To do this, the air cleaner is removed and a small amount of oil is poured in. The oil goes through the carburetor and then into the combustion chamber. When dark blue exhaust is produced the engine can be stopped.

The crankcase should be drained while the engine is warm so the oil drains more readily. In some cases the crankcase is then refilled; on most engines it need not be refilled until the engine is returned to service.

Any unpainted surfaces that might rust should be lightly coated with oil before storage. Linkages should be oiled. The radiating fins and the engine in general should be cleaned. Do not put too much oil on the engine as it will collect dirt and be difficult to remove.

Finally, wrap the engine in a canvas or dry blanket and store the engine in a dry place. A garage, dry shed, or dry basement is good. Do not allow the engine to be left outside, exposed to the weather.

RETURNING THE ENGINE TO SERVICE

Upon returning the engine to service refill the gas tank with new gasoline. Before filling the gas tank check to see if any water has condensed in the tank. If so, drain the water completely.

Check for condensation in the crankcase and refill the crankcase with new oil of the correct type, figure 8-16. Also, clean the air cleaner and refill the bowl with oil if it is an oil-bath type.

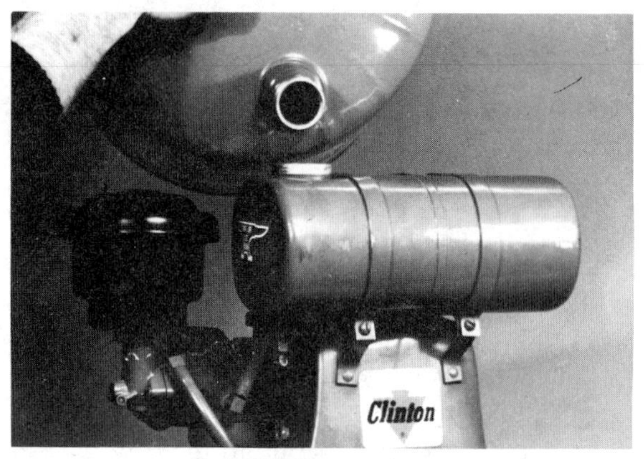

(A) Fill the gas tank

(B) Check the oil in the crankcase

(C) Fuel shut-off valve open

(D) Choke closed

(E) Spark plug shorting bar off spark plug

(F) Air cleaner clean

Fig. 8-16 Things to check before returning the engine to service (Courtesy Clinton Engines Corporation)

Check the spark plug; clean and regap it. Many manufacturers of small engines recommend that a new plug be installed at the beginning of each season to ensure peak performance throughout the season.

SPECIAL CONSIDERATIONS FOR OUTBOARD MOTORS

Winter Storage

Generally, winter storage for outboard motors is similar to that for other engines. One notable difference is the cooling system. Being water cooled, all water must be removed from the cooling system. If a water-filled cooling system on an idle engine is exposed to freezing weather, cracked water jackets and other major damage can result.

To remove the water from the system, remove the engine from the water, set the speed control on stop (to prevent accidental starting), and turn the engine over several times by hand. This allows the water to drain from the water jackets and passages.

Salt Water Operation

Although most engines are treated with anticorrosives, not all engines are free from corrosion. If an engine normally used in salt water is not to be used for a while, its cooling system should be flushed with fresh water to prevent any possible corrosion.

Use in Freezing Temperatures

If the engine is being used during freezing temperatures, care must be taken not to allow the cooling water to freeze in the engine or lower unit during an idle period. While the engine is operating there is no danger of freezing.

REVIEW QUESTIONS

1. Explain what routine care and maintenance is and why it is important.
2. Where can routine care and maintenance information be found?
3. What are the four basic points to be considered in routine care and maintenance?
4. How often should the oil in the crankcase be checked?
5. How often should the oil in the crankcase of a Briggs and Stratton engine be changed?
6. What are the two types of cooling systems?
7. How often should the air-cooling system be cleaned?
8. How is an air-cooling system cleaned?
9. What care should be given a water-cooling system?
10. How often should spark plugs by cleaned? Why do some manufacturers recommend no cleaning of spark plugs?
11. What is the purpose of cleaning a spark plug?

12. List several types of spark plug fouling.
13. What are the steps in cleaning a spark plug?
14. What is the purpose of the air cleaner?
15. How often should air cleaners be cleaned?
16. What are the two basic types of air cleaners?
17. Discuss the cleaning of the several types of air cleaners.
18. List the general steps in winter storage of engines.
19. List the general steps in returning the engine to service.
20. What special considerations must be made for outboard motors?

SECTION 3
OTHER INTERNAL COMBUSTION ENGINES

Unit 9 The Automobile Engine

OBJECTIVES **After completing this unit, the student should be able to:**

- **Recognize the parts of an automobile engine and discuss their function.**
- **Discuss the operation of the fuel, ignition, lubrication, cooling and exhaust systems.**
- **Explain pollution reducing designs of the automobile engine.**
- **Recognize parts of the drive system and discuss their function.**
- **Recognize and avoid unsafe conditions when working on an automobile.**

The automobile engine is a larger version of the small gasoline engine, figures 9–1 and 9–2. The engine has the same basic parts, the same basic systems, and operates on the same four-stroke cycle principle. Every engine is designed and engineered for a specific purpose, but the basic parts and operating principles remain almost unchanged, figures 9–3 and 9–4. A thorough knowledge of any one engine makes it easier to understand other engines.

The foundation of the engine is the block, figure 9–5. This casting contains all of the cylinders, usually four, six, or eight. Four-cylinder engines normally have their cylinders in a straight line and are called an in-line engine, figure 9–6. Six-cylinder engines may be either in-line or "V." The V-6 engine has three cylinders on each side and the block, when viewed from the front, takes on a "Y" shape, figure 9–7. Today's eight-cylinder engines are V-8s, having four cylinders on each side.

The lower part of the block is called the crankcase. It provides mounting for the crankshaft and space for the crankshaft to revolve. Bolted to the bottom of the block is the oil

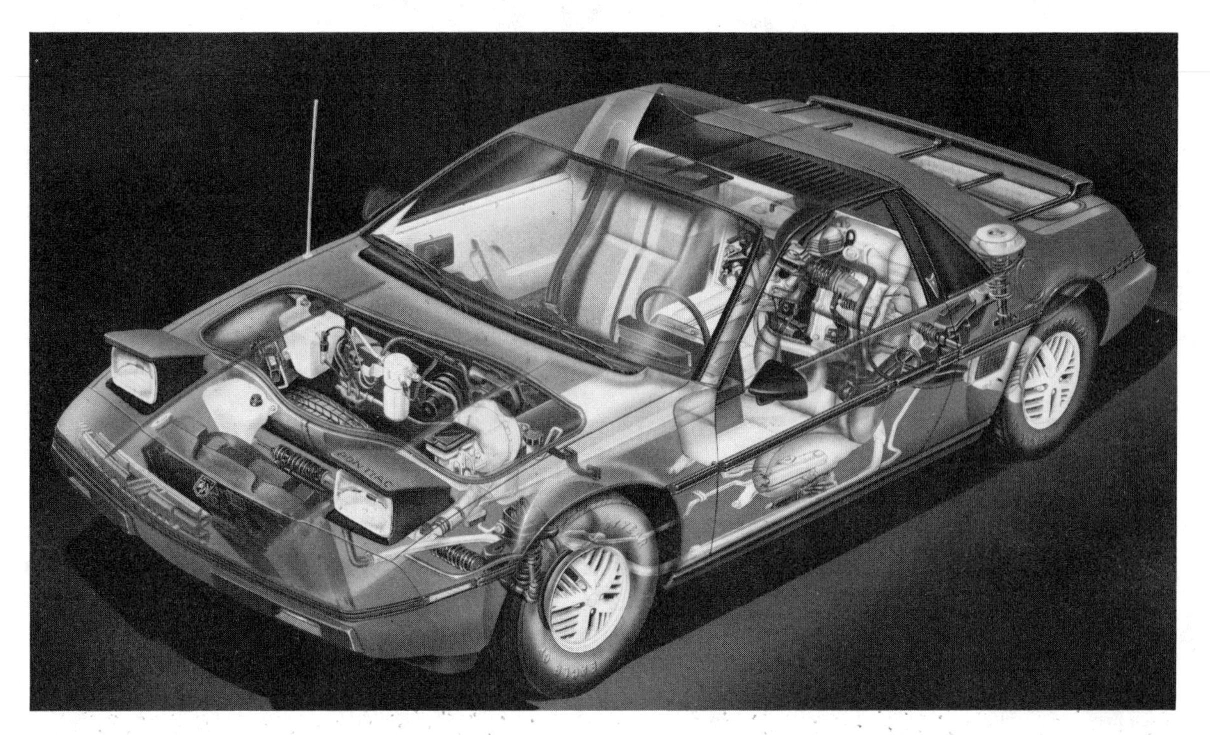

Fig. 9-1 Pontiac Fiero cutaway view (Photo courtesy Pontiac Motor Division, General Motors Corporation)

pan, figure 9-8. This part is usually a stamped-steel pan that acts as a reservoir for the engine's oil supply. It is called the sump, base, or crankcase.

The cylinder head is bolted to the top of the block; one head for six-cylinder engines and two heads for V-8 engines, figure 9-9. Most modern engines have overhead valves or valves mounted in the cylinder head, called valve in-head. The valve system and cylinder head are more complicated on overhead valve engines, figure 9-10. Instead of simply pushing the valves up for opening, like most small engines, the valves must be pushed down for opening. Since more parts are needed, rocker arms are added to make this change in direction, figure 9-11. Some rocker arms mount on a shaft, while others are separately bolted to the head. The cylinder head, therefore, contains the intake and exhaust valves, rocker arm shaft, rocker arms, valve springs, and other valve parts.

The camshaft is usually in the engine block. The cam movement is transferred to the rocker arm through a push rod. The train is: camshaft, tappet, push rod, rocker arm, valve, as seen in figure 9-12. When the nose of the cam pushes up on the tappet, push rod, and rocker arm; the rocker arm pushes the valve down or open, figure 9-13.

Engines are also manufactured with overhead cams, figures 9-14 and 9-15. The camshaft is located in the head, with the cams pushing directly on the rocker arms, figure 9-16.

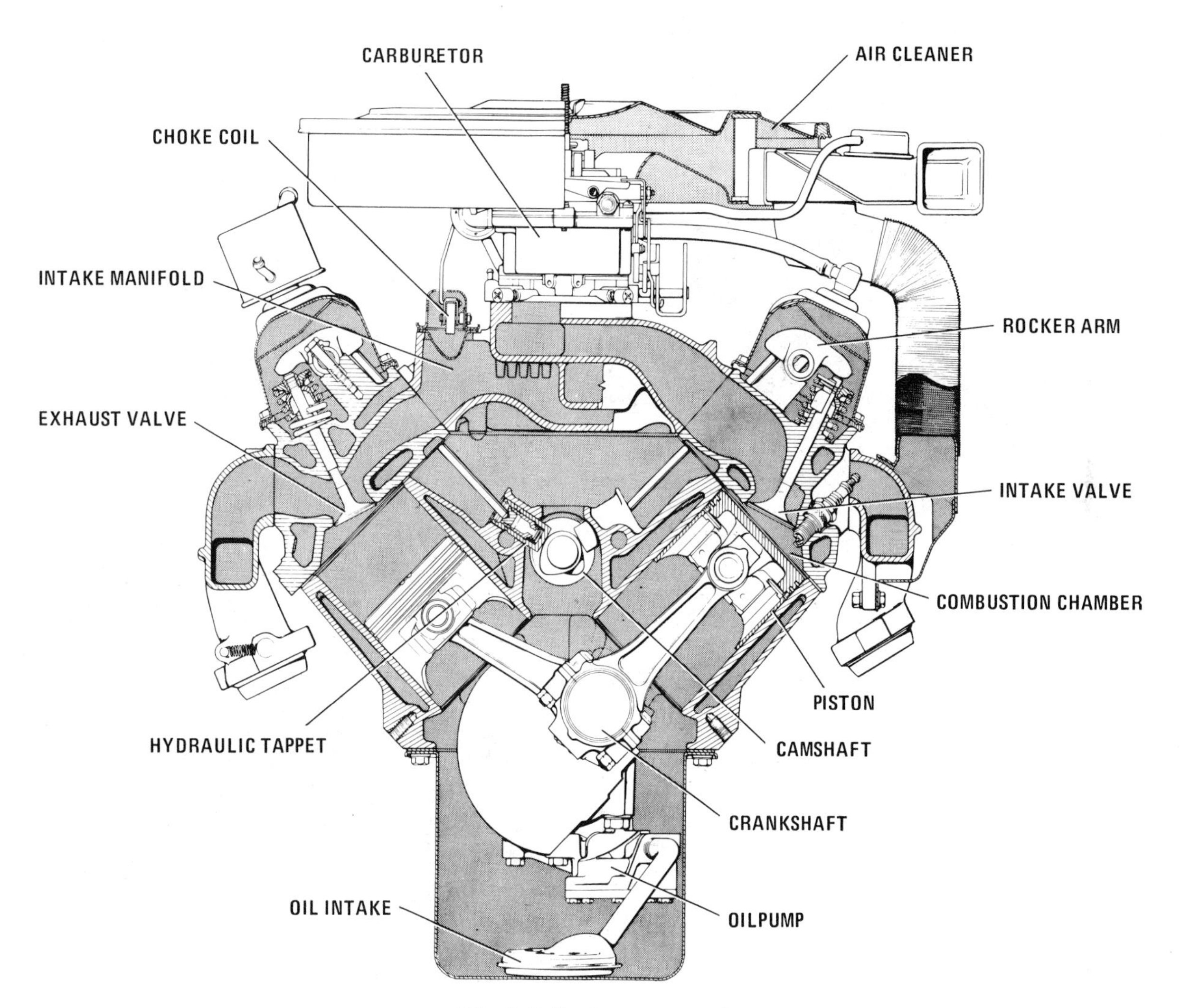

Fig. 9-2 Engine cross section

Fig. 9-3 Pontiac 2.5 liter longitudinal engine (Photo courtesy of Pontiac Motor Division, General Motors Corporation)

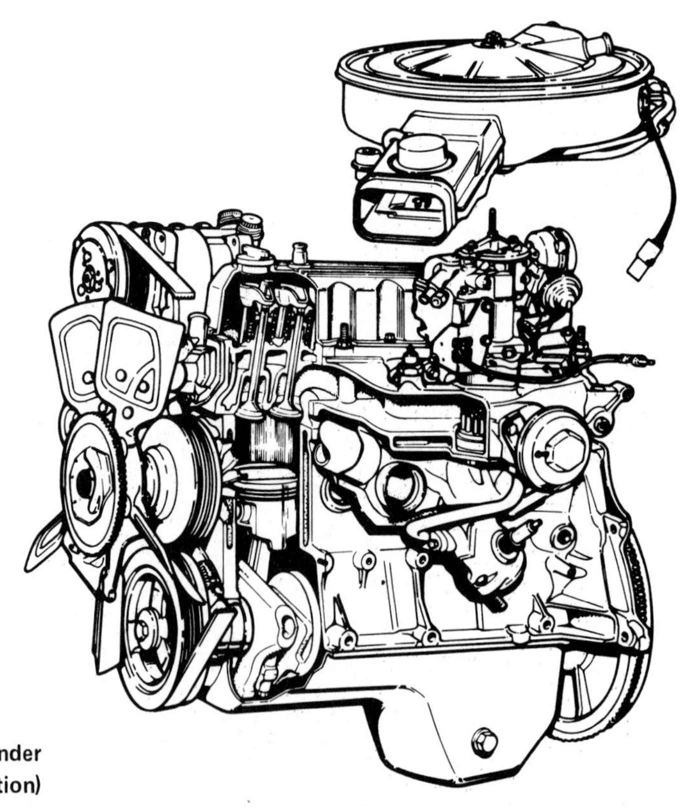

Fig. 9-4 American Motors 2.5 liter four-cylinder engine (Courtesy of American Motors Corporation)

Fig. 9-5 Cylinder block, top view, 350 CID V-8 engine

Fig. 9-6 2.5 liter engine block

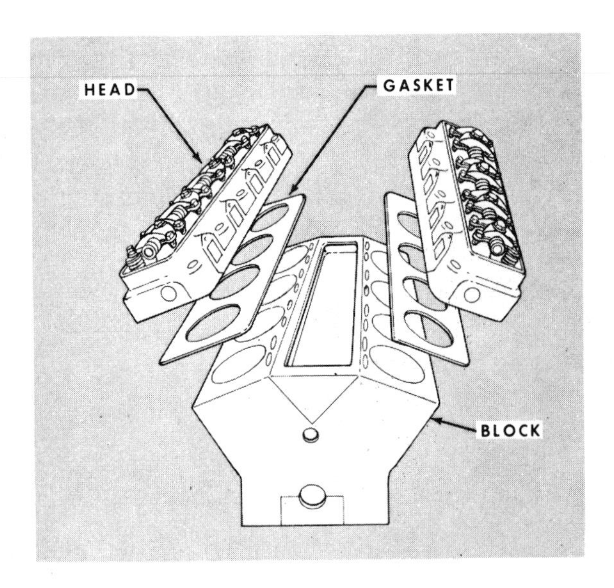

Fig. 9-7 Block, head, and gasket of a V-8 engine

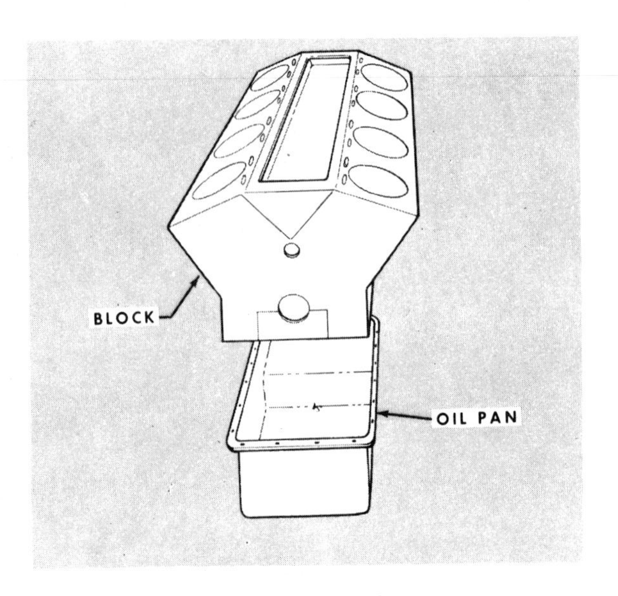

Fig. 9-8 Block and oil pan

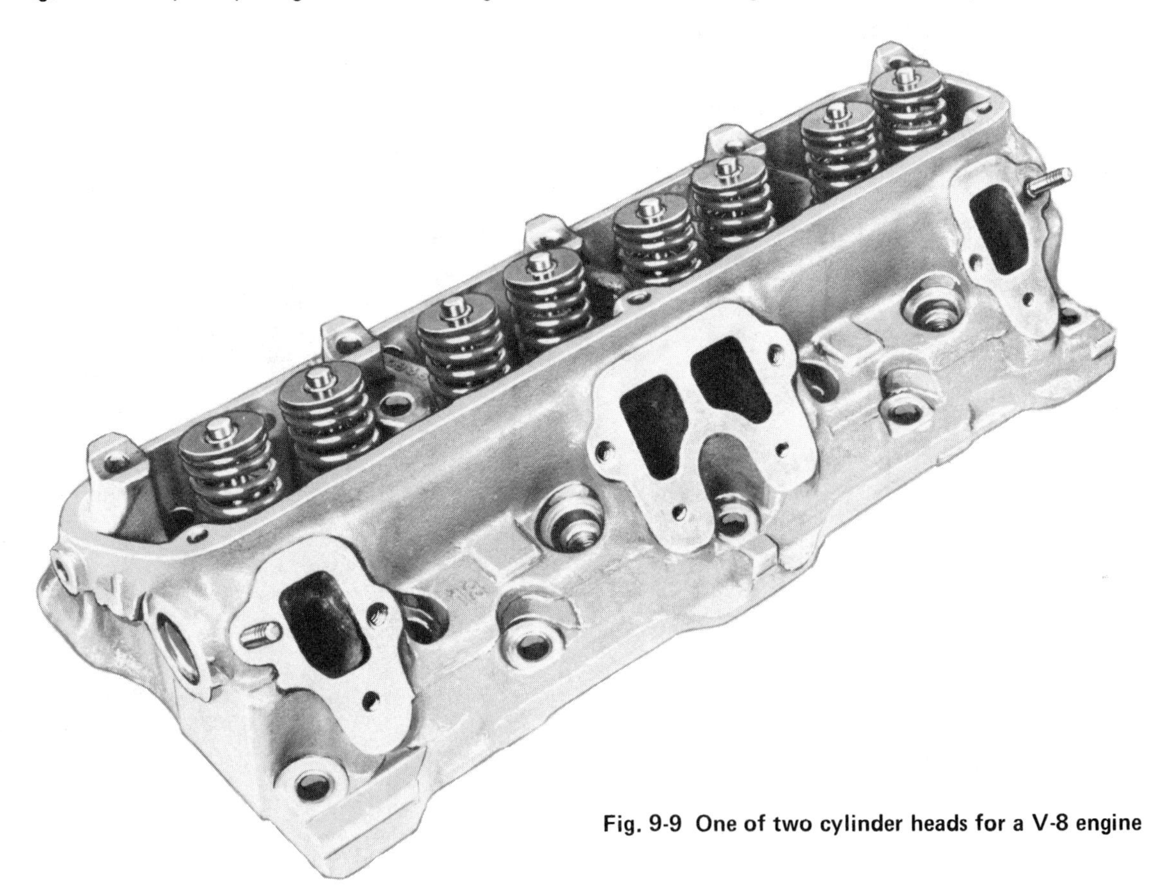

Fig. 9-9 One of two cylinder heads for a V-8 engine

Fig. 9-10 Valve in-head arrangement of a six-cylinder engine

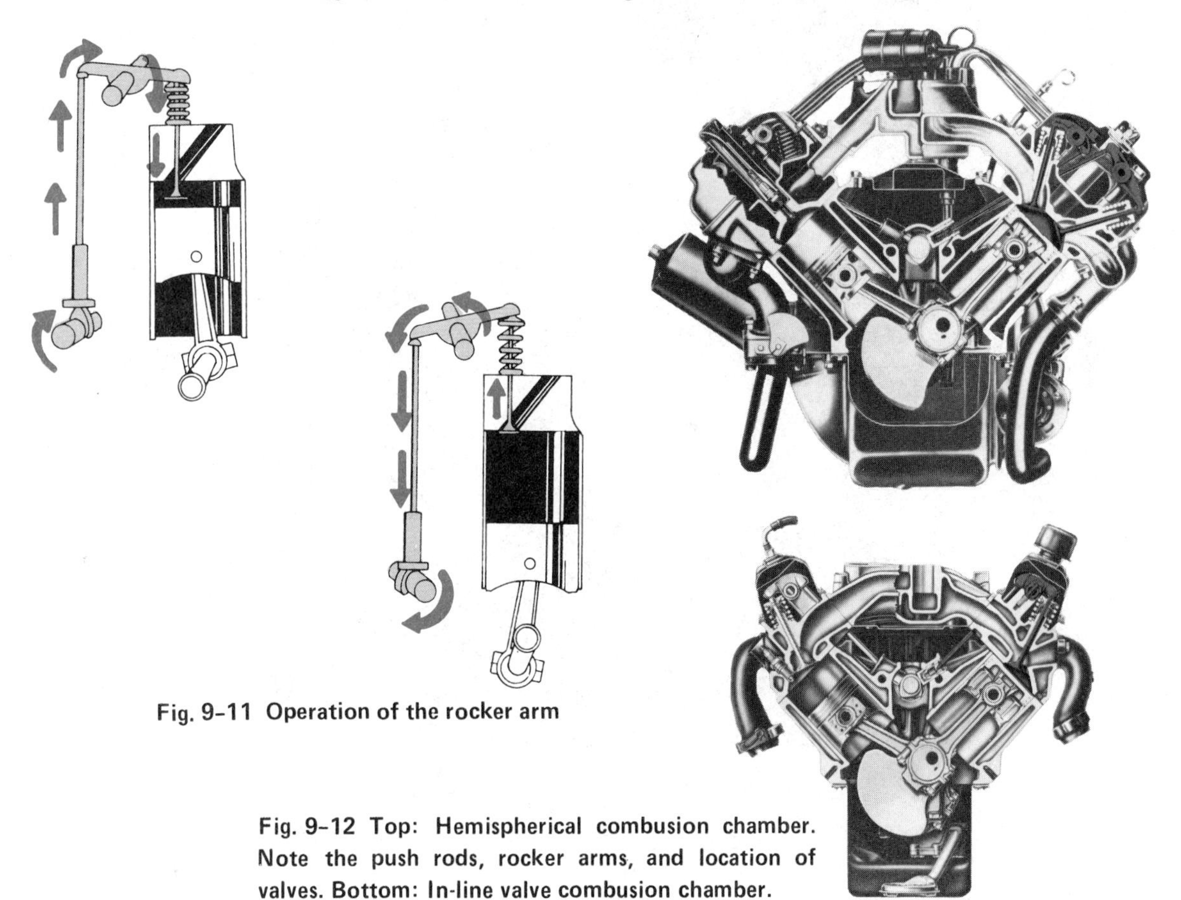

Fig. 9-11 Operation of the rocker arm

Fig. 9-12 Top: Hemispherical combustion chamber. Note the push rods, rocker arms, and location of valves. Bottom: In-line valve combustion chamber.

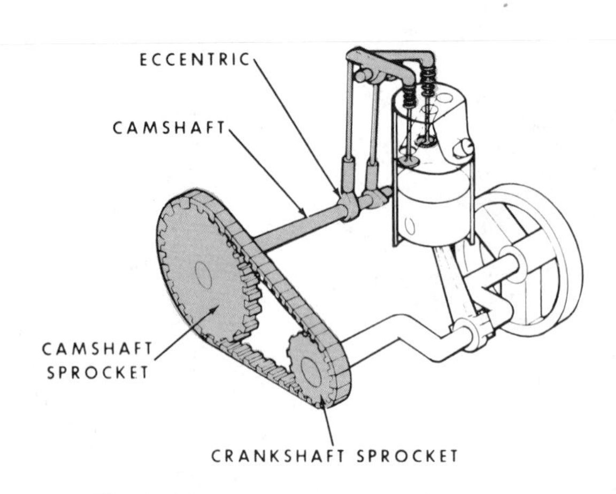

Fig. 9-13 Valve in-head arrangement

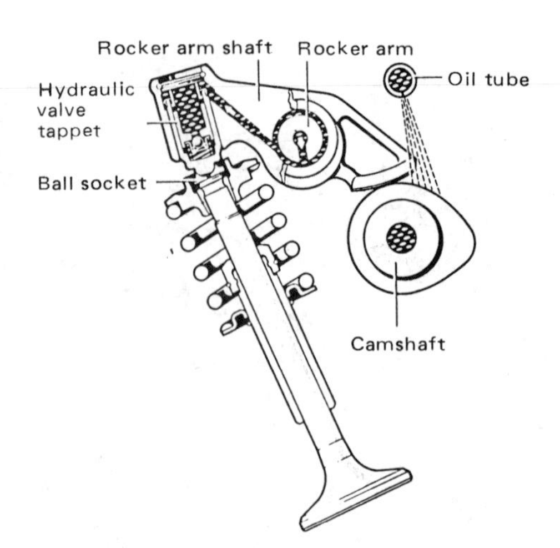

Fig. 9-14 Overhead camshaft (Courtesy of Mercedes-Benz of North America, Inc.)

Fig. 9-15 1.8 liter overhead cam, four-cylinder engine (Photo courtesy of Pontiac Motor Division, General Motors Corporation)

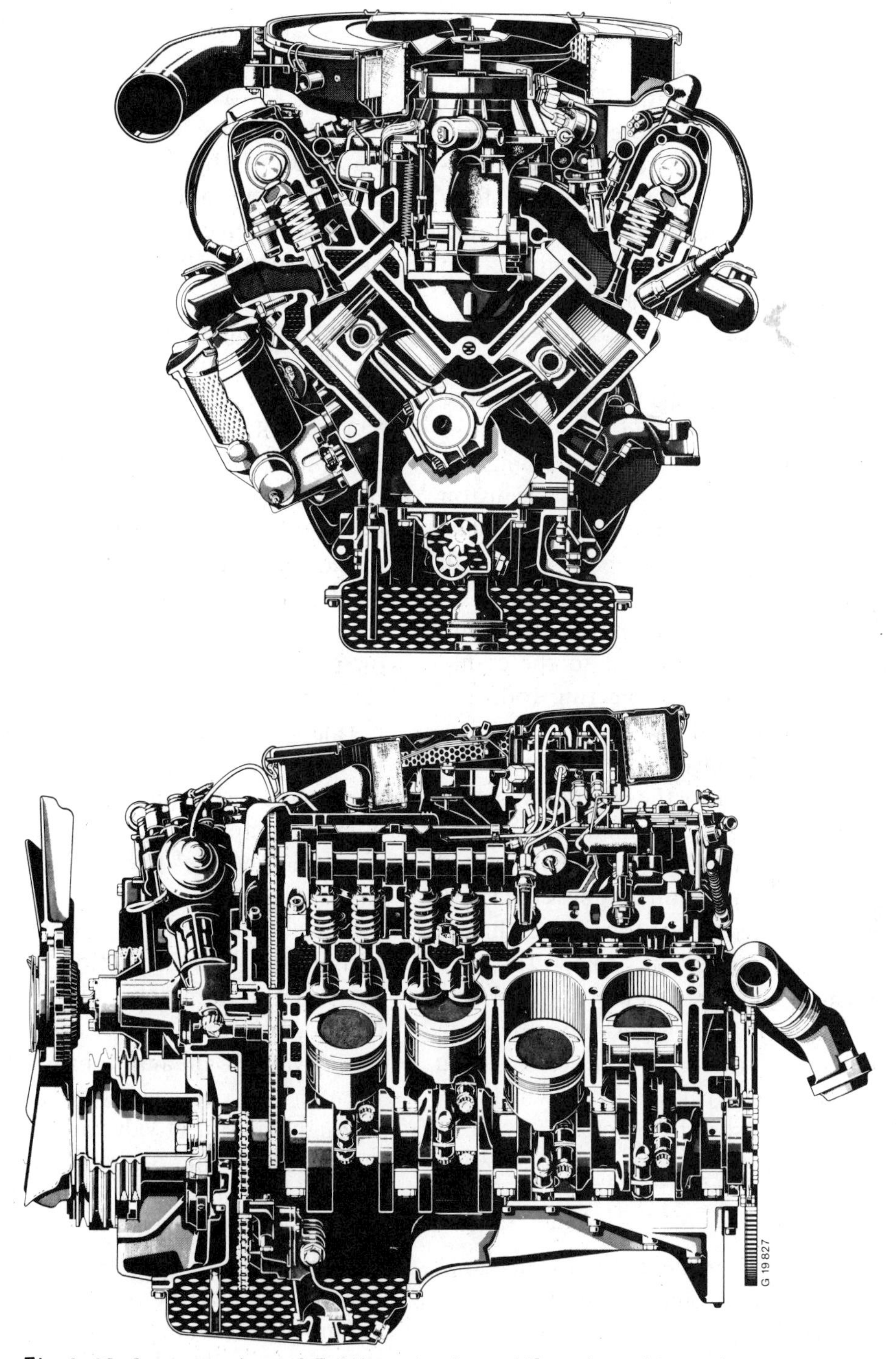

Fig. 9-16 Cutaway view of 5.0-liter aluminum V8 engine with overhead cams (Courtesy of Mercedes-Benz of North America, Inc.)

Correct valve timing is vital to engine performance. The cams must open and close the valves at the correct instant. The crankshaft and camshaft must, therefore, be perfectly coordinated. Timing marks on the gears indicate the proper alignment, figure 9-17. Many engines use a timing chain between the crankshaft and camshaft gears. If the timing chain becomes worn, improper timing and sluggish engine operation can result. Some engines eliminate the chain completely and use direct gear drive. Still others use a reinforced rubber belt between the camshaft and crankshaft.

Pistons are usually made of an aluminum alloy to reduce weight, figure 9-18. The piston is equipped with cast iron piston rings. The compression rings reduce power lost through blow-by. The oil rings scrape away excess oil and spread out an even oil film on the cylinder walls. The piston is linked to the crankshaft by the wrist pin, connecting rod, and connecting rod cap, figure 9-19.

The crankshaft converts the reciprocating motion of the piston to the necessary rotary motion. The six-cylinder crankshaft has six crank throws, the offset sections, evenly spaced around the axis of rotation to deliver smooth power. The V-8 engine has four crank throws spaced around its axis, one throw for each two opposite cylinders, figure 9-20. When one piston is down, the opposite piston is up. Two connecting rods are fastened to each throw. The crankshaft has journals which rest in the main bearings of the block.

FUEL SYSTEM

The fuel system is made up of the fuel tank, fuel line, fuel pump, and carburetor, figure 9-21. These same basic parts are also found on most small engines.

The fuel tank is a metal container that stores the gasoline to be used by the automobile engine. Capacity ranges from 6 to 25 gallons (23 to 95 liters), depending upon the automobile. Inside the tank are metal baffles that prevent the fuel from sloshing. The fuel gauge sensing unit is mounted inside the tank. It connects electrically or mechanically to the fuel gauge on the dashboard to register the level of fuel. A filter screen prevents foreign matter that may enter the gasoline from leaving the tank and entering other parts of the fuel system.

The fuel line carries the fuel between the major parts of the fuel system. It is usually steel tubing. Flexible sections of the fuel line use neoprene or plastic hose. Rubber hose cannot be used to carry gasoline.

The fuel pump is operated by a cam located on the camshaft, figure 9-22, or an eccentric on the front of the camshaft gear. Some engines use electrically-operated fuel pumps, normally located near or in the gas tank. Electric fuel pumps do not take power from the engine like a mechanical pump. They also eliminate vapor lock, a situation in which gas vapors in the fuel line limit the flow of liquid gasoline to the carburetor. The pump brings gasoline from the tank, delivering it to the carburetor or fuel injection system. The main parts of the pump are: body, sediment bowl, diaphragm, pump rocker arm, inlet valve, and outlet valve. Cam action moves the diaphragm in and out; at one time forcing fuel out through the outlet valve and up to the carburetor, and another time drawing fuel into the pump from the gas tank. In actual operation the pump delivers fuel only when the carburetor calls for it, that is, when the fuel inlet valve in the carburetor is open. If the float chamber is full and the fuel inlet valve on the carburetor is closed, the pump will not deliver fuel.

Fig. 9-17 Timing chain

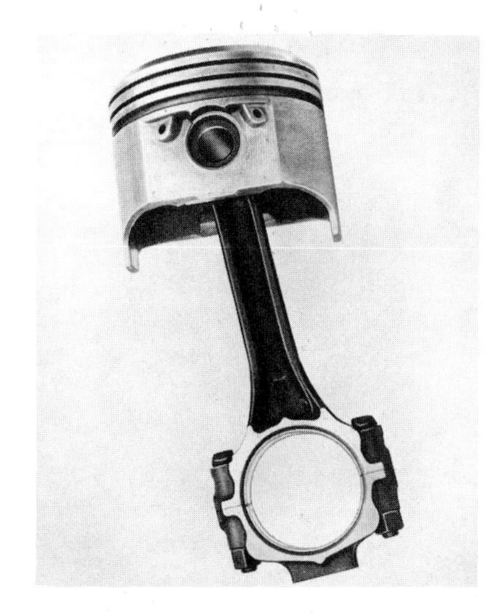

Fig. 9-18 Piston, connecting rod, and cap for 2.5 liter engine

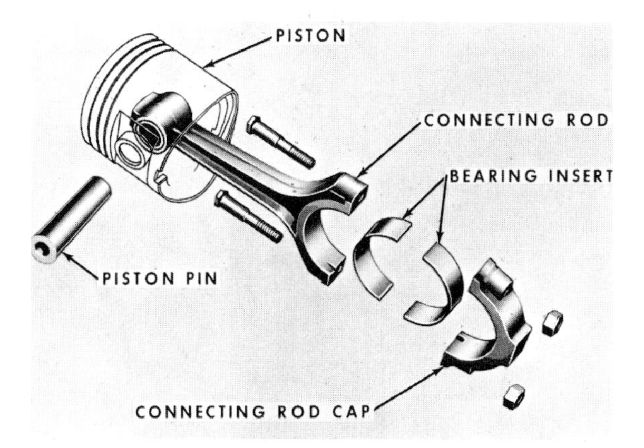

Fig. 9-19 Piston, piston pin, connecting rod, bearing insert, and connecting rod cap

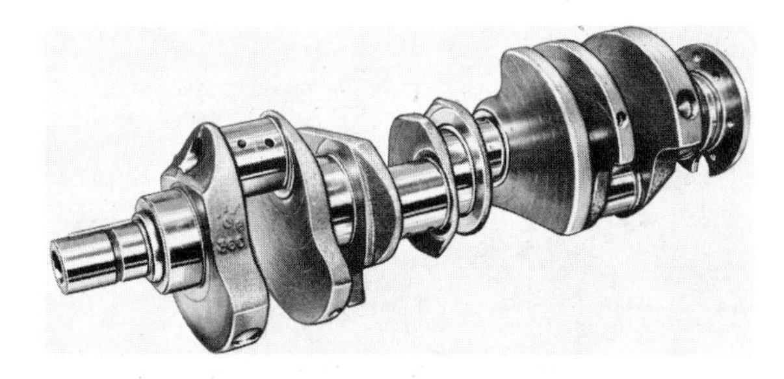

Fig. 9-20 Crankcase 360 CID V-8 engine

Automobile carburetors are more complex than small engine carburetors. The operating principles are the same for each, but the larger carburetors have more refined systems. In either case, the carburetor's job remains the same: to mix air and gasoline in the proper proportions and to deliver the mixture to the combustion chambers. A typical carburetor would consist of the idle system, main system, power system, choke system and accelerating system. In discussing the systems, it is assumed that the student already has gained an understanding of carburetor fundamentals.

The idle system supplies fuel for idle or slow speeds, figure 9-23. When a relatively small amount of air is drawn through the venturi section, the venturi vacuum is low. Therefore, little fuel from the main discharge system enters the venturi section to mix with the air. The mixture would be too lean without the addition of the idle system.

Fuel from the main well travels up the idle tube and down to the idle jet during idle speeds. Idle discharge jets are located just behind the throttle plate and, with the throttle closed down for idle speed, the pressure be-

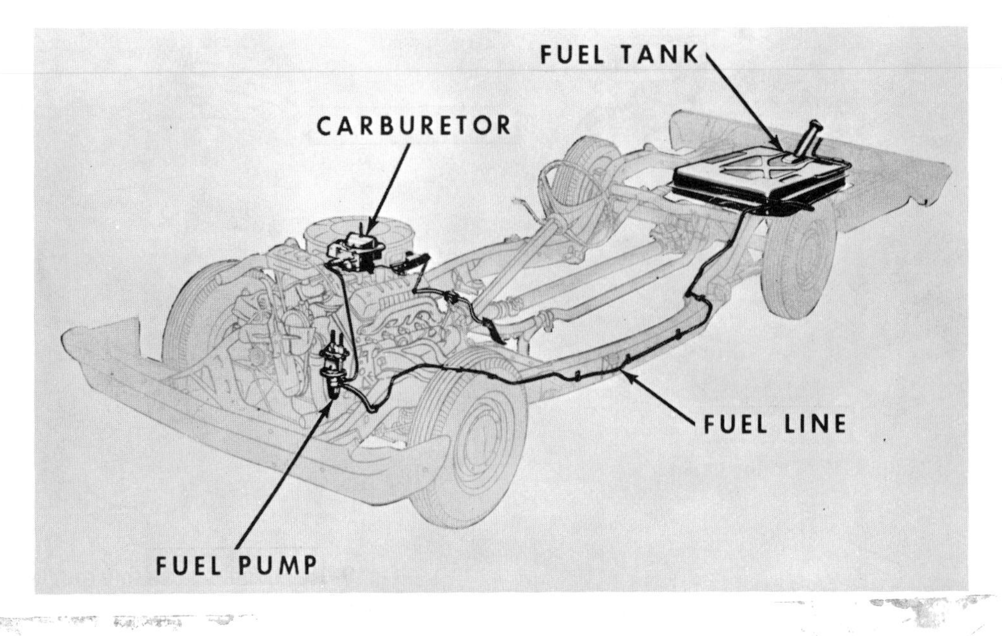

Fig. 9-21 Fuel system

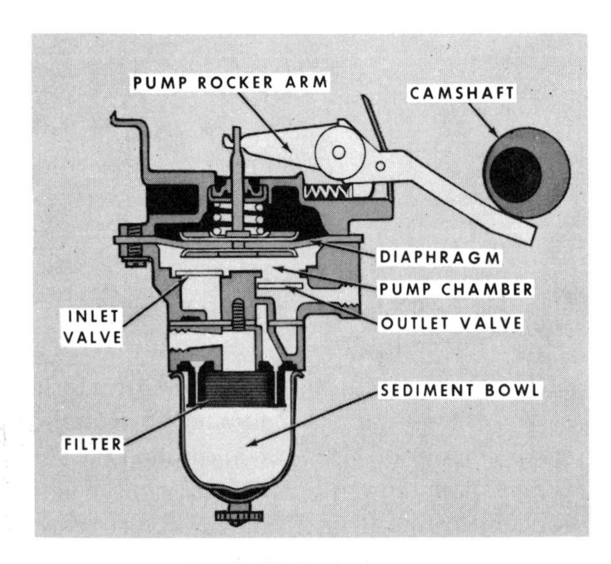

Fig. 9-22 Fuel pump

hind the throttle is very low. Atmospheric pressure forces the gasoline to this low pressure area. As the throttle is opened and the engine speeds up, the fuel supply shifts from the idle system to the main system, figure 9-24.

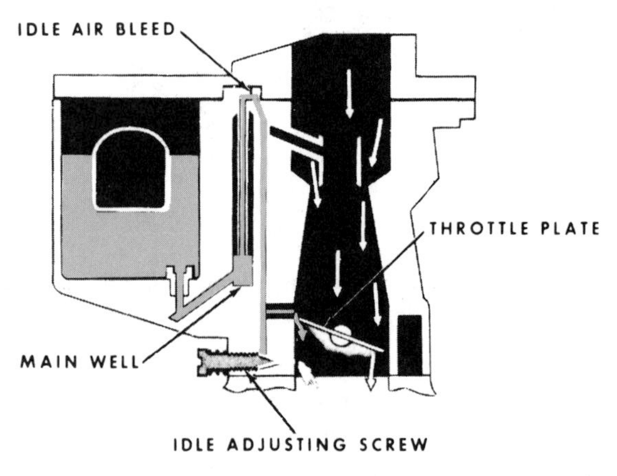

Fig. 9-23 Idle system

During normal driving speeds the gasoline is delivered by the main fuel system. High vacuum is attained in the venturi section and atmospheric pressure pushes the fuel into the airstream.

The power system supplies extra fuel for the richer mixture necessary for heavy loads or high speeds. A valve, operated by intake manifold pressure, in the bottom of the float bowl opens to allow additional fuel to enter the main well, figure 9-25. At low speeds, with the throttle closed, intake manifold pressure is very low and the spring-loaded power valve is held closed. However, at high speeds, with throttle opened wide, the intake manifold pressure is relatively high, enabling the spring tension to open the power valve, admitting extra gasoline. The power system will stay in operation until the throttle is partially closed.

The acceleration system operates when the throttle is opened. Because good acceleration requires a richer fuel mixture, the system sprays additional gasoline into the carburetor. The system, figure 9-26, has a diaphragm rod between the throttle plate and the diaphragm. When the throttle is opened, its motion is transferred to the diaphragm, moving the diaphragm inward. The fuel in the diaphragm chamber is then forcefully sprayed into the venturi section of the carburetor.

Most carburetors used on automobiles are equipped with an automatic choke. Unlike the manual choke, the automatic choke cannot be adjusted from the driver's seat. The function of the choke, automatic or manual, is to provide a rich fuel mixture for starting the engine. When the engine is first started the choke is in a closed position, figure 9-27. As the engine warms up, the choke automatically moves into an open position, figure 9-28. The movement of the choke from a closed position to an open position is caused by a thermostatic spring. When the spring is cold, it holds the choke in a closed position. As the spring begins to warm up, it unwinds. This unwinding moves the choke from a closed position to an open position. Heat is supplied to the spring from the exhaust manifold or

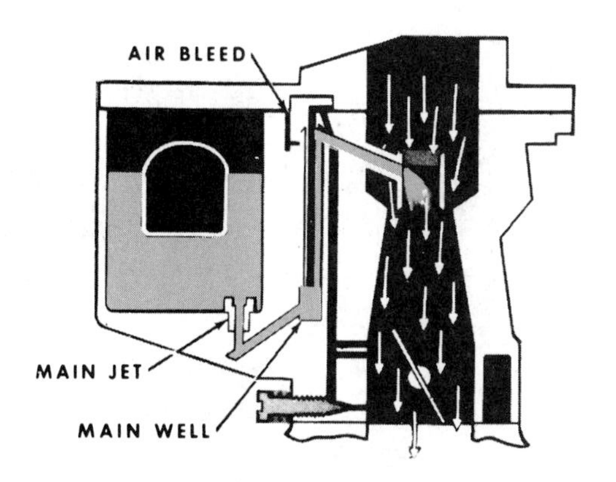

Fig. 9-24 Main system

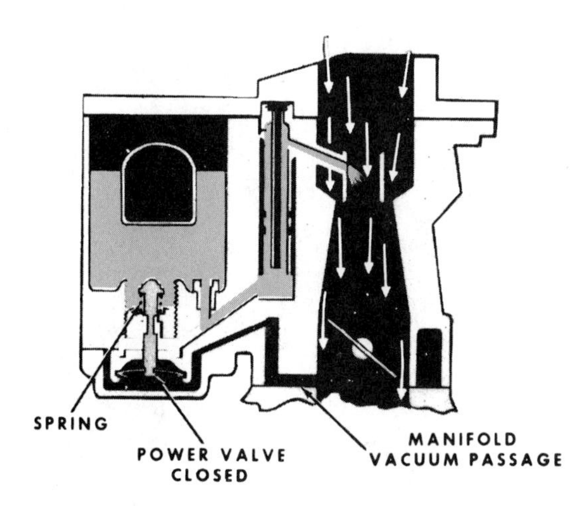

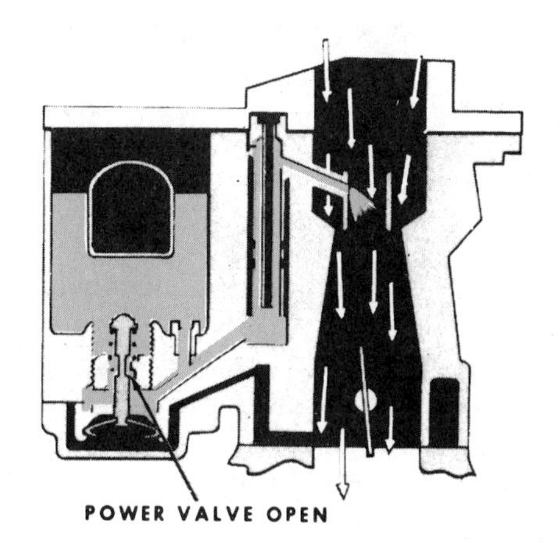

Fig. 9-25 Power system

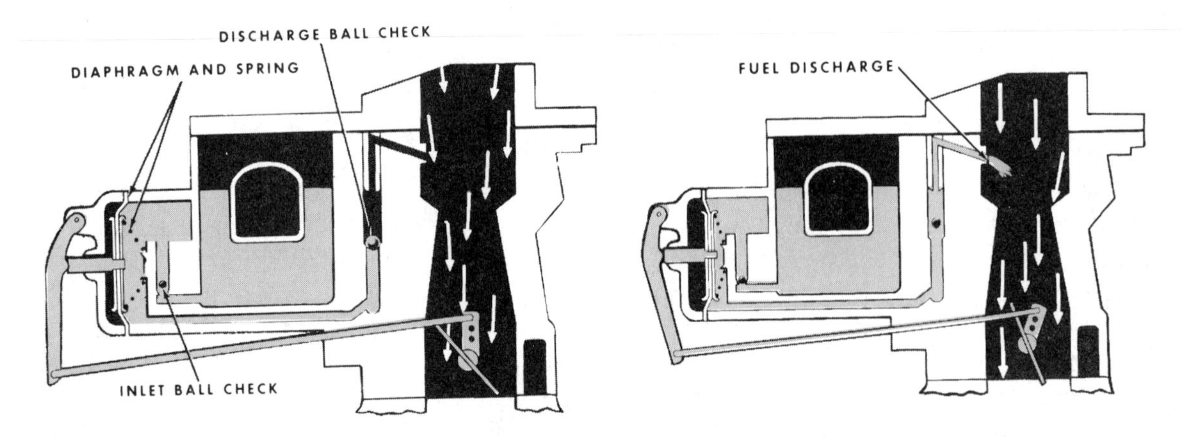

Fig. 9-26 Acceleration system

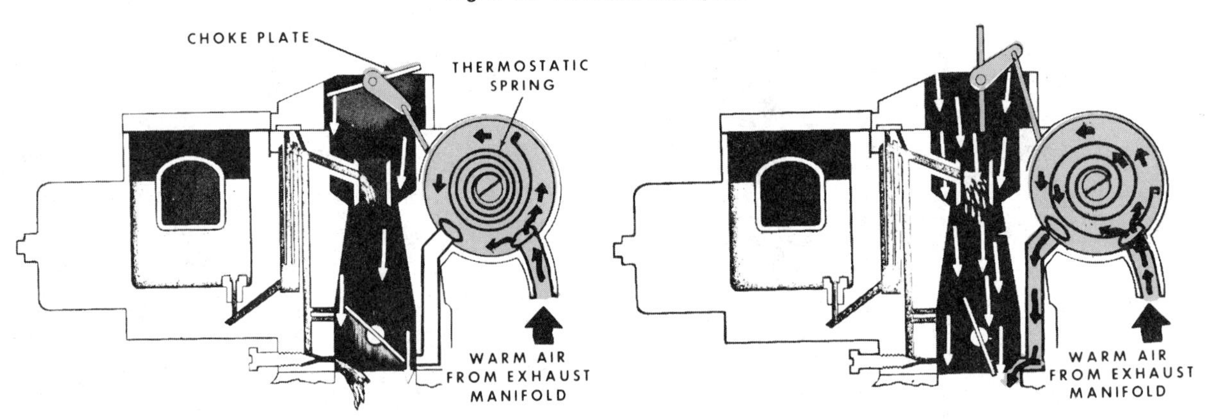

Fig. 9-27 Automatic choke, cold

Fig. 9-28 Automatic choke, hot

electrically from a heating element built into the side of the carburetor.

The carburetor is not bolted directly to the head. Between the two is the intake manifold, figure 9-29. Gas and air are carried from the carburetor, through the passages in the intake manifold, to the combustion chamber.

AIR CLEANER

An air cleaner is located on top of the carburetor. The large drum made of a fibrous paper material removes dust and grit that would be harmful to the carburetor and the

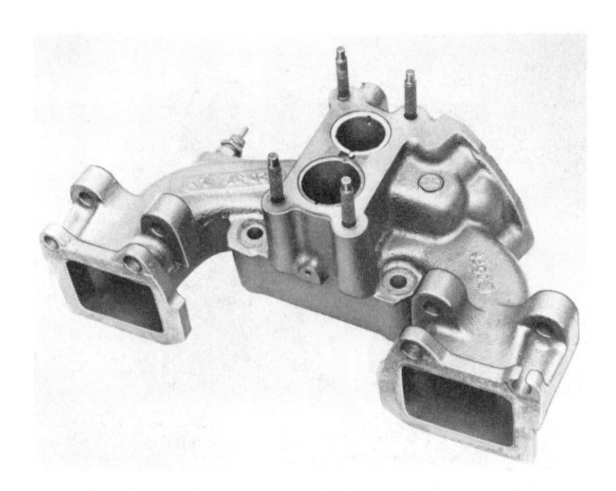

Fig. 9-29 Intake manifold—2.5 liter engine

cylinder walls. The air cleaner also muffles the sound that would be caused by air rushing directly through the carburetor, and acts as a flame arrester in case a flame from engine backfiring occurs as a result of late ignition timing.

TURBO-CHARGED ENGINES

Turbo charging of auto engines is becoming a more popular option as the need for smaller, lighter, more powerful, and more efficient engines continues, figure 9-30. The horsepower of a turbo-charged engine can be 35% greater than that of a naturally-aspirated engine. The turbo charger will greatly increase the pressure of the incoming air so that air and fuel are *forced* into the combustion chambers rather than just being drawn in by the piston action.

The turbine wheel is driven by exhaust gases and does not rob the engine of otherwise useful power, figures 9-31 and 9-32. The centrifugal compressor on the other end of the shaft gathers and forces air into the engine.

ELECTRONIC FUEL INJECTION

Electronic fuel injection will deliver precise, equal amounts of gasoline to each of the cylinders. This increases engine efficiency and reduces pollution problems. On the General Motors system, injectors that are located in the intake manifold will spray fuel toward the intake valve; this is the port injection system, figure 9-33. The operation of the injectors must be timed to the motion of the pistons in the engine. As engine speed increases the injectors are held open for a longer time to allow more atomized gasoline to be sprayed. Sensors detect the rate of air flow and coordinate the amount of gasoline to provide the proper air/fuel ratio.

Fig. 9-30 1.8 liter tubocharged engine (Photo courtesy of Pontiac Motor Division, General Motors Corporation)

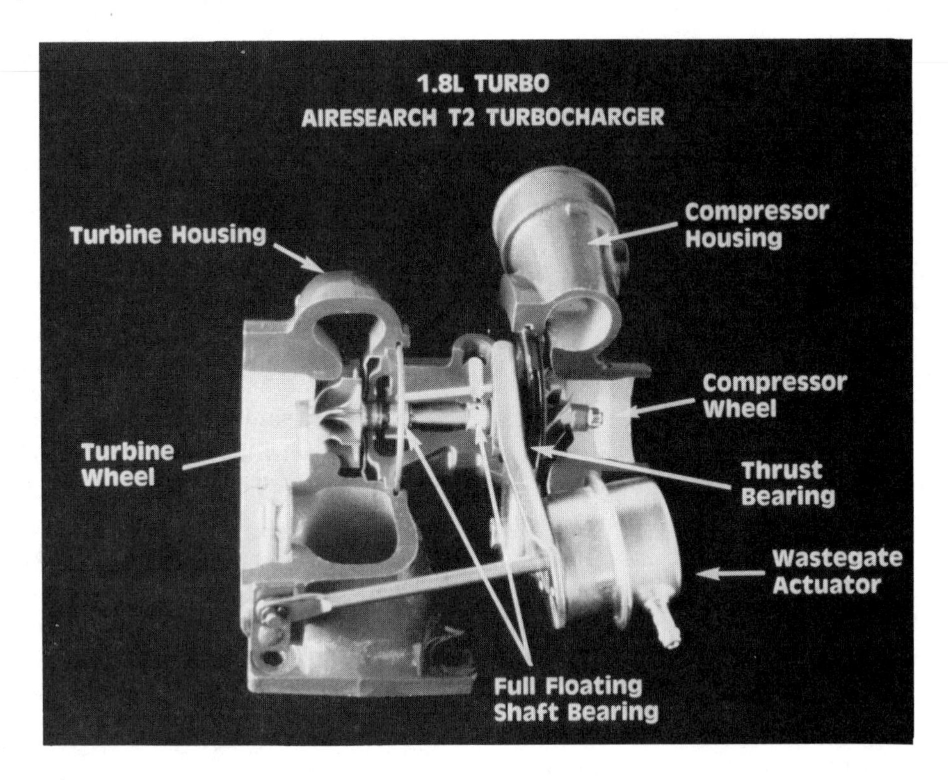

Fig. 9-31 Cutaway view of a 1.8 liter turbo Airesearch T2 turbocharger (Photo courtesy of Pontiac Motor Division, General Motors Corporation)

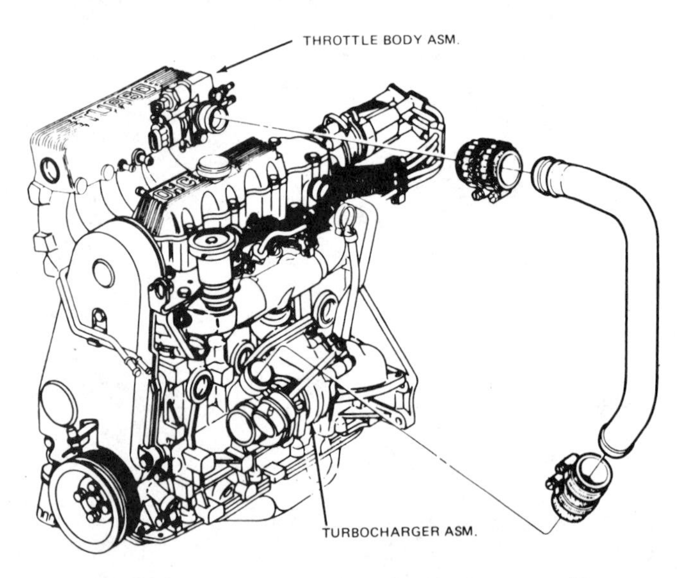

Fig. 9-32 Turbocharger induction tube and hoses (Courtesy of Pontiac Motor Division, General Motors Corporation)

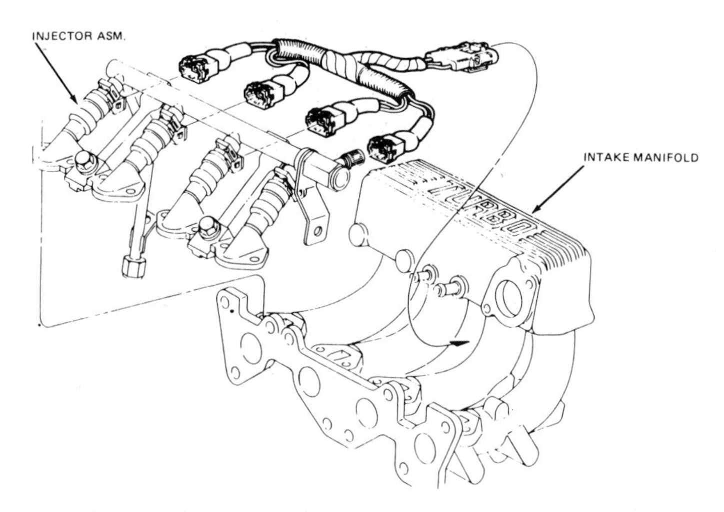

Fig. 9-33 Fuel injector assembly (Courtesy of Pontiac Motor Division, General Motors Corporation)

EXHAUST SYSTEM

The exhaust system consists of the exhaust manifold, inlet pipes, muffler, catalytic converter and tailpipe, figure 9-34. The muffler is designed to reduce noise without letting the exhaust pressure build up to a point that would be harmful to engine efficiency. The manifolds carry the exhaust gases from the engine to the inlet pipes and then through them to the muffler. After passing through the muffler, the gases are routed through the catalytic converter. In the converter, the exhaust pollutants (hydrocarbons and carbon monoxide) are changed into harmless carbon dioxide (CO_2) and water. The remaining gases and the carbon dioxide are then discharged through the tailpipe.

IGNITION SYSTEM

The automobile ignition system is very similar to the small engine ignition system. Except for the magneto, many of the parts used in the two systems are the same, figure 9-35.

The parts of the ignition system are:

- *Battery:* Provides the energy source for the primary circuit, starter circuit, and accessories.
- *Alternator or Generator:* Keeps the battery charged.
- *Ignition Switch:* Opens and closes the electrical circuit to stop and start ignition.
- *Ignition Coil:* Consists of a primary and a secondary coil. The primary coil contains relatively few turns of wire and produces low voltages. The secondary coil contains thousands of turns of wire and produces high voltage.
- *Distributor:* Delivers high voltage to the correct spark plug at the correct time. It consists of the:

  1. Distributor Rotor: Spins around, mov-

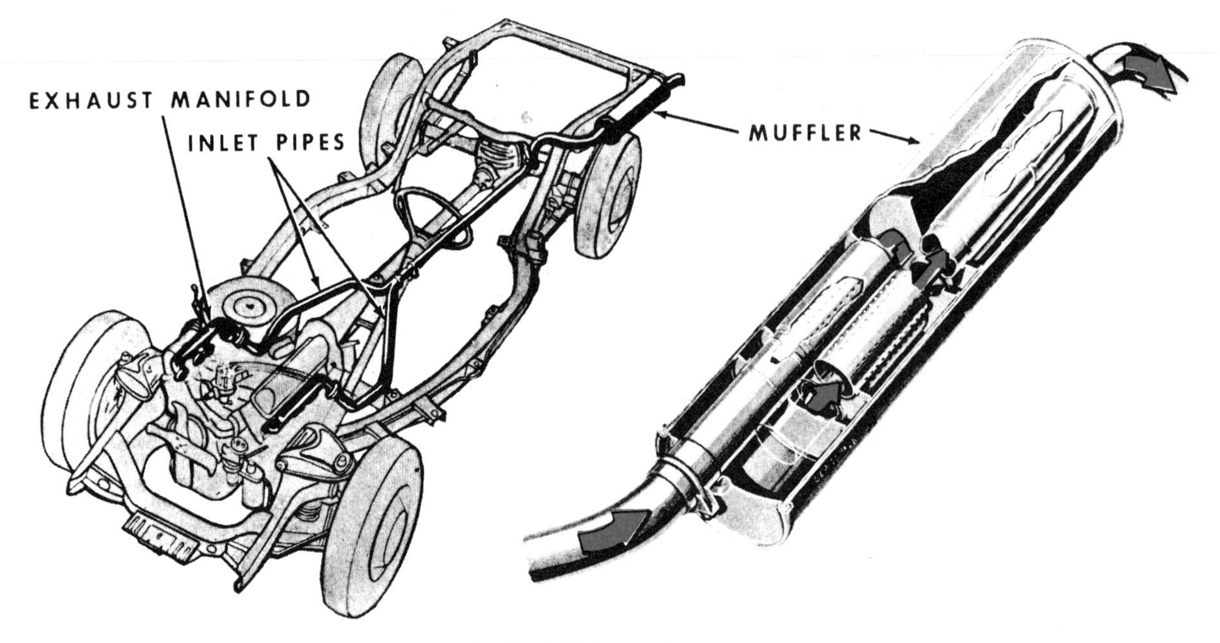

Fig. 9-34 Exhaust system

ing across the contacts for the spark plug leads.

2. Breaker Points: Act as a switch in the primary circuit.

3. Condenser: Protects the breaker points from arcing by acting as an electrical storage tank.

4. Breaker Cam: Opens and closes the breaker points.

- *Spark Plugs:* Ignite the fuel mixture in the cylinder as high voltage jumps the spark gap.

- *Spark Plug Leads:* Carry high voltage from the distributor to the spark plugs.

When the distributor breaker points are closed and the ignition switch is on, current travels from the battery through the ignition coil low voltage circuit, through the closed breaker points, and back to the battery through a ground or a return wire. This is known as the primary circuit, figure 9-36.

With the flow of electrons in the primary circuit, energy is stored in the coil in the form of a magnetic field. As the breaker cam rotates, it opens the beaker points just as a piston is reaching the top of its compression stroke and as the distributor rotor is passing the contact point for that cylinder. When the breaker points open, current immediately stops and the magnetic field that was built up collapses. This collapse of the magnetic field transforms the energy in the coil into a high-voltage surge large enough to create a spark at the spark lug. This voltage is usually between 15,000 anu 20,000 volts. The high voltage is created in the secondary circuit and follows this path: high-tension wiring, distributor rotor, spark plug lead, and spark plug, figure 9-37.

The breaker cam continues to rotate with the rotor, delivering high voltages to the spark

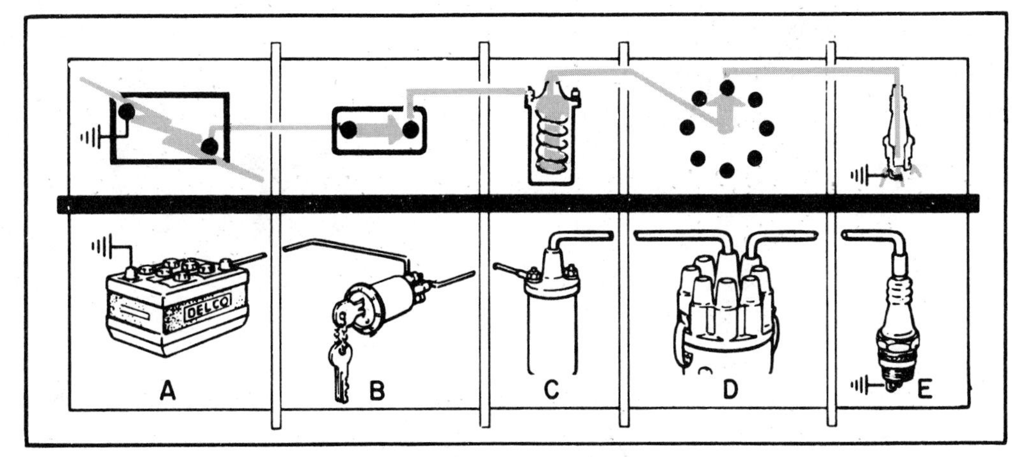

Fig. 9-35 Ignition system

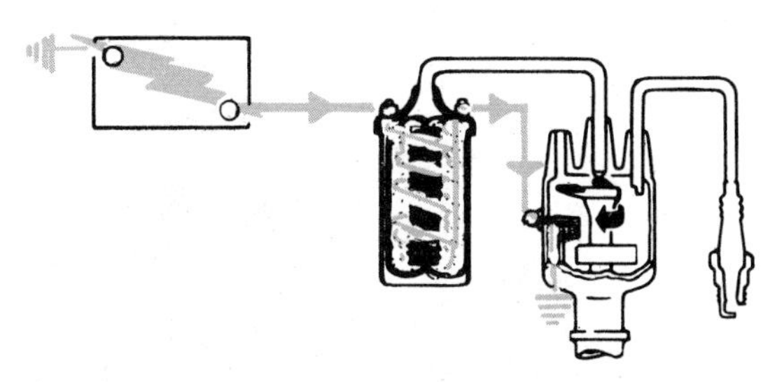

Fig. 9-36 Primary circuit with points closed. Energy is stored as magnetic field.

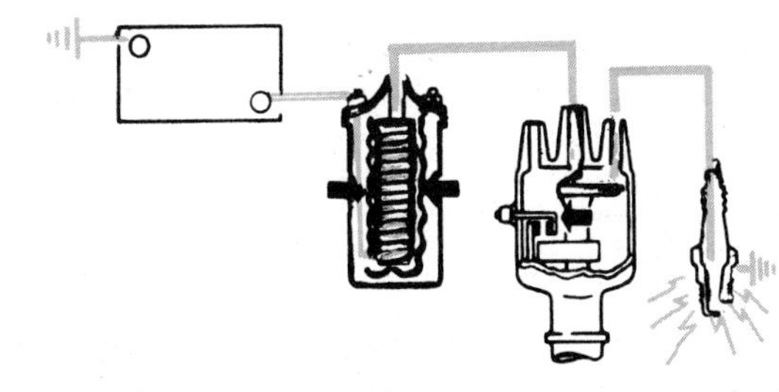

Fig. 9-37 Secondary circuit with points open. Coil produces high voltage.

plugs at the correct time. Each time the breaker points open, a spark occurs.

The spark must be timed to occur as the piston reaches the top of its compression stroke. Engine speed and load determine the exact time that the spark should occur. A centrifugal and a vacuum spark advance mechanism accomplish this task.

Solid-state ignition is sometimes called breakerless ignition or electronic ignition, figure 9-38. This system allows more precise timing of the spark, increasing the efficiency of the engine. It also eliminates the breaker points and condenser, parts that periodically wear out and need replacing. In the case of an eight-cylinder engine, an eight-toothed iron reluctor is located where the breaker cam used to be, figure 9-39. As this reluctor rotates, it passes a permanent magnet and pick-up coil, figure 9-40. The change in the magnetic field triggers the control unit to stop current flow in the primary winding of the ignition coil. The collapse of this magnetic field induces high voltage in the secondary of the ignition coil and this voltage fires the plug.

In automobiles with a 12-volt electrical system, a modification of the simplified system is made by using a resistor, figure 9-41. This allows a parallel circuit to be set up which bypasses the resistor during the cranking interval. This resistor may be installed in the solenoid switch or in the ignition switch. By using this bypass circuit, full battery voltage is supplied to the coil to insure quick starting. When the ignition switch is released after

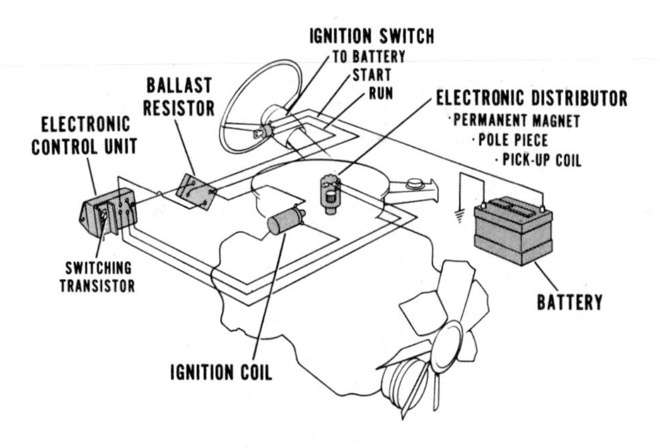

Fig. 9-38 Electronic ignition system

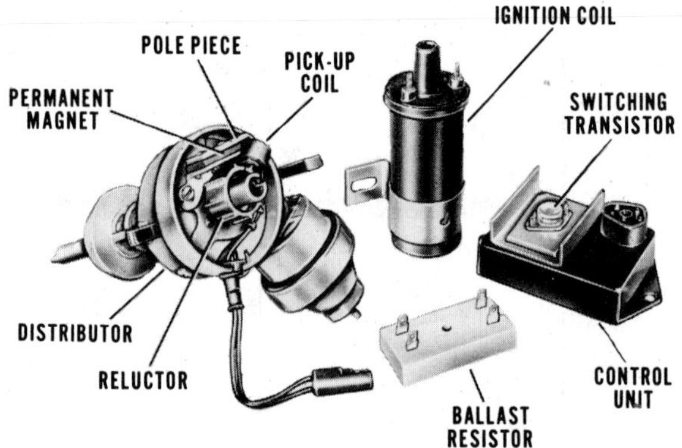

Fig. 9-39 Electronic ignition system components

starting, the resistor is automatically added to the primary circuit. This protects the coil from overload while the engine is running.

COOLING

Auto engines are usually water cooled. Cool water is circulated through water passages or jackets found in the cylinder head and block where the greatest concentrations of heat are found. The system will consist of the radiator, fan, water pump, thermostat, and radiator pressure cap.

Heated water from the water jackets is delivered to the top of the radiator. As the water moves through the radiator it is cooled by the fan drawing cool air past the radiator fins. The thermostat will hold the water in the engine where it circulates until the correct engine operating temperature is attained. (It might be noted that modern automobile engines require that antifreeze be mixed with the water.) At this time the thermostat will open to allow heated water to flow to the radiator. The water pump is mounted behind the fan and its centrifugal action moves the water through the system. The radiator pressure cap contains a blow-off valve that will

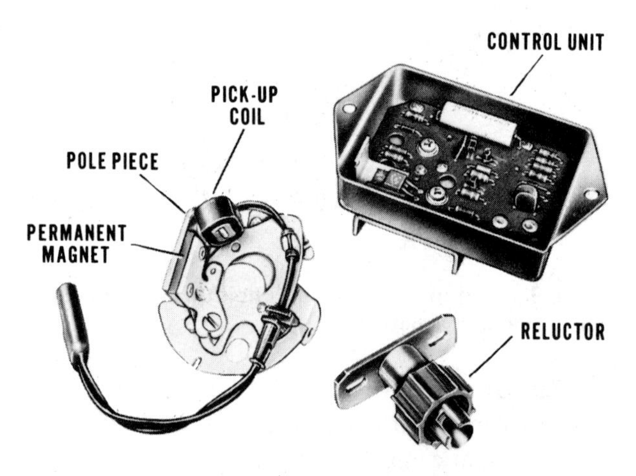

Fig. 9-40 Magnetic pick-up components and control unit

open if pressure in the system becomes dangerously high.

LUBRICATION

The chief function of the lubrication system is to reduce friction, which causes damaging heat and excessive wear of engine parts. Lubricating oils also perform the auxiliary jobs of (1) providing a better power seal, (2) cleaning by washing away small bits of metal and dirt, and (3) cooling by improving heat transfer between parts, figure 9-42.

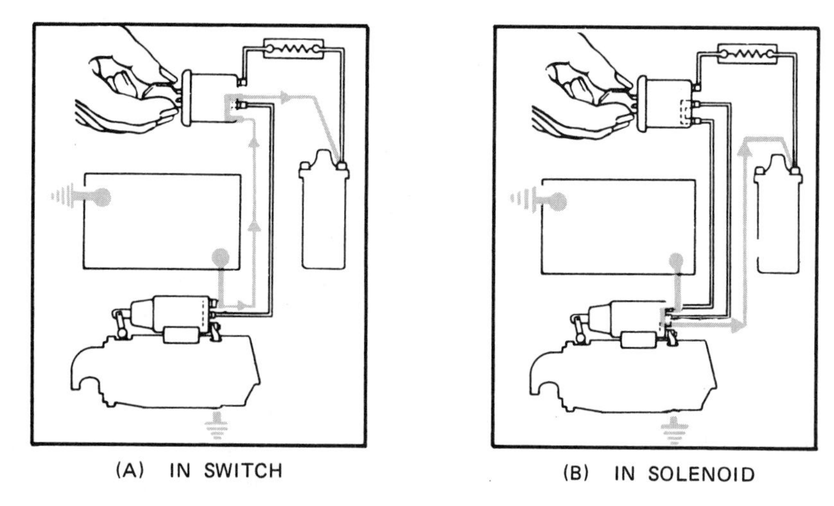

Fig. 9-41 Resistor bypass during starting

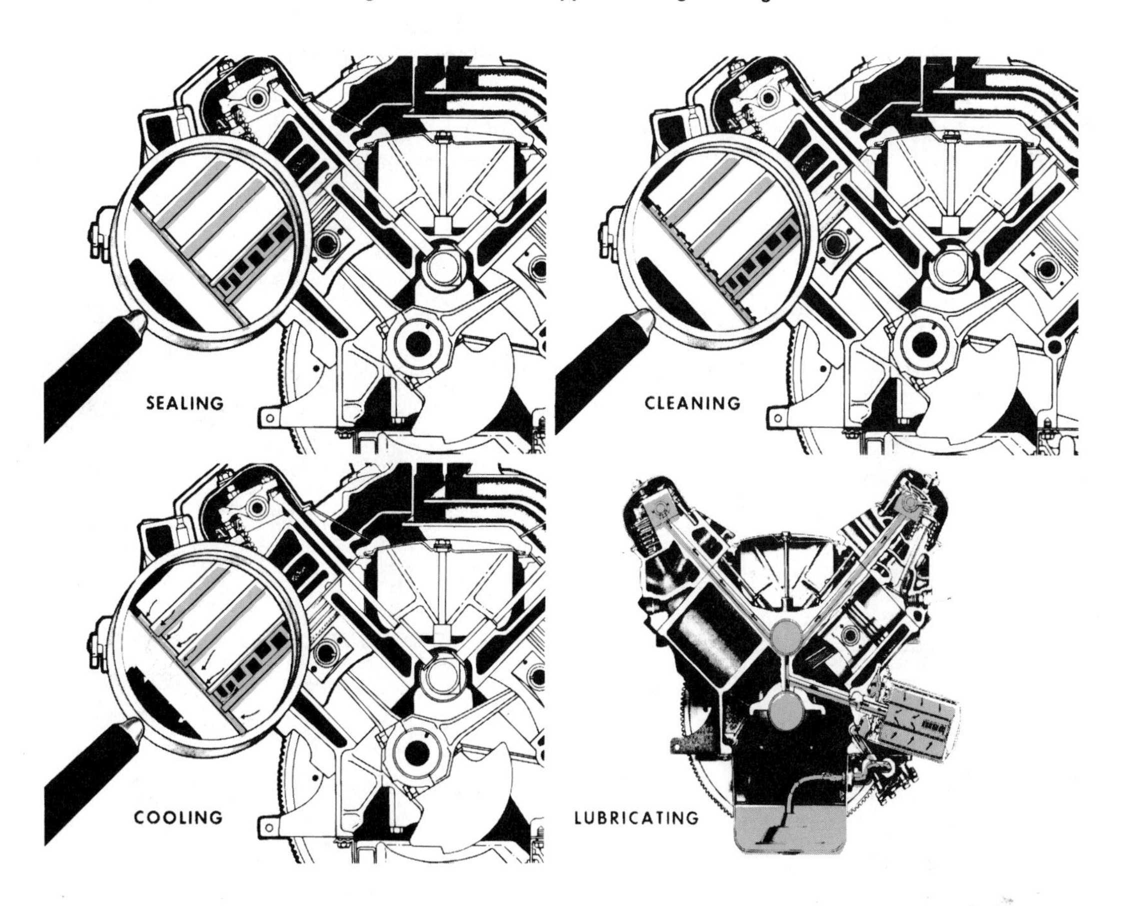

Fig. 9-42 Lubricating oil seals, cleans, cools, and reduces friction

An oil pump moves the oil throughout the engine. Generally, either some or all of the oil from the pump goes first to the oil filter, then through passages drilled in the engine block to the crankshaft and camshaft. Oil is also sent to the rocker arm shaft where it lubricates the rocker arms, valve stems, pushrods, and, as it drains back to the oil pan, the timing sprockets, chain, and distributor drive gear.

Oil to the crankshaft, camshaft, rocker arm shaft, connecting rod bearings, main bearings and other necessary parts is delivered through small drilled passages within the parts themselves. This system is usually referred to as full-pressure lubrication.

Cylinder walls and the parts inside the crankcase are splashed or sprayed with oil. Whether the part requires a continuous oil bath or a moderate one, the system is designed to meet the specific needs of all parts.

A gear-type oil pump is capable of delivering more oil than the system can distribute. To prevent excessive pressures from building up in the lubrication system, an oil pressure relief valve is installed. If the maximum desired pressure is passed, the spring-loaded relief valve opens allowing the excess oil to drain back to the crankcase and oil pan.

The oil filter continually cleans the lubricating oil. Dirt, soot, small bits of metal, and any foreign matter is trapped in the filter. The filter may be made of cotton waste, cellulose, inert earth, paper, or any material that will effectively filter the oil.

On some engines all the oil is passed through the oil filter as it comes from the pump. This is referred to as the full-flow system, figure 9-43. Other engines have the bypass system in which only a portion of the oil is filtered at any given moment, figure 9-44. The bypass

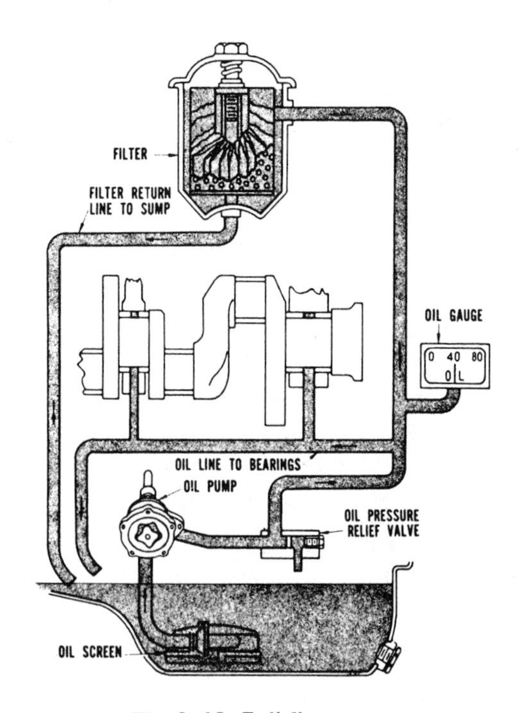

Fig. 9-43 Full flow system

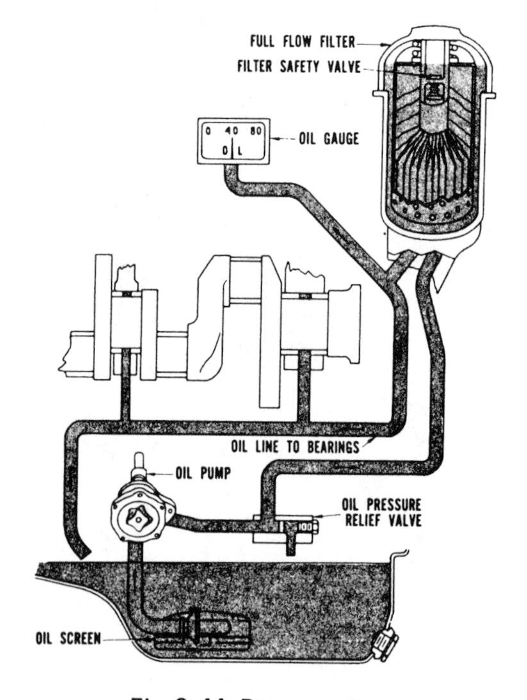

Fig. 9-44 Bypass system

system may filter 5% to 20% of the oil at a time. However, over a period of time all the oil will be filtered. The full-flow system is not used in many automobiles because, if the filter becomes clogged, oil cannot flow to the moving parts of the engine. This problem cannot happen in the bypass system.

The engine crankcase has a ventilation system designed to help keep the engine oil pure and reduce air pollution. By-products of combustion may leak down into the crankcase, gasoline may leak into it, or moisture in the air may condense in the crankcase. These conditions result in the formation of sludge and acids and the dilution of the oil. The ventilation system keeps these conditions to a minimum by providing a draft of clean air through the crankcase. Filtered air enters the crankcase, flows through it, and then is routed through the air cleaner and carburetor to the combustion chamber, figure 9–45.

POLLUTION CONTROL

Over the years several government acts have concerned themselves with air pollution. However, it was not until 1970 that a special agency, the Environmental Protection Agency (EPA), was set up to determine dangerous pollution levels, set standards, and then to monitor and regulate control of pollution. Pollutants from the automobile engine come under the jurisdiction of the EPA, so all United States automobile manufacturers, and those cars imported to the United States, must have engines that meet certain pollution standards.

Automobile pollution comes from three sources: vapors from the crankcase, fuel vaporization from the gas tank and carburetor, and combustion by-products from the tailpipe.

Crankcases were vented directly to the atmosphere until it was discovered that the emissions from the crankcases were pollutants; about 20% of the total problem. These pollutants are unburned hydrocarbons from blowby along the cylinder walls that pass into the crankcase. Since 1963 the crankcase pollution problem has been brought under control with positive crankcase ventilation systems. Positive crankcase ventilation recirculates crankcase gases through the carburetor and back into

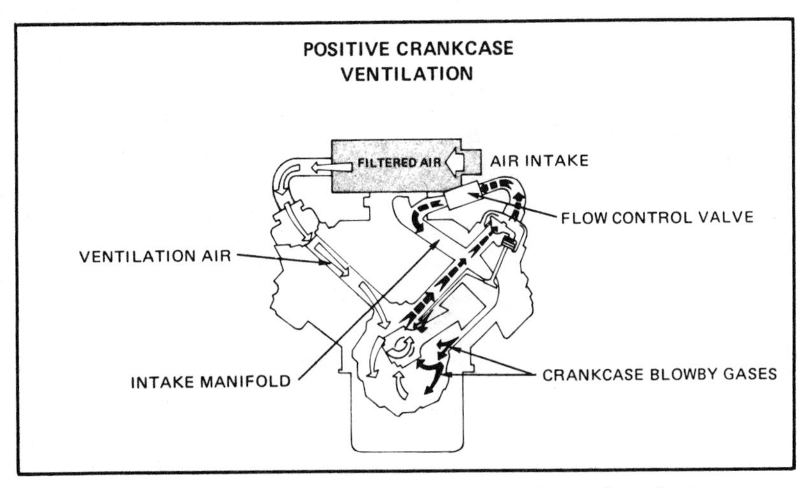

Fig. 9-45 Positive crankcase ventilation reduces air pollution

the combustion chamber to be reburned. This way the gases do not escape directly into the atmosphere.

About 20% of the hydrocarbon pollution has been traced to the carburetor and gas tank. Gasoline is an ideal fuel because it vaporizes easily, but unfortunately, pollutants are created when the gas evaporates from the gas tank and carburetor. Tanks and carburetors must be vented to prevent the buildup of pressures. Before pollution controls, they too were vented directly to the atmosphere. Most vaporization occurs just after the engine has stopped, when the hot engine evaporates the gasoline left in the carburetor. This kind of evaporation is now controlled, by one of two ways, figure 9-46. In one system, vapors are collected and stored in the crankcase to be used when the engine is restarted. In another system, the vapors are collected in a canister of activated charcoal granules. The vapors remain in the charcoal until the engine is started. Then air purges the canister and sends the vapors to the base of the carburetor to be burned.

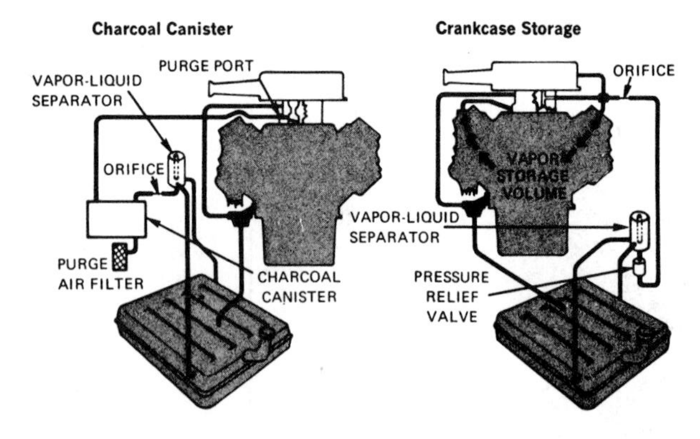

Fig. 9-46 Evaporate control systems eliminate pollution from the carburetor and the gas tank.

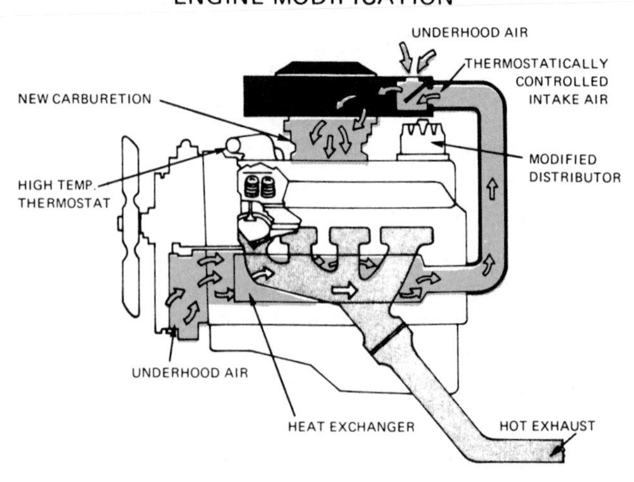

Fig. 9-47 Modification of the engine's fuel injection and cooling system reduces hydrocarbon emission from the tailpipe

Exhaust emissions contain hydrocarbons, oxides of nitrogen, carbon monoxide, lead particles, and other pollutants. With this in mind it's easy to understand that the more completely the fuel is burned, the less pollution there will be. Controls for exhaust emissions are already in use, but they must continue to improve to meet standards of the future. Manufacturers are designing combustion chambers, fuel induction systems, ignition systems, and other modifications and parts that improve the combustion process, figure 9-47. Two currently-used methods of reducing emissions are shown in figure 9-48, the afterburner and the catalytic converter.

Afterburners are used to insure more complete combustion. In an afterburner, air is pumped into the exhaust manifold to burn the exhaust more completely before it is released to the atmosphere.

Some manufacturers are approaching emission control with the catalytic converter. The catalytic converter chemically changes the

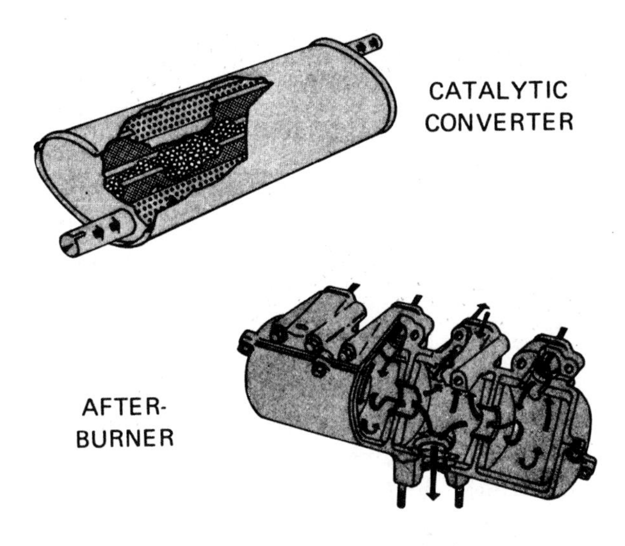

Fig. 9-48 The catalytic converter and the afterburner are methods of further reducing tailpipe emissions

hydrocarbons and carbon monoxide emissions into water vapor and carbon dioxide. This device depends on the use of lead-free gasoline. Remember that tetraethyl lead is now blended into gasoline to improve its octane rating. Lead does not hurt the engine, in fact, metallic lead acts as a lubricant for valves. Also, while lead is harmful as a pollutant, experts have not established the extent of lead hazard in emissions. It has been found, however, that lead is very destructive to catalytic converters. If leaded gasoline goes through the converter, the platinum elements in it lose their effectiveness and have to be replaced. Gasoline service stations and auto manufacturers have tried to prevent this mistake by using smaller gas nozzles and smaller gas filler holes.

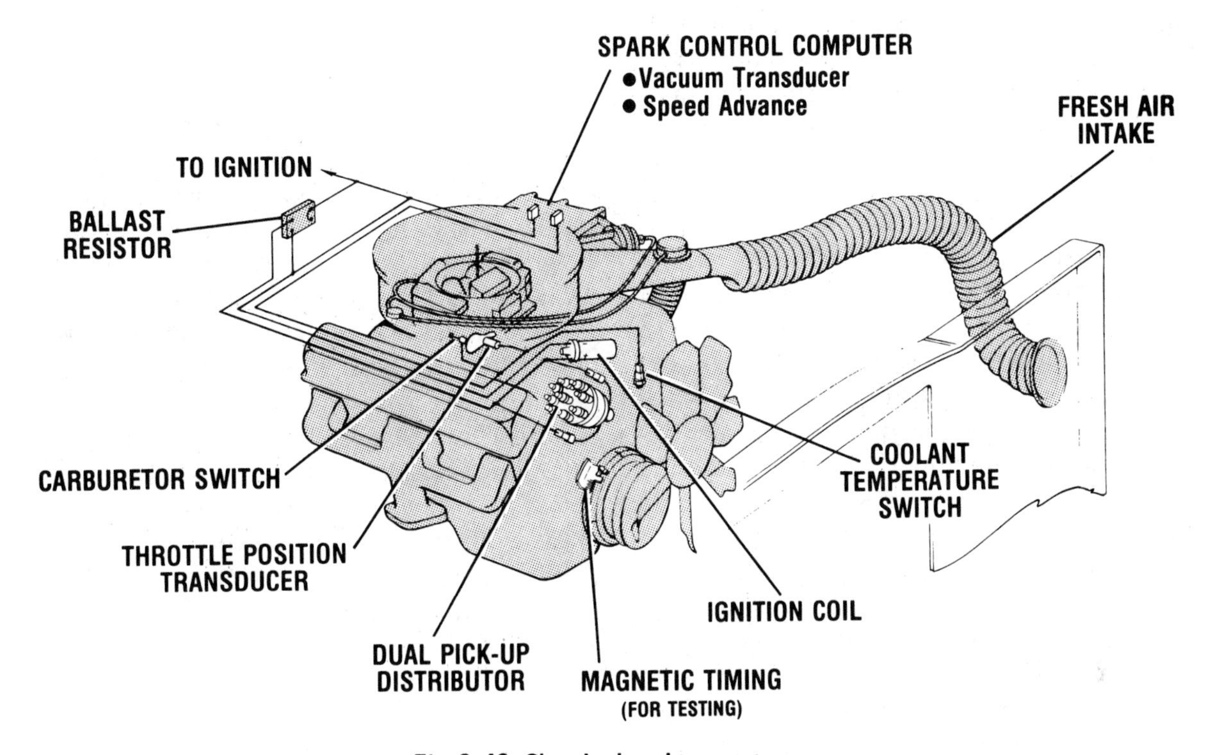

Fig. 9-49 Chrysler lean burn system

Electronic Lean Burn

A further step in pollution control is Chrysler's electronic lean burn system which controls pollution in the combustion chamber as the fuel is being burned, figure 9-49. Lean fuel mixture, near 18:1, burn with low hydrocarbon, HO, and carbon monoxide, CO, emissions. When lean mixtures are used, the oxides of nitrogen, NO_x, emissions are also very low due to the presence of excess air and cooler combustion temperatures.

Electronic lean burn uses a computerized electronic spark advance control system combined with the lean fuel mixture. This system improves the efficiency of the combustion process without sacrificing engine performance. The system responds to starting, throttle position, coolant temperature, rate of throttle position change, and length of operating at open or closed throttle. Sensors feed information on these changing conditions to the spark control computer which determines the correct ignition timing for the conditions at that moment.

Stratified Charge

Stratified charge engines are in limited use today. The basic idea of this engine is to start combustion with a rich mixture and finish with a lean mixture; this completely burns the gasoline. The combustion chamber is divided into a pre-chamber and the larger main chamber. A rich fuel mixture is introduced into the pre-chamber where it is ignited. The flame then moves into the main chamber where it ignites a much leaner mixture. Fuel can be delivered by a fuel injector or a carburetor depending on the engine design. This system is used in the Honda CVCC.

The stratified charge programmed combustion PROCO engine, figure 9-50, injects a precisely measured amount of fuel into the combustion chamber. A specially shaped piston controls the air motion as a "shaped" cloud of fuel is ignited, figure 9-51. The initial rich burning produces little oxides of nitrogen, NO_x, while the complete burning of the fuel reduces CO and HO emission. At present this engine still needs a catalytic converter to reduce pollutants below government standards.

Stirling Engine

Robert Stirling, a Scottish churchman, devised this principle in 1816 and, although the engine has not gained wide acceptance, it is being considered as an alternative power plant for the automobile. The engine has potential since it is a low emitter of pollutants, very smooth and quiet, and offers good gas mileage.

The engine, figure 9-52, uses an enclosed working fluid such as hydrogen which is alternately heated to 1300°F (704°C) and cooled to 175°F (79°C). Fuel is burned in a combustor outside the engine; the combustion is continuous. The engine is capable of using gasoline, diesel fuel, or kerosene. The current design uses four pistons arranged around a shaft and swash plate. This plate mounts at an angle on the shaft and is used instead of a crankshaft. The hot expanding gases push the piston down, transferring its linear motion into rotary motion at the swash plate. Tubing connects the cool end of the cylinder to the hot end of the next cylinder and the gas is alternately displaced from one cylinder to the next. A regenerator stores heat, picking up heat from the fluid as it is cooled and releasing heat as the fluid is cycled through for reheating and expansion.

TUNEUP AND DRIVING TIPS

In addition to pollution control devices, a properly-tuned engine will emit fewer pollutants and improve engine efficiency, thus in-

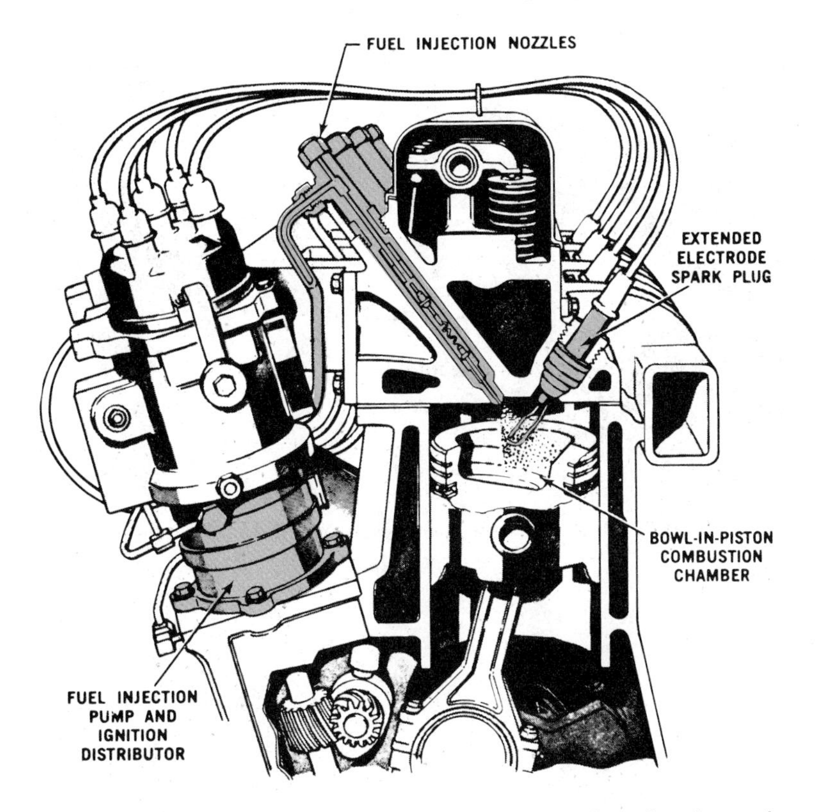

Fig. 9-50 Fuel-injected stratified charge internal combustion engine

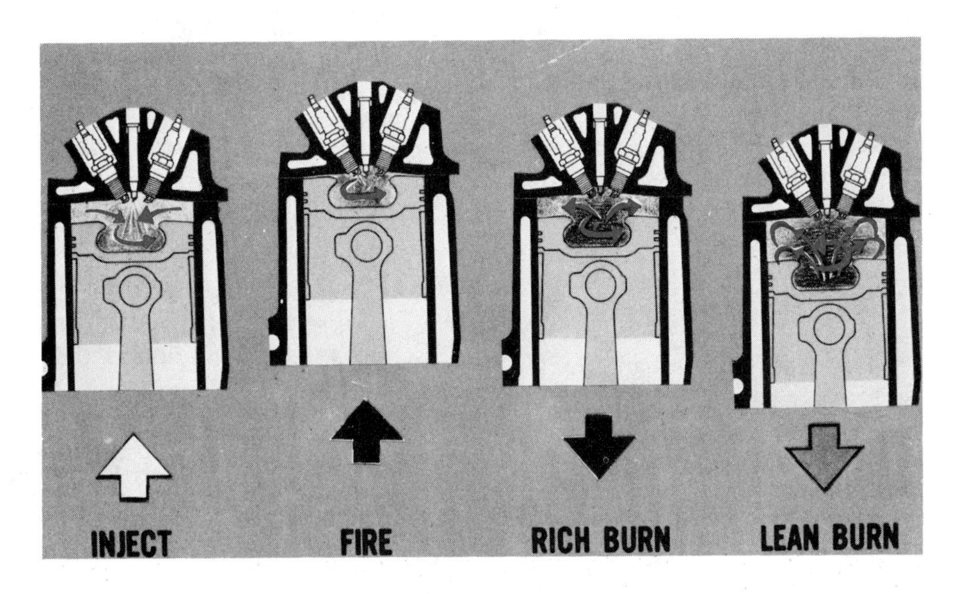

Fig. 9-51 PROCO burning cycle

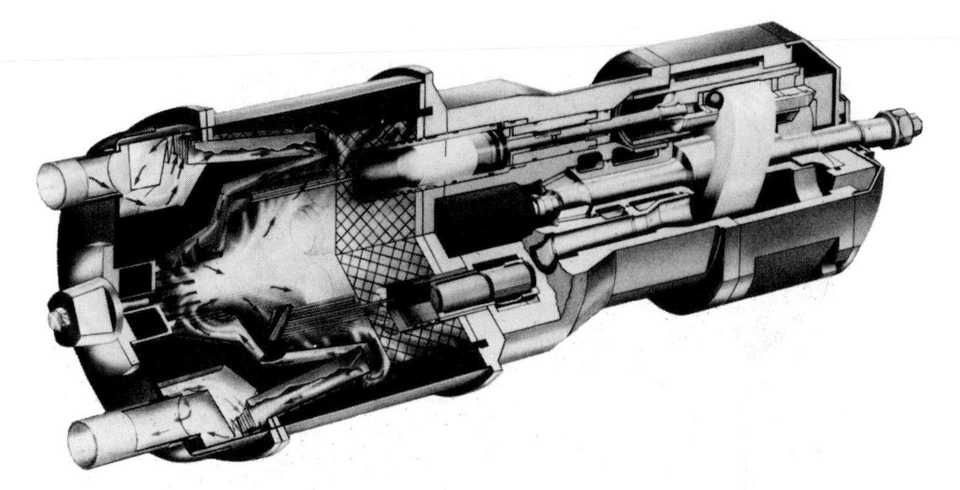

Fig. 9-52 Stirling engine (Courtesy Ford Motor Company)

creasing gasoline mileage. There are several other things that can be done also. These include:

- Idle speed should be adjusted to the manufacturer's specifications.
- Ignition timing should be correct.
- Spark plugs should be clean and the breaker points should be clean and set correctly.
- The carburetor should be adjusted correctly.

To improve gasoline mileage, consider these items.

1. Avoid high speed driving. Fuel economy can be reduced up to 25% as a result of high speed driving.
2. *Jack rabbit* starts and hard braking should be avoided. Eliminating these may increase fuel economy up to 18%.
3. Inflate tires properly.
4. Use the air conditioner sparingly at low speeds. An air conditioner may cost you 7% in fuel economy.
5. Avoid long engine warmups. Start driving slowly after thirty seconds of warmup.

DRIVE SYSTEM

The drive system transmits the power to the wheels which propel the automobile. The system contains four basic parts: clutch or torque converter, transmission, drive shaft and universal joints, and differential.

- The clutch engages or disengages the engine from the rest of the drive system. When the automobile is first started or standing still, the clutch is disengaged. When moving, the clutch is engaged. The torque converter is an automatic, hydraulically operated device that replaces the clutch.
- The transmission changes the speed of the wheels in relation to the speed of the engine: low speed, high torque for starting up; medium speed, medium torque for

intermediate speed; and high speed, low torque for rapid travel. The transmission also provides a reverse gear for backing.

- The drive shaft transmits the power to the differential. It is sometimes called the propeller shaft. The universal joints enable the transmission and differential to be out of alignment. It has the effect of being a flexible shaft.
- The differential changes the direction of the drive, since the drive shaft and rear axles are at right angles to each other. Also, the differential allows the wheels to rotate at different speeds for turning corners.

Clutch

The single plate clutch is the most common type, figure 9-53. It is mounted directly behind the engine. Its main parts are the flywheel, the friction disc, and the pressure plate. The polished flywheel, which is fastened to the end of the crankshaft, is always turning as long as the engine is running. The friction disc is faced with a material similar to brake lining material and mounted on the input side of the transmission. The end of the transmission shaft is splined so the disc can slide back and forth. The polished pressure plate is behind the disc and revolves with the flywheel. When the clutch pedal is out, strong springs push the pressure plate against the disc and up to the flywheel. The disc and transmission input shaft rotates with the flywheel; the clutch is said to be engaged. When the clutch pedal is pushed in, the spring pressure is released and the friction disc loses contact with the flywheel; the clutch is said to be disengaged.

Transmission

The transmission changes the relationship of the engine speed to the rotating speed of

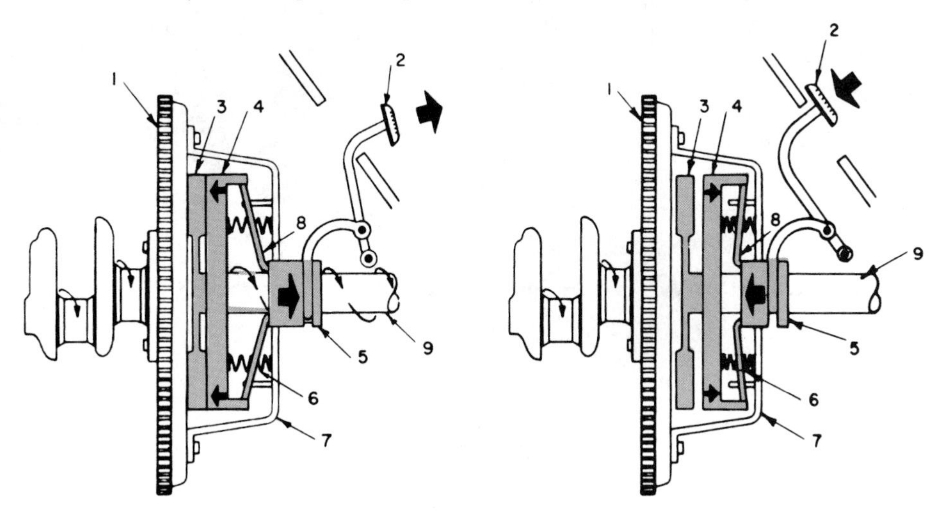

Fig. 9-53 Clutch operation

the wheels. This is done through the use of gears, figure 9-54. By shifting the power to the various gears, the proper torque and speed can be obtained. For example, a large gear driving a small gear increases speed but reduces torque, while a small gear driving a large gear reduces speed but increases torque.

The sliding gear transmission is the conventional type manual transmission providing low gear, intermediate gear, high gear, and reverse, figure 9-55. To engage first or low gear, gear (F) is slid forward. The power follows these gears: (A) from the clutch, (B) on the counter shaft, (D) on the counter shaft, (F) on the transmission main shaft. This arrangement produces a gear reduction ratio of 3:1. Accompanied by a 4:1 reduction in the differential, the crankshaft turns twelve times for one turn of the wheels. Gears (E), (C), (G), and (H) are arranged to deliver no power in low gear.

In second or intermediate gear, (F) is returned to its middle position, out of mesh. The sliding collar "X" locks gear (E) to the output shaft. The power follows these gears: (A) from the clutch, (B) on the counter shaft, (C) on the counter shaft, (E) on the transmission main shaft. This arrangment produces a gear reduction ratio of 2:1. Again accompanied by a 4:1 reduction in the differential, the crankshaft turns eight times for one turn of the wheels.

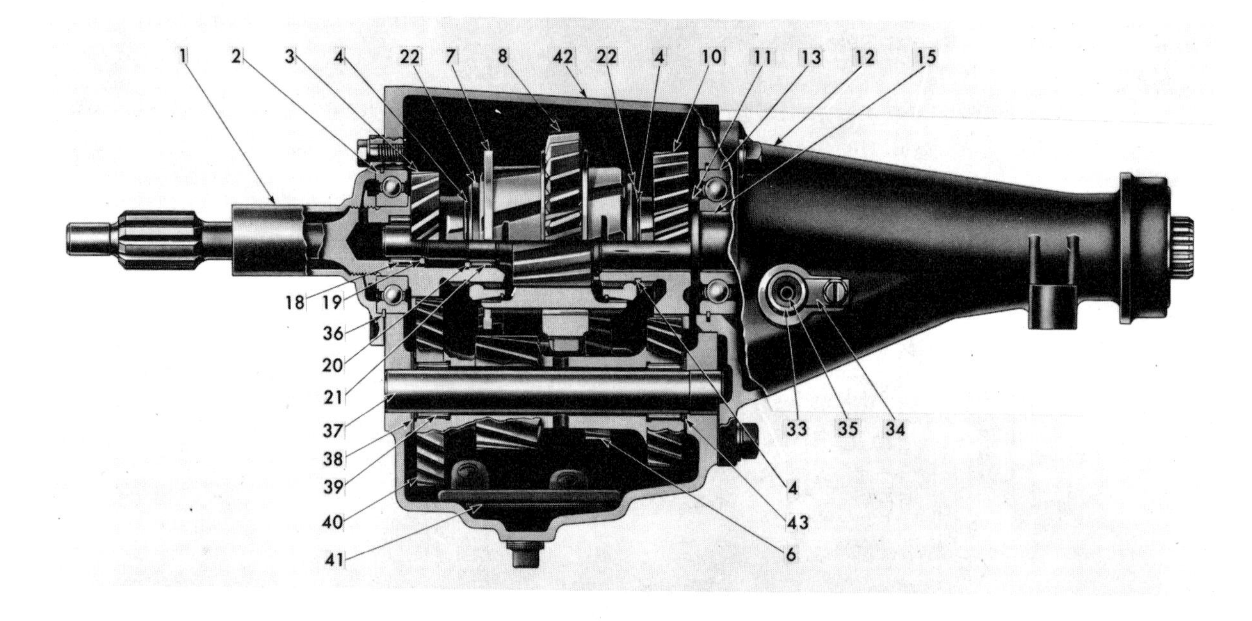

1. Clutch Gear Bearing Retainer
2. Clutch Gear Bearing
3. Clutch Gear
4. Energizing Spring
6. Reverse Idler Gear
7. Second and Third Speed Clutch
8. First and Reverse Sliding Gear
10. Second Speed Gear
11. Thrust Washer
12. Case Extension
13. Mainshaft Rear Bearing
15. Mainshaft
18. Front Pilot Bearing Rollers
19. Thrust Washer
20. Thrust Washer
21. Rear Pilot Bearing Rollers
22. Synchronizer Ring
33. Speedometer Shaft Fitting
34. Lock Plate
35. Speedometer Driven Gear Shaft
36. Snap Ring
37. Countershaft
38. Thrust Washer
39. Roller Bearing
40. Countergear
41. Collector
42. Transmission Case
43. Roller Thrust Washer

Fig. 9-54 Cross section of transmission

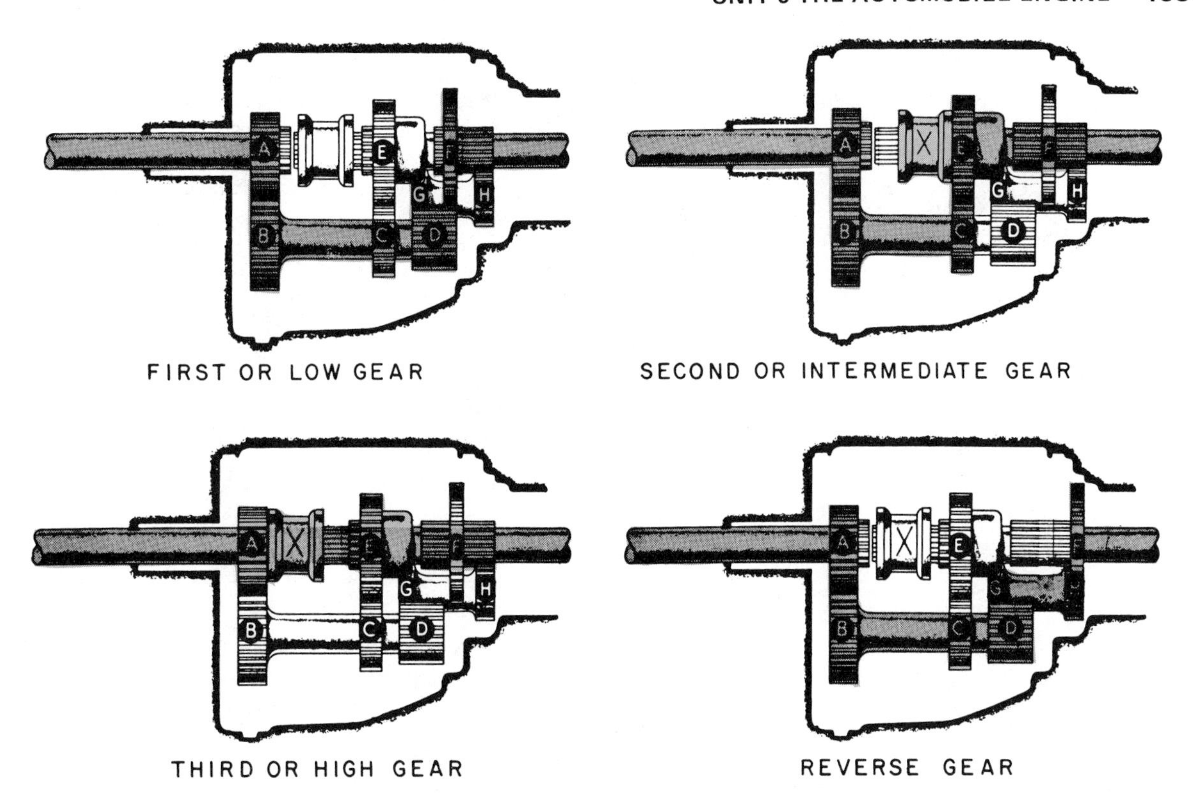

Fig. 9-55 Conventional constant mesh transmission

In third or high gear, the power from the clutch goes directly to the main transmission shaft. The sliding collar "X" is moved to the left, engaging the clutch shaft and disengaging gear (E). Gear (F) is also disengaged. This direct connection provides a ratio of 1:1. However, it is accompanied by the 4:1 reduction in the differential. The crankshaft turns four times for one turn of the wheels.

In reverse, the power follows these gears: (A) from the clutch, (B) on the counter shaft, (D) on the counter shaft, (G) on the reverse idler, (F) on the transmission main shaft. Gear (E) revolves but delivers no power. When gear (G) is used, the direction of the output transmission shaft rotation is reversed. The gear ratios are the same as for low gear if gears (D), (G), and (H) are the same size.

Drive Shaft, Universal Joints, and Differential

The drive shaft and universal joints carry the power to the differential. The differential transmits the power to the axles and also enables the wheels to revolve at different speeds when a corner is turned. The wheel on the inside of the turn needs to travel less distance than the wheel on the outside of the turn, just as a column of marching soldiers do in turning a corner.

The parts of the differential are the case, drive pinion, ring gear, differential side bevel gears on the end of the axles, and pinion gears mounted in the differential case, figure 9-56.

The drive pinion brings power into the differential and meshes with the ring gear. These two gears transfer the power from the drive

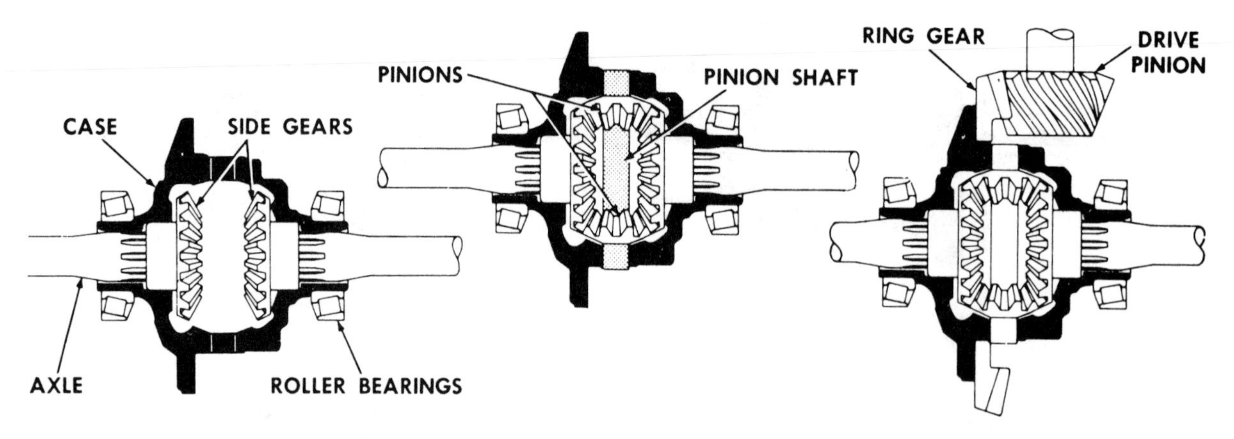

Fig. 9-56 Differential changes the direction of power

shaft to the axles, which are at right angles to each other. The ring gear has about four times as many teeth as the drive pinion, the effect being a gear reduction ratio of 4:1. If the automobile always traveled in a straight line, these would be the only parts necessary. However, to compensate for different rear wheel speeds when turning, the differential side bevel gears and pinion gears are added. These gears are mounted in the case. The ring gear is also mounted on this case. When the auto travels straight ahead, the ring gear, case, and differential gears all move as a unit. But when a corner is turned, the differential gears turn relative to each other; one faster, the other an equal amount slower. On slippery ice with one wheel spinning, this wheel's differential side gear would be turning at twice ring gear speed while the other differential side gear is stationary.

Overdrive and Automatic Systems

Overdrive offers a great savings in fuel economy at highway speeds, and once was a popular feature, but the public's desire for automatic transmissions almost eliminated overdrive transmissions. The overdrive transmission, figure 9-57, has a gear ratio of 0.73:1 in overdrive. Assuming the same speed, this has the effect of reducing the engine's revolutions per minute in third gear by 27%. This decreases engine wear, noise level, and oil consumption while getting more miles to the gallon of gasoline.

Many automobiles are equipped with automatic transmissions, figure 9-58. The conventional clutch is eliminated, being replaced by a type of hydraulic drive, the simplest of which is fluid coupling. It consists of two wheels fitted with blades. One is connected to the crankshaft and called the pump or driver. The other is connected to the transmission shaft and called the turbine or driven member. The two are fitted close together but do not touch. They might be visualized as a doughnut sliced in two discs. These parts are in a sealed, oil-filled case. The driver blades throw the oil, or exert pressure, against the driven blades when the engine is running. At idle speed, the pressure exerted is very small and the driven member does not move. The driven member is slipping 100%. However, as the engine speeds up, the oil pressures against the driven blades increase and the driven member begins to revolve, slowly at first, then faster and faster.

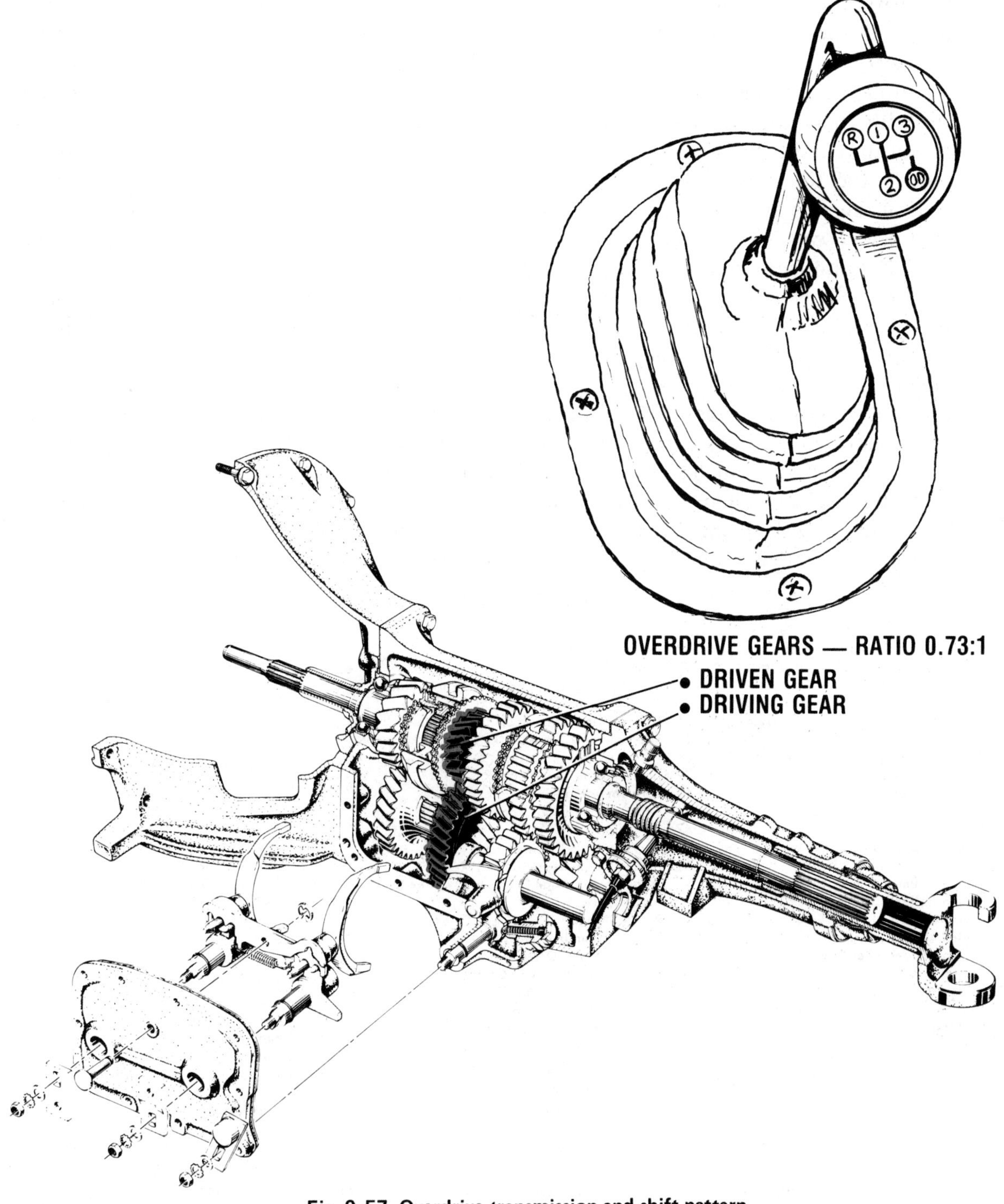

Fig. 9-57 Overdrive transmission and shift pattern

Soon the driver and driven member are at almost the same speed, and there is almost no slippage.

The automatic transmission works in conjunction with the hydraulic drive. The driver does no shifting, rather shifting is done automatically in the transmission, controlled by the speed of the auto and how far the accelerator is pushed down. The planetary gear is the heart of the automatic transmission. Often there are three planetary gear sets in the transmission; two are used to give four forward speeds and one is for reverse. The planetary gears are always in mesh.

The planetary gear is made of three main parts: the center or sun gear, planet pinion gears, and internal or annulus gear, figure 9-59. The three planet pinion gears are held 120 degrees apart by the planet carrier.

The correct gear reduction is attained by holding one of the gear elements, allowing the power to be transmitted in either direction between the other two elements. For example, if the internal gear is held stationary and power is applied to the center gear, the planet pinion gears revolve inside the internal gear. Power is taken from the planet pinion's planet carrier at a certain gear reduction ratio.

Also, if the center gear is held stationary and power is applied to the internal gear, the planet carrier will revolve around the center gear. Gear reduction is again attained, but at a different ratio.

To obtain reverse rotation in the transmission, the planet carrier is held stationary and power is applied to the center gear. The planet pinion gears spin on their bearings acting as idlers and transfer the power to the internal gear. Power is taken from the internal gear which is rotating in reverse direction.

LEGEND – Fig. 9-58

1. Transmission Housing
2. Converter Cover "O" Ring Seal
3. Turbine Assembly
4. Stator Assembly
5. Converter Housing & Pump Assembly
6. Converter Pump
7. Converter Pump Thrust Washer
8. Front Oil Pump Body Oil Seal
9. Front Oil Pump Body
10. Front Oil Pump Body "O" Ring Seal
11. Stator Support
12. Transmission Valve Body
13. Input Shaft Oil Seal Ring
14. Clutch Drum Oil Seal Rings
15. Clutch Relief Valve Ball
16. Low Brake Band
17. Clutch Drum
18. Clutch Piston Inner Seal
19. Clutch Hub
20. Clutch Hub Thrust Washer
21. Low Sun Gear & Clutch Flange Assembly
22. Parking Lock Gear
23. Planet Short Pinion
24. Planet Input Sun Gear
25. Planet Input Sun Gear Thrust Washer
26. Planet Carrier
27. Reverse Brake Band
28. Output Shaft
29. Transmission Case
30. Rear Oil Pump Gasket
31. Rear Oil Pump Cover to Body Attaching Screw
32. Rear Oil Pump Cover
33. Rear Oil Pump Body
34. Rear Bearing Locating Front Snap Ring
35. Transmission Rear Bearing Assembly
36. Transmission Rear Bearing Retainer
37. Rear Bearing Locating Rear Snap Ring
38. Transmission Extension "O" Ring Seal
39. Transmission Extension
40. Speedometer Drive Gear
41. Extension Rear Oil Seal
42. Extension Bushing
43. Speedometer Driven Gear
44. Transmission Rear Bearing Retainer Screw
45. Transmission Rear Bearing Retainer Screw Lockwasher
46. Rear Oil Pump Drive Gear Drive Pin
47. Rear Oil Pump Assembly Attaching Screw
48. Rear Oil Pump Drive Gear
49. Rear Oil Pump Driven Gear
50. Governor Drive Gear
51. Governor Driven Gear
52. Transmission Case Bushing
53. Reverse Drum Thrust Washer
54. Planet Long Pinion
55. Reverse Band Lever & Link Assembly
56. Low Sun Gear Thrust Washer
57. Planet Pinion Shaft Lock Plate
58. Reverse Drum & Ring Gear
59. Clutch Flange Retainer
60. Clutch Flange Retainer Ring
61. Clutch Spring Seat
62. Clutch Spring Snap Ring
63. Clutch Spring
64. Clutch Drive Plates
65. Clutch Driven Plates
66. Clutch Piston
67. Clutch Piston Outer Seal
68. Clutch Drum Thrust Washer (Selective)
69. Manual Valve
70. Converter Housing Dowel Pin
71. Converter Housing-to-Case Gasket
72. Front Oil Pump Drive Gear
73. Front Oil Pump Driven Gear
74. Sump Baffle
75. Oil Pickup and Suction Screen
76. Access Hole Plug
77. Converter Pump Housing Bolt
78. Converter Pump Housing Nut
79. Stator Retaining Rings
80. Stator Thrust Washers
81. Over-Run Cam Roller
82. Stator Race
83. Converter Cover Hub Bushing
84. Input Shaft
85. Turbine Thrust Washer
86. Over-Run Cam Roller Spring
87. Over-Run Cam
88. Converter Cover Assembly
89. Flywheel to Transmission Anchor Nut Assy.

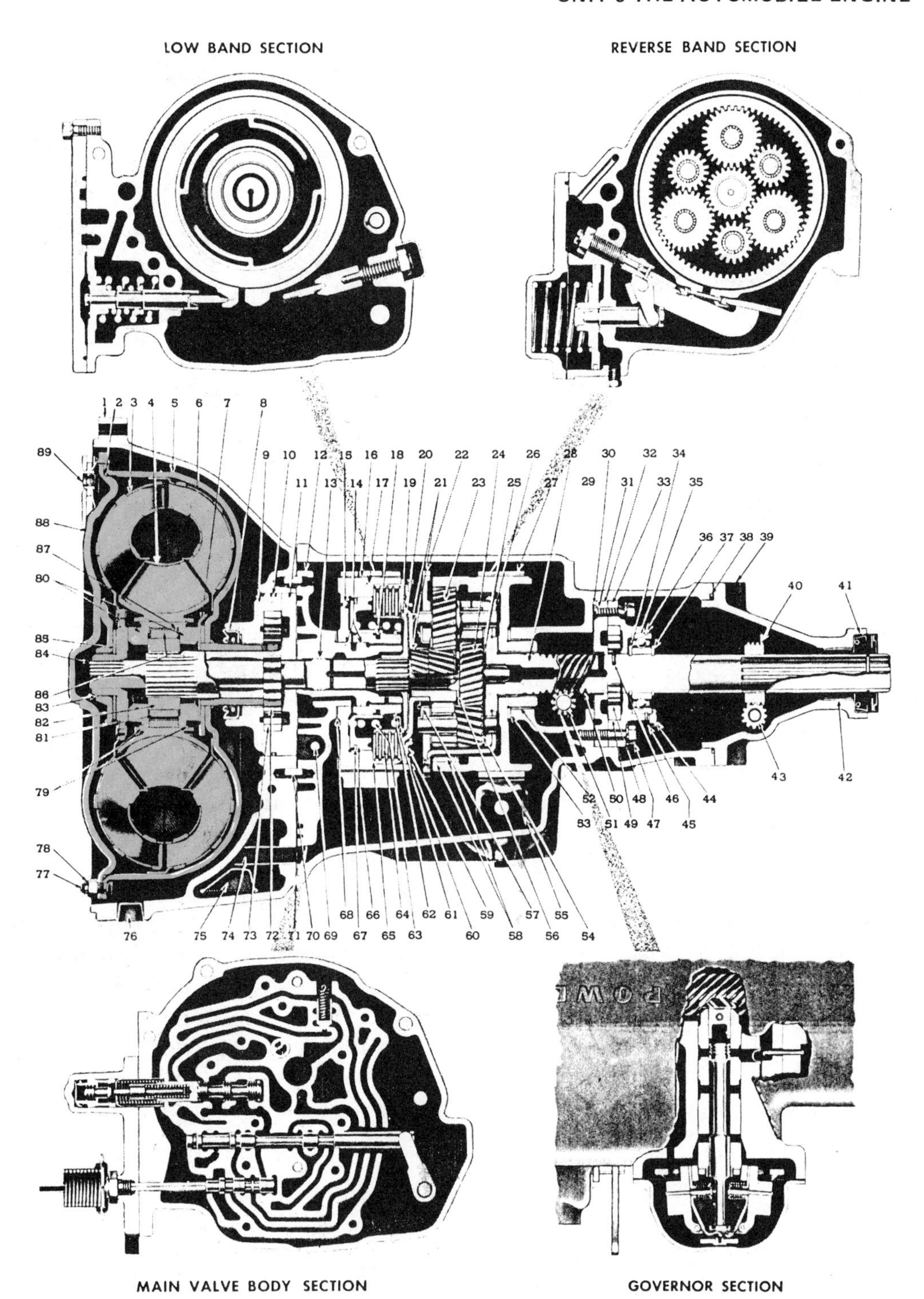

Fig. 9-58 Cross section of torque converter and transmission

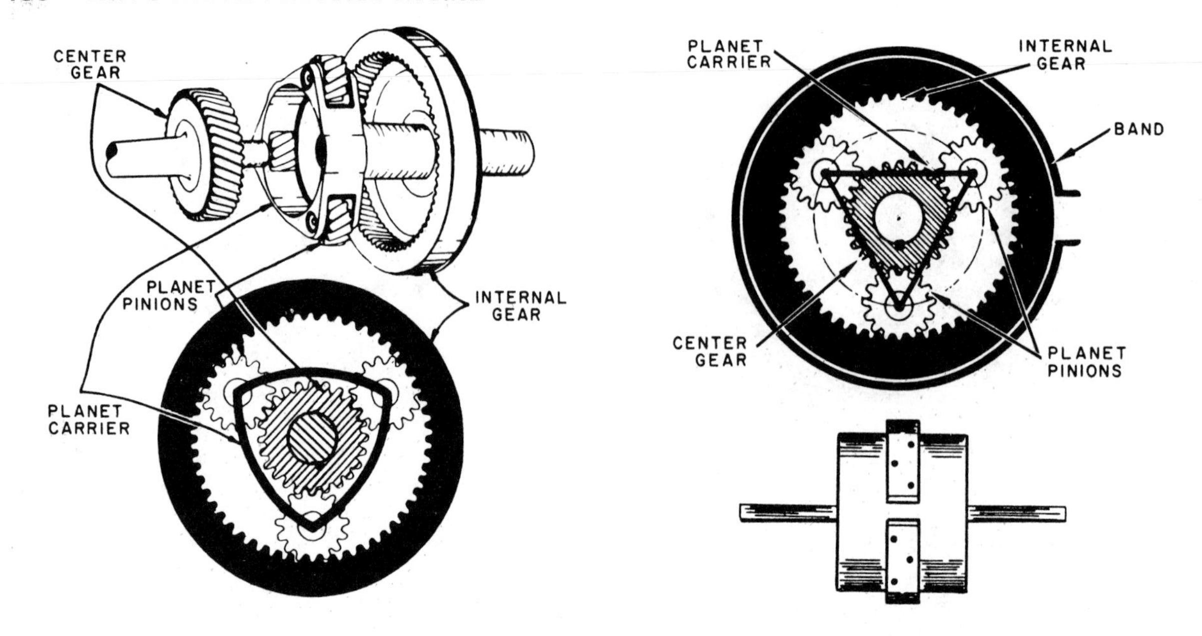

Fig. 9-59 Planetary gear system

If none of the planetary gear elements are held stationary, no power is transmitted; the transmission is in neutral. The planetary gear transmission is operated by oil pressure which tightens bands and clutches onto the gears at the proper time.

The hydraulic torque converter is another type of automatic transmission. It is very similar to fluid coupling but it has the ability to multiply the torque or twisting force produced. The torque converter has a driver and driven member, pump and turbine, but, in addition, it has a set of stationary blades that receive the oil from the turbine and change its direction of flow. This change of direction helps the pump to pump harder when the oil re-enters, thereby multiplying the torque up to 2⅓ times under certain conditions. Once the automobile has attained speed, the torque converter acts as a simple fluid coupling. Planetary gears are usually used with the torque converter to obtain reverse or to obtain a special low gear for heavy pulling.

SAFETY IN AUTOMOTIVE MAINTENANCE AND REPAIR

The repair and maintenance of auto engines and the associated parts of the automobile present conditions that are a safety hazard if safety rules are not followed in an intelligent manner. Accidents can quickly convert a satisfying activity into a personal tragedy.

Before discussing safety rules as such, every student mechanic or professional should be aware of the responsibility involved with auto repair work. There is absolutely no room for halfway or makeshift work. Errors can place a life in jeopardy.

The mechanic's attitude must be one that demands perfection.

- No mistakes.
- Every nut and bolt properly tightened.
- Every part in its proper place.
- Every part functioning correctly.

Burns

- Take extreme care in draining radiators, being careful to protect hands and face when removing the cap from a boiling radiator.
- For protection against battery acid burns, wash the hands immediately after coming in contact with battery acid.
- Avoid contact with exhaust pipes, manifolds, and other parts of an engine which are hot.

Falling Objects

- Be sure that chain hoists are fastened securely and that chain slings are not overloaded.
- Do not depend on jacks or hoists alone. Always use strong wooden blocks or a substantial steel or wooden stand made for the purpose. Never use concrete blocks.
- Allow no one under the car while it is being lifted.
- When two wheels are off the floor, keep the other wheels blocked to prevent the car from rolling.
- When working under a car, keep feet and legs clear of passageways.

Fire

- Splashing gasoline may start a fire on a hot motor. Never run a car above idle speed when the float bowl cover is removed.
- Store gasoline in an approved safety can.
- Be sure that all liquid containers in the shop are labeled properly.
- Do not use gasoline to clean tools, machine parts, or clothing. Use kerosene for cleaning oil and grease from metal parts and tools.
- Never wash hands or arms in gasoline.
- If clothing has been soaked with gasoline, remove it as it may irritate the skin.
- Put gasoline-soaked rags in safe, covered metal waste cans only.
- When pouring gas from one container to another, keep containers in contact to prevent static electricity.
- Place fire extinguishers where handy. Use only foam or carbon dioxide on gas or oil fires.
- Engines operated indoors should be provided with flexible exhaust tubes attached to the tailpipe to carry dangerous carbon monoxide gases outdoors.

Shocks

- When working on a car in which the motor need not be in operation, disconnect the battery and insulate the connections.
- Check extension cords to see that they are well insulated.
- Check portable electric machines to see that they are grounded properly.

Lifting

- Lift with the legs, not the back. Use hoists wherever possible. Get help on heavy lifting.

Cuts

- Be careful when removing bulbs and

lenses from the headlights. Use proper tools to prevent cuts in case of breakage.

- Use great care in replacing broken glass in the car.

Moving Parts

- Do as much work as possible on the engine when it is stopped.
- Do not lubricate the engine while it is in motion.
- Do not use a wiping cloth on moving parts of the engine.
- Do not get the hands in door jambs when cleaning windshields or when performing other similar tasks.

Hoists

- Always use a safety device on a hoist designed to keep it from falling.
- Make sure the car is out of gear, with the motor turned off, but the brakes set.
- Check the wheel blocks before lifting.
- Keep from under the hoist while the car is being raised or lowered.

Lubricating Rack

- Have frequent inspections of the air hose. Replace worn hose because it may burst under pressure.
- Allow no fooling with air hoses or with high pressure lubrication equipment. Air hoses are dangerous when used in horseplay or for dusting off clothing or hair.
- Do not allow students to point the gun at any person even though it is grease clogged. High pressure lubrication equipment can cause injury to face or body.

Tire Inflation

- Check pressure gauges frequently to see that they are working properly.
- Never face a tire while it is being inflated. Excessive pressure may cause the tire to explode.

Unsafe Tools

- Have the person at the tool crib inspect all tools as they are issued and as they are returned. Unsafe tools are not allowed in the shop.
- Use goggles on all jobs that have an eye hazard, such as grinding or chipping.
- Have students report any unsafe tool or equipment to the instructor at once.

Horseplay

- Do not allow running, scuffling, or throwing tools and materials in the shops. Thoughtless acts such as these can result in serious injury.

Some General Rules for Auto Shops

- Have keys removed from all cars in the school shop and turned over to instructor.
- Allow only the instructor to start the engine in any car in the shop.
- Allow only one student to work on the inside of a car at a time.

REVIEW QUESTIONS

1. Explain the job of the rocker arms.
2. Explain why correct engine timing is essential.
3. The fuel pump is located between what two parts of the fuel system? Explain the operation of the fuel pump.
4. Why are the idle system, power system, and accelerator system necessary on a carburetor?
5. What is the key part of the automatic choke?
6. What parts make up the exhaust system?
7. List the main parts of the ignition system.
8. Explain the difference between full-flow and bypass lubrication.
9. Why is a ventilation system necessary?
10. Why is the differential necessary?
11. On many automatic transmissions, what replaces the conventional clutch?
12. What gears are the heart of the automatic transmission? List their parts.
13. What is a hydraulic torque converter?

Unit 10 Diesel Engines

OBJECTIVES After completing this unit, the student should be able to:

- Explain how a diesel engine operates.
- Discuss the advantages of the diesel.
- Explain fuel injection and how it works.
- Discuss supercharging and turbocharging.

For centuries, the natives of the South Pacific have employed an ingenious method for lighting fires. Using a bamboo cylinder in which dry tinder is placed, a close-fitting plug is driven sharply into the open end. The air entrapped in the cylinder is compressed and heated to the point where the tinder ignites. This spontaneous combustion from the heat of compression is the principle of *compression ignition,* the basis on which the diesel engine operates.

The characteristic, then, of the diesel engine which distinguishes it from other internal combustion engines in the **absence of any external ignition device**, figure 10-1. The gasoline engine uses an outside spark source to ignite the fuel mixture. This type of engine is thus known as the *spark-ignition engine.* The diesel, however, uses only the heat generated by the compression of air in the cylinder to ignite the fuel. Thus, the diesel is classified as a *compression-ignition engine.*

The inventor of the diesel engine, Dr. Rudolph Diesel, was born in 1858 of German parents. He became interested in engines during his college years from which he graduated when he was 21 years old with an engineering degree. He then immediately set out to build an engine that would be more efficient than the low-efficiency steam engines of the time. His first engine was patented in 1892, but when built, it exploded as he tried to start it. The tragedy did not cause him to give up. In 1897, Dr. Diesel produced a successful single-cylinder engine capable of developing 25 horsepower (18.6 kilowatts).

The first commercial diesel engine was put into service a year later in St. Louis, Missouri. It was a two-cylinder, 60-horsepower (44.8 kilowatts) engine. The use of diesel engines spread rapidly, in a few years, thousands of engines were in use.

The first engines were heavy, weighing as much as 250 pounds per horsepower (152 kilograms per kilowatt), were used mostly to replace steam engines in large installations such as stationary power plants or on large boats. They could not compete with the gasoline engine used on land vehicles where weight was a more important factor. Today, because diesel engines are much lighter in weight—as little as 7 pounds per horsepower (4.3 kilograms per kilowatt)—they are commonly found on all types of trucks, heavy road machinery, railroad locomotives, and in some automobiles.

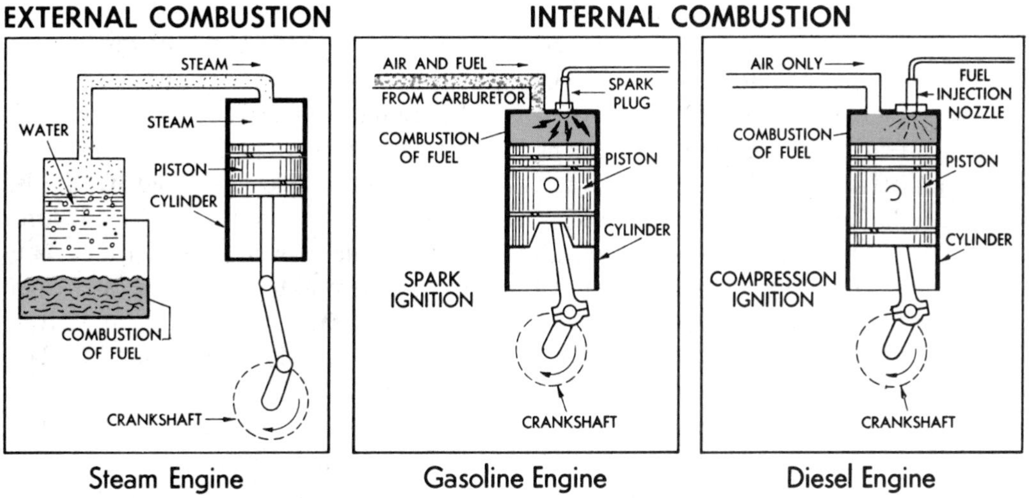

EXTERNAL COMBUSTION

INTERNAL COMBUSTION

STEAM
WATER
STEAM
PISTON
CYLINDER
COMBUSTION OF FUEL
CRANKSHAFT

Steam Engine

AIR AND FUEL
FROM CARBURETOR
SPARK PLUG
COMBUSTION OF FUEL
PISTON
CYLINDER
SPARK IGNITION
CRANKSHAFT

Gasoline Engine

AIR ONLY
FUEL INJECTION NOZZLE
COMBUSTION OF FUEL
PISTON
CYLINDER
COMPRESSION IGNITION
CRANKSHAFT

Diesel Engine

Fig. 10-1 Comparison of combustion systems

A major difference between diesel and gasoline engines is the *compression ratio* for which each engine is designed. Compression ratio is defined as the ratio of the cylinder and clearance volume with the piston on bottom dead center to the clearance volume alone with the piston on top dead center, figure 10-2.

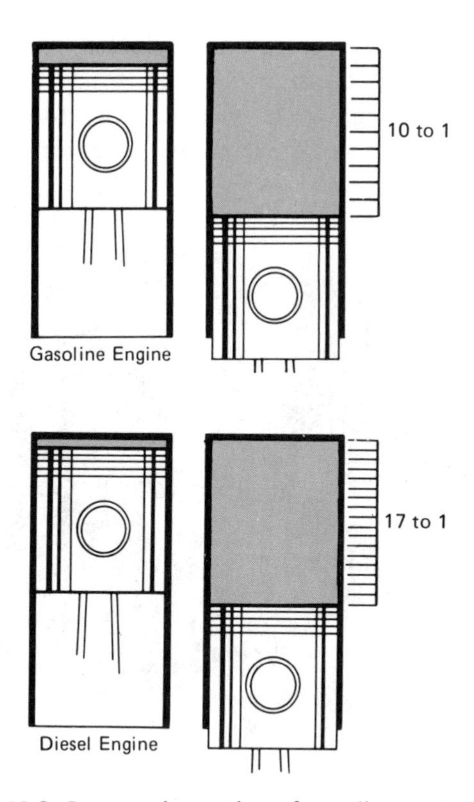

Fig. 10-2 Compression ratios of gasoline and diesel engines

Diesel engines are all high-compression engines, usually in the range of 17:1 versus 8.5:1 for gasoline engines. Compression ratios in the area of 17:1 are not possible in a carburator-fed gasoline engine even using the best anti-knock gasoline since a mixture of air and gasoline would ignite long before the piston reached top dead center because of the intense heat caused by the high compression.

The high compression ratio is what gives the diesel its efficiency advantage, because more of the potential energy in the fuel is put to work. When the stored energy of a fuel is released or burned, in other engines, all of the energy does not do useful work. Much of the heat energy is lost either in waste exhaust gases, to friction, or in heating up the engine. That is why, of the common engine types, the diesel engine is the most efficient. The simple steam engine has an efficiency of 6 to 8%. The steam turbine and condensing engine has an efficiency of 16 to 33%. The gasoline engine has an efficiency of 25 to 32%. The diesel engine has an efficiency of 32 to 38%. Remember that, the more efficient the engine, the less fuel it will have to consume to do the same amount of work.

The construction of gasoline and diesel engines is similar in many respects: crankcase, crankshaft, cylinder, piston, etc., figure 10-3. The main differences between the engines lie in four points: (1) how the air enters the cylinder, (2) how the fuel enters the cylinder, (3) how the fuel is ignited and, (4) how the exhaust gases leave the cylinder.

The diesel can be either a four-stroke or a two-stroke cycle engine. Therefore, the differences also depend on which engine is discussed.

In a two-stroke cycle diesel the air enters the cylinder by being blown in through intake ports that are located near the bottom of the cylinder. The fuel enters the cylinder at the top of the compression stroke through a fuel injector, and is ignited by the high temperature of the compressed air. Upon completion of the power stroke, the exhaust gases leave when exhaust valves are opened. At about the same time, new air is forced into the cylinder scavenging out the last bit of exhaust.

The system illustrated in figure 10-4 is a uniflow system with intake ports all around the cylinder and an exhaust valve at the top of the cylinder. One side of the cylinder has ports for incoming air while the other side has ports for the exit of the exhaust. The principle used is the same as that for small two-stroke cycle gasoline engines that use loop or cross flow scavenging. Hot exhaust gases leave the combustion chamber on their own, then a new charge of air rushes in behind to scavenge

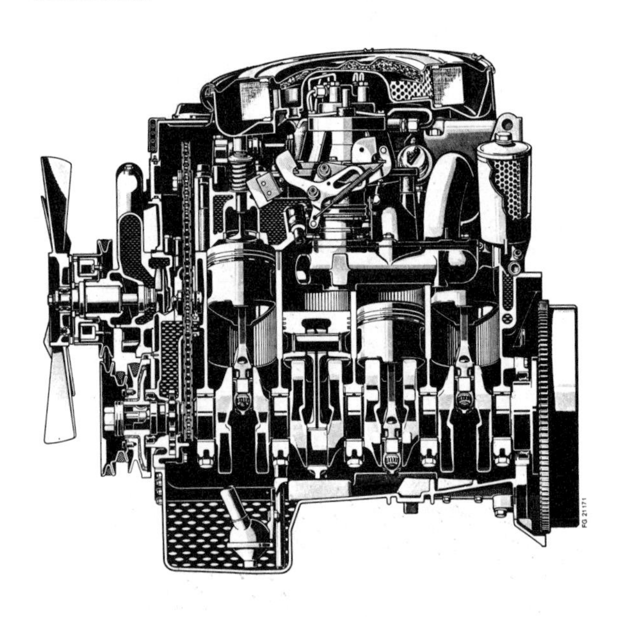

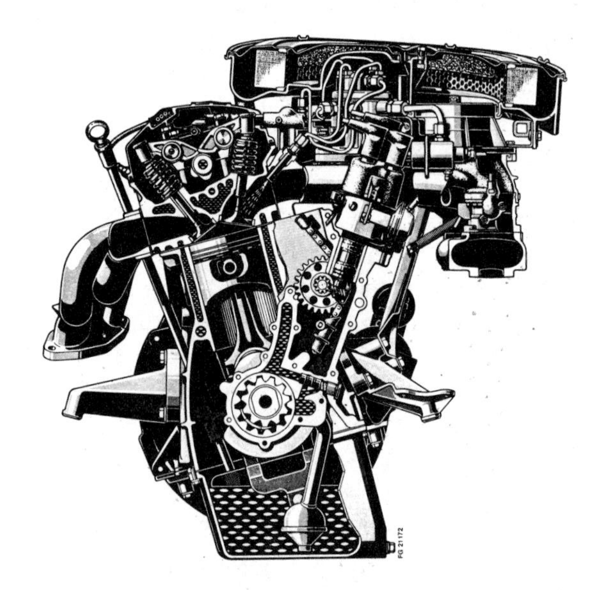

Fig. 10-3 Cross-section view of diesel engine for automative use (Courtesy of Mercedes-Benz of North America, Inc.)

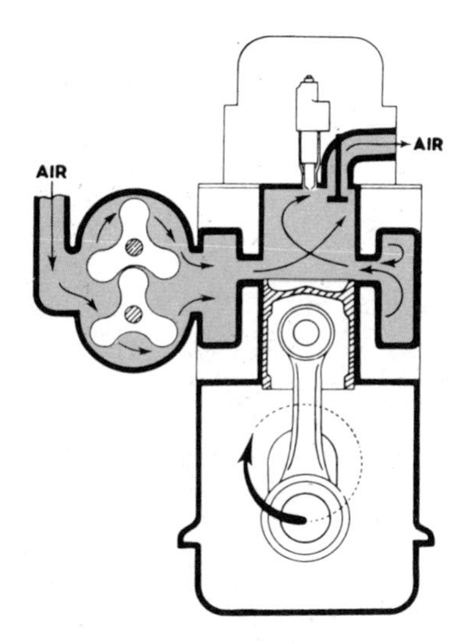

AIR ENTERING THROUGH PORT TO COMBUSTION CHAMBER

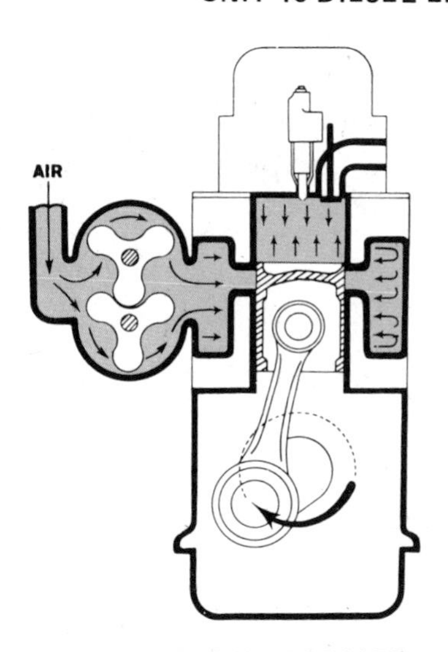

AIR BEING COMPRESSED WITH EXHAUST VALVE CLOSED

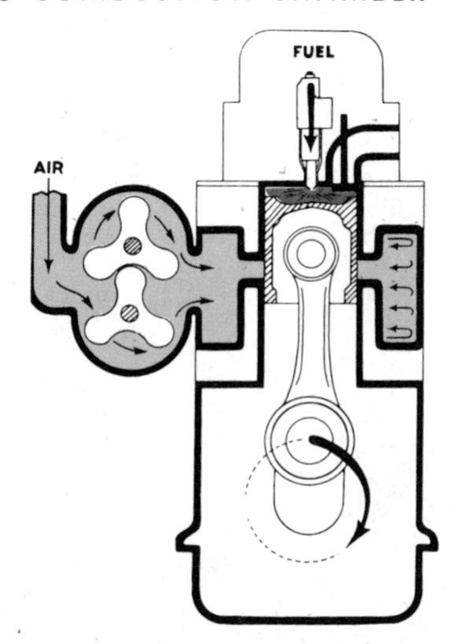

CHARGE OF FUEL BEING INJECTED INTO COMBUSTION CHAMBER

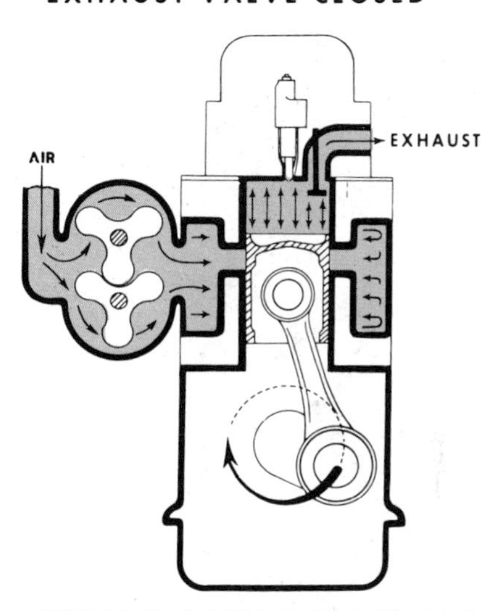

EXHAUST TAKING PLACE AND CYLINDERS ABOUT TO BE SWEPT CLEAN WITH SCAVENGING AIR

Fig. 10-4 Two-stroke cycle diesel engine

out the last of the exhaust. Many two-stroke cycle diesels are loop or cross flow scavenged.

In a four-stroke cycle diesel, the air enters the cylinder through an intake valve. Sometimes there are two intake valves per cylinder to increase airflow, figure 10-5. On most modern four-stroke cycle engines, a turbocharger brings large quanitiies of air into the cylinder. However, the air does not pass through a carburetor, figure 10-6. Fuel is

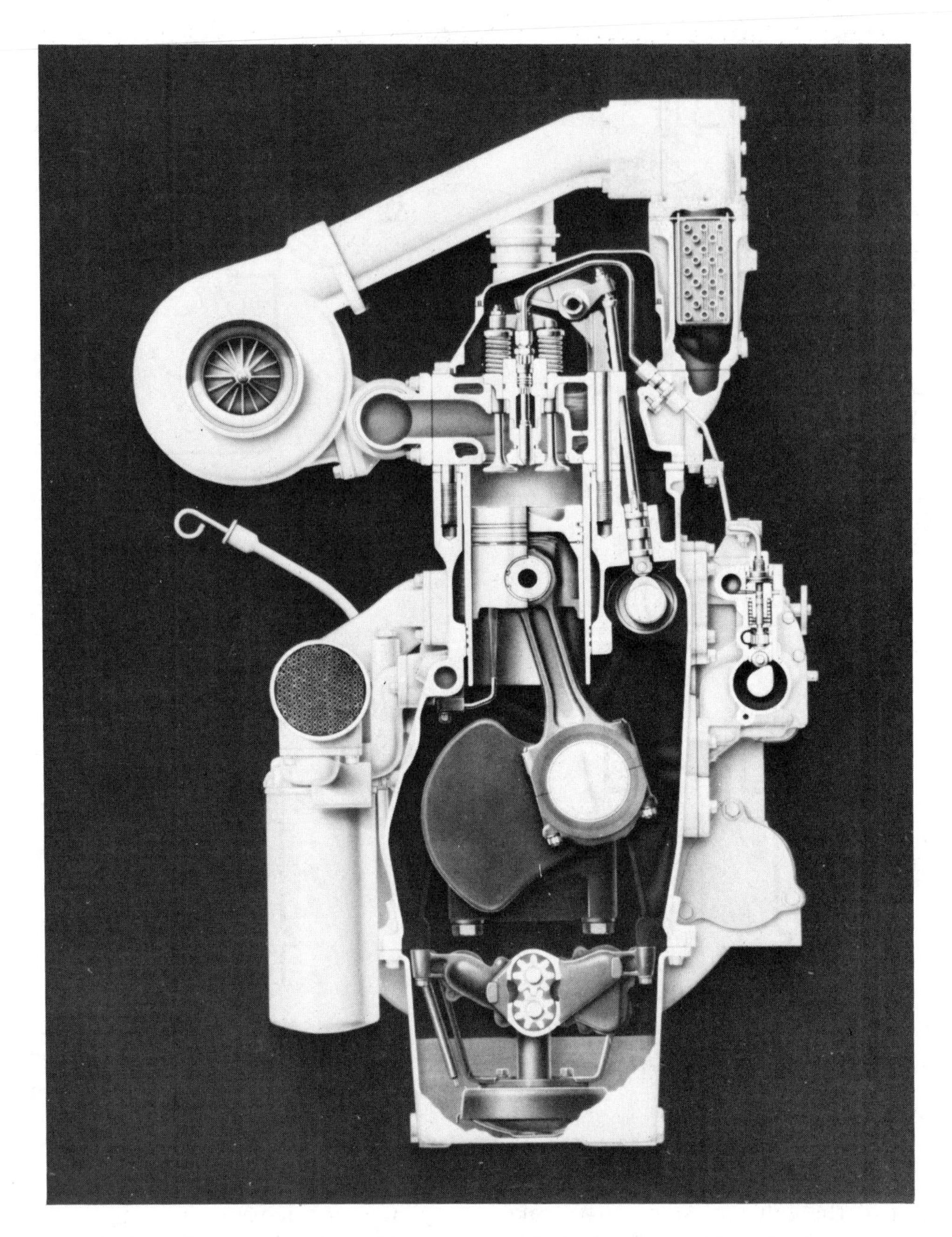

Fig. 10-5 Cross-section view of four-stroke cycle, in-line diesel with turbocharger

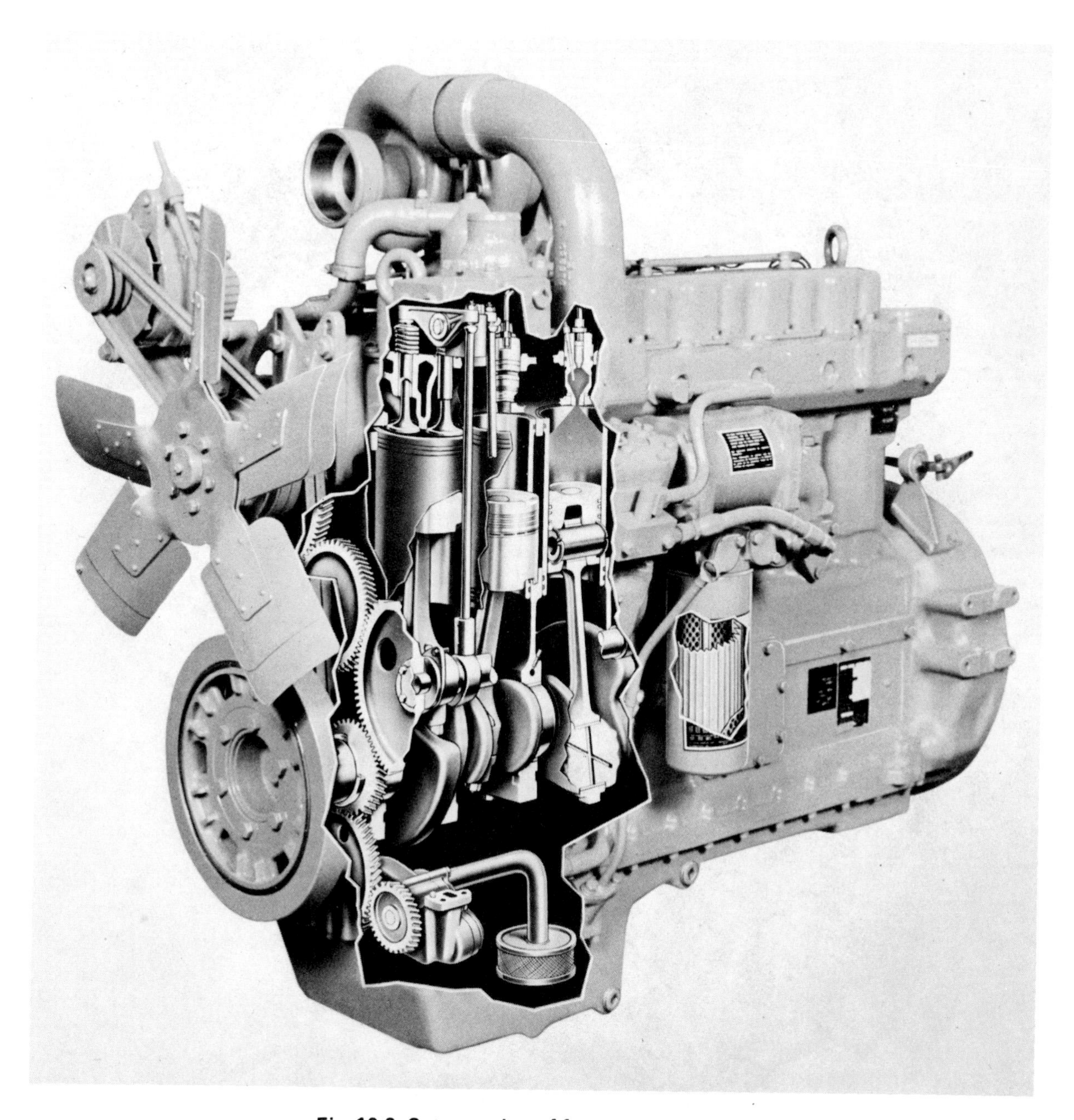

Fig. 10-6 Cutaway view of four-stroke cycle diesel

Fig. 10-7 Four-cylinder, 2.0 liter, turbocharged diesel engine. Note the fuel injectors (Courtesy of Ford Motor Company)

introduced in the cylinder when the piston reaches the top of the compression stroke, through the tip of a fuel injector, figure 10-7. The fuel is ignited by the high temperature of the compressed air. The exhaust gases leave as they would in a gasoline engine, being pushed out by the upward motion of the piston on the exhaust stroke and by the scavenging effect of the new incoming air.

FUEL INJECTION

The fuel injector eliminates the need for a carburetor on the diesel. With the air tightly compressed and at a high temperature, the fuel injector sprays diesel oil into the cylinder as a mist or fog, figure 10-8, and begins burning immediately upon contact with the superheated air.

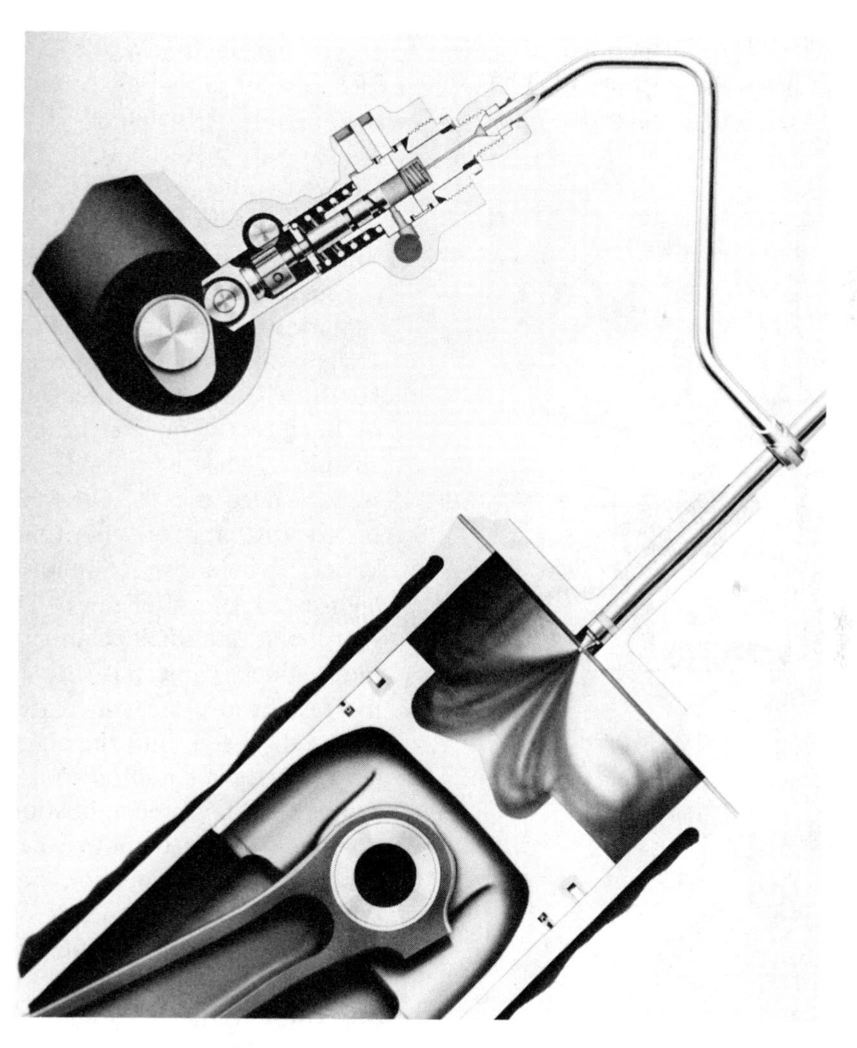

Fig. 10-8 Direct or open charging showing injector and injection pump

The fuel injector parts are machined to extremely exacting tolerances, figure 10-9. The unit is a precision machine in itself. For example, on a unit injector, the injection pump piston is fitted to the pump cylinder to an accuracy of 30 to 60 millionths of an inch (762 to 1524 micrometers). Such accuracy is necessary since the fuel is forced through the injector nozzle in a split second and at a tremendous pressure. The fuel may be forced through the several needlepoint-sized holes at the nozzle tip with a pressure of 3000 to 30,000 pounds per square inch (20.7 to 207 megapascals). There cannot be leakage between the plunger and cylinder. The injector must meter out the exact amount of fuel needed. Usually each cylinder of the engine is fitted with its own fuel injector. Sometimes each cylinder is equipped with its own fuel pump, however, fuel pumps can be made to supply several injectors in much the same way that a car's distributor channels high voltage to the proper cylinder at the correct instant.

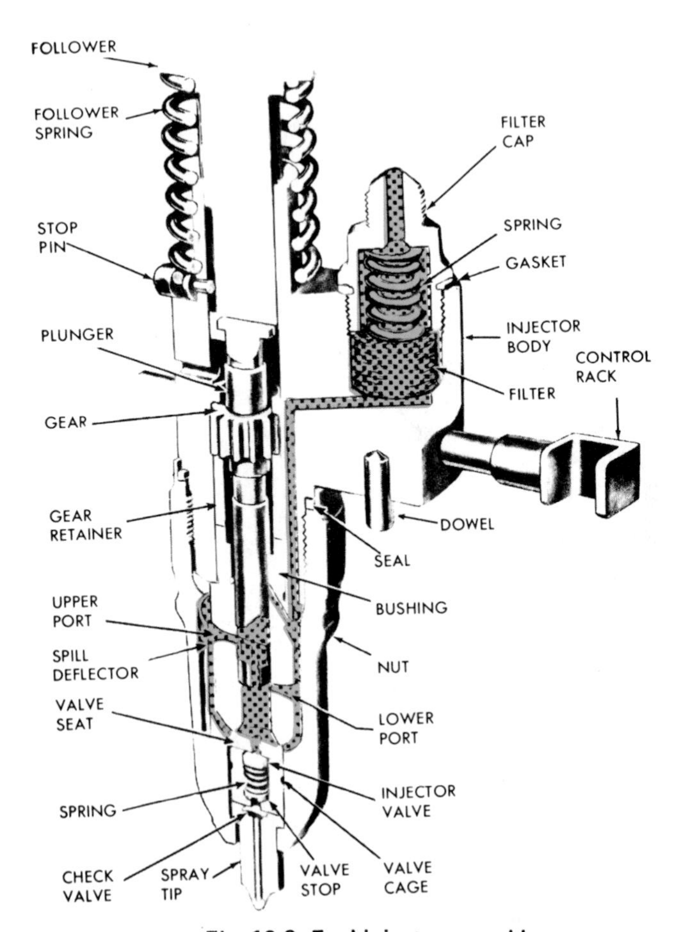

Fig. 10-9 Fuel injector assembly

Another factor in the complexity of the fuel injector is that as the engine speed varies, the amount of fuel sprayed from the injector is changed; just the right amount of fuel is metered out for the desired speed. The amount of air taken into the cylinder and compressed remains the same throughout the speed range, however.

Various types of combustion chamber designs are used in diesel engines, figure 10-10. The *open* or direct design allows fuel to be directly injected into the combustion chamber. This is suitable for all large and slower speed diesels. Other designs are being tested to improve smoothness and prevent knock, which are common problems with high-speed diesel engines. The *auxiliary* design includes a small chamber to the main cylinder. Fuel is injected into the chamber and the restricted passage increases the turbulence of the discharge into the main chamber. Similarly, in the *precombustion* design, burning begins in the precombustion chamber and explodes violently into the main cylinder for complete combustion, figure 10-11. The *air cell* design is a means of increasing the force of the combustion. The energy cell, with its chamber opposite from the fuel injector, produces a smooth buildup of pressure in the main chamber rather than a violent burst.

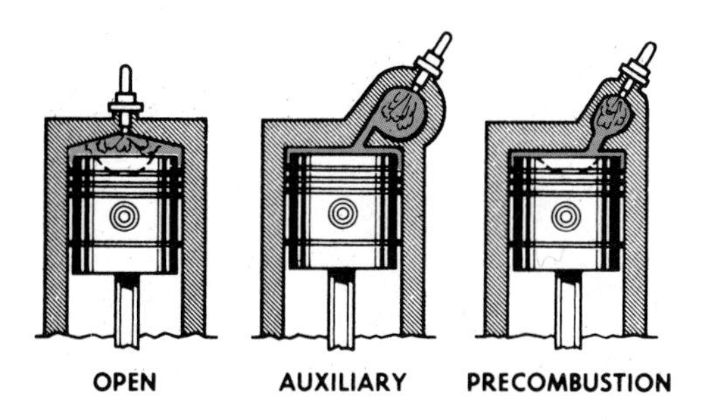

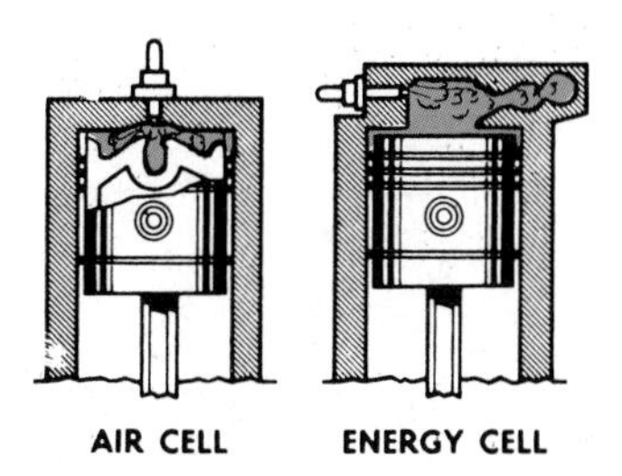

Fig. 10-10 Combustion chamber designs

SUPERCHARGING

The more air that can be packed into the cylinder, the better. Superchargers supplement the pumping action of the pistons in four-stroke cycle engines and provide rapid air movement in two-stroke cycle engines, greatly increasing the efficiency of both.

One common type of supercharger is the Roots blower, figure 10-12. It consists of 2 three-lobe, intermeshing spiral gears which trap air and force it to the opposite side. It turns constantly and pumps air under pressure to the cylinder. Its action might be compared to a revolving door.

Turbocharging is also very common, figure 10-13. The turbocharger is a centrifugal air pump driven by the exhaust gases of the diesel, figure 10-14. This device does not rob power from the engine in order to operate, but uses energy that would otherwise be wasted. Turbocharging can increase the power de-

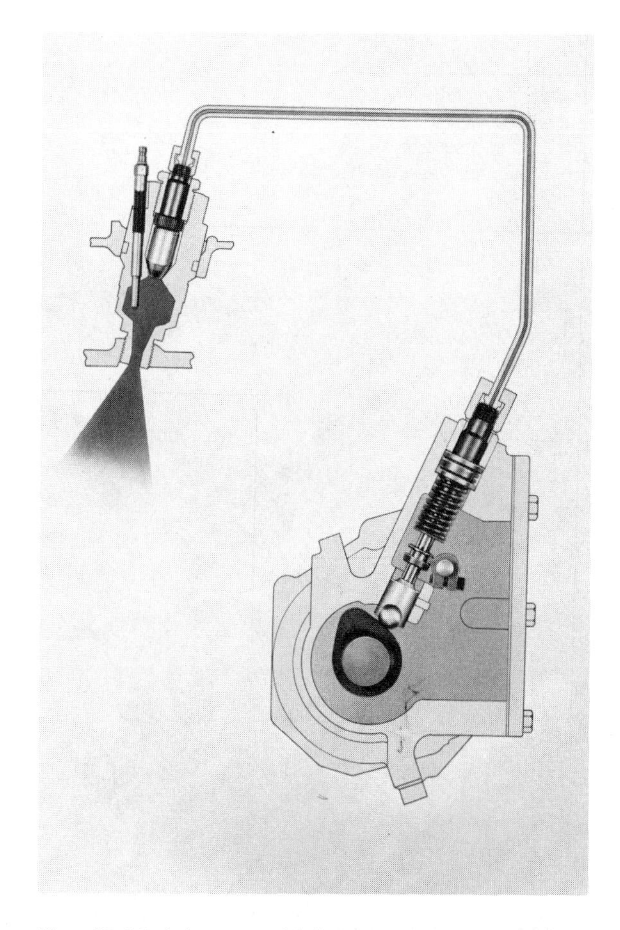

Fig. 10-11 Injector and injector pump using precombustion design

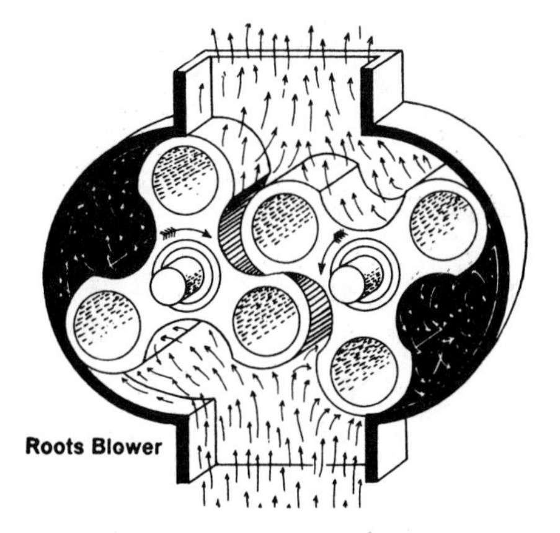

Fig. 10-12 The Roots blower

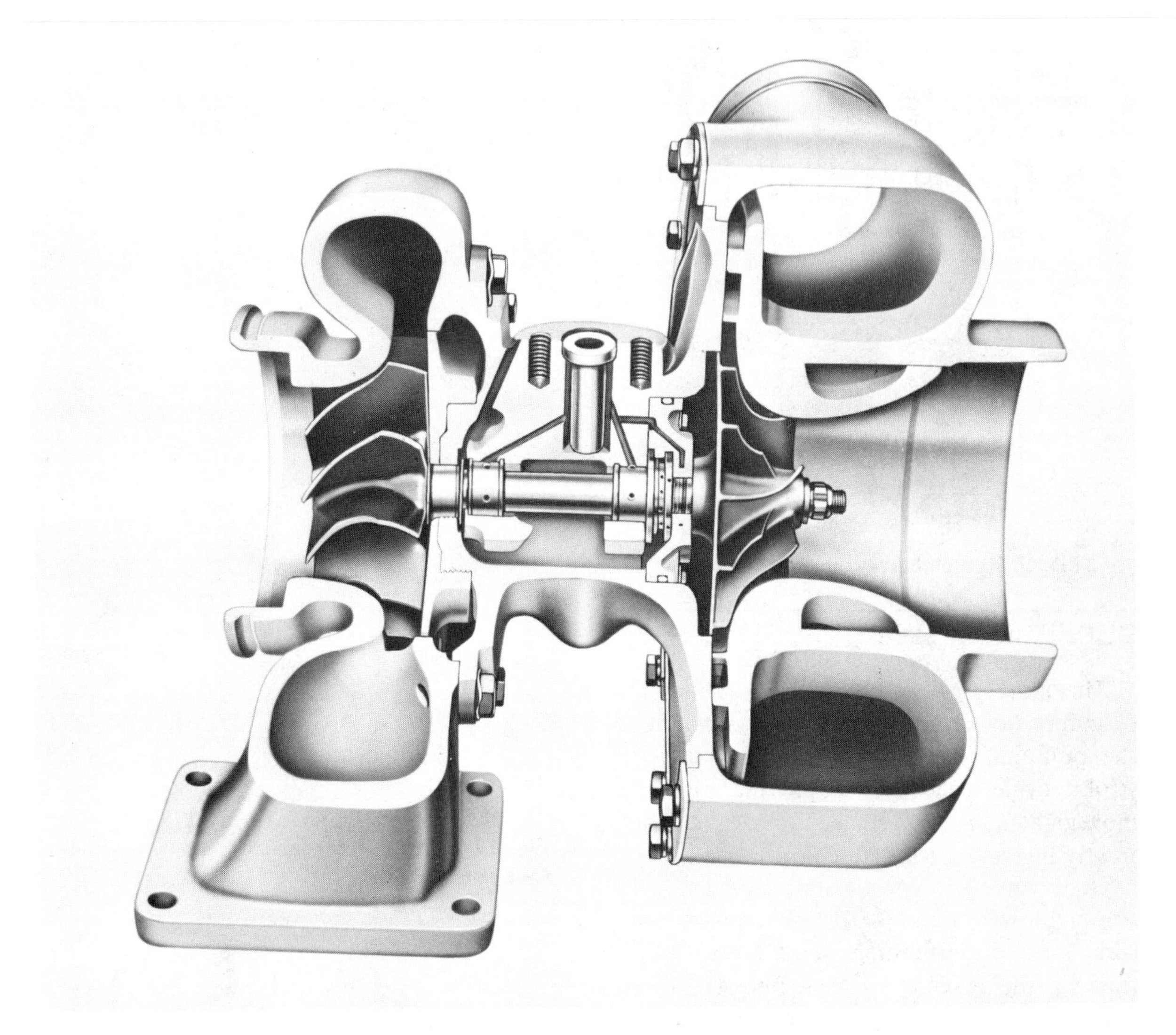

Fig. 10-13 Cutaway of turbocharger shows exhaust-driven turbine on left and centrifugal blower on right

veloped by an engine by 50 to 150%. Supercharging and turbocharging have done much to reduce the size, weight, and cost of diesel engines.

Other Design Features

High speed diesels have a crankshaft speed range of 1200 to 4000 revolutions per minute, and can develop 482 horsepower (360 kilowatts) at 2100 revolutions per minute. To save space and keep them light in weight, most of these engines are designed using the "V" block, figure 10-15. They are used in farm tractors, highway vehicles, and earth moving equipment.

Medium speed diesels have a speed range of

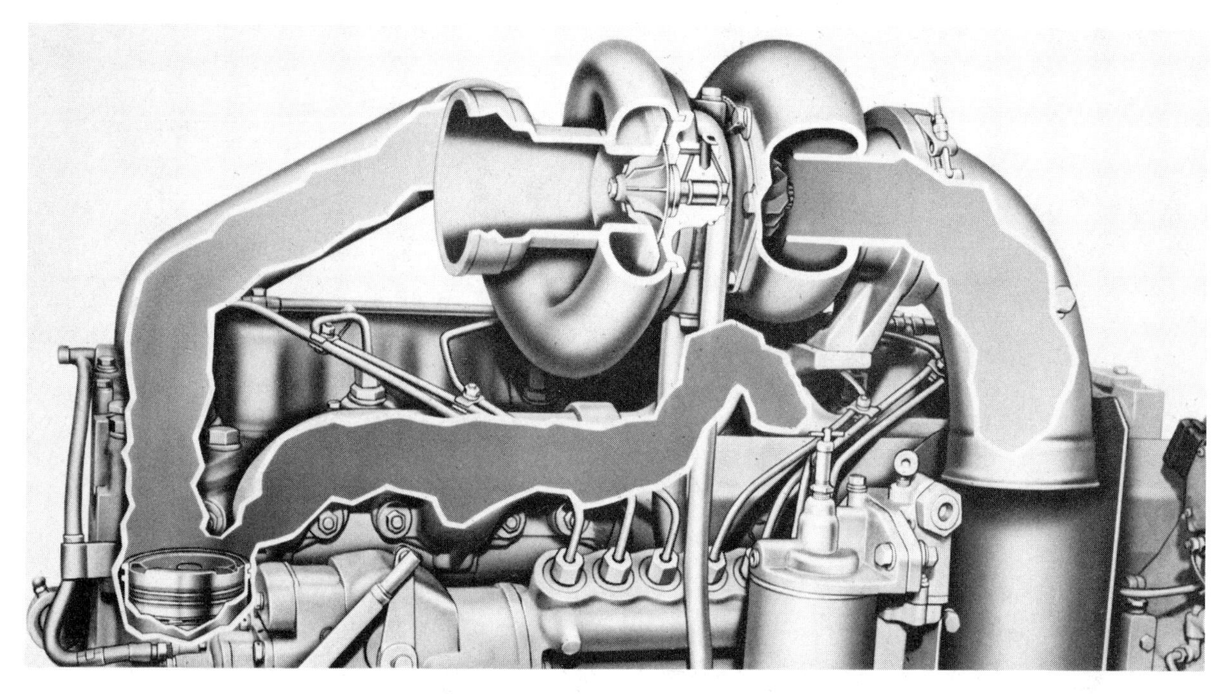

Fig. 10-14 Turbocharger mounted on engine

500 to 1200 revolutions per minute and will be 2000 horsepower (1.5 megawatts). They are used mostly for locomotives, ships, and stationary equipment. Cylinder bores range from 8 to 9 inches (203 to 229 millimeters) in diameter.

Low speed diesels operate as low as 115 revolutions per minute, developing up to 12,500 horsepower (9 megawatts). These huge engines, with cylinder bores measuring up to 29 inches (737 millimeters), are used in power plants to drive electric generators.

It is common for people to think of diesel fuel as being cheap or of low quality. Nothing could be further from the truth. It is distilled petroleum fuel oil produced in much the same manner as gasoline. The quality of the fuel is high and it is produced to exacting standards. It must burn leaving little or no deposit, be very clean, and ignite properly. Surprisingly, the price of diesel fuel and gasoline is almost the same at the refinery. Any cost difference may be due to greater taxation on one or the other. The economy of operating a diesel does not come from burning cheaper fuel, rather it comes from burning the fuel in a very efficient engine.

It will be helpful to follow one cycle of a two-stroke cycle diesel engine in order to understand this fact better, figure 10-17. At the top of the compression stroke, the air is tightly compressed and held at a temperature of 1000°F (538°C). Now the fuel injector sprays an oil fog into the combustion chamber. The oil begins burning immediately and the extreme pressures push the piston rapidly back down the cylinder. When the piston nears the bottom of its stroke, two things happen: (1) intake ports are uncovered and air is blown into the cylinder, (2) the exhaust valves

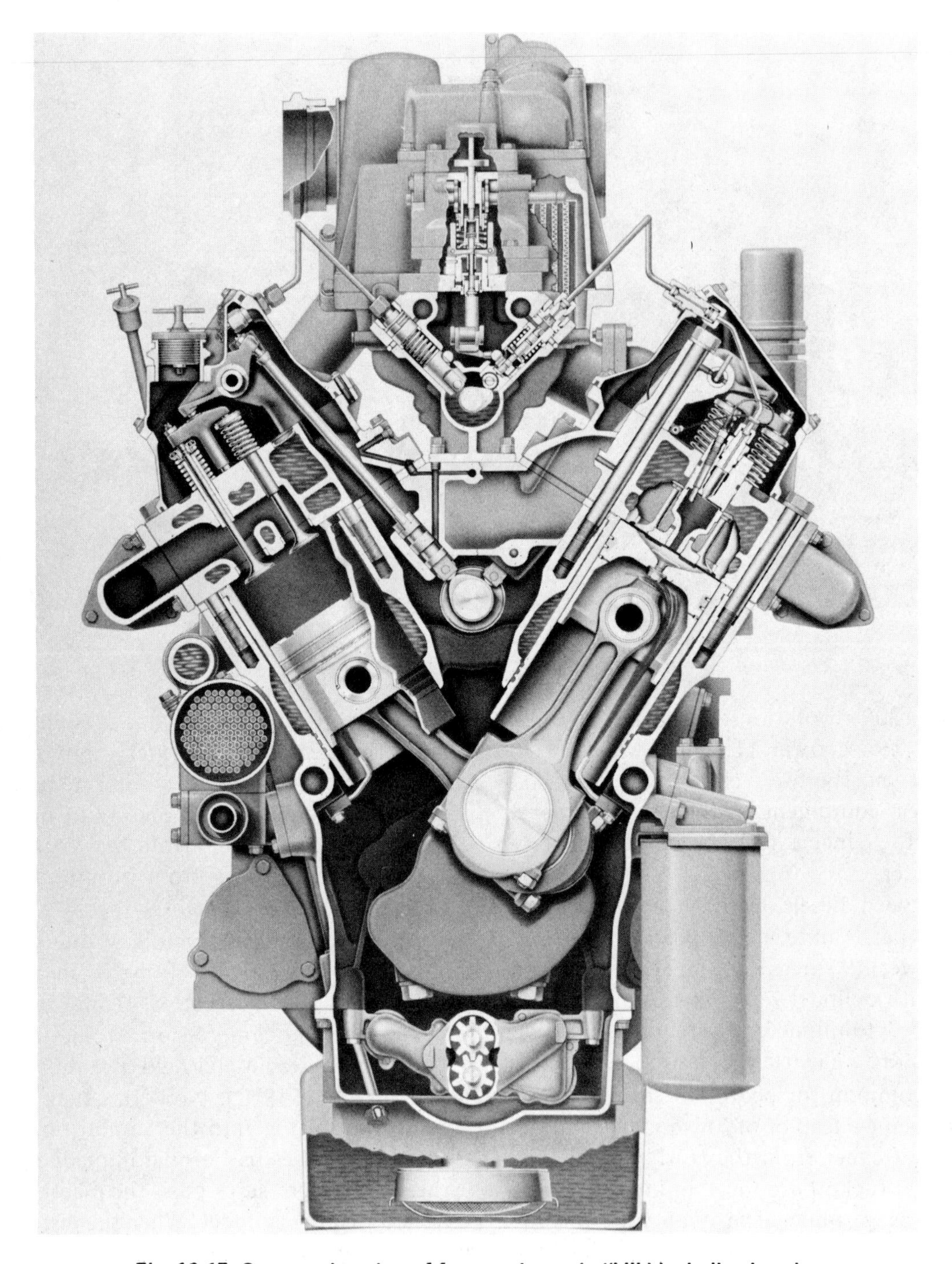

Fig. 10-15 Cross-section view of four-stroke cycle "V" block diesel engine

Fig. 10-16 International DT-446C turbocharged, in-line, six-cylinder medium diesel engine for highway use (Courtesy of International Harvester)

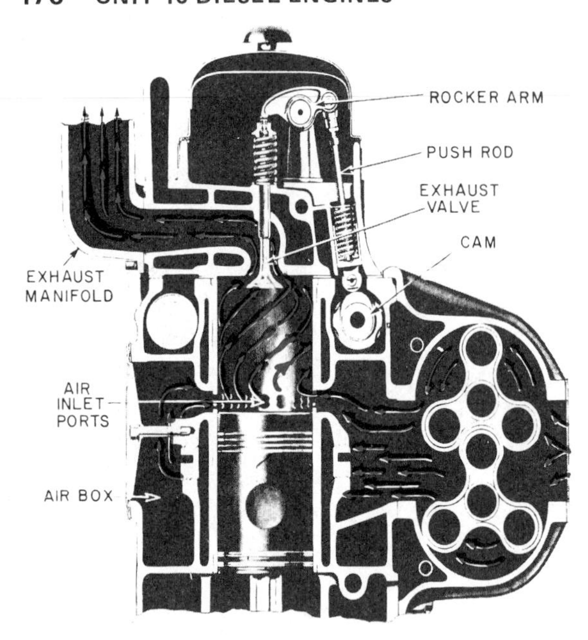

Fig. 10-17 Intake and exhaust system of two-stroke cycle diesel

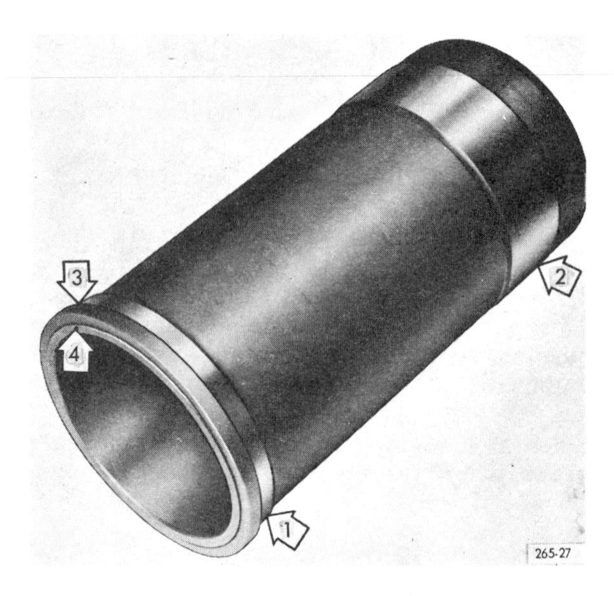

Fig. 10-18 Diesel cylinder sleeve can be removed from the engine block

located at the top of the cylinder open, allowing exhaust gases to leave. The last amount of exhaust is scavenged from the cylinder by the incoming air. Exhaust valves now close, the piston moves upward sealing off the intake ports, and the air charge is trapped and compressed.

Since the diesel requires heat to ignite the fuel mixture, it is a difficult engine to start in very cold weather. There are several methods used to pre-heat the cylinder or make the fuel easier to ignite:

- Glow plugs are used to heat the combustion chamber.
- Special electric heaters warm the oil or water before starting.
- Ether is even injected into the air intake to improve the burning of the fuel.

Because it is difficult to start when cold, it is not unusual to leave a diesel truck running

Fig. 10-19 Long hours under heavy load do not strain the diesel (Courtesy International Harvestor Company)

even if it is not to be used again for several hours, or even overnight.

The diesel is also difficult to stop once it is running. A gasoline-powered engine has an ignition system which can be shut off or shorted out to stop it; the diesel does not. Instead, it is stopped by a decompressing device that holds open the intake valves to prevent the pistons from building up enough pressure to ignite the fuel. Although the engine could be stopped by shutting off the fuel supply, this would only drain the system and leave no fuel in the pump or injectors for restarting.

Diesel Uses

The world of the diesel is constantly expanding. One of the most important applications of the diesel is the diesel locomotive; this has made the steam locomotive nearly obsolete. Diesel engines for boats, trucks, road machinery, buses, offshore oil drilling rigs, and transportable electric power units are other applications where the diesel is very effective. It is an engine known for durability and ease of maintenance, figure 10-18. Because of its slower speeds, it can operate continuously for many hours without requiring major repairs, figure 10-19. The diesel is enjoying expanded popularity as an automobile engine because of its fuel efficiency, figure 10-20, and its ability to meet government pollution control standards with little difficulty, figure 10-21.

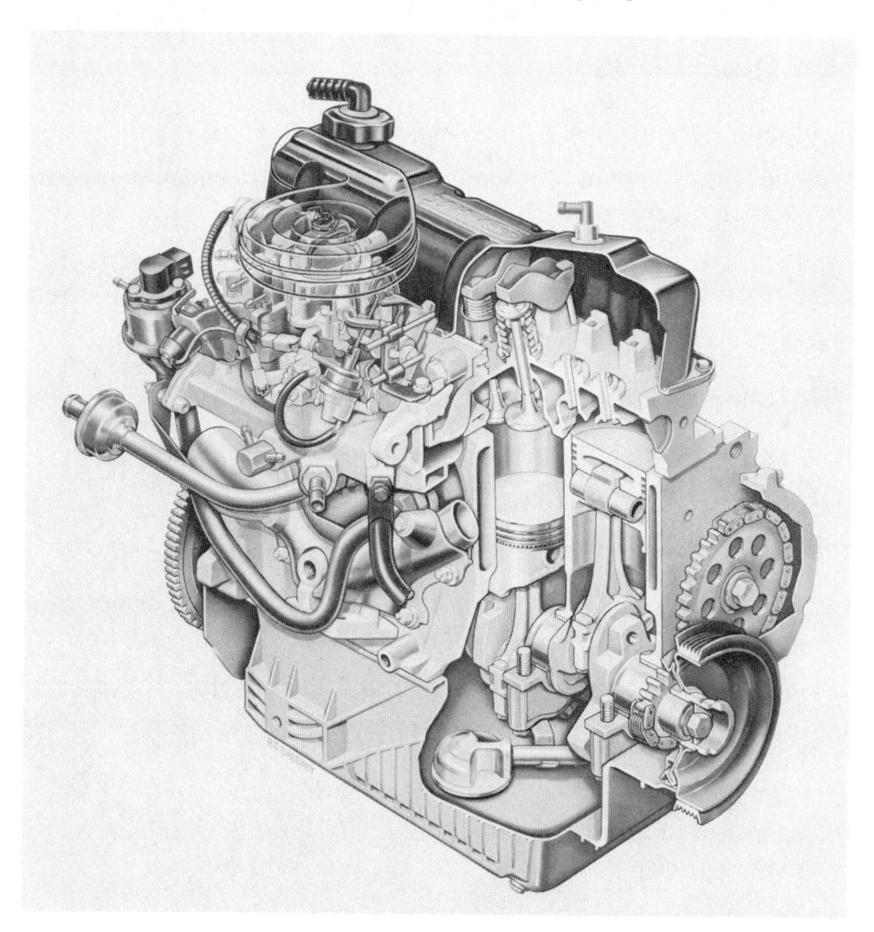

Fig. 10-20 Four-cylinder, 2.3 liter diesel engine for automative use (Courtesy of Ford Motor Company)

Fig. 10-21 Diesel-powered automobile (Courtesy Mercedes-Benz of North America Inc.)

REVIEW QUESTIONS

1. What were early diesel engines used for?
2. Why does the diesel produce more power with a given amount of fuel energy than other engines?
3. What does the diesel's high compression ratio have to do with its success?
4. Why doesn't the diesel use a carburetor?
5. What temperature does the air reach under compression?
6. What causes the fuel to ignite in the combustion chamber?
7. How does the fuel get into the cylinder?
8. What is the purpose of an air blower on a diesel engine?
9. When engine speed is changed, what is varied: fuel, air, or both fuel and air?
10. Explain why diesel fuel is not cheap or low-grade fuel.

Unit 11 Jet Engines

OBJECTIVES After completing this unit, the student should be able to:

- Explain the principle of jet propulsion.
- Compare rocket and jet engines.
- Identify and explain the operation of:
  A. ramjets
  B. pulsejets.
  C. turbojets.
  D. turboprops.
  E. turbofans.

Most of us think of jet propulsion as a new idea, one developed during this century. While the most notable application of jet propulsion, the jet aircraft, figure 11-1, has come about during the past few decades, the theory and use of jet propulsion stretches back for

Fig. 11-1 The jet engine powers modern commercial and military aircraft. (Photo courtesy of Electric Power Research Institute)

centuries. Hero of Alexandria invented the aeolipile 2000 years ago, figure 11-2. His machine illustrates the *principles* of jet propulsion.

The aeolipile is a covered vessel containing water. Two tubes mounted on top of the vessel act as an axis for a hollow sphere. On the sphere are two nozzles with their opening 90° to the axis of rotation and opposing each other. A fire built below the vessel boils the water and the steam travels up the tubes, into the sphere, and out the nozzles. The pressure of the escaping steam causes the sphere to rotate rapidly. The idea was regarded as a novelty, a toy, and was destined not to be developed into a useful machine for centuries.

Newton's third law of motion explains why a jet engine works, and why Hero's aeolipile spins. This law states: For every acting force there is an equal and opposite reacting force. In the aeolipile, the force of the escaping steam causes an equal and opposite reacting force in the spinning of the sphere.

A toy balloon also illustrates the jet principle, figure 11-3. Blown up and held closed, the balloon has no motion. Pressure inside the balloon is equal in all directions. However, when the balloon is released, its elasticity causes the air to be forced out rapidly. This action is accompanied by an equal but opposite reaction that can be seen as the balloon is propelled forward.

Fig. 11-2 Hero's aeolipile illustrates the principles of jet propulsion

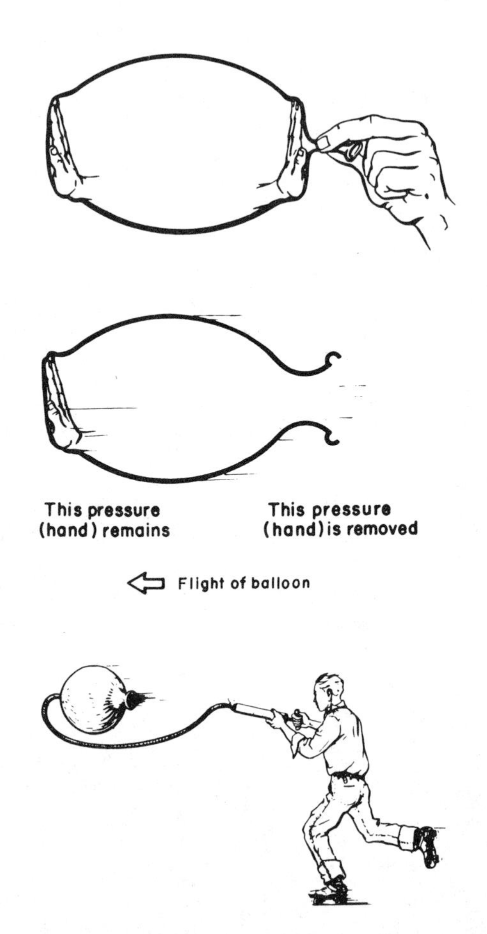

Fig. 11-3 A toy balloon also illustrates jet propulsion principles

In a jet engine, tremendous pressures build up inside the engine during combustion. This pressure is exerted in all directions, but the only possible escape route is out the rear of the engine. In simple terms, the action of the gases rushing rearward produce the reaction of forward motion.

The rapid ejection of gases from inside the engine produces *thrust,* the driving force that pushes the engine forward. Jet propulsion power plants are, therefore, classed as reaction engines, figure 11-4. This type of engine produces a forward thrust by the forceful ejection of matter from within itself. Jet engine power is measured in pounds of thrust produced.

It might be noted that almost all applications of jet engines are in the aviation field: commercial and military aircraft, drones, among others. Speed is the answer to the jet's popularity. Today, travel to any corner of the earth is a matter of hours.

AIR REQUIREMENTS: ROCKET VS JET VS PROPELLER

All jets burn oxygen from the earth's atmosphere with their fuel. This makes it an *air-dependent engine;* it cannot leave the atmosphere and still operate. A rocket engine is an air-independent engine; the relatively thin layer of the earth's atmosphere does not limit its range. The rocket engine carries its own oxidizer for its fuel. These two things are really the main distinction between jet and rocket engines since they both operate on the same physical principles, action and reaction.

The air mass around the earth is dense close to the earth, and thins out as the altitude increases. Eventually, the earth's layer of atmosphere becomes nonexistent. The comparison is easier to understand if you think of

Fig. 11-4 A modern jet engine, General Electric's J79-11A (Courtesy of General Electric Company)

the air nearest the earth as being 100 and declining by ones the further away from the surface you get. The thinning atmosphere limits the altitude of the propeller-driven aircraft, because its propellers need to bite into atmosphere, the thicker the better; let us say its best operation is in the 100 to 70 range. The jet goes through this layer to the 70 to 30 range, at which time performance is maximum. The rocket, on the other hand, goes through both these ranges to its optimum performance area of 30 to 0 and beyond.

RAMJET

The ramjet is the simplest of the jet engines—it has no moving parts. Technically, it is an aero-thermodynamic duct. It is a cylinder, open at both ends, having a fuel injection system inside, figure 11-5.

Air enters the front end of the cylinder and burns with the fuel being sprayed into the combustion section. The action of the high pressure exhaust gases leaving the cylinder produces the thrust that pushes the jet forward.

As the name ram implies, air is pushed into the engine when the jet speeds through the air. Air is literally rammed into the open cylinder.

It is impossible for this jet to take off under its own power. It must be accelerated to high speed by an auxiliary power source before the ram effect can take over. That means that if the pressure of the incoming air is too small due to slow speed, the burning fuel and air will tend to blow out both ends of the jet. The pressure of the ram air must be enough to keep the combustion gases traveling out the rear only.

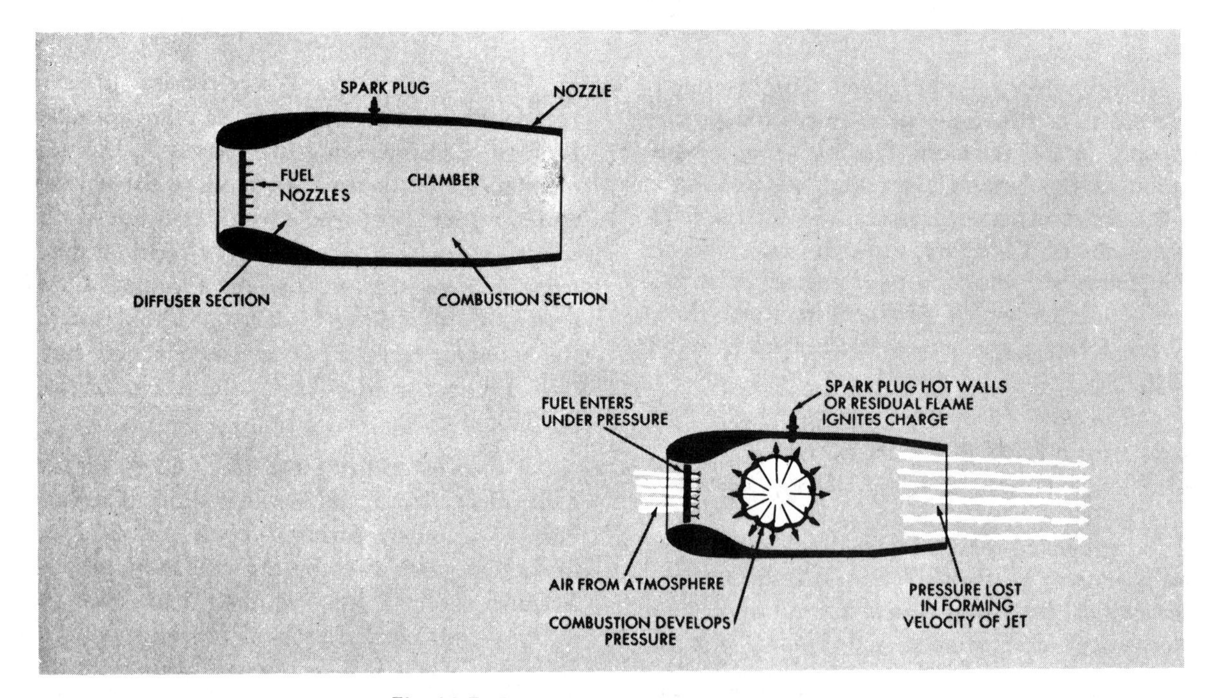

Fig. 11-5 Operating principles of the ramjet

The diffuser section is designed to get the correct ram effect. Its job is to decrease the velocity and increase the pressure of the incoming air. To do this, it is designed with a small entrance diameter sloping back to a larger diameter. This design increases the air pressure to the point where the combustion gases will not overcome the ram air pressure.

Fuel injectors spray fuel into the incoming air continuously. This incoming stream of fuel and air is then ignited by the hot gases in the combustion chamber. Initial starting is by a spark plug but once the engine starts there is no interruption of power. A flame holder toward the rear of the jet retards the gases so they will not be swept too far to the rear of the engine.

The ramjet's high power-to-weight ratio make it an excellent high altitude power plant. Its performance at high speed and high altitude is the best of the four jet engines. Its characteristics make it useful as an auxiliary power source mounted on the wing tips of high altitude jet aircraft. Also, it is suitable for jet helicopters, the ramjet being mounted on the rotor blade tips. The only problem it has is very high fuel consumption.

PULSEJET

The pulsejet was developed into an operational engine by the Germans during the 1930s. Their application of the engine was the V-1 or Buzz Bomb which was used primarily against England during World War II. The Buzz Bomb was unmanned and, once aloft, it maintained an altitude of about 10,000 feet (3000 meters) with a preset gyroscopic system and barometric unit. When the bomb reached target area, a measuring device cut off the fuel and the bomb went into a spiraling dive to the earth. Speed was 400 miles per hour (645 kilometers per hour). Although the bomb had many drawbacks, it was effective for bombing a large metropolitan area such as London.

During World War II a pulsejet missile was also developed by the United States, known as the JB-2. However, the war ended before it could be put into use.

The pulsejet is quite simple and economical to construct. Today, it is used primarily for drones and test vehicles. Most pulsejets need to be boosted into operation. Once in operation, their speeds are moderate, being limited by the design of the engine.

As the name implies the pulsejet has a pulsating power, that is, the combustion is broken and not continuous. The pulses of power are very close together, for example 200 combustions per second. Other jets have continuous burning of the fuel.

The main parts of the pulsejet, figure 11-6, are: the diffuser, the grill assembly, the combustion chamber, and the tailpipe.

The ram air comes into the engine through the diffuser section. This high velocity air travels into an expanding cross-sectional area, the result being to decrease the velocity and to increase the air pressure.

The grill assembly is made up of a number of flapper units or shutters. The flapper unit has spring steel strips similar to a reed valve on two-stroke cycle engines. These strips spring shut upon combustion. Ram air pressure causes the flapper unit to open, but when combustion takes place, the combustion pressures overcome the ram air pressure and the flappers spring shut. These flapper units are the key to the pulsating operation of the engine.

The grill assembly also contains the fuel

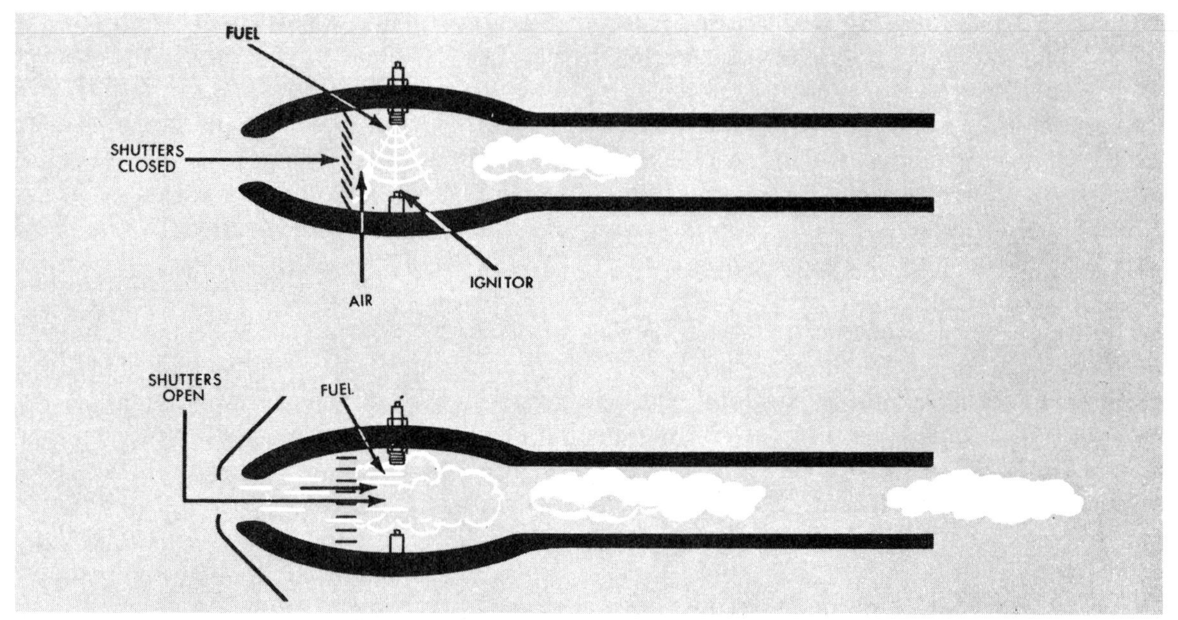

Fig. 11-6 Operating principles of the pulsejet

injectors and venturies, both located just behind the flapper units. The venturies insure proper mixture as fuel is sprayed into the airstream.

The combustion chamber provides the area for the burning gases to expand. The initial combustion is begun by a spark plug but, once the engine is operating, the spark ignition is shut off. Flame from the tailpipe is drawn back up to ignite each succeeding fuel charge.

The tailpipe section with its smaller cross section accelerates the exhaust gases. The length of the tailpipe determines the number of cycles per second; the shorter the tailpipe, the greater the frequency of cycles.

TURBOJET

The principle of the turbojet engine goes back several centuries, but active development of the engine began during World War II. Both England and Germany developed high-speed turbojet-powered fighters. Germany began work in 1936 and the world's first turbojet-powered aircraft was flown in 1939; it was the HE-178. A later German aircraft, the ME-262, was capable of speeds up to 500 miles per hour (805 kilometers per hour). Over 1600 were produced near the end of World War II.

Air Commodore Frank G. Whittle was the guiding force behind the English effort, where jet engine work ran much the same course as that in Germany. In 1936 design and development began and in 1937 the engine was ground tested. In 1941 England's first jet-powered aircraft, the *Gloucester Pioneer,* was flown. Its successor, the *Gloucester Meteor,* was the only allied jet aircraft to become operational during the war.

The United States entered the jet aircraft field in 1941 when the General Electric Company was awarded the contract to improve and produce the English Whittle engine. The jet fighter XP-80 was the first operational U.S. fighter. It was built by Lockheed and powered by a General Electric GE-J-33 engine. This aircraft, later designated the F-80 Shooting Star, was capable of speeds over 600 miles per hour (966 kilometers per hour).

In a short time, jet fighters replaced propeller-driven fighters among all major nations. The constant improvement of turbojet engines soon led the way to their use on bombers. Now bombers such as the B–52 are jet-powered. Perfected aerial refueling techniques along with engine improvements have made the intercontinental bomber a mainstay in our nation's defense structure.

Of the jet engine types the turbojet, figure 11-7, is the most complicated because of the inclusion of a gas turbine as a major engine component. With the turbine included, this engine can take off on its own power. Auxiliary boosters are not necessary as they are in the ram and pulsejets.

The air entering a turbojet goes through the following sections of the engine, figure 11-8: air inlet, compressor, combustion chambers, turbine, exhaust cone, tailpipe, and jet nozzle.

The compressor increases the pressure of the incoming air and delivers that air to the

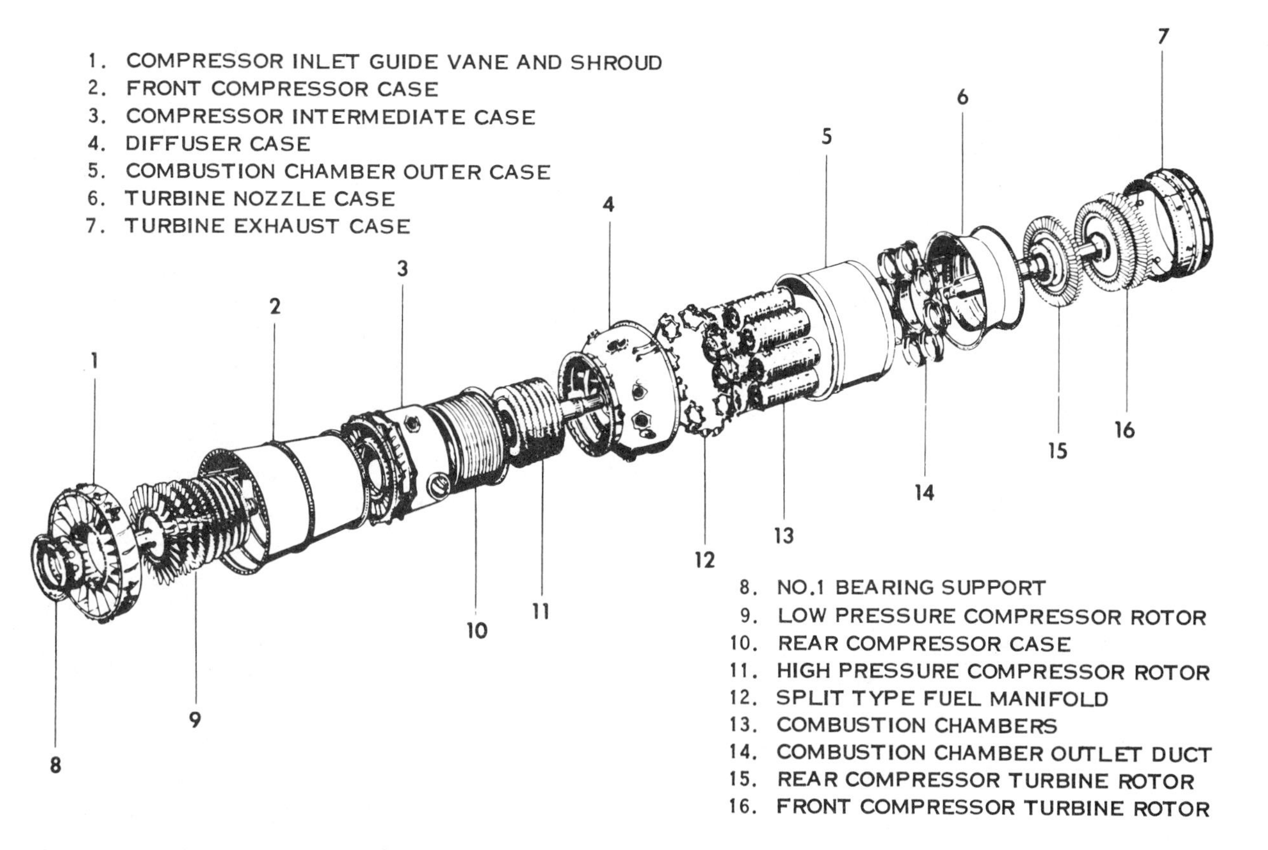

Fig. 11-7 Major sections of Pratt & Whitney JT3C nonafterburning engine (Courtesy Pratt & Whitney Aircraft Group/United Tech.)

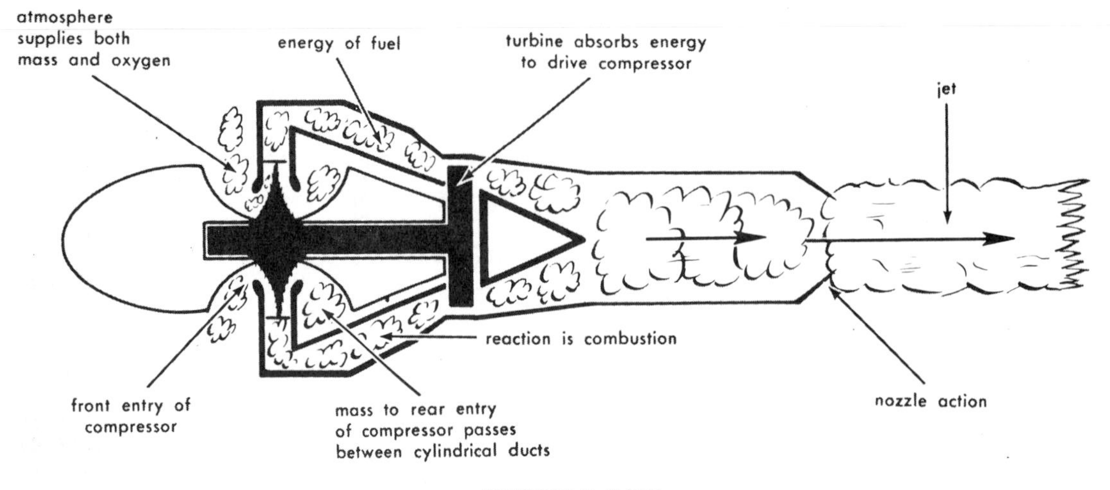

CENTRIFUGAL FLOW

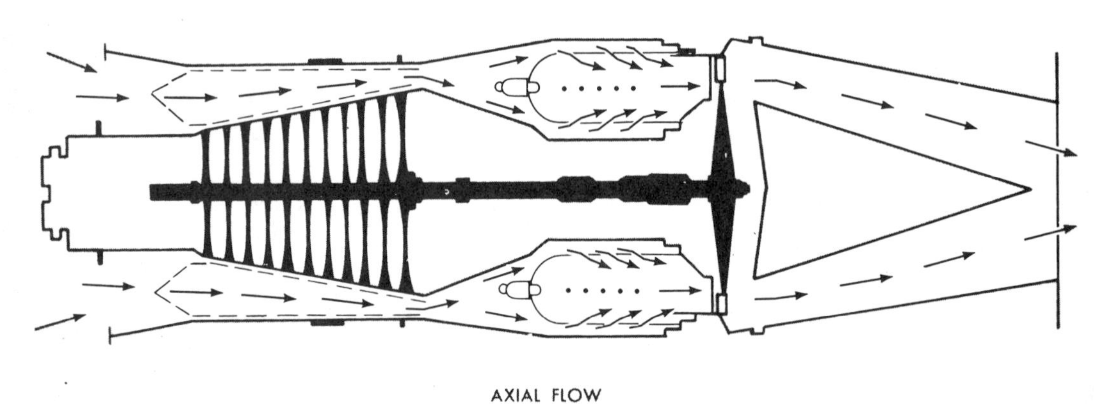

AXIAL FLOW

Fig. 11-8 Operating principles of the turbojet

combustion chambers, where air is compressed from three to twelve times its original volume. As the amount of air delivered to the combustion chambers increases, the thrust produced increases. The amount of air in the chamber depends on three factors: the speed at which the compressor is turning, the aircraft speed, and the density of the air. Increasing any or all of these will put more air through the engine.

The compressed air supports the combustion of fuel—at temperatures as high as 1800°F (982°C)—that is being sprayed into the combustion section. The expanding gases rush rearward to the turbine blades. As the gases travel through the turbine section, a considerable amount of their energy is spent in turning the turbine. The turbine's motion is applied directly to turn the compressor.

About two-thirds of the engine's power is used to rotate the compressor. The remaining

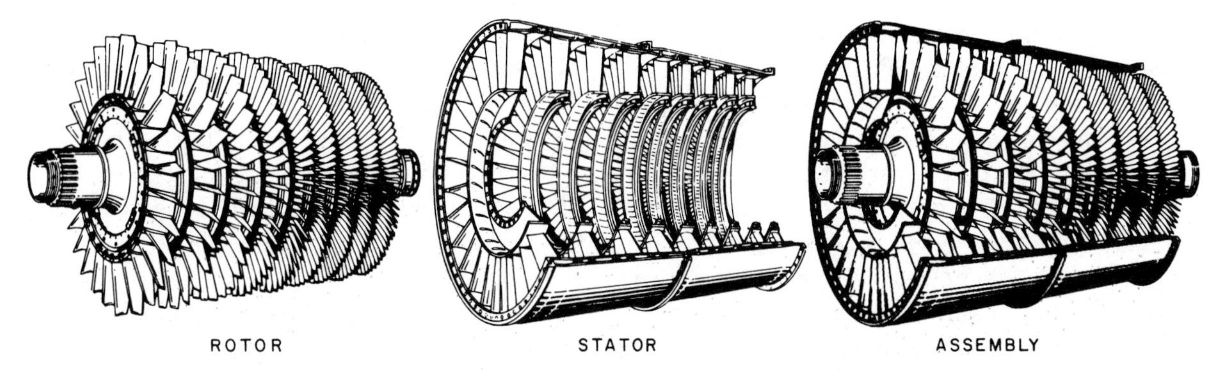

Fig. 11-9 Components and assembly of an axial-flow compressor

one-third goes out the exhaust. This one-third produces the thrust that pushes the aircraft forward.

There are two types of compressors, axial flow and centrifugal flow. The *axial flow compressor,* figure 11-9, is comprised of a series of rotor blades and stator blades. The air moves rearward along the axis of the rotating shaft. The spinning rotor blades scoop the air and force it rearward. The stator blades are stationary and pick up the air, deflecting it on to the next rotor blade. The rotor and stator stages get smaller and the angle of the blades scooping the air gets sharper. Air pressures and air acceleration, therefore, increase with each successive compressor stage.

The *centrifugal flow compressor,* figure 11-10, consists of three main parts: an impeller, a diffuser, and a compressor manifold. Air

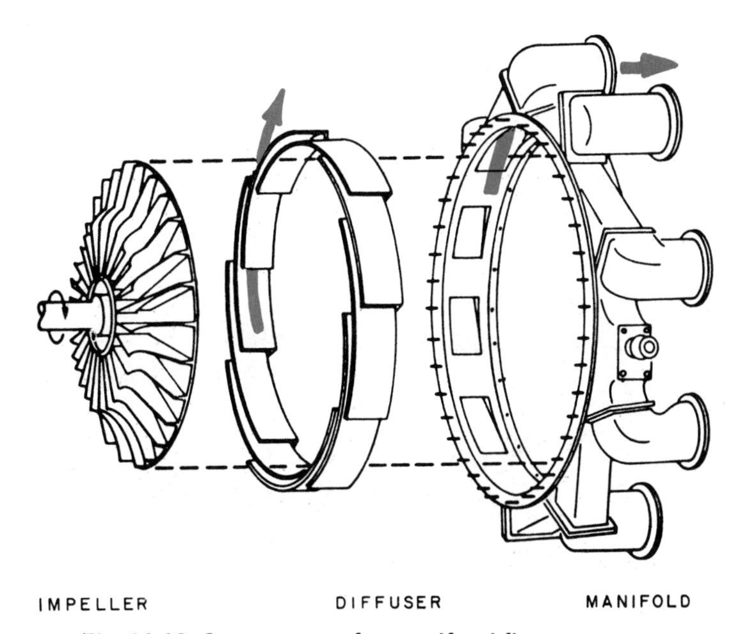

Fig. 11-10 Components of a centrifugal-flow compressor

is scooped up by the impeller blades as it enters the compressor and is thrown to the outside rim of the impeller by centrifugal force. The impeller is designed in such a way that the speed and pressure of the air increase as the air is thrown outward. As the air comes off the impeller rim, it is picked up by the diffuser and channeled to the compressor manifold. The manifold delivers the air on to the combustion chamber.

The combustion section contains the combustion chambers, spark plugs, and fuel nozzles. Within the combustion chambers, the air and fuel are burned, the hot gases being used to power the turbine and to produce the required thrust. The size and number of combustion chambers may vary from engine to engine. However, the combustion chambers themselves are usually made up of these parts, figure 11-11: outer combustion chamber, inner liner, inner liner dome, flame crossover tube, and fuel injector nozzle.

The outer chamber receives the high-pressure air and surrounds the inner chamber much like a large tin can with a small can inside. The inner liner has many holes in its surface. Air enters through these holes to be burned with the jet fuel. The flame sweeps rearward down the inner liner. Incoming air all along the liner keeps the flame from touching the metal. Combustion is completed just before the hot gases reach the turbine. Ignition is started by spark plugs but, once started, the plugs are turned off and burning is continuous. The flame crossover tubes carry the flame to the other combustion chambers which are not equipped with spark plugs.

The turbine section harnesses much of the kinetic energy of the hot gases, using the power to drive the compressor and engine accessories. The turbine is basically a dynamically balanced disc with steel alloy blades or buckets fastened to its perimeter, figure 11-12. A stationary turbine nozzle vane assembly directs the hot gases against the rotating turbine blades. The turbine may be a single rotor, or it may be designed with several turbine rotors to increase its output.

The exhaust section is designed to discharge the exhaust gases in a way that will produce maximum thrust. The turbulent gases from the turbine are straightened and concentrated by an inner cone and expelled through the tailpipe.

There is always a surplus of oxygen in the

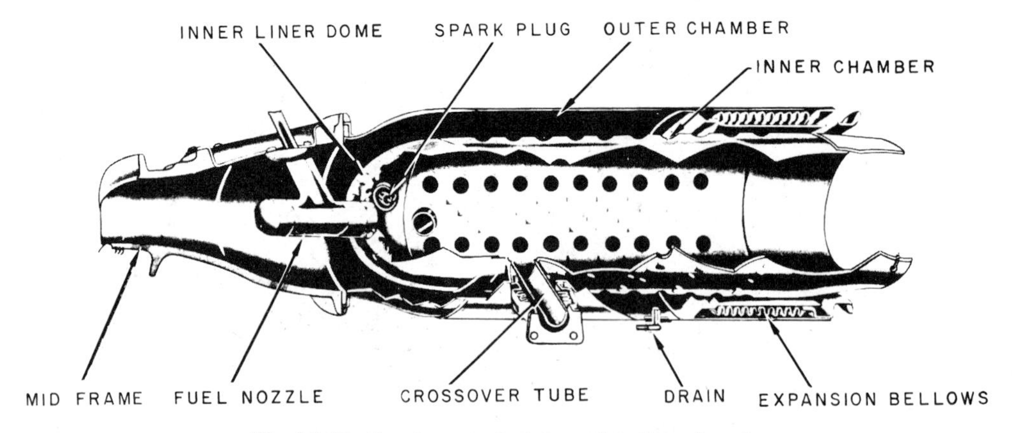

Fig. 11-11 Can-type turbojet combustion chamber

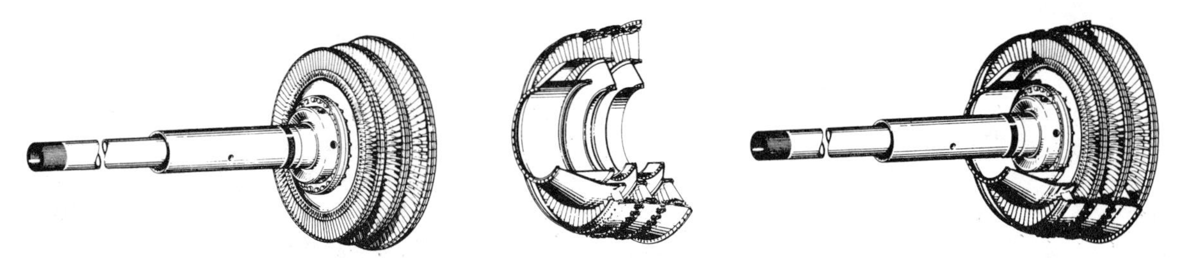

Fig. 11-12 Turbine elements

jet engine. The oxygen in the air mass serves only to cool the engine parts, then goes out the exhaust. Some engines are equipped with afterburners to utilize this oxygen to supply additional engine thrust, figure 11-13. The afterburner is behind the combustion and turbine section. More fuel is introduced in the afterburner, increasing the force of the expelled gases.

Turbojet noise suppressors have become a necessity in the vicinity of public airports, figure 11-14. The basic idea of a noise sup-

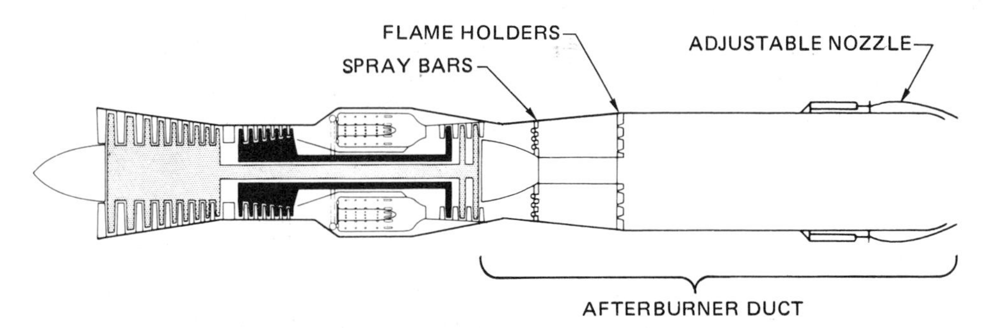

Fig. 11-13 Dual axial flow compressor turbojet with afterburner

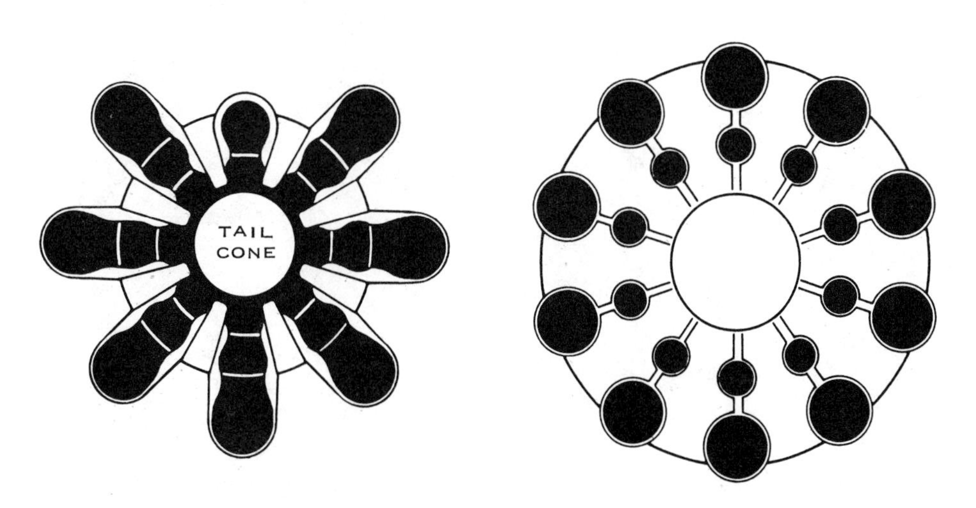

Fig. 11-14 Left: Corrugated perimeter-type noise suppressor, rear view. Right: Multitube-type noise suppressor, rear view

pressor is to split up high volume, low frequency sound into high frequency sound that can be absorbed better by the surrounding atmosphere.

TURBOPROP

Essentially the turboprop engine consists of a turbojet engine that drives a propeller, figure 11-15. There are some differences in the designs of turboprop engines, but the basic engine components are present. Figure 11-16 shows the inlet, diffuser and duct, compressor, burner, turbine, exhaust duct, and exhaust nozzle. The principles of operation are virtually the same as those of the turbojet and, therefore, will not be repeated.

The engine turbine, however, is designed to extract more power from the hot gases since the turbine must drive the propeller. The propeller is responsible for 90% of the engine's thrust, the remaining thrust being produced by the gases leaving the exhaust nozzle. The shaft speed of the engine is far greater than the speed range of a propeller, therefore, a reduction gear assembly is used to obtain the desired propeller speed.

The turboprop aircraft is especially suited for commercial cargo and passenger service where flights are relatively short and turbojet speeds are not necessary. Runways may be shorter for these aircraft since the engine's performance is excellent at low speeds. The turboprop combines the power of the turbojet with the efficiency of the propeller. The turboprop engine can produce about twice as much power as a conventional piston engine of equal weight.

Fig. 11-15 Avco Lycoming LTP 101 turboprop engine (Courtesy AVCO Lycoming Div.)

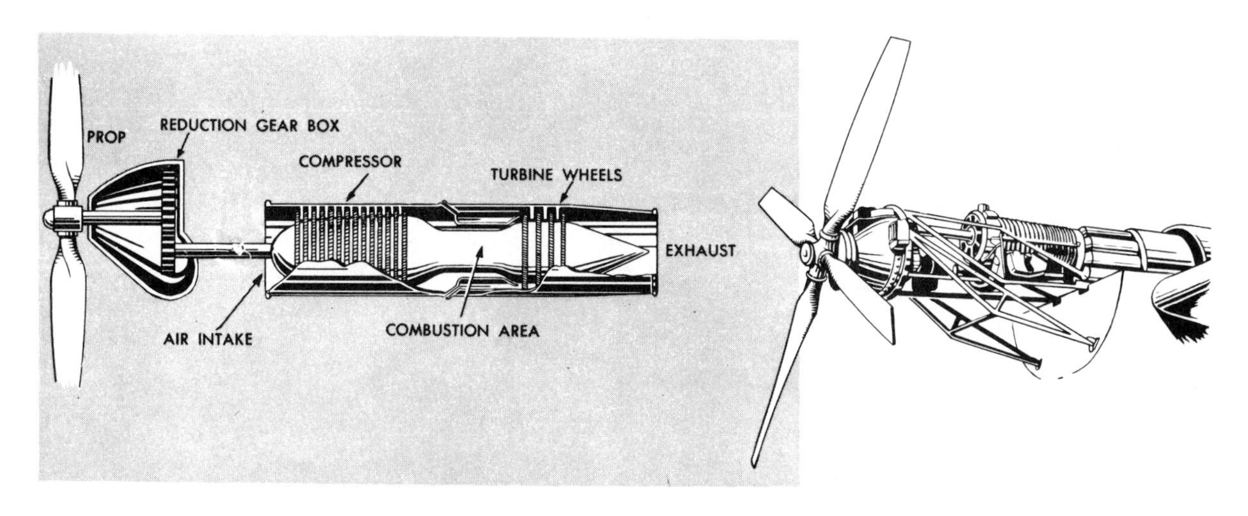

Fig. 11-16 Operating principles of the turboprop

TURBOFAN

Another variation of the turboprop engine is the turbofan engine, figure 11-17. On this type of engine, a ducted fan replaces the propeller. The large fan is really the first section of the compressor, figure 11-18. However, much of the air is released before the final compressor stages, burner, and turbine. The fan accelerates the air mass much the same as the propeller would. The fan is driven at shaft speed and is designed to operate more efficiently at higher air speeds than is possible with a propeller. The fan is responsible for 30 to 60% of the propulsive force. The turbofan engine is also lighter and simpler than the turboprop engine, figure 11-19.

The turbofan engine provides a large amount of thrust for take-off, and good fuel economy at cruising speeds. Turbofan engines are well suited for commercial transport aircraft that travel just below the speed of sound. Noise problems are also reduced with the turbofan engine.

The higher speeds of jet aircraft have created special landing problems. Brakes and tires wear more rapidly trying to slow down from such high speeds. To help reduce this wear, and lower stopping distances, thrust reversers are fitted to turbojet and turbofan engines, figure 11-20. These reversers turn the exhaust gases around to point forward. They are most effective at high speeds and are used during the first part of the landing roll.

JET FUELS

During the early development of jet engines, it was commonly thought that the engine could operate on almost any petroleum fuel. Soon it was found that kerosene was a desirable fuel. Since kerosenes cover quite a range of fuel characteristics, very fine fuel specifications have been laid down for jet fuels.

Most jet fuels are not pure kerosene, but

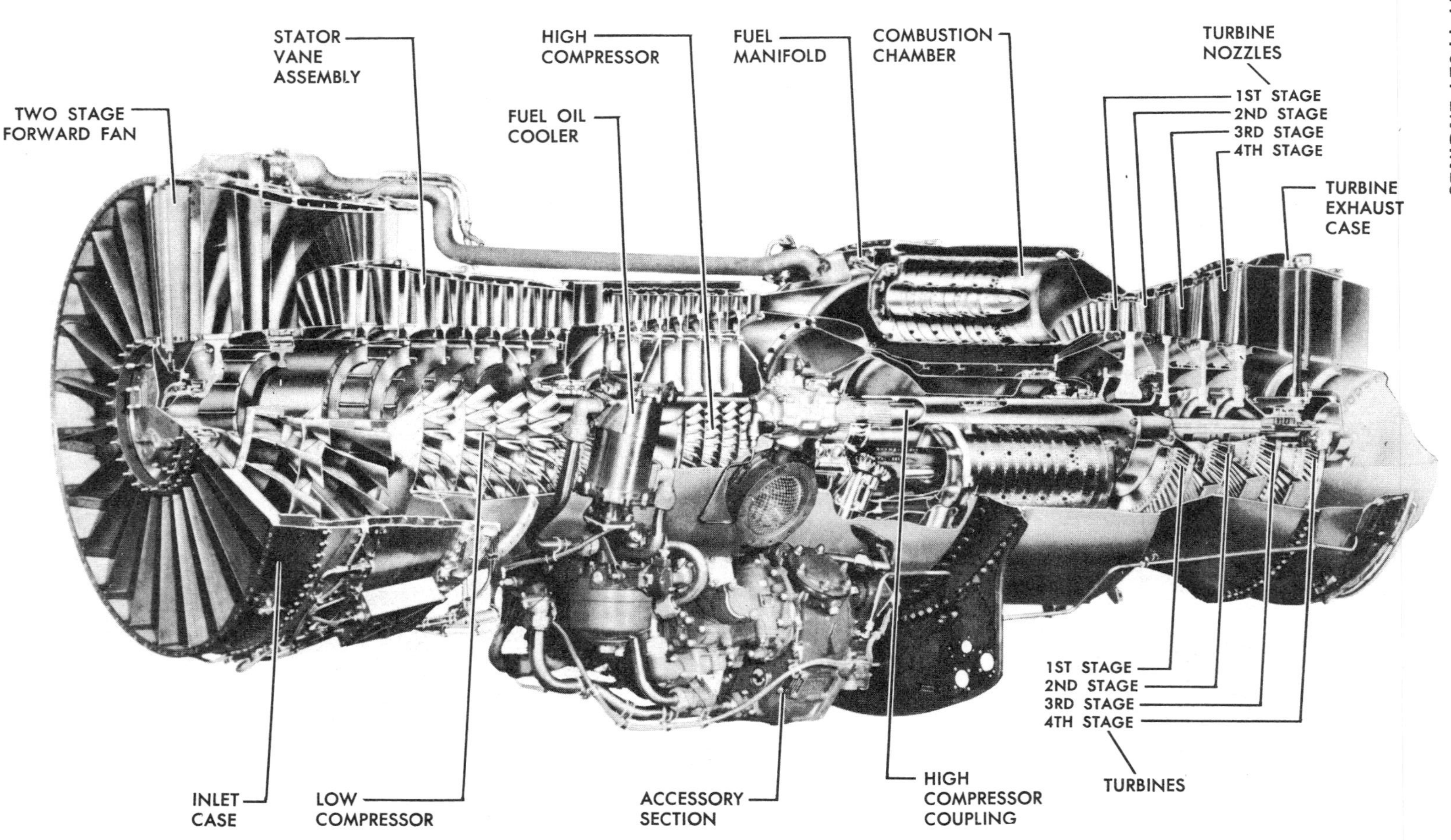

Fig. 11-17 Pratt & Whitney JT3D turbofan engine cutaway (Courtesy Pratt & Whitney Aircraft Group/United Tech.)

rather a blend of kerosene and gasoline. A common ratio is 65 to 70% gasoline to 30 to 35% kerosene. Military jet fuels are designated by JP-1, JP-2, JP-3, etc. Jet fuel JP-4 and JP-5 are the most common, both for military and commercial use. Jet fuel must be high quality and made to meet rigid specifications. Often, jet engines are designed to operate on one specific fuel type.

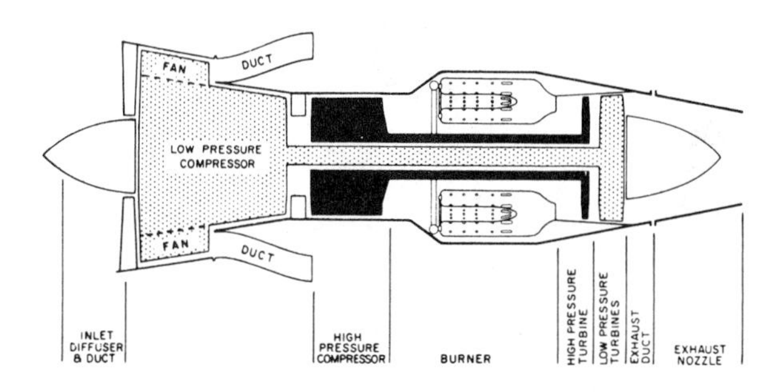

Fig. 11-18 Dual-compressor turbofan

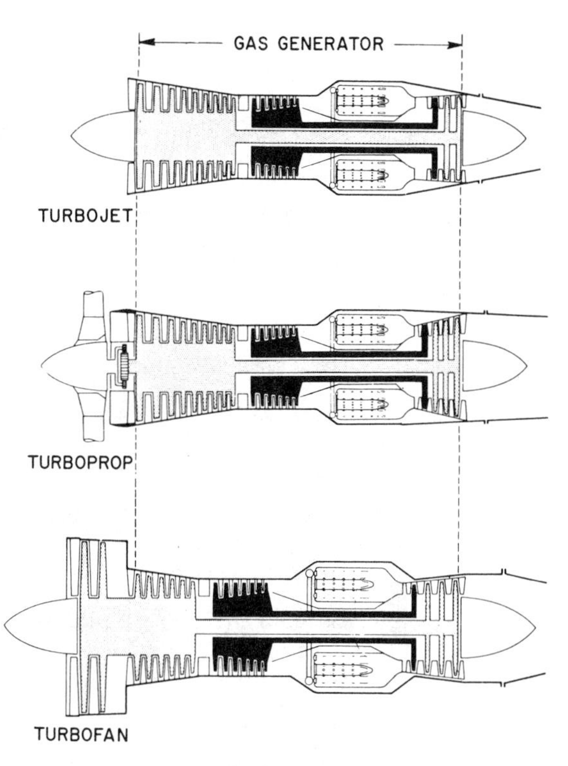

Fig. 11-19 Gas generator sections of turbojet, turboprop, and turbofan engines

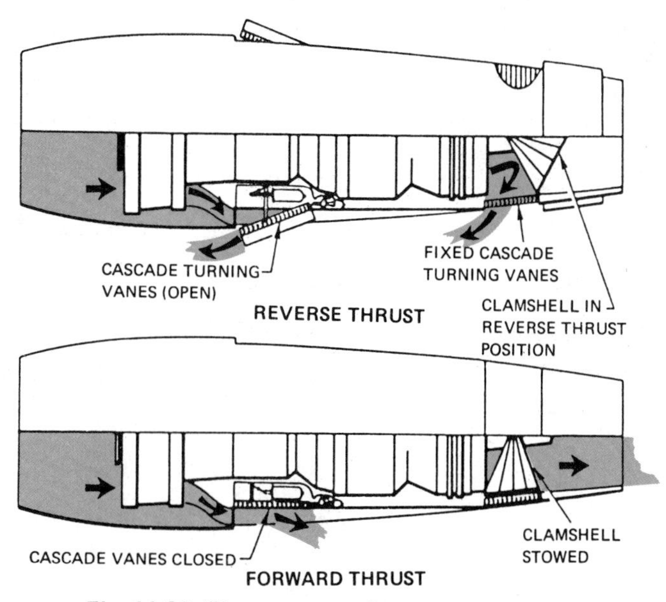

Fig. 11-20 Thrust reverser for turbofan engine

REVIEW QUESTIONS

1. Is the theory of jet propulsion new?
2. What does Newton's third law of motion state?
3. What term is used to describe the propulsive force of a jet engine?
4. How are jet engines and rocket engines alike?
5. How are jet engines and rocket engines different?
6. Does the ramjet engine have any moving parts?
7. Can the ramjet take off on its own power?
8. Does the ramjet have any practical use today? What?
9. Explain the pulsing operation of the pulsejet.
10. Is the pulsejet capable of very high speeds?
11. What two countries took the lead in developing the turbojet?
12. What are the main sections of the turbojet engine?
13. What amount of energy is needed to operate the compressor?
14. What is the purpose of the compressor?
15. What is the purpose of the turbine?
16. Why is the design of the exhaust system important?
17. Does the turbojet burn all the air that it takes in?
18. How is a turboprop engine different from a turbojet engine?
19. What are the advantages of a turboprop engine?
20. What are the advantages of a turbofan engine?

Unit 12 Rocket Engines and the Exploration of Space

OBJECTIVES

After completing this unit, the student should be able to:

- Discuss the development of rockets throughout history.
- Recognize four types of trajectories and state their use.
- Explain the operation of a rocket engine.
- Discuss and compare liquid and solid propellent rockets.

Rockets have been used by the military for several centuries. The Chinese of the thirteenth century used rockets for warfare. The Indians used rockets against the British in India in the late eighteenth century. "The rockets' red glare" in our National Anthem refers to the rockets used by the British in the War of 1812. Rockets served as signals and flares in World War I. It wasn't until World War II, however, that rockets were used as weapons on a large scale.

The Germans spent 40 million dollars to develop the V-2 rocket, and used about 2675 of these rockets against the Allies in World War II. Though inaccurate, the V-2 rocket could travel up to 200 miles (322 kilometers) at a speed of 3600 miles per hour (5806 kilometers per hour). The V-2 was huge: 47 feet (14.3 meters) long, 5.5 feet (1.7 meters) in diameter, and 27,000 pounds (12 270 kilograms) at takeoff.

Dr. Robert H. Goddard of the U.S. was an early rocket pioneer who developed the forerunner of the famous bazooka of World War II, an antitank missile. His primary interest, however, was the development of high altitude meteorological rockets.

Rocket development gained national attention after World War II. The United States developed the Viking and Aerobee rockets for meteorological research. The Honest John, Redstone, and Nike rockets were designed to strengthen our defense network.

The United States and Russia soon directed their efforts to the intercontinental ballistic missile, ICBM, capable of traveling over 5000 miles (8050 kilometers) and yet striking a target with pinpoint accuracy. Fortunately, the ICBMs have never been used, but the technology of these rocket weapons has provided a base for peaceful space exploration.

SPACE EXPLORATION

Before space exploration programs could be started, it was necessary to develop an engine with a thrust powerful enough to overcome the earth's gravitational pull. Acceleration is the key to escape, and in the last twenty-five years rockets have been developed that are able to achieve astounding acceleration rates. As they consume their fuel, they become increasingly lighter, and so they can accelerate at higher and higher rates of speed.

It was discovered early in the study of rockets that staging also aids rapid acceleration, and so three stages are assembled for escape from the gravity of the earth and into orbit. When the first stage propellants are exhausted, the first stage of the rocket drops off. Already at a high velocity and now considerably lighter, the second stage ignites and the rocket's acceleration increases. When the second stage burns out it drops off, and the third stage is then capable of developing the final velocity and acceleration necessary for leaving the earth's gravitational field and traveling into space.

The rocket's path in space is called its *trajectory*. Four general trajectories are now used: sounding, earth orbit, earth escape, and planetary.

Sounding rockets are used for research of the upper atmosphere. They are two-stage rockets which go straight up and then fall back into the ocean or desert. Scientific data is radioed or parachuted back to earth. The Argo, Aerobee, and Astrobe are sounding rockets.

Earth orbit is attained by launching the rocket vertically and then tilting the trajectory so that it becomes parallel to the earth's surface. When the desired altitude of 125 miles (201 kilometers) is reached and orbital velocity of 18,000 miles per hour (28 980 kilometers per hour) is attained, the final-stage rocket shuts off. The gravitational pull of the earth and the centrifugal pull of the rocket balance each other, and the spacecraft coasts into orbit around the earth. The earth-orbiting rocket is usually used for some type of research and is commonly referred to as a satellite. The first of these was the Russian satellite Sputnik 1, launched on October 4, 1957. Explorer 1, launched on January 31, 1958, was a close second for the United States. Our first satellite weighed 30.8 pounds (14 kilograms) and orbited the earth each hour and fifty-five minutes. Several hundred research satellites have been launched since that time, figure 12-1.

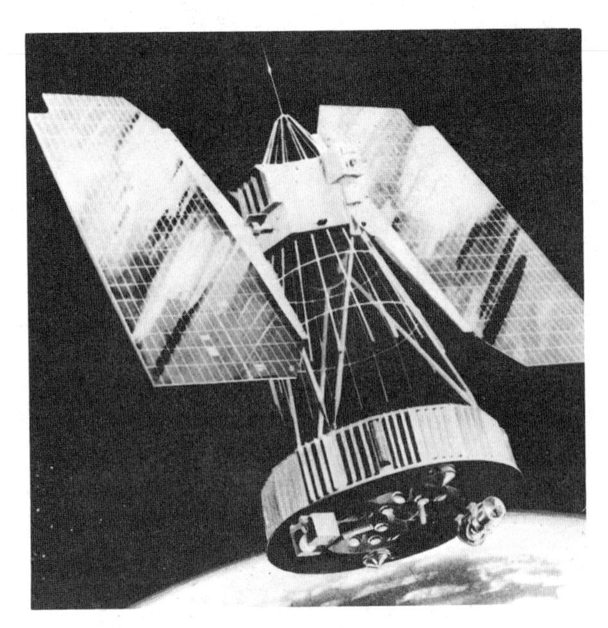

Fig. 12-1 Nimbus 1 weather satellite

MANNED SPACECRAFT

The next goal in rocketry was to place a manned spacecraft into orbit around the earth and return it safely. The Russian Vostok and American Mercury programs were directed toward this goal. On April 12, 1961, Russian cosmonaut Yuri Gagarin became the first human to orbit the earth. Alan Shepard, when he followed on May 5, 1961, rode a ballistic flight into space and traveled 300 miles (483 kilometers) downrange on the Freedom 7. John Glenn Jr., was the first American to orbit the earth, as he flew the Friendship 7 for three orbits around the earth before returning safely home. His spacecraft was powered by a recently developed Atlas booster rocket.

The United States Gemini Program and the Russian Voskhad Program went beyond the previous flights and proved that humans could function in space and return to earth. The Gemini Program involved walking in space and making a rendezvous with a target vehicle and docking with it. Extreme precision was necessary to accomplish the rendezvous. The Gemini spacecraft had to be launched into the identical orbit that the target vehicle, launched a day earlier, had used. The Gemini astronauts of 1965 and 1966 accomplished their objectives. Edward White II became the first American to walk in space on June 3, 1965. Astronauts Frank Borman and James Lovell performed a rendezvous between Gemini 6 and 7. The Titan 2 rocket was the launch vehicle for the Gemini Program, figures 12-2 and 12-3. This rocket is a two-stage rocket, 109 feet (33.2 meters) high, delivering 430,000 pounds (1.9 meganewtons) of thrust at sea level and able to place the Gemini space capsule, which weighed more than 7000 pounds (3200 kilograms), in orbit.

MAN TO THE MOON

Earth escape is a step beyond orbital travel. The craft must gain enough velocity to escape the gravitational influence of the earth, about 25,000 miles per hour (40 250 kilometers per hour). After its escape, the spacecraft eventually goes into orbit around the sun, like a tiny planet, unless it comes under the gravitational influence of another celestial body. The Apollo Program, using the powerful Saturn 5 launch vehicle, was designed to escape the earth, orbit the moon, land on the moon, and then return to the earth, figure 12-4.

Apollo 11 put the first human being on the surface of the moon. The 365-foot (111-meters), 3-stage Saturn rocket began its journey

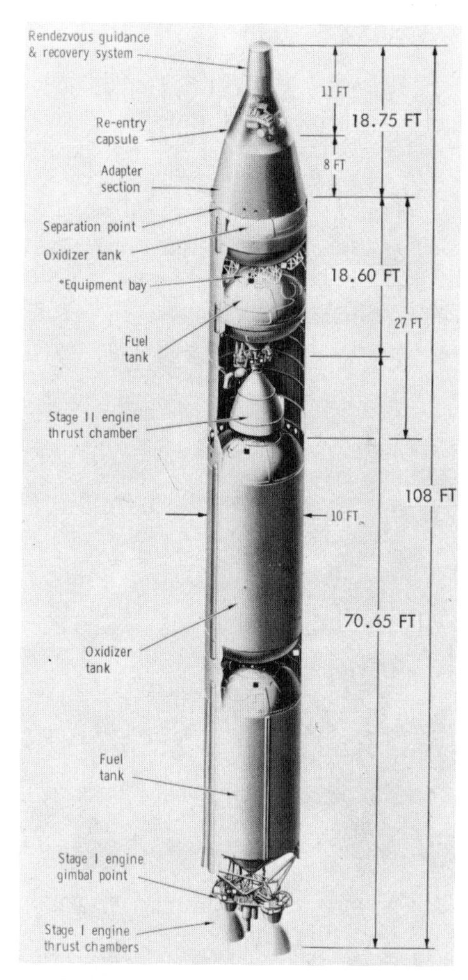

Equipment bay contains:

- Batteries
- Malfunction detection system (MDS) units
- Range of safety command control system
- Programmer
- Three-axis reference system (TARS)
- Radio guidance system (RGS)
- Autopilot.
- Instrumentation and telemetry system

Fig. 12-2 Titan 2 launch vehicle for the Gemini spacecraft (Courtesy NASA)

Fig. 12-3 Gemini 4 spacecraft lifts off, powered by a Titan 2 rocket. (Courtesy NASA)

with the ignition of the first stage, burning 214,000 gallons (811 060 liters) of kerosene and 346,000 gallons (131 340 liters) of liquid oxygen in 2 1/2 minutes, figure 12-5. At the completion of this stage, the rocket was traveling at the rate of 6100 miles per hour (9821 kilometers per hour), 38 miles (61 kilometers) above the earth. The second stage burned liquid hydrogen and liquid oxygen and attained

Fig. 12-4 The Apollo Saturn vehicle generates a thrust of 1 1/2 million pounds (6.7 meganewtons) at liftoff. (Courtesy NASA)

Fig. 12-5 The H-1 engine, generating up to 188,000 pounds (.84 meganewtons) of thrust, is used in a cluster of eight to provide the 1,500,000 pounds (6.7 meganewtons) of thrust for the Saturn booster.

a speed of 15,300 miles per hour (24 633 kilometers per hour) and an altitude of 117 miles (188 kilometers). The third stage, using the same kind of fuel, placed the spacecraft in orbit around the earth. While in orbit the spacecraft was checked out thoroughly before the third stage was fired again. A final firing of the third stage accelerated the spacecraft to 24,200 miles per hour (38 962 kilometers per hour) and the trajectory of the craft was changed for the moon, figure 12-6.

The spacecraft consists of three modules: the command module, which carries the astronauts; the service module, which consists of the rocket and fuel tanks; and the lunar module, which is equipped to land and take off from the moon.

As the spacecraft enters the moon's gravitational field, it is slowed down with retro-rockets and put into lunar orbit. Two astronauts transfer from the command module to the lunar module, separate the lunar module from the others, and descend to the moon's surface. Once on the moon, the astronauts collect rocks and perform scientific experiments. Then the lunar module, using its landing section as a launch pad, blasts off at exactly the correct time to rendezvous with the orbiting command and service modules. After docking, the rocket in the service module is burned to power the craft back to earth. The lunar module is left to orbit the moon, and the service module is disconnected outside the earth's atmosphere, leaving only the command module to return to earth. Neil Armstrong, Michael Collins, and Edwin Aldrin accomplished the historic Apollo 11 mission in July 1969. Further Apollo flights, following the same pattern of operations, permitted further exploration and observation of the lunar surface.

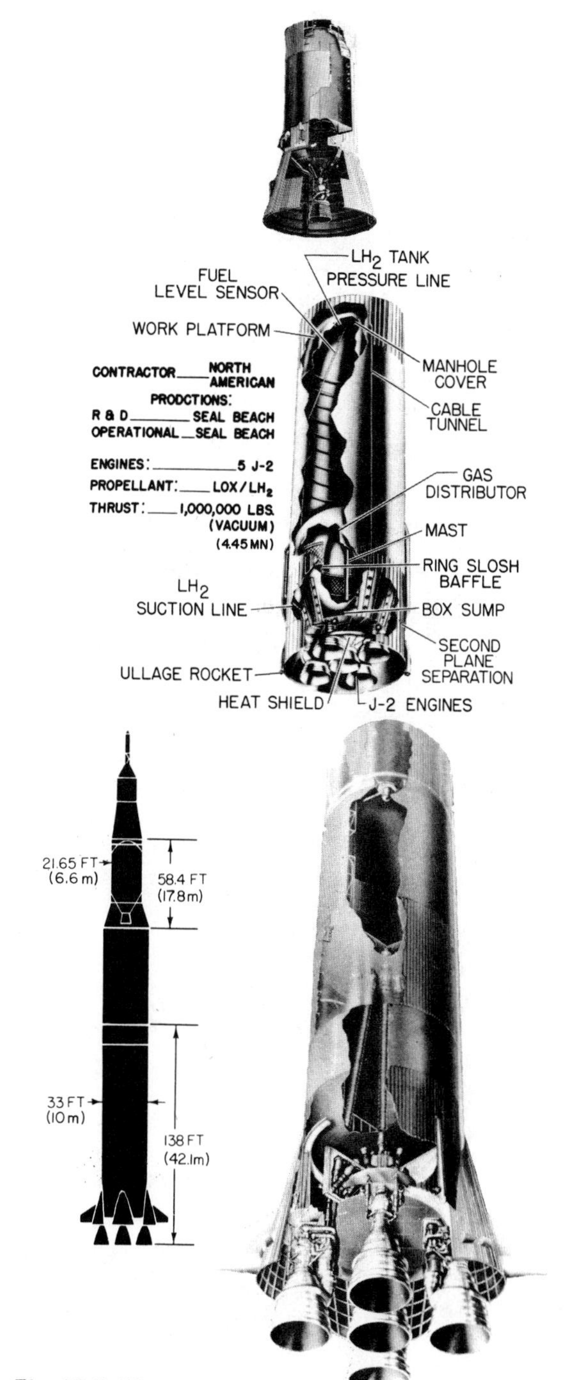

Fig. 12-6 First, second, and third stages of the Saturn 5 launch vehicle (Courtesy NASA)

SKYLAB

The Skylab missions of 1973 and 1974 accumulated a large amount of scientific information. Skylab was launched unmanned on May 14, 1973. Shortly thereafter three crews of astronauts were sent into orbit to perform experiments aboard Skylab. The first crew spent 28 days, the second crew 59 days, and the third crew spent 84 days aboard the spacecraft. The missions proved that it is possible to spend extended periods of time in space without suffering ill effects.

NATIONS COOPERATE IN SPACE

The next manned space flight for the United States was the joint U.S.–Soviet Apollo/Soyuz Test Flight in 1975. Three U.S. astronauts aboard an Apollo spacecraft successfully docked with the Russian Soyuz spacecraft. This mission consisted of five scientific experiments: an exchange of micro-organisms among crewmen and their spacecraft, melting various metals and semiconductors, studying cultures of a fungus, spectrometer experiments to determine the atomic oxygen and nitrogen in the upper atmosphere, and photography of the solar corona.

SPACE SHUTTLE

The space shuttle program began in 1972 and is now a successful, ongoing program. The basic concept is to provide a space transportation system that will deliver payloads of satellites and test equipment into space and then return to earth in perfect condition so it can be used again. The space shuttle takes off like a rocket, maneuvers in orbit like a spacecraft, and then returns to earth and lands on a

Fig. 12-7 Space shuttle orbiter *Enterprise* (Courtesy NASA)

runway like an airplane. The spacecraft itself is called the orbiter. After return to earth the orbiter can be ready to go up again in two weeks.

The orbiter is about the size of a commercial DC-9 aircraft—122 feet (37 meters) long and has a wing span of 78 feet (24 meters), figure 12-7. The first orbiter named the Enterprise was used for testing and did not fly in space. The first orbiter to travel into space and return to earth was the Columbia—piloted by John Young with Robert Crippen as the co-pilot. The maiden flight was in April 1981. This flight proved the flight worthiness and reliability of the craft. Other orbiters in the program are the Challenger, Discovery, and the Atlantis.

The main units for the space shuttle are two solid rocket boosters, the orbiter, and the external tank, figure 12-8. At lift off five rockets are firing, two solid propellent rocket boosters with a thrust of 2.6 million pounds (11.6 meganewtons) and three liquid propellent rocket engines supplied with liquid oxygen and liquid hydrogen and producing 1,250,000 pounds of thrust.

The boosters, 12 feet in diameter and 150 feet tall (3.7 meters x 45.5 meters) are the largest solid fuel rockets ever flown. Even at this great size, the boosters will burn out after two minutes of flight, but by this time the spacecraft is 31 miles in space (50 kilometers). Being spent, the boosters are cut loose from the external tank and allowed to return to earth, figure 12-9. They first drift up several more miles into space and then descend into the ocean, their fall being slowed by parachutes, where they are picked up by waiting ships. Upon return to land they are cleaned, refitted, and refueled for further use.

The orbiter's three rocket engines continue

Fig. 12-8 Space shuttle launching with all five engines firing (Courtesy NASA)

to fire, being fueled by the external tank, figures 12-10 and 12-11, until orbit is just about achieved. From lift off only eight minutes has elapsed. Just after the rocket engines are shut down the orbiter is separated from the external tank, figure 12-2. The large tank will fall back to earth in a ballistic trajectory and disintegrate in a safe ocean area.

Now in orbit, in weightless and airless space, the orbiter uses thrusters to orient it in any direction desired. The flight deck and cabin are spacious and supplied with ordinary air and are temperature and humidity con-

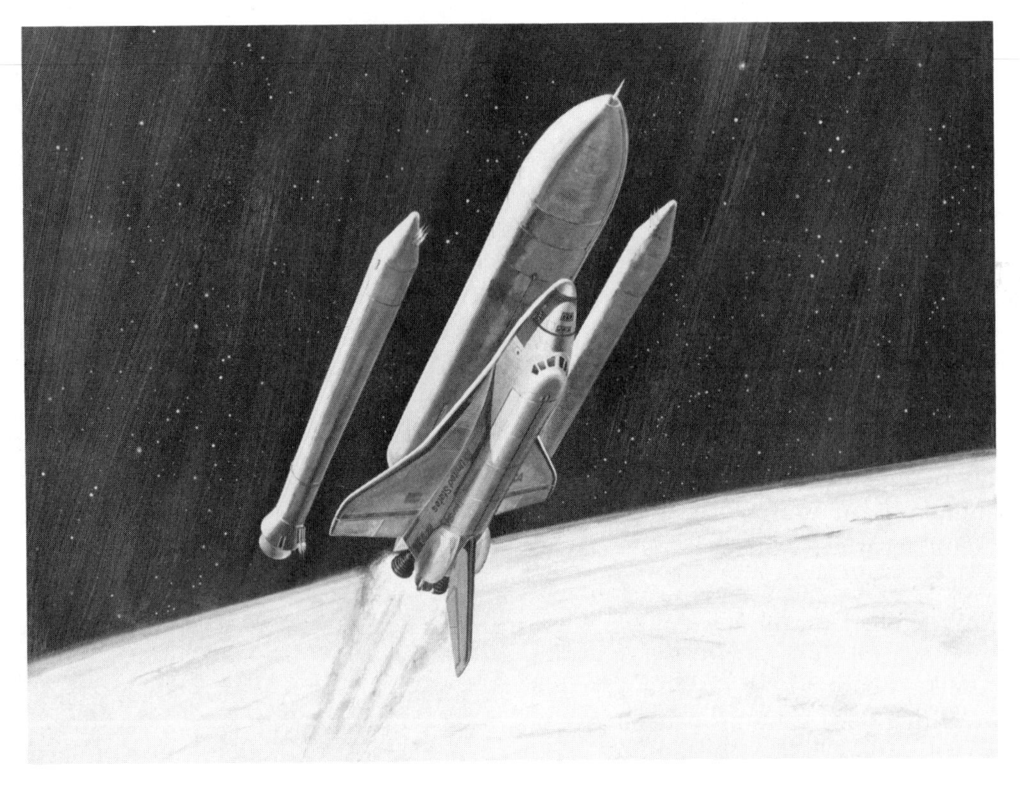

Fig. 12-9 Solid rocket boosters being jettisoned (Courtesy NASA)

trolled. Since space suits are not necessary, astronauts can work and live in comfort.

The large cargo bay can carry either a large satellite or up to five smaller satellites. A pressurized spacelab can also be in the cargo bay, figure 12-13, which is adaptable for whatever the mission requires. A remotely-controlled manipulator arm can lift satellites from the cargo bay and place them into space where they will orbit the earth, figure 12-14. If astronauts wish to enter the cargo bay they must enter through an air lock chamber and wear space suits since the bay is not pressurized. Sometime in the future the payload of the orbiter will be modules that will fit together to make up a large permanent space station that will be habitable for humans.

When the mission is complete the orbiter heads for home. Attitude control thrusters turn the orbiter so that it is tail first when the craft is halfway around the earth from the landing area. Orbital maneuvering engines fire for about two minutes to slow the orbiter down so that it starts its slow curve toward the earth. In about half an hour the craft is flying nose first and starting to enter the earth's atmosphere. Within a few minutes the spacecraft must change into an aircraft, slowing from 18,000 to 225 miles per hour for landing, figure 12-15. There is no opportunity for a fly by or a second chance. The pilot, aided by computers, adjusts the control surfaces and fires appropriate rockets to aim this "heavy glider" at the runway. Tremendous

Fig. 12-10 Rocketdyne engine #0004. Three such engines power the space shuttle orbiter. (Courtesy Rocketdyne Division, Rockwell International)

heat builds up on the spacecraft during reentry but some 32,000 ceramic tiles on the underside of the orbiter prevent the craft from burning up. Seconds before touchdown the landing gear comes down and the orbiter lands safely on earth.

PLANETARY FLIGHTS

Planetary flights are more difficult than lunar flights. The moon is in the earth's gravitational field and is traveling around the sun at the same rate as the earth. The orbits of the earth and the planets around the sun make the problem of interplanetary flight complex. Favorable opportunities to launch rockets for Venus and Mars occur about every two years. These periods for favorable launching are called *launch windows*. Much planning is required to develop a transfer trajectory, which is the elliptical trajectory used for launching a rocket from the earth to another planet, figure 12-16.

Planetary space probes from the United States include the Viking program toward Mars, figure 12-17, the Mariner program toward Mercury, the Voyager mission to Jupiter and Saturn, and the Pioneer flights to Venus.

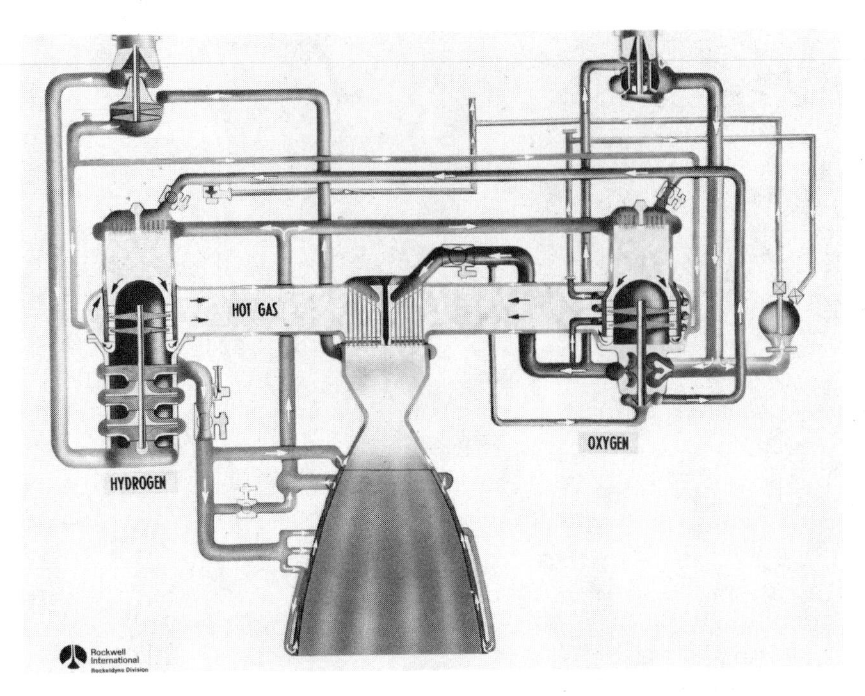

Fig. 12-11 Schematic drawing of the space shuttle orbiter's rocket engine. Three are used. (Courtesy Rocketdyne Division, Rockwell International)

All of these space probes contribute to the total knowledge of the planets by transmitting thousands of pictures and data to earth.

In 1976 Viking 1 and 2 landed on Mars, figure 12-18. The cameras sent out pictures of a rocky, sandy, reddish landscape with a pink sky, figure 12-19. The Viking discovered that winds on Mars average 19 miles per hour (30.6 kilometers per hour), and the atmosphere consists of 95% carbon dioxide, 2.7% nitrogen, 1.6% argon, and 0.15% oxygen; there is no evidence of life on Mars. The Soviet Venera 9 and 10 landings on Venus revealed that the rock there is similar to the basalts on the earth.

The Pioneer missions to Venus have found that the temperature on the surface was 900°F. This is due to the atmospheric greenhouse effect in which radiation from the sun is trapped by the planet's atmosphere. Tests also show high concentrations of Argon 36, Krypton, and Xenon—higher than the earth's concentrations. The first maps of the surface of Venus were provided by radar scanning devices on board Pioneer.

Voyager missions 1 and 2 surveyed Saturn and discovered that the well-known rings around the planet are made up of literally thousands of individual rings. Also, it was discovered that Saturn has at least 15 moons. In a trip that took two years, Voyager 1 traveled 1,250,000,000 miles and passed within 125,000 miles of Saturn. Voyager 1 is now on a path that will take it out of the solar system.

Fig. 12-12 Liquid propellant tank being jettisoned just before orbit (Courtesy NASA)

SATELLITES

In the relatively few years that nations have been placing man-made craft into space with rockets some 14,000 have been launched into orbit. Most of these satellites have eventually fallen back toward earth and been burned up as they re-entered the earth's atmosphere. At last count nearly 5,000 satellites are still in space but most are not operational. Of the 290 operational satellites 180 belong to the United States.

The life of a satellite is usually just a few years even under the best conditions. In addition to falling out of orbit, the satellite can run out of fuel, or fail electronically. Even when these major failures do not occur, satellites do not always remain in orbit perfectly. They are affected by solar flares, atmospheric conditions, and the competing gravitational fields of the earth, moon, and sun. On the ground, crews can sometimes signal the satellite to fire small fuel jets to correct their orbit, but this can only be done so many times before the fuel runs out. If the satellite is about to become useless, ground crews will usually propel it into a higher orbit to get it out of the way. In other cases the

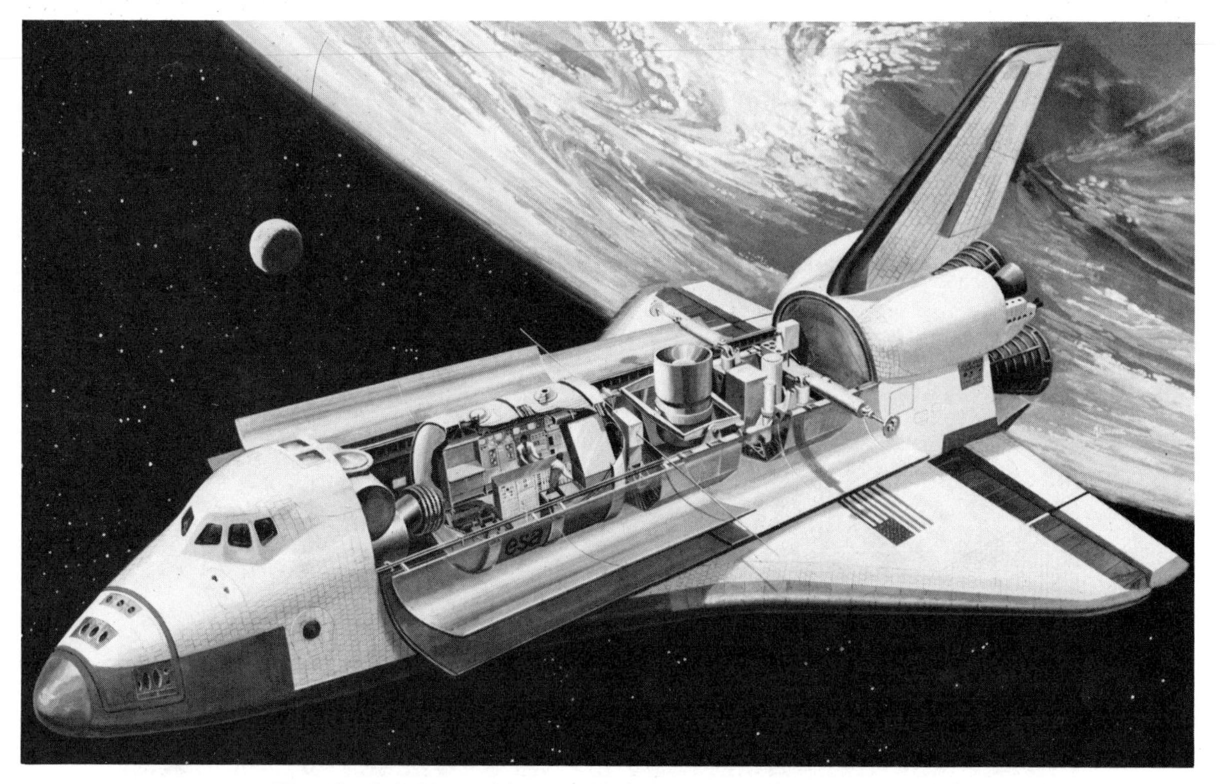

Fig. 12-13 Orbiter with space lab hatches open (Courtesy (NASA)

satellite will be allowed to slowly drift toward earth and eventually burn up in the earth's atmosphere.

Satellites perform a vast number of scientific experiments and much of the information gathered and transmitted to earth has been of great benefit to mankind. Several uses are:

- Weather information to warn of storms
- Mapping snow cover to predict spring runoff
- Mapping timer and grassland areas of the earth
- Mapping potential mineral deposits to aid in exploration
- Mapping glaciers and their movement
- Mapping floods
- Providing navigational information for ships.

Satellites such as Landsat orbit the earth at 570 miles and carry electronics equipment for mapping and mineral exploration. A TV camera that is capable of distinguishing objects the size of a football field sends pictures to earth. Another camera, a multispectral scanner, is capable of recording the reflective color of the earth and is far more sensitive than the human eye.

Communications satellites can relay radio signals from distant points on the earth in-

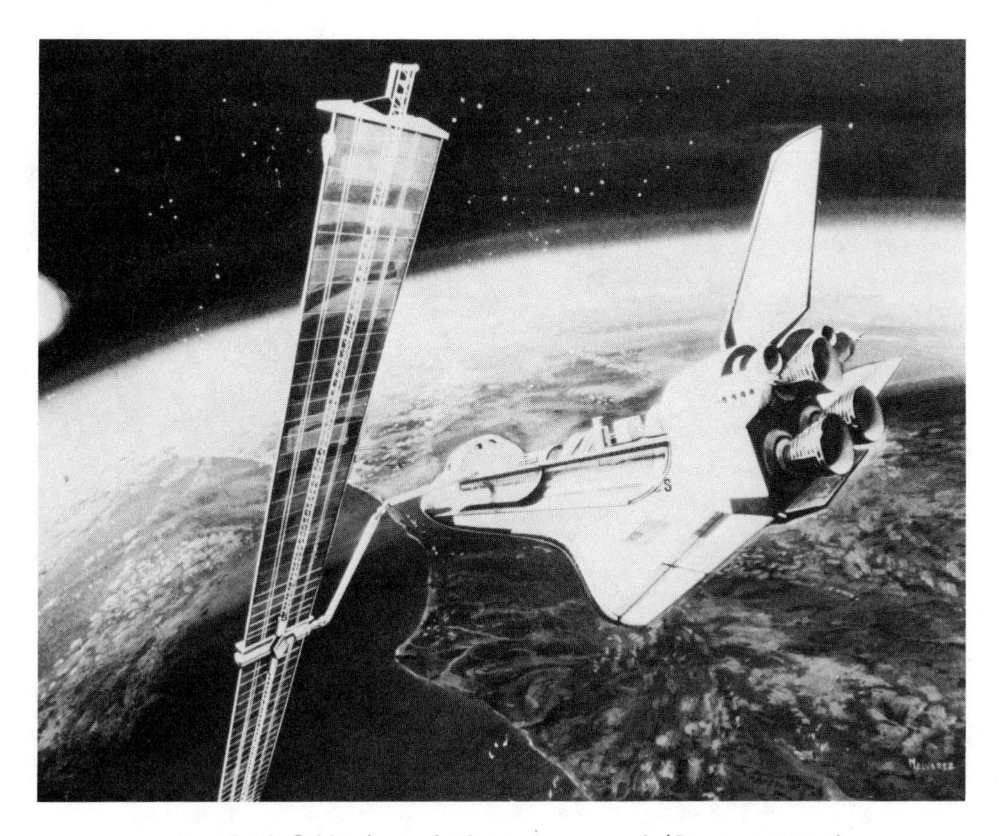

Fig. 12-14 Orbiter's manipulator arm at work (Courtesy NASA)

stantaneously; Echo I was the first. It was a hugh balloon made of Mylar that reflected radio signals from coast to coast. It was launched in 1960 and, when inflated in space, it was the height of a 10 story building.

Telstar was also an early communications satellite and was launched in 1962. Telstar carried transponders that amplified the strength of radio signals. Telstar could relay signals only when its orbit placed it between the sender and receiver–it could not be used all the time.

There was a need to put satellites into an orbit timed to the earth's rotation. This would have the effect of having the satellite fixed right over a given spot on the earth; this is called a geostationary orbit, and is 22,300 miles above the equator. In 1965 the Early Bird communications satellite was launched in this geostationary orbit. It could relay 240 telephone calls between North America and Europe and also carried a TV channel. The words "Live Via Satellite" became commonplace as a result of this event.

The International Telecommunications Satellite Organization (Intelsat) was formed by 109 member nations to use satellites orbiting in the geostationary path for worldwide communications. Fifteen satellites are in use and the current models have 12,000 voice circuits and two color TV channels. The costs of telecommunications have been reduced

Fig. 12-15 Orbiter returning to earth for runway landing (Courtesy NASA)

tremendously by their use, even though such satellites are not cheap—the price tag is about $34 million each.

In addition to Intelsat communications satellites many nations and private companies are putting communications satellites into orbit, but in spite of this, the demand for space communications satellites is greater than the supply.

The effect of these satellites can be seen directly as they are used in cable and commercial TV. Signals from TV companies are sent to the satellite and returned to local TV stations for local distribution. There has been a great increase of individuals who now own dish antennas that pick up these TV signals and bring them into their homes directly.

ROCKETS ARE AIR-INDEPENDENT

Rockets are unique as a major power producer because they are air-independent. That doesn't mean that air is not necessary for fuel combustion—it is—it just means that since the rocket carries its own oxidizer or liquid oxygen *within* the engine, it doesn't need an *external* supply. Rockets can operate equally as well in the heavy air near the earth's surface—the troposphere—as they move through the stratosphere, ionosphere, or in the vacuum of outer

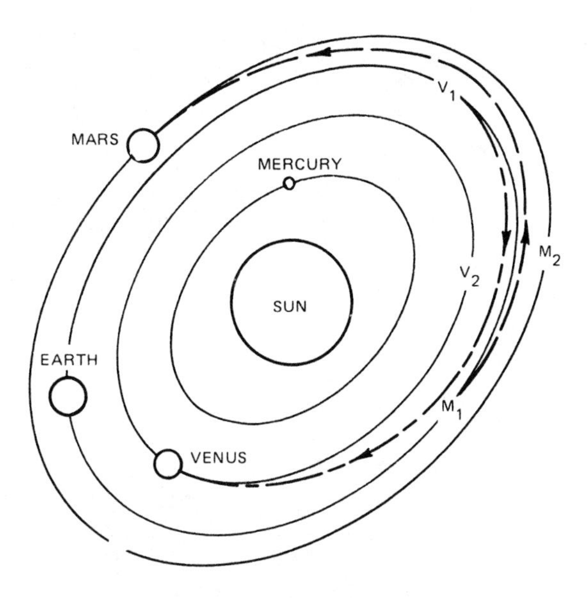

M_1 = POSITION OF EARTH WHEN ROCKET IS LAUNCHED FROM EARTH FOR TRAVEL TO MARS
M_2 = POSITION OF MARS WHEN ROCKET IS LAUNCHED FROM EARTH FOR TRAVEL TO MARS
V_1 = POSITION OF EARTH WHEN ROCKET IS LAUNCHED FROM EARTH FOR TRAVEL TO VENUS
V_2 = POSITION OF VENUS WHEN ROCKET IS LAUNCHED FROM EARTH FOR TRAVEL TO VENUS

Fig. 12-16 Transfer trajectory for launching rockets to Mars and Venus

space. Actually, the thinner the air around the rocket, the faster the rocket travels since there is less resistance from air friction.

The rocket engine burns fuel, just as any other engine does, by changing potential chemical energy into useful kinetic energy. However, because it finally travels in conditions where there is no air present, the rocket can in no way be dependent on oxygen in the atmosphere for maintaining its combustion. Therefore, the rocket carries its own supply of oxygen or oxidizer, as just mentioned.

Just as in jet propulsion, the thrust of the rocket comes from within the rocket itself. As the fuel ignites, the rapidly expanding gases of combustion rush toward the rear of the rocket. Here again, Newton's third law of motion, for every action there is an equal and opposite reaction, comes into play. The action of the gases leaving the rear of the rocket creates an equal but opposite reaction by pushing the rocket forward.

TYPICAL ROCKET OPERATION

It might be helpful to examine the operation of a typical liquid propellant rocket before studying more specific rocket construction, figure 12-20.

The propellant, or oxidizer and fuel, is at rest in tanks. The propellant represents untapped chemical energy.

During rocket operation, a feed system forces the propellant from the storage tanks into the rocket engine's injector. This feed system applies great force, greater than

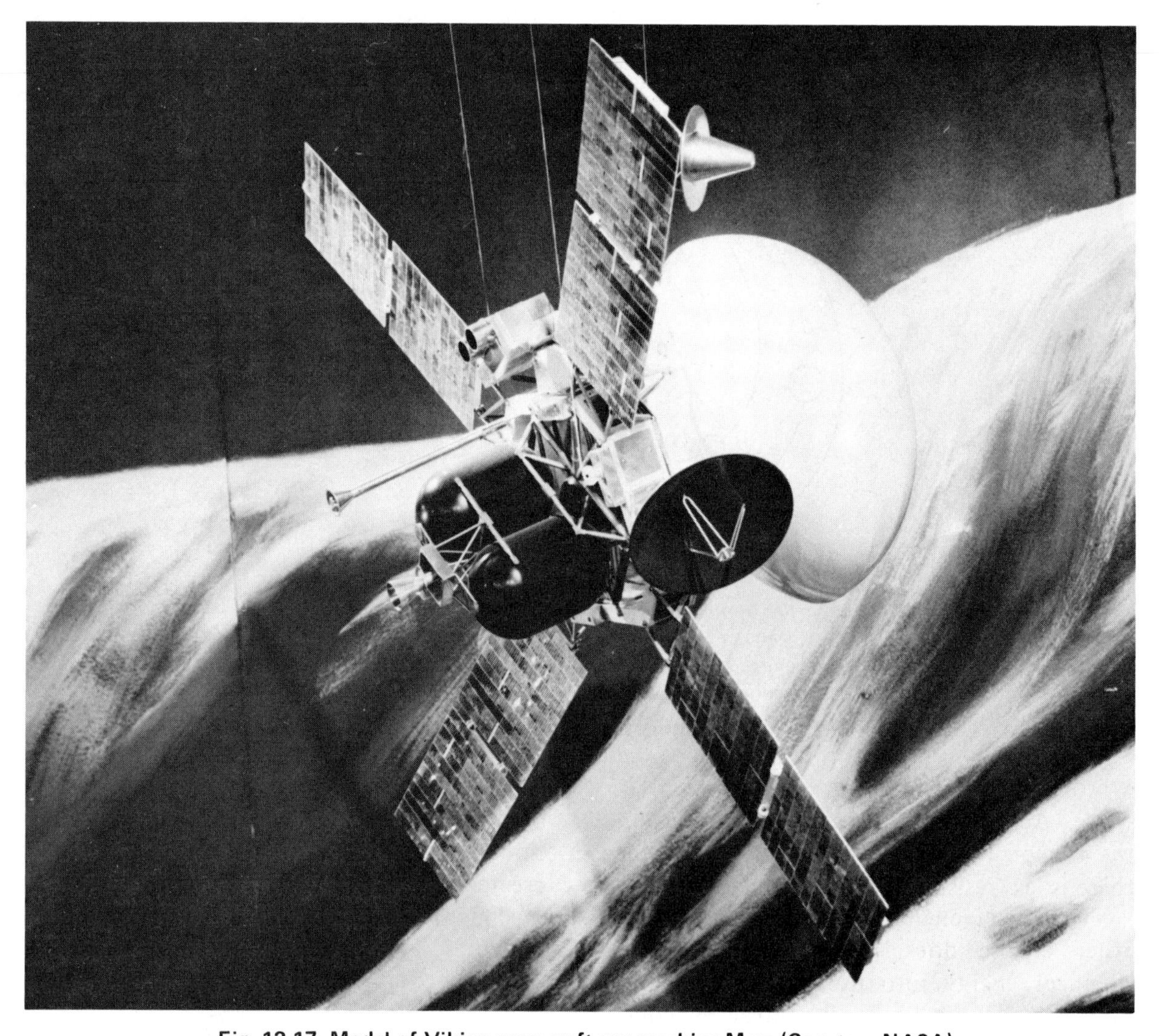

Fig. 12-17 Model of Viking spacecraft approaching Mars (Courtesy NASA)

the high pressures usually found in the combustion chamber. The injector thoroughly mixes the oxidizer and fuel in the proper proportions.

Ignition begins at the face of the injectors as the propellant enters the thrust chamber; burning is continuous. The liquids are converted into hot burning gases that travel back to the nozzle and out of the rocket.

Most of the combustion takes place before the gases reach the nozzle, but some combustion continues even as the gases leave the rocket.

Most rocket engines use the convergent–divergent nozzle for accelerating the gases and producing the thrust, figure 12-21. As the

Fig. 12-18 Scale model of Viking lander (Courtesy NASA)

name suggests, the cross-sectional area of the nozzle is at first small, diverging to a larger cross-sectional area at the rear of the engine.

The speed of the burning gases in the combustion chamber accelerates to 4000 feet (1200 meters) per second as they travel rearward in the convergent section. In the divergent section of the nozzle, the gases continue to expand, increasing the velocity to 6000 to 8000 feet (1830 to 2440 meters). The diverging section of the nozzle actually increases the thrust produced. A simile of the divergent section's action would be a balloon which is forced into the large end of a funnel. When released, the balloon moves away, exerting pressure to thrust the funnel forward.

Rocket engines are classified into two major groups:

- Liquid propellant

Fig. 12-19 The Mars landscape from Viking 1 (Courtesy NASA)

- Solid propellant

Generally, liquid propellants are used where greater impulse and range are necessary. Solid propellants are used for short range missiles.

LIQUID PROPELLANT ROCKETS

The liquid propellant rocket may have a *monopropellant,* which is a mixture of two or more compounds. The oxidizer and fuel are blended together to form a single substance. This has the advantage of requiring only one storage tank and one feed system. An example of a monopropellant would be hydrogen peroxide and alcohol mixed together.

A rocket may have the *bipropellant* system which has the oxidizer and fuel stored in separate tanks. This propellant system requires two storage tanks and a more complicated feed system. However, the bipropellant system has been more successful than the monopropellant system and currently is in wide use. A common system would use liquid oxygen and alcohol.

Feed Systems

The feed system's job is to force the liquid propellant into the combustion chamber. The pressure feed system is relatively simple. It uses gas under high pressure to displace the oxidizer and fuel, figure 12-22. The high-pressure gas is led through the necessary accompanying pressure regulator, feed lines, and control valves and into the propellant tanks. Propellant is delivered to the combustion chamber at the proper rate.

This system is simple, trouble free, and reliable, but there is one limitation. The weight of the high-pressure gas tank is too great for many applications. However, the pressure feed system is used on assisted takeoff units and on rocket sled engines.

The pump feed system uses a centrifugal pump to deliver the propellant. The pump is driven by a turbine, and the system is commonly referred to as the turbopump feed

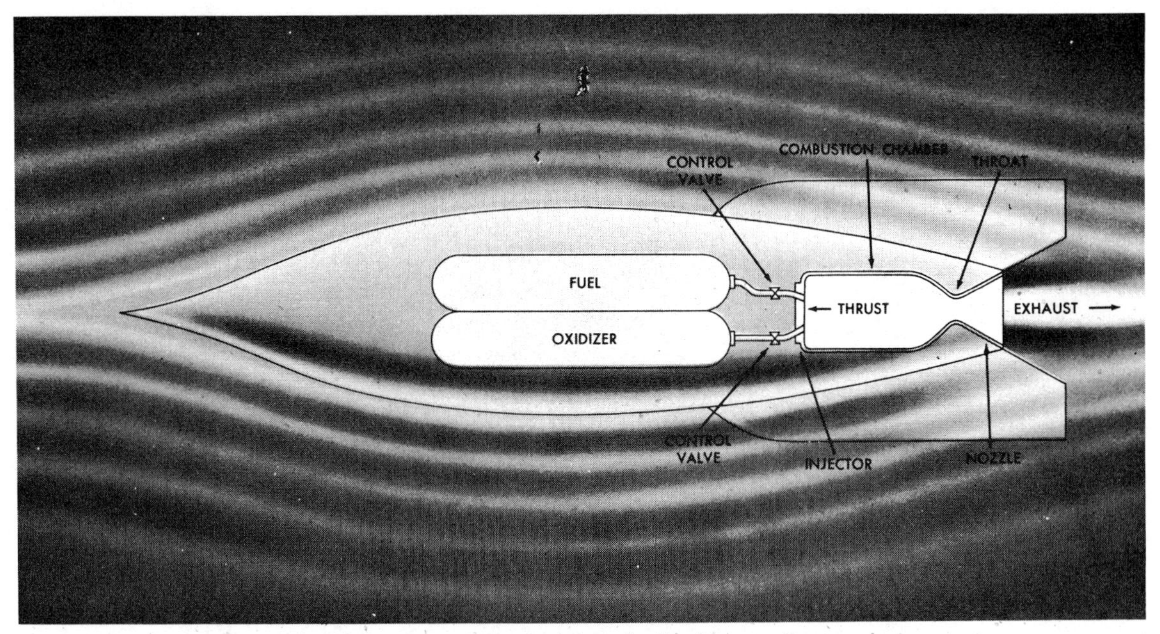

Fig. 12-20 Operating principles of a liquid propellant rocket

system, figure 12-23. The turbine, in turn, may be driven by several sub-systems.

The *gas bleeding* or bootstrap system takes a portion of the hot, high-pressure gases from the main combustion chamber, cools them somewhat, and then uses this power to drive the turbine.

The *gas generator* generates high-pressure gases by a chemical reaction of the propellants. The reaction is similar to that in the

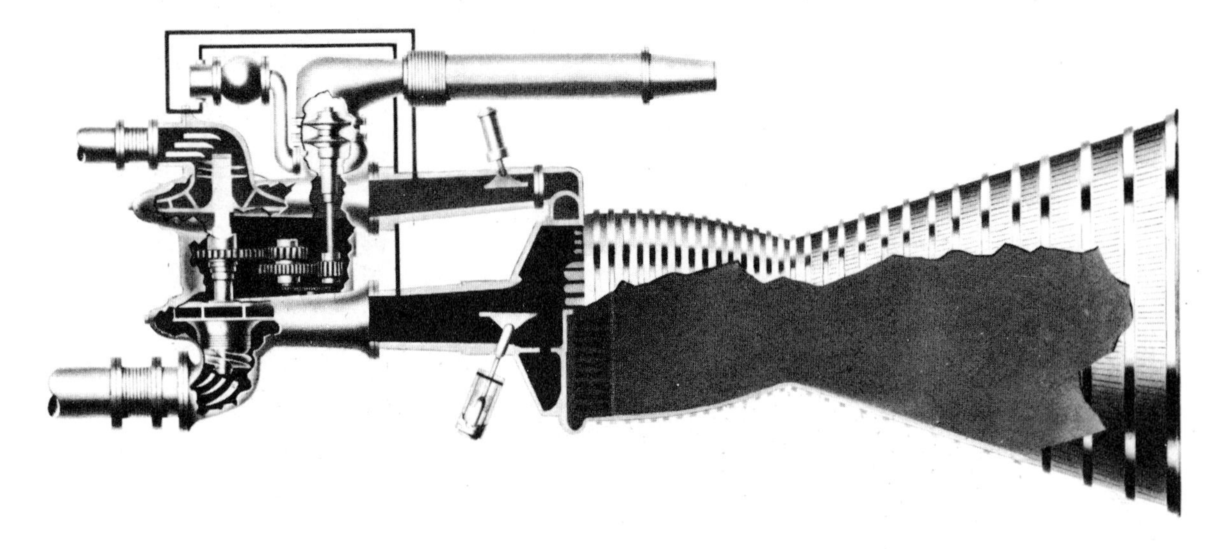

Fig. 12-21 Liquid propellant rocket engine

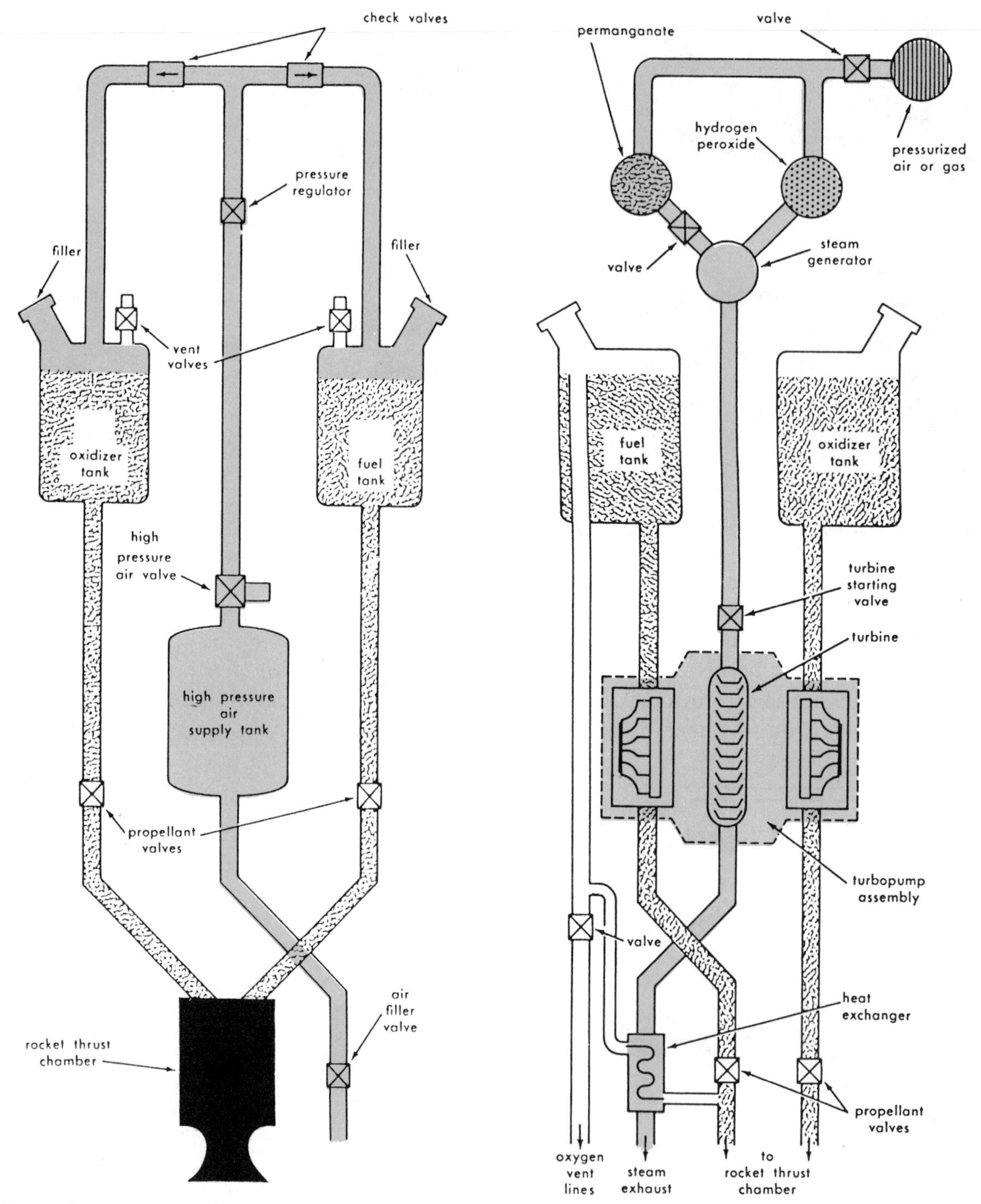

Fig. 12-22 Gas pressure feed system for liquid propellant rockets

Fig. 12-23 Turbopump feed system

main combustion chamber but temperatures are lower. The V-2 rocket used this type of generator to drive its turbine. Potassium permanganate and hydrogen peroxide were sprayed together, the high temperature gases produced by the chemical reaction being used to drive the turbine.

The *high-pressure gas system* is equipped with storage tanks filled with high-pressure air or nitrogen. The controlled flow of the gas from these tanks is directed against the turbine blades which drive the turbopump. A gas generator drives the centrifugal pump which delivers the propellant to the injectors and thrust chamber.

The large, liquid propellant rocket engine for use in the Atlas, Thor, Jupiter, and Redstone missile programs consists primarily of a propellant feed system, control system, and thrust chamber.

Figure 12-22 shows the small spherical gas generator, the power source for the turbopump. The turbine rotates twin pumps forcing fuel and oxidizer through the injector into the combustion chamber. The thermal energy of expanded gases resulting from the combustion of propellants within the chamber is converted to direct kinetic energy as the gases are expelled through the nozzle and produce thrust.

SOLID PROPELLANT ROCKETS

The principal action of a solid propellant rocket is the same as that of a liquid propellant rocket. Both produce thrust as hot gases are discharged through the rocket engine's nozzle. Solid propellant rockets are simpler in design and construction, present fewer servicing problems, and are believed to be more reliable than liquid propellant rockets, figure 12-24. Since they generally produce thrust for a relatively short time, they are used extensively for missile and aircraft boosters, guided missiles, and rocket projectiles.

The propellant is usually made up of a

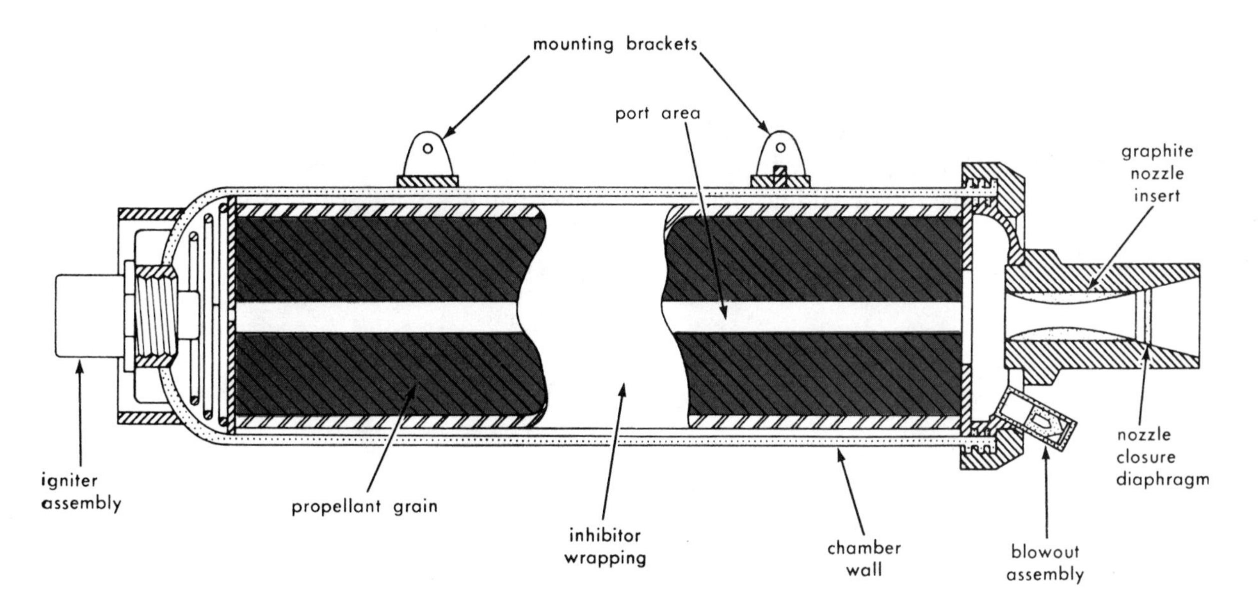

Fig. 12-24 Typical solid propellant rocket

hydrocarbon and an oxidizer. The two are mixed, forming a liquid or plastic mass which is poured into a mold and allowed to solidify or set. In some cases the rocket casing is the mold. The finished unit of propellant is referred to as a grain.

One of the first solid propellants used was black powder. Carbon and sulphur (fuel), saltpeter (oxidizer) and glue or oil (binder) were mixed to form a mass that could be shaped, and then allowed to solidify. A common propellant in current use is Ballistite. It has two chemical bases, nitroglycerine and nitrocellulose. Galcit, composed of an asphalt-oil mixture and potassium perchlorate, is another solid propellant.

The rate of burning, thrust, operating pressure, and duration of burning are determined by the construction of the grain, its size, shape, and exposed burning surface. The burning rate of a propellant is expressed in inches per second, and is composed of the number of inches burned through the propellant each second measured perpendicular to the burning surface. Most propellants burn at a rate of 0.03 to 2.5 inches (0.76 to 63.5 millimeters) per second.

Solid propellant rockets are further classified as restricted burning and unrestricted burning. Generally, the restricted burning propellant will deliver a smaller amount of thrust for a longer time while the unrestricted burning propellant will deliver more thrust but for a shorter time. The restricted burning propellant has some of the exposed propellant surface covered with an inhibitor which controls the burning rate. The unrestricted propellant allows all of the exposed surfaces to burn at the same time. Thrust and time range of typical restricted burning grains of propellant are 100 to 10,000 pounds (445 to 44 500 newtons) thrust for 4 to 120 seconds. The range for typical unrestricted burning grains of propellant is 500 to 100,000 pounds 2220 to 445 000 newtons) thrust for 0.05 to 10 seconds.

Solid propellants are molded into different shapes to produce the desired burning characteristics. *Progressive burning* means that more surface is exposed as the propellant burns, thus increasing thrust. *Regressive burning* means that less surface is exposed as the propellant burns, decreasing thrust. *Neutral burning* means that the same amount of surface is exposed as the propellant burns, maintaining constant thrust.

Rocket engine power may never become a power unit for the average consumer to use, but it has already affected recent history. Our nation's defense rests more and more on weapons powered by rocket engines. The present and future exploration of outer space and other planets is and will be carried on with the aid of rockets that were only dreams a few years ago.

REVIEW QUESTIONS

1. Briefly discuss the rocket's early history.
2. What country was responsible for the V-2 rocket, the forerunner of modern rockets?
3. List some modern uses for rockets.
4. Explain the term air-independent.
5. What are the two major classifications of rockets?
6. Explain the terms monopropellant and bipropellant.
7. What is the function of the rocket feed system?
8. What are the advantages of a solid propellant rocket?
9. What is the solid propellant rocket's major limitation?
10. What is the difference between restricted and unrestricted burning solid propellants?

Unit 13 Gas Turbine Engines

OBJECTIVES After completing this unit, the student should be able to:

- Explain the gas turbine's basic principles of operation.
- List the advantages of gas turbines.
- Discuss the development of the gas turbine in automobiles, trucks, and aircraft.

WHY GAS TURBINE?

The term gas refers to the method that is used to drive the turbine wheel. Various fuels can be used to create the gas: natural gas, naphtha, distillate oil, residual fuels, jet fuels, diesel fuel, and kerosene. Gas, then, is the combustion product of fuel and air that rushes through the turbine, potential chemical energy converted into kinetic thermal energy.

The thermodynamic cycle of the gas turbine is similar to other types of fuel-burning engines. Air and fuel are brought together, or ignited, and the resulting hot gases are directed against a surface that can respond to the pressure–surfaces such as a turbine blade or a reciprocating engine's piston. However, the turbine is a steady-flow process while reciprocating engines are a batch process; a batch of fuel mixture enters on the intake stroke and is processed as a unit.

Gas turbines have a large appetite for air which can be satisfied only by drawing air into the turbine with a compressor. A portion of the output gases are used by the gas turbine to drive the compressor. The compressors themselves may be axial flow, centrifugal flow, or a combination of both.

GAS TURBINES AND JET ENGINES

This introduction may bring to mind the jet engine principles and components that were discussed in Unit 11. It is important to distinguish between the two prime movers, jet engines and gas turbines. The turbojet discussed in the other unit is clearly a turbine type of engine, but the primary purpose of the engine is that of the reaction of the gases producing thrust as they *leave* the engine. And, even though the vast portion of the power produced by a turboprop aircraft is used to power the propeller, still about 10% of the propulsive force is derived from the jet principle as gases leave the engine.

The gas turbines discussed in this unit do not rely on jet thrust principles. Their construction and application is strictly that of a turbine shaft driving a mechanism.

GAS TURBINE APPLICATIONS

Gas turbines have found many uses even though their development, like that of the turbojet, has been underway for a relatively short time. In some applications their use has been rather firmly established, such as in electrical generation and powering helicopters. However, in other applications, such as passenger automobiles, research and development continues.

Applications of turbines can be classified into several areas:

- Gas turbines for trucks, buses, large military vehicles, and off-the-road equipment
- Gas turbines for automobiles
- Gas turbines for helicopters and small propeller aircraft
- Gas turbines for marine use
- Gas turbines for process industries
- Gas turbines for electrical generation
- Gas turbines for multipurpose use

BASIC PRINCIPLES

The basic principle of the gas turbine engine is converting the kinetic energy of hot expanding gases into mechanical rotary energy of a turbine wheel. The typical gas turbine first brings large quantities of air into the engine with the aid of a compressor. Fuel enters the combustion chambers where it is combined with the air and ignited. The hot, expanding gas is directed against the buckets or blades mounted on the turbine wheel. In addition to the gasifier turbine that drives the compressor, there is a power turbine that provides shaft power to the driven machine. It is at the turbine wheel that most of the kinetic energy of the expanding gases is converted to mechanical rotary energy. An operating cycle of this type is referred to as a simple *open cycle* because the gases are exhausted directly to the atmosphere, figure 13-1.

It is in the area of exhausting gases that heat and efficiency can be lost in any engine. The waste involved in the reciprocating piston engine is obvious. Hot gases are removed from

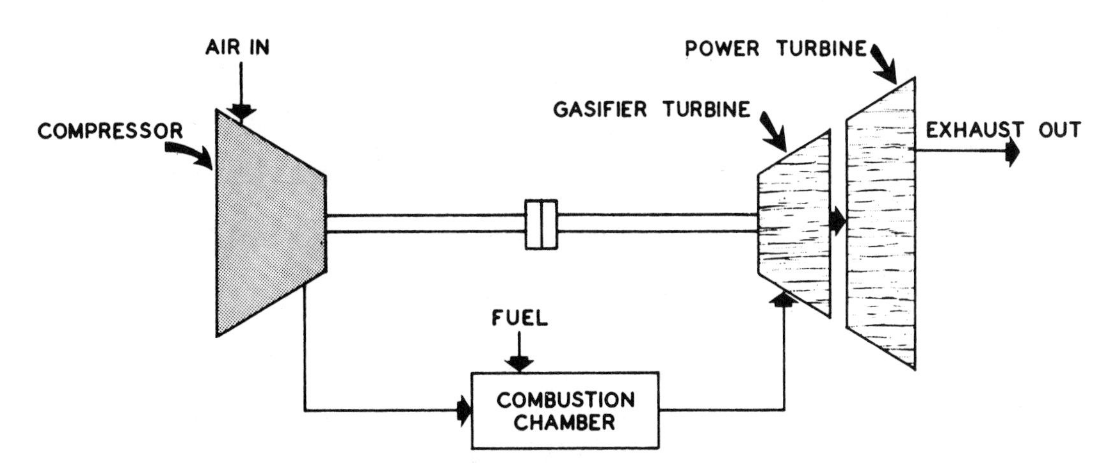

Fig. 13-1 Simple open cycle gas turbine

the engine even though much heat energy remains. Gas turbines are frequently designed to minimize this type of waste. Passing the gases through several stages of turbine wheels uses the energy to a greater degree than a single turbine wheel. Also, a *regenerator* that uses exhaust heat to preheat the incoming air increases efficiency, figure 13-2.

Another technique that can be used to increase thermal efficiency is to exhaust the gases from the gas turbine to a steam turbine system forming a combined cycle, figure 13-3.

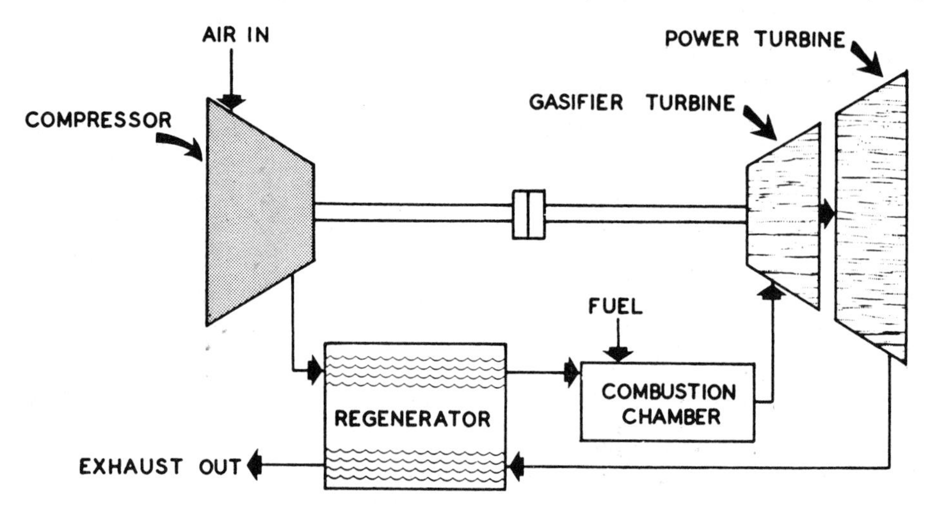

Fig. 13-2 Regenerative open cycle gas turbine

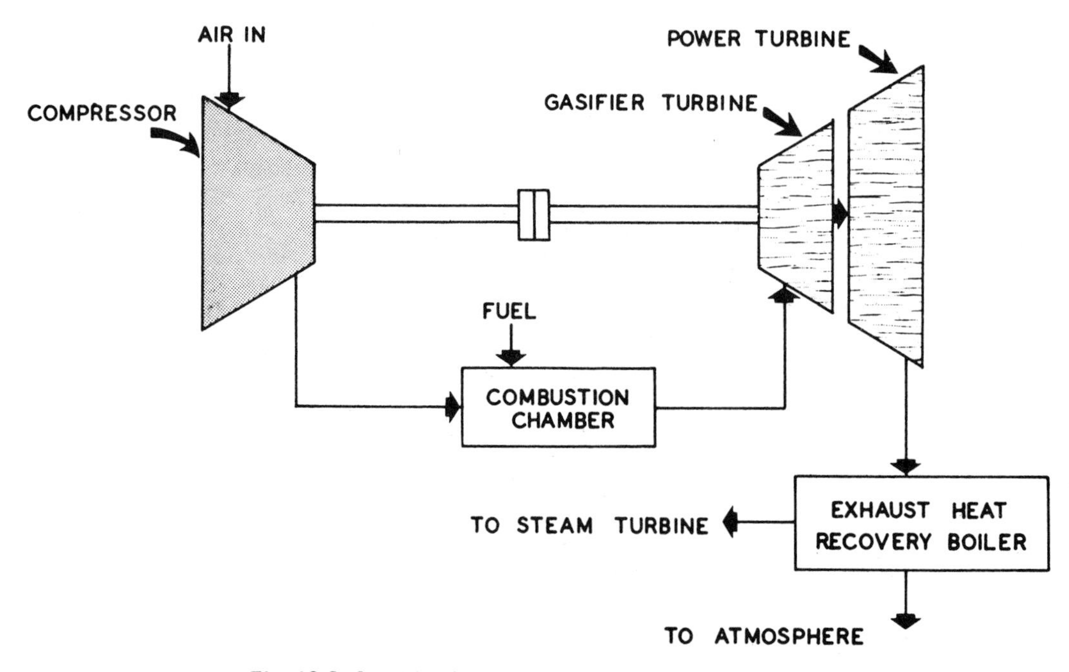

Fig. 13-3 Combined cycle gas turbine with steam turbine

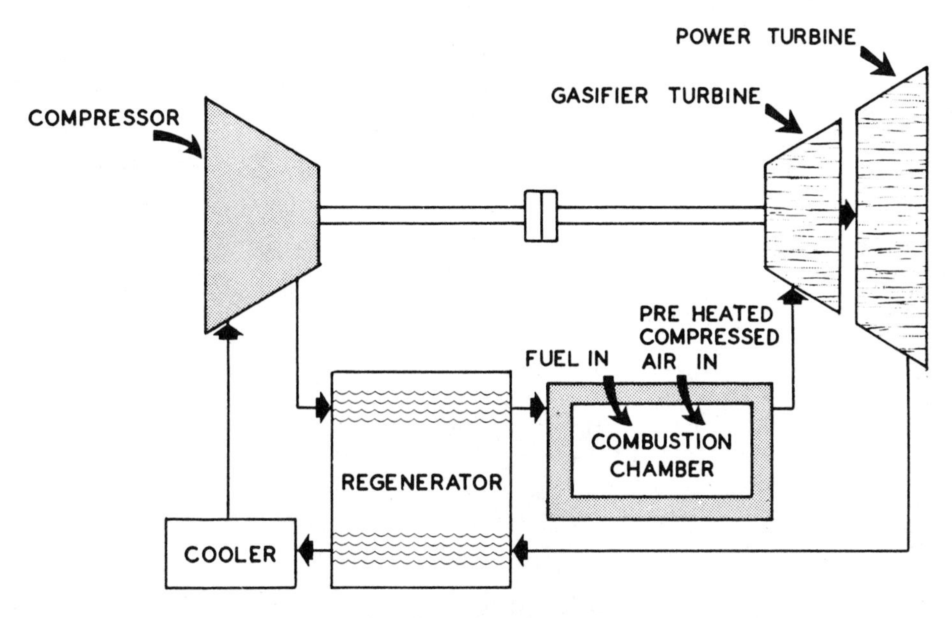

Fig. 13-4 Closed cycle gas turbine

Gas turbines can also operate on the *closed cycle* principle, figure 13-4. In this case, hot combustion gases superheat, but do not contact, the working gas which is expanded through the gasifier and power turbines, channeled to the regenerator where it gives up heat, and then to the cooler for further temperature reduction. The gas is then compressed and ready for expansion through the system. Since the working gas does not come in direct contact with the combustion products, friction losses are reduced as well as turbine wear. The incoming combustion air also has the benefit of the regenerator. Any fuel may be used: oil, gas, or solid.

ADVANTAGES OF GAS TURBINES

Gas turbines have many advantages. High on the list is the smooth, vibration-free power that is produced. There are no reciprocating parts, just continuous rotary motion. Gas turbines require little warmup time, so they can be brought up to full operating power in a short time. Of course, advantages vary with the application but it could be further said

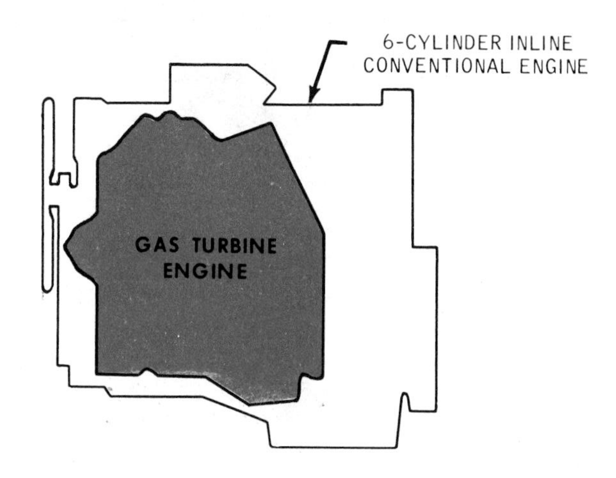

Fig. 13-5 Showing comparison of sizes between gas turbine and standard 6-cylinder engines

that gas turbines:

- compete very well with other prime movers on size and weight per horsepower, figure 13-5.
- operate well on a wide variety of relatively inexpensive fuels.
- present no significant maintenance problems and, in fact, are considered to be easy to maintain.

In the case of vehicular applications, there are additional advantages. Gas turbines:

- are air-cooled; no radiator freezing problems.

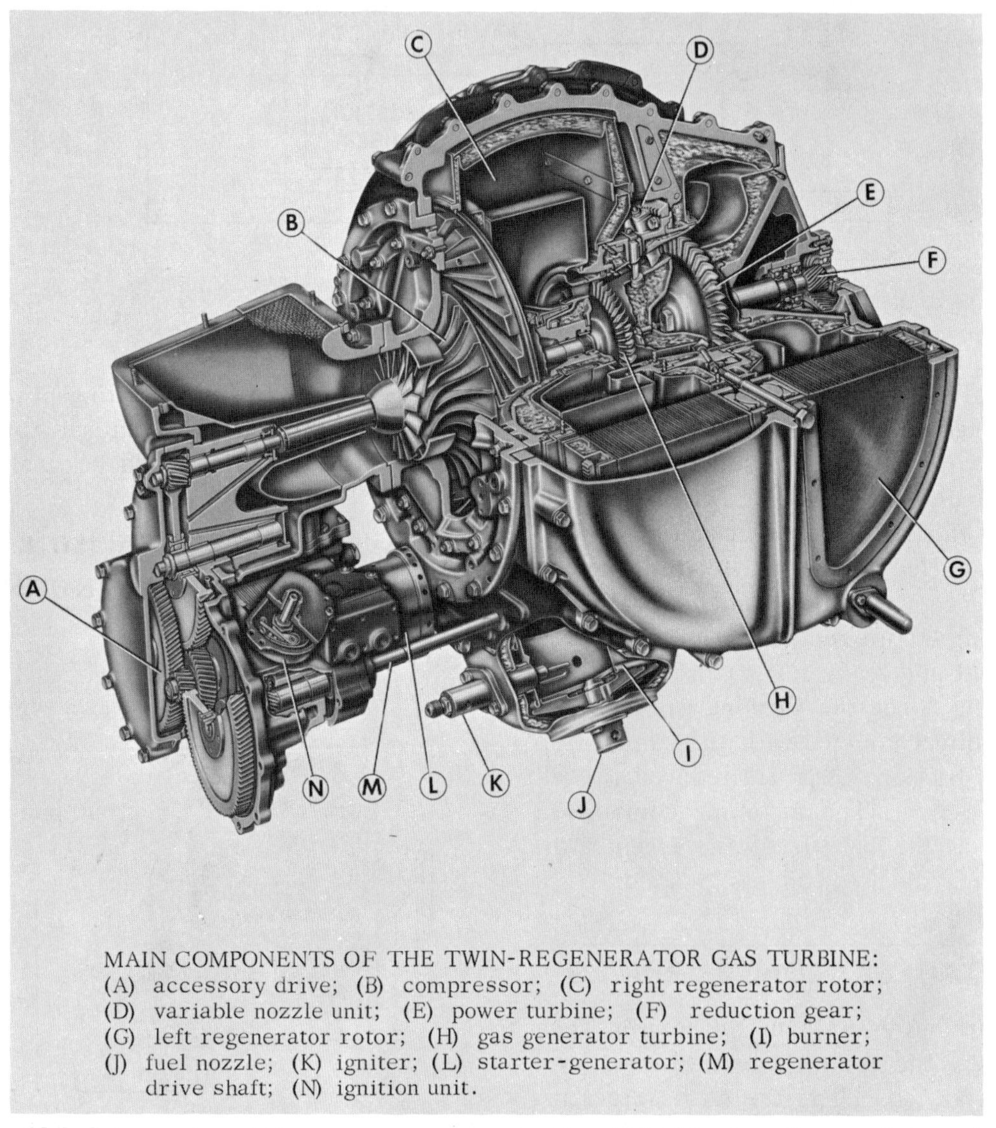

Fig. 13-6 Cutaway view of Chrysler regenerative gas turbine A-831 (Courtesy Chrysler Corporation)

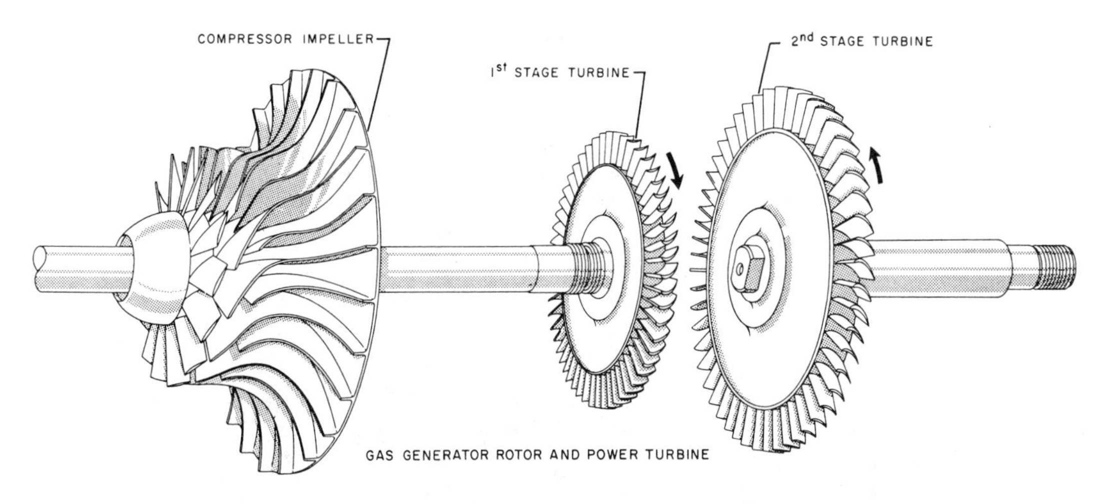

Fig. 13-7 Basic parts of the two-shaft turbine

- have fewer moving parts than reciprocating engines.
- start easily in cold weather.
- combustion is very complete. Exhaust is cleaner than that of other engines and, therefore, the gas turbine represents less of an air pollution problem.
- do not stall under overload.
- supply heat for passengers instantly.

SPECIAL CONSIDERATIONS

Gas turbines do require some special considerations in certain areas. One is their appetite for clean air. The air must be well cleaned or it will damage the precision parts within the turbine. Turbines also operate best at top speed. To slow a turbine down by reducing the fuel available lessens the efficiency of the turbine. These problems have been difficult for engineers and researchers to solve, especially in vehicular applications.

VEHICULAR DEVELOPMENT

Vehicular turbine applications must solve the problems of fuel consumption, wide variety of speed requirements, exhaust temperature, and engine braking. The regenerative gas turbine is able to do this, figure 13-6.

The turbine developed by Chrysler engineers is a two-shaft regenerative type. Figure 13-7 illustrates the first stage gas generator rotor which drives the compressor, and the second stage power turbine which provides power for propelling the auto. The regenerators are rotating heat exchangers that extract heat from the exhaust gases and then use this heat to preheat air as it comes from the compressor to the burner or combustion chamber, figure 13-8. The regenerators serve to increase fuel economy and reduce temperature.

Air is drawn into the turbine and compressed by a centrifugal compressor. The compressed air picks up heat as it passes through the regenerator. Arriving at the burner, the air is burned with the fuel. Through the combustion process, the energy level of the fuel is

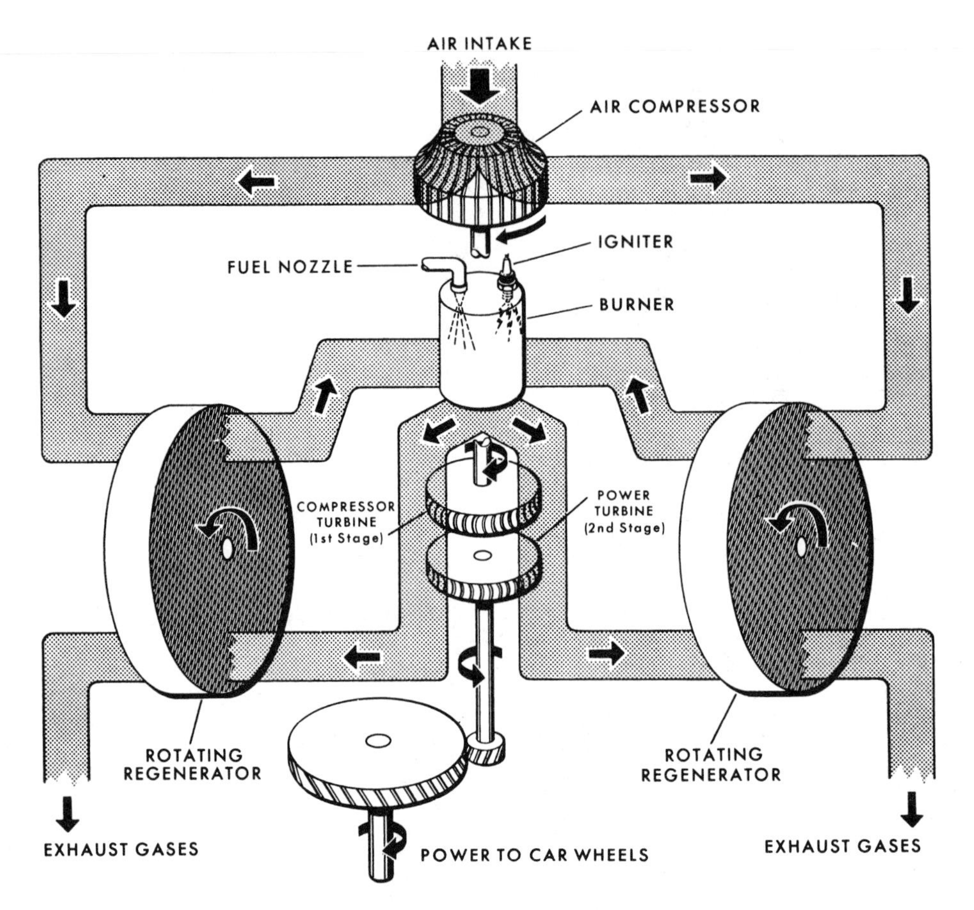

Fig. 13-8 Diagram of gas turbine operation

increased and this gas is applied to the first stage turbine which drives the compressor. The hot gases are then applied to the second stage turbine or power turbine which drives the auto through transmission. Before entering the exhaust ducts, the hot gases pass through the honeycomb regenerator drums.

The first stage turbine always rotates whenever the turbine is operating. This can be anywhere from 18,000 revolutions per minute to idle to 44,600 revolutions per minute at full power. The power turbine speed may range from zero when the auto is standing still to a maximum of about 45,700 revolutions per minute.

Fixed nozzles direct the hot gases against the first stage turbine wheel. However, at the second stage or power turbine wheel, the variable nozzle is used, figure 13-9. This is a feature of the Chrysler turbine. The variable nozzles permit engine braking or acceleration at all speed ranges. The nozzles are linked to the accelerator, which changes nozzle direction. At starting or idle, the nozzles are in an axial position, but as the accelerator pedal is depressed, the variable nozzles are turned toward the direction of turbine wheel rotation. The variable nozzles are also used to provide engine braking. If the auto is traveling at more than 15 miles per hour (24 kilometers per

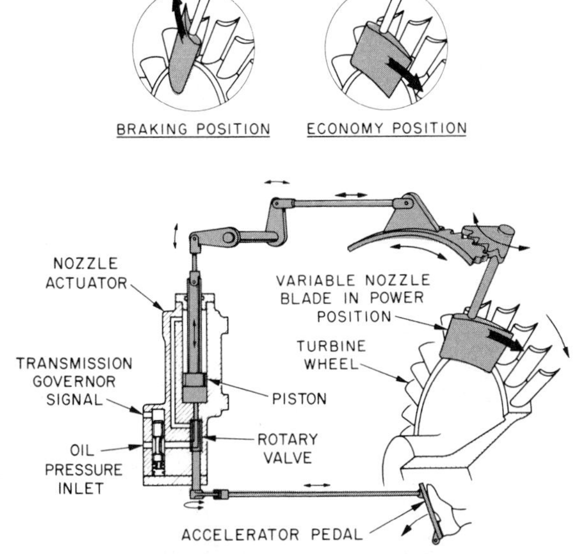

Fig. 13-9 Chrysler variable nozzle system (Courtesy Chrysler Corporation)

hour) and the accelerator is released, the nozzles are moved to such a position that the gases will be directed against the turbine wheel's direction of rotation. At speeds of less than 15 miles per hour, release of the accelerator will cause the nozzles to assume an axial position or open idling position.

Fuel control must be variable to insure a high degree of operating economy. Operating at full power for all auto speed requirements would be quite wasteful. This fuel control system consists of a fuel pump, governor, pressure regulator, and metering orifice. At constant speeds, the governor will provide fuel to the burner in response to the position of the accelerator. Fuel flow will shut off when the accelerator pedal is released, but fuel will resume flowing when idle speed is reached.

This engine has excellent acceleration and flexibility as shown by its horsepower and torque curves, figure 13-10. It is difficult to stall this turbine because even if torque requirements become very great, the power turbine will merely slow down. The compressor will continue to function normally.

General Motors has been researching the

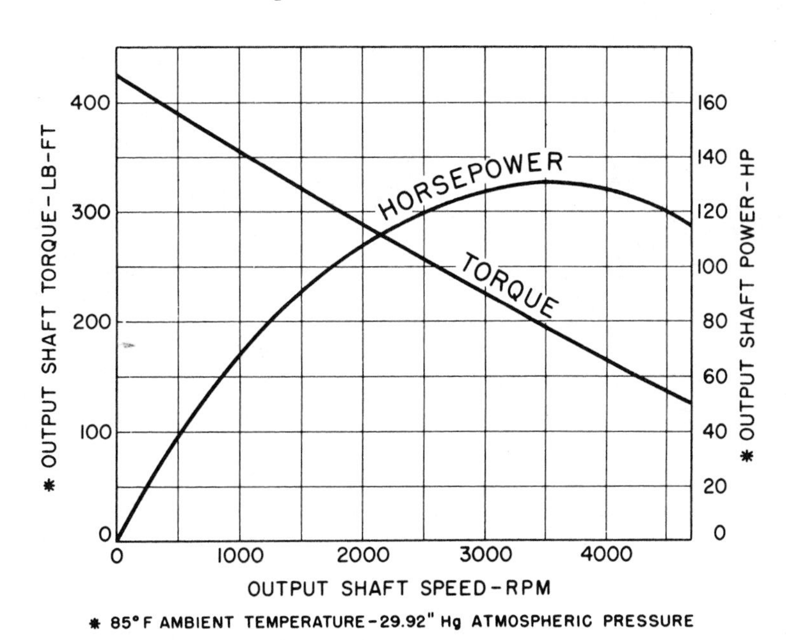

Fig. 13-10 Horsepower and torque curves for Chrysler turbine (Courtesy Chrysler Corporation)

use of the gas turbine for automobiles since the early 1950s. Their GT series is used on the Firebird experimental automobiles, figure 13-11. The GM engineering staff's latest work on gas turbines designed for passenger cars is the GT-225, a small size regenerative turbine, figure 13-12. It is being used to test compressors, combustors, regenerators, controls, and new materials. This turbine can burn kerosene or diesel fuel. Hydrocarbon

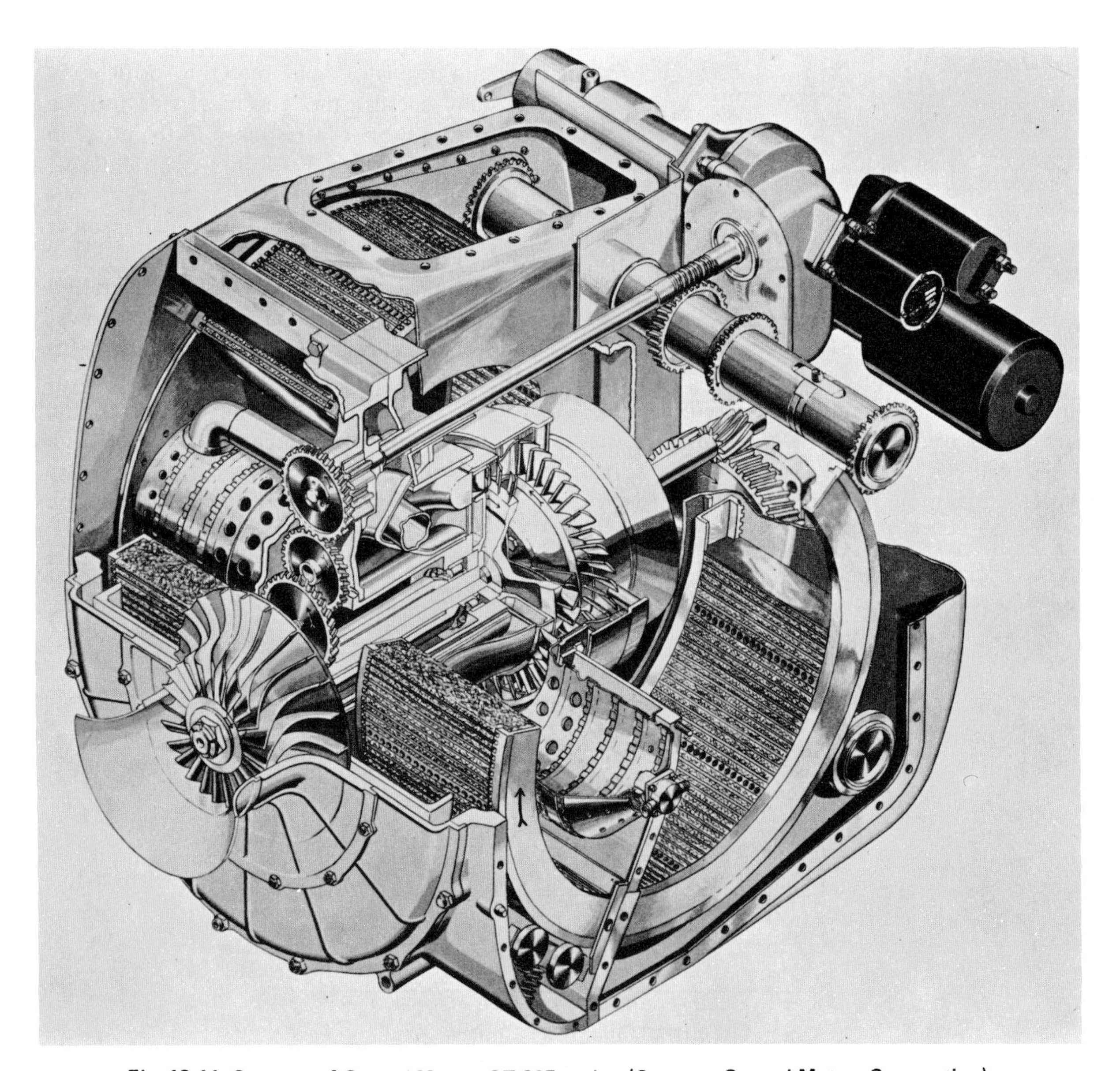

Fig. 13-11 Cutaway of General Motors GT-305 engine (Courtesy General Motors Corporation)

and carbon monoxide emissions are low, but the high emission of oxides of nitrogen is a problem with turbines. However, the GT-225 engine equipped with a pre-chamber combustor was able to meet the 1978 Federal emission standards. Research continues on fuel

Fig. 13-12 General Motors AGT-1 gas turbine and the latest, smaller AGT-5 (Courtesy General Motors Corporation)

Fig. 13-13 General Motors GT-225 experimental gas turbine (Courtesy General Motors Corporation)

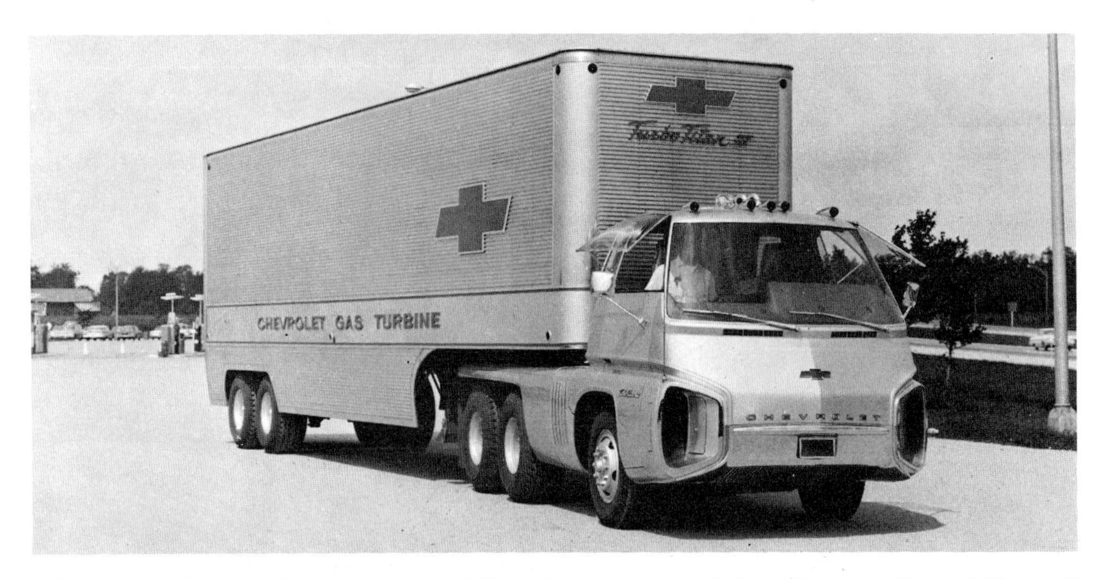

Fig. 13-14 Note that the air scoops on this GM truck turbine installation. (Courtesy General Motors Corporation)

economy and engine durability, figure 13-13.

High noise level and high exhaust temperatures have led research away from the automobile toward development of engines capable of competing directly with the diesel in the heavy-duty vehicle market, figure 13-14. The GT-309 turbine by GM has developed into the 404 and 505 industrial gas turbine engines, figure 13-15.

General operating principles for this regenerative gas turbine are similar to smaller automobile units. A centrifugal compressor compresses air to 3.8 times atmospheric pressure and directs it through the regenerative drum to the combustors. High energy gases from the combustion process are directed against the gasifier turbine and then against the power turbine.

The General Motors turbine employs a power transfer system, figure 13-16, to insure high efficiency at part throttle and to provide for engine braking. The independent gasifier turbine and the power turbine shafts are equipped with a multiplate slipping clutch which is the heart of the power transfer system. At part load, a portion of the gasifier turbine power is transferred to the power turbine side of the engine. At less than full load, the system slows the power turbine as a slight amount of pressure is applied to the clutch. The governor on the gasifier turbine responds by increasing fuel output to increase speed. The net result is increased turbine inlet temperature and the higher turbine efficiency that can be gained with high temperatures. With power transfer, the gasifier turbine temperature is maintained at a constant high level, temperatures that will permit high thermal efficiency. Clutch pressure is controlled by a pressure programmer which senses gasifier turbine torque and speed as well as temperature and altitude.

Fig. 13-15 General Motors GT-309 turbine engine for heavy-duty vehicular applications (Courtesy General Motors Corporation)

Engine braking is accomplished by locking the power transfer clutch which causes the output of the power turbine to drive the compressor, dissipating the heat energy as air is pumped through the engine.

Efforts at the Ford Motor Company are in the area of passenger car applications for turbines, figure 13-17. Research is directed at finding materials and manufacturing processes that can produce turbine parts inexpensive enough for passenger car use. These parts must be able to withstand high turbine temperatures while maintaining exacting engine tolerances. Ceramic components are being considered to replace the expensive high-temperature metal parts. This permits an increase in engine temperature, allowing the engine to be made smaller and more economical to operate. Fuel economy is expected to be better than a comparable internal com-

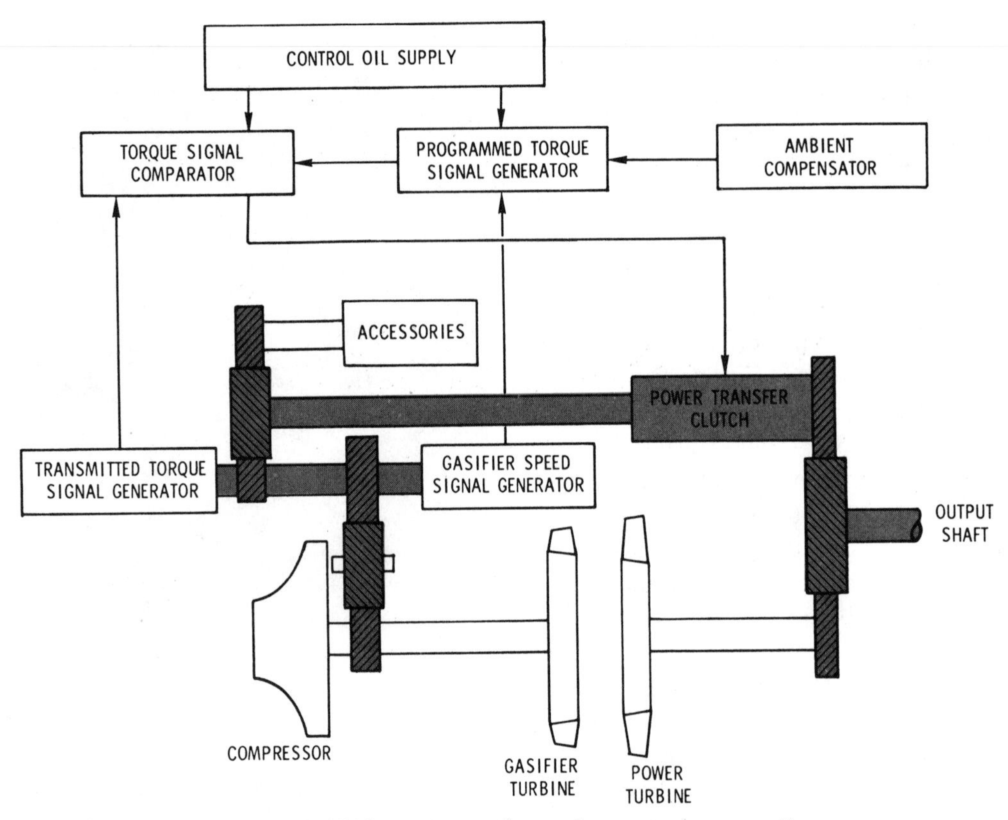

Fig. 13-16 GT-309 power transfer engine control system diagram

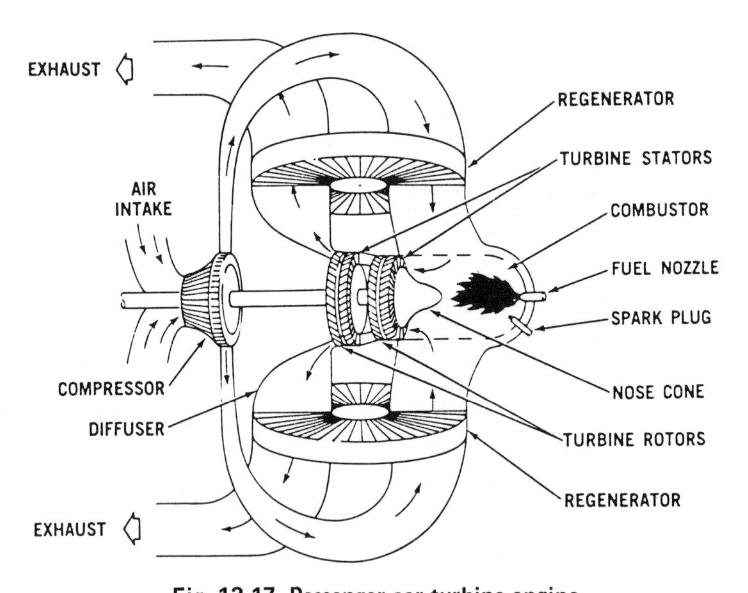

Fig. 13-17 Passenger car turbine engine

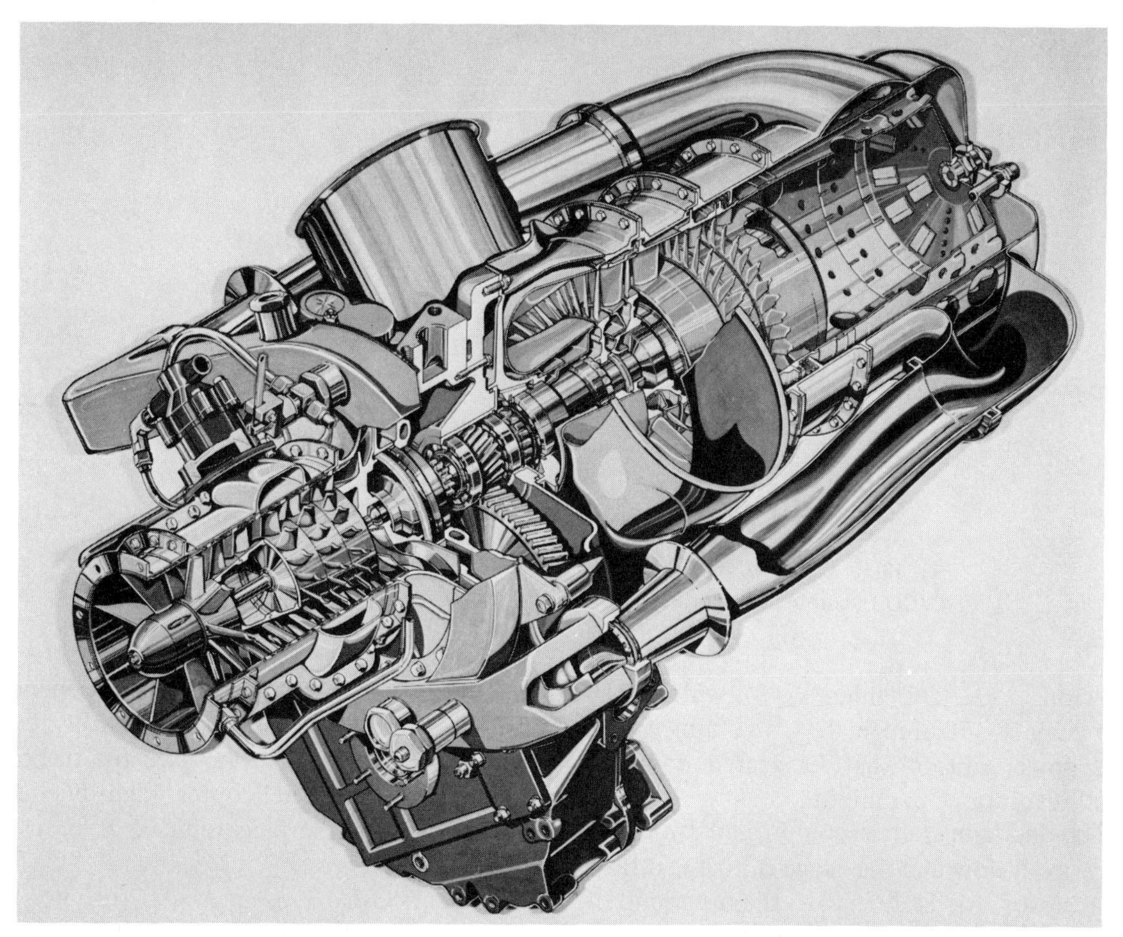

Fig. 13-18 Cutaway view of a Model 250 turboshaft engine (Courtesy of Allison Gas Turbine Division of General Motors)

bustion engine. New combustion systems are also being developed to reduce the emissions of nitrogen oxides. Turbines are already low in hydrocarbon and carbon monoxide emissions. It can take several years to determine if a competitive passenger car turbine is possible.

AIRCRAFT DEVELOPMENT

Jet engines revolutionized the field of commercial and military aviation with turbojet and turboprop engines. Gas turbine manufacturers developed gas turbine engines for smaller fixed-wing, propeller-driven aircraft and helicopters. In these applications, shaft power is used rather than jet thrust. The Allison Model 250 turboshaft light engine falls into this category, figure 13-18. The model 250 was designed specifically for the Army's new observation helicopter. The 250-C18 weighs only 136 pounds (62 kilograms), yet can develop 317 horsepower (236 kilowatts). The engine dimensions are a compact

Fig. 13-19 Cutaway of the AVCO Lycoming T-53 gas turbine (Courtesy AVCO Lycoming Division)

40″ long x 22 1/2″ high x 19″ wide (1016 mm long x 572 mm high x 483 mm wide). The power output shaft is geared down to 6000 revolutions per minute.

The incoming air is compressed by a six-stage axial flow and one-stage centrifugal flow compressor. Air is ducted to the other end of the engine and into the single large combustor. Gases pass through a two-stage gasifier turbine which drives the compressor, then to a two-stage power turbine which delivers power to the output shaft. Exhaust gases are ducted from the engine upward through twin ducts.

In the future, engines such as these may be equipped with regenerative devices which will increase the engine's thermal efficiency and the potential applications for the engine.

One of the outstanding universal-purpose turbines that has been developed is the T-53 by AVCO Lycoming, figure 13–19. Depending on the gearing, accessories, and hardware, the basic T-53 turbine can be used for turbofan, turboprop, geared helicopter, high-speed helicopter, marine, surface vehicle, or industrial applications.

The T-53-L-13 is a shaft turbine helicopter engine that develops 1400 shaft horsepower (1 megawatt) at 6300 revolutions per minute, yet weighs only 530 pounds (241 kilograms). The overall dimensions of the engine are 47.61 inches (1209 millimeters) along with a diameter of 23 inches (584 millimeters). The AVCO AFT-1500 is a vehicular version that develops 1500 brake horsepower (1.1 megawatts) at 3000 revolutions per minute. Again, the size is quite small. It is 28″ high x 60″ long x 40″ wide (711 mm high x 1524 mm long x 1016 mm wide). It weighs 1600 pounds (427 kilograms).

The T-53 turbine has a five-stage axial compressor followed by a single centrifugal compressor. The compressed air is directed to the rear of the turbine and into the combustion chamber. Fuel atomizers ring the chamber and, after initial spark ignition, the burning is continuous. Gases pass through two compressor turbines and two power turbines. The power turbine shaft extends forward and concentrically through the compressor shaft, thus delivering shaft power at the forward or cold end of the engine.

ELECTRICAL POWER GENERATION

The field of large horsepower prime movers has long been dominated by the steam turbine. However, the gas turbine has in recent years been making its place in this important field, figure 13-20. Large, heavy-duty gas turbines may range in size from several thousand horsepower to more than 20,000 horsepower. Except for the fact that they use gas directly from combustion, these turbines closely resemble those using steam, figures 13-21 and

Fig. 13-20 Complete installation of Westinghouse gas turbine power plant (Courtesy Westinghouse Electric Corp. Power Generation Group)

Fig. 13-21 Axial flow compressor and reaction turbine rotor for gas turbine power plant

13-22. The prime applications of these turbines is electrical power generation where, depending on the installation of the turbine, they may be used in base load generation or in peaking. *Peaking* refers to supplying the needed generation requirements during times of heavy demand on the system. In this application, the base load may be delivered by a conventional steam generator with a smaller gas turbine supplying peaking requirements. Here the gas turbine brings its advantages of quick start and low maintenance into play. Base load installations may be of a combined nature, steam turbine and gas turbine, using the gas turbine exhaust gases to preheat the air before it enters the boiler. Such a combined cycle may increase thermal efficiency as much as nine percent.

OTHER APPLICATIONS

Large gas turbines are used as the prime mover to supply the many phases of an industrial complex, providing electricity, hot water, and heat for the various industrial processes of the industry. By integrating the gas turbine into the process, industries such as chemicals, petroleum, petro-chemical, paper, and steel, among others, a complete power package can be provided. These industries require large amounts of heat and power and are natural targets for gas turbine installations.

Fig. 13-22 Split casing for gas turbine shaft and turbine wheels

REVIEW QUESTIONS

1. Explain how the gas turbine's appetite for air is satisfied.
2. List several fuels that gas turbines are capable of using.
3. Explain the difference between a turboshaft engine and a turbojet engine.
4. Can a gas turbine use both turboshaft and turbojet principles? Explain.
5. List several applications for gas turbine engines.
6. Explain several of the advantages of gas turbine engines.
7. Explain the regenerative principle and the advantage of its use.
8. What is the function of the gasifier turbine?
9. What is the function of the power turbine?

10. Describe the systems that can be used to insure efficiency of the gas turbine at slow speeds and also provide for engine braking.

11. Explain how a gas turbine can be used in a combined cycle with a steam turbine.

SECTION 4 EXTERNAL COMBUSTION ENGINES

Unit 14 Steam Engines and Steam Turbines

OBJECTIVES

After completing this unit, the student should be able to:

- **Explain the difference between internal and external combustion engines.**
- **List the parts of a typical steam engine.**
- **Describe how an impulse turbine is different from a reaction turbine.**
- **Discuss the uses of steam engines and turbines.**

The steam engine is classified as a heat engine. The chemical energy of a fuel is released in the form of heat energy which the steam engine is designed to harness. Unlike most engines, the steam engine can be made to operate on almost anything that will burn: wood, coal, petroleum, or their many variations. The steam engine cares little what fuel is used, just as long as the water is boiled and steam is produced. The reason why steam engines can operate on many different fuels is that the fuel is not burned within the engine itself. Fuel is usually burned in a fire box under a boiler, outside the working parts of the engine. The engine is called an *external combustion engine*.

The energy path of a typical steam engine starts with the chemical energy of the fuel being converted into heat energy. This heat boils the water to produce steam, which is confined in the boiler. High steam pressures build up and, when released and channeled,

Fig. 14-1 Hero's aeolipile—steam put to work more than 2000 years ago

are able to move the engine's piston or turbine parts. It should be remembered that when water is converted into steam it expands at the rate of 1600 to 1. It is not difficult, therefore, to see how steam pressures can be built up in a closed boiler.

It is almost impossible to credit any one person with the invention of the steam engine. Many scientists, engineers, and inventors contributed their ideas down through the years, each person building upon the accomplishments of forerunners. Modern, powerful, efficient steam engines and turbines evolved from many crude, inefficient, and often unsuccessful efforts.

Hero's aeolipile which was devised 2000 years ago—and which was used in Unit 11 to illustrate jet propulsion can also be used here to describe the power of steam—the two illustrations are equally valid. Remember, the aeolipile was basically a hollow sphere that was spun around by jets of steam coming from within the sphere itself, figure 14-1. A boiler fed steam to the sphere. The action of the aeolipile can also be compared to water coming from a spinning lawn sprinkler. This novelty of the time could have been one of the first doors opened to the possibilities of steam power.

FIRST MAJOR DEVELOPMENTS

Further progress was almost nonexistent until about 300 years ago when Lord Worcester, Captain Thomas Savery, and Thomas Newcomen began to work on engines to pump water from coal mines in England, where the water, seeping into and flooding the mines, had created a need for a good pumping device. Thomas Newcomen built several hundred steam engines that were used throughout England's coal mines, figure 14-2. His pump had a rocking action: a large seesaw beam with the pump on one end and the steam engine on the other. Coal burned under the boiler, boiled the water and produced steam. Steam was let into the cylinder as the piston moved upward. The steam valve was shut off. Cold water was sprayed into the cylinder to condense the steam. Atmospheric pressure pushed the piston back down the cylinder; the pump on the other end raised water out of the mine. Since the power delivered was actually produced when the steam was condensed and atmospheric pressure, which was greater, pushed the piston down the cylinder, Newcomen's engine was more an atmospheric engine than a steam engine.

Newcomen's engines consumed an enormous amount of coal and, in spite of the large number of them in use, they were very

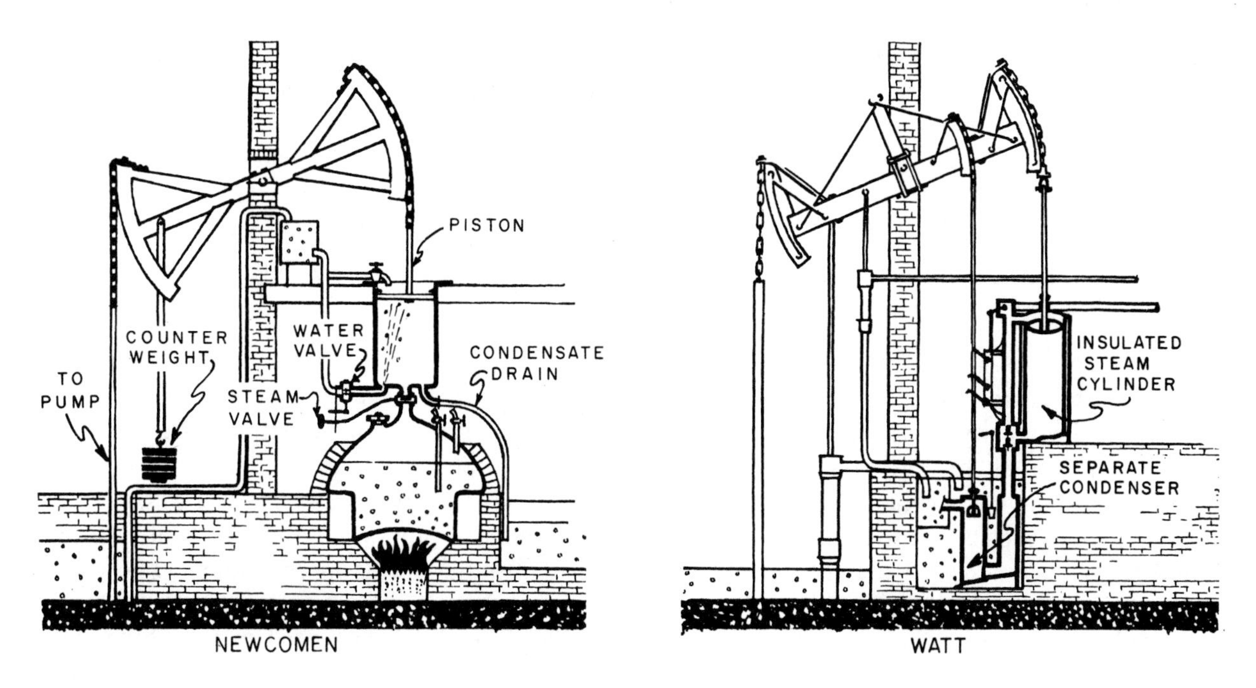

Fig.. 14-2 Early steam engines produced by Newcomen and Watt

inefficient. As coal prices rose, the cost of operating these engines became a problem.

James Watt studied Newcomen's engine and detected some of its faults. He reasoned that the principal cause of inefficiency was that the cylinder was first heated by the steam, then cooled by the condensing water. This alternate heating and cooling wasted much power. His solution was to connect a separate tank for condensing the steam. With this condensing tank connected to the cylinder, the cylinder could be kept hot all the time. Watt also devised a way to make the engine double-acting, that is, it would operate on both the push and pull strokes. The engines that Watt developed were more efficient than Newcomen's and widely used in England

Methods were developed to convert the up and down motion of the engine into rotary motion. Engines became more versatile and were soon found powering the wheels of many industries: paper mills, corn mills, cotton spinning mills, breweries, and iron works. The steam engine was a contributing cause of the Industrial Revolution.

Watt's steam engines were heavy, so they were built to remain in one location. The eventual answer to reducing the size and weight of steam engines lay in the use of high-pressure steam, but higher pressures required boilers that could withstand these pressures without bursting. Richard Trevithick, an Englishman, successfully developed high-pressure boilers that operated at 60 pounds per square inch (408 kilopascals) pressure.

Lighter, high-pressure steam engines were adaptable to vehicles; the locomotive lent itself best to the use of steam. Trevithick built a steam locomotive and, though it was not entirely successful, it did work. George

Stephenson was the engineer who brought together all of the best steam propulsion ideas to build a practical steam locomotive. His locomotive, the *Rocket*, sparked the rapid development of the railroad industry.

MODERN DEVELOPMENTS

The modern steam engine has changed little in its *basic* components as it has developed. Two important factors, however, have contributed to the steam engine's increasing *efficiency:* higher steam pressure and higher steam temperature. A small amount of steam placed under conditions of high pressure and high temperature is capable of producing considerable power as it expands, pushing the piston down the cylinder. Steam locomotives that produce as much as 6500 horsepower (4850 kilowatts) have been built, for example. Steam pressures on locomotives have increased to the range of 300 pounds per square inch (2 megapascals).

Even though the steam locomotive has been replaced by the diesel-electric locomotive, it is worthwhile to consider what takes place within a typical railroad locomotive, figure 14-3.

Steam Locomotive

Fuel is burned in the firebox, and the hot gases of combustion are drawn through the boiler, which consists of many steel pipes surrounded by water. As the hot gases sweep through the pipes, some of the heat is transferred to the water in the boiler. The water boils and the steam rises to the top of the boiler where it is collected in the dome. The steam next travels through superheaters where its temperature is increased. The superheated steam is then introduced into the steam chest

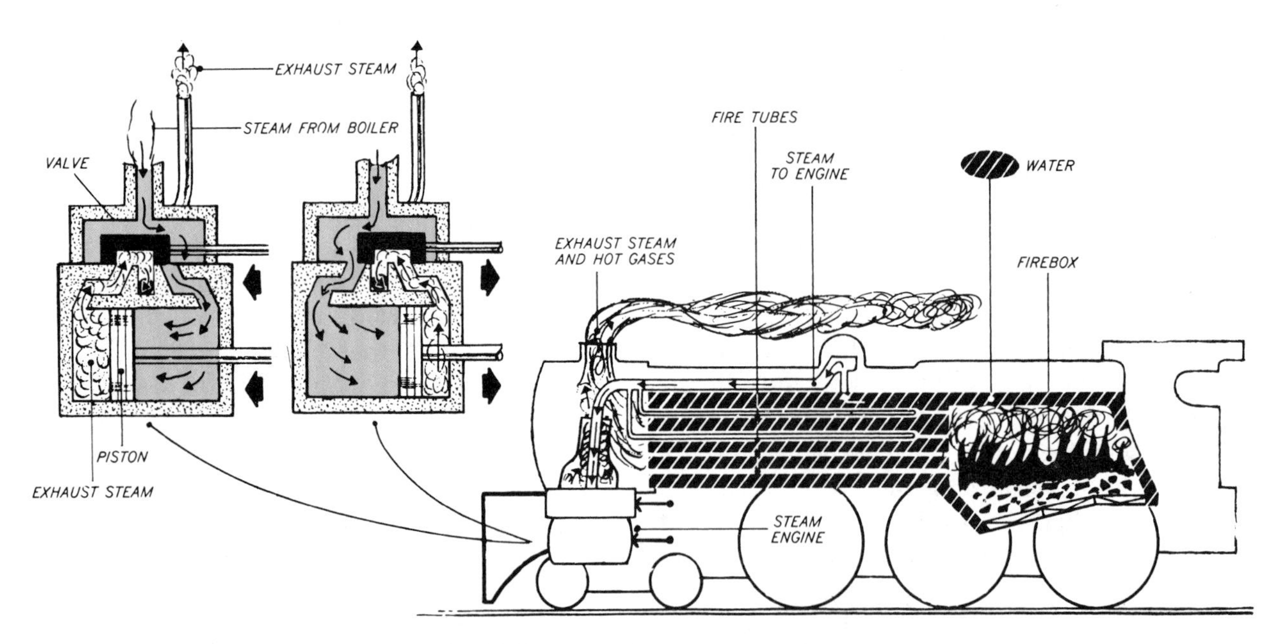

Fig. 14-3 A typical steam locomotive

and is ready to deliver its energy. The sliding valve within the steam chest is designed to deliver steam to one side of the piston. When the piston has completed its stroke, the sliding valve admits steam to the other side of the piston. Power is, of course, produced on both strokes of the piston. The steam is exhausted up the exhaust stack of the locomotive, helping to create a good draft through the engine.

Variations

Some steam engines are compound engines, that is, the exhaust steam from one cylinder is led into a second or third cylinder to expand further, extracting more of the steam's energy. On compound engines, each succeeding cylinder must be larger. Most steam locomotives are not of the compound type. Instead, they have one piston on each side of the locomotive.

Some steam engines are equipped with condensers to conserve the boiler's water supply. The steam engines on ships need condensers since salt water cannot be used in a boiler without building up large deposits in the fire tubes. Because it would not be practical to carry enough fresh water to replace the water that is boiled away, the used steam must be condensed back into water and returned to the boiler. The railroad locomotive does not need a condenser. It is more practical to stop periodically and take on more water for the boiler, replacing the water that is lost as steam.

STEAM TURBINE

The idea of a steam turbine can be traced back to 1628 when Giovanni Branca designed a steam turbine. Though his ideas are on record, it is not known if his turbine was actually built. Little was done with the turbine until some years after Watt's reciprocating steam engine was developed and in wide use. At that time, engineers and scientists again began to think in terms of steam turbines. They reasoned that much power was being wasted in driving the reciprocating engine's many cranks and connecting rods, and in the start-stop, back-and-forth motion of the piston.

The first person to build a successful turbine was a Swedish engineer, Carl DeLaval. DeLaval's turbine, built in 1882, consisted of a wheel with vanes mounted on the rim. Four nozzles directed jets of steam against the vanes to drive the turbine. The action was much the same as wind driving a windmill. The kinetic force of the steam struck the blades, giving up a portion of its energy to revolve the turbine. This type of turbine is called an *impulse turbine,* figure 14-4.

DeLaval designed a nozzle that takes advantage of the behavior of steam as it is released from a boiler. The steam expands,

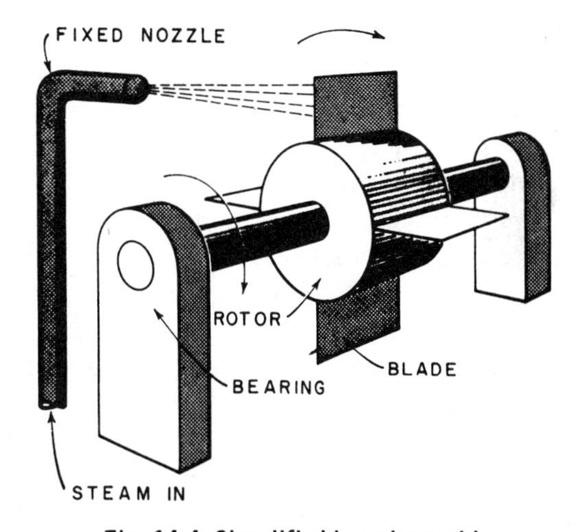

Fig. 14-4 Simplified impulse turbine

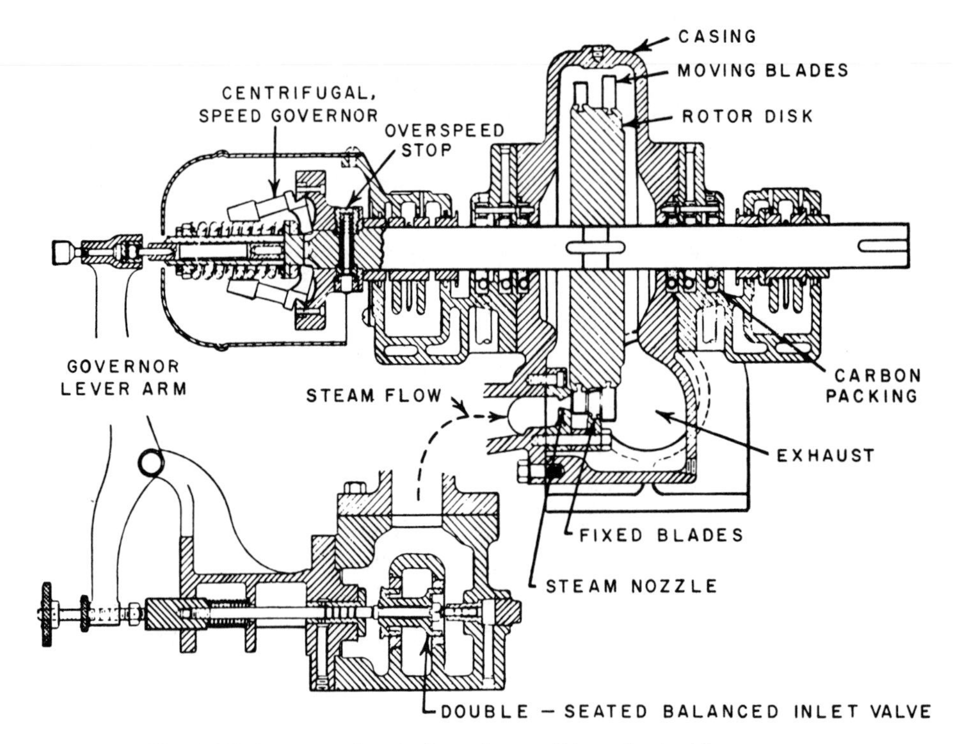

Fig. 14-5 Cross-sectional view of a small impulse turbine

but it also increases in speed or velocity as it is released. The DeLaval nozzle helps to increase the velocity and kinetic force of the steam.

In order to be efficient, the impulse turbine must rotate at about one-half the velocity of the incoming steam. This means that the turbine rotor turns 10,000 to 30,000 revolutions per minute. Such speeds create problems in themselves since the speed must be geared down to operate machinery. The characteristics of DeLaval type turbines make them best suited for installations that require a relatively small amount of power, figure 14-5.

C. G. Curtis, an American inventor, used DeLaval's ideas to produce a more powerful impulse turbine. Curtis constructed his turbine with a series of revolving and stationary blades, figure 14-6. In today's Curtis turbine, the steam hits the first rotor vanes, imparting motion to the rotor as the steam changes its direction of flow. The set of stationary vanes changes the steam direction back again and the steam strikes a second set of rotor vanes, giving up more of its kinetic energy. The next set of stationary blades reverses the steam

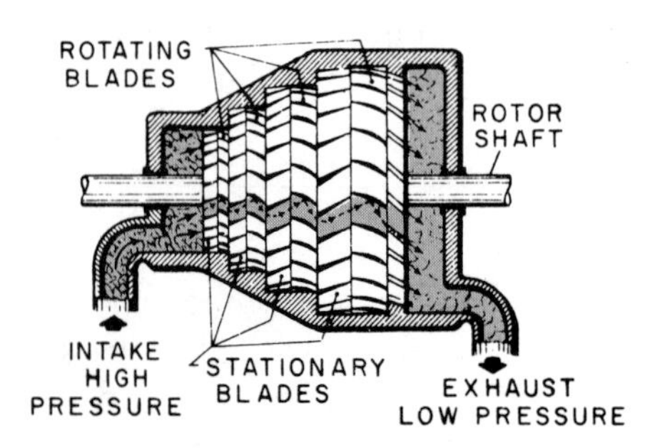

Fig. 14-6 A simplified steam turbine

direction again and the process continues as the steam travels through the turbine.

Often turbines are divided into two or more pressure sections to gain increased efficiency. Each diminishing pressure section has larger turbine rotors to accommodate the expanded steam. These turbines are the compound impulse type, figure 14-7.

Another type of impulse turbine in use today is the helical flow impulse turbine, figure 14-8. In this turbine the steam follows a spiralling path as it delivers its kinetic energy to the rotor, figure 14-9.

The *reaction* turbine was invented by C. A. Parsons of Ireland. His turbine is similar to Curtis's in that it has alternating and fixed blades, sometimes called diaphragms, but the design of the blades is different. Instead of having a symmetrical cross-sectional path to follow, the steam follows a converging cross-sectional path between the blades, figure 14-10. As a result, the steam is accelerated. All of the many turbine blades that surround the wheel produce this reaction. The principle is identical to that of a jet or rocket engine. The action of the accelerated steam causes the reaction of pushing the rotor forward, figure 14-11. Steam that enters a turbine at 75 miles per hour (120 kilometers per hour) under pressure leaves the turbine at 300 miles per hour (483 kilometers per hour) under low pressure. The steam passes through a 13-foot (4-meter) long turbine in 1/16 of a second.

In actual practice the reaction turbine does not derive 100% of its power from the reaction principle. About half comes from im-

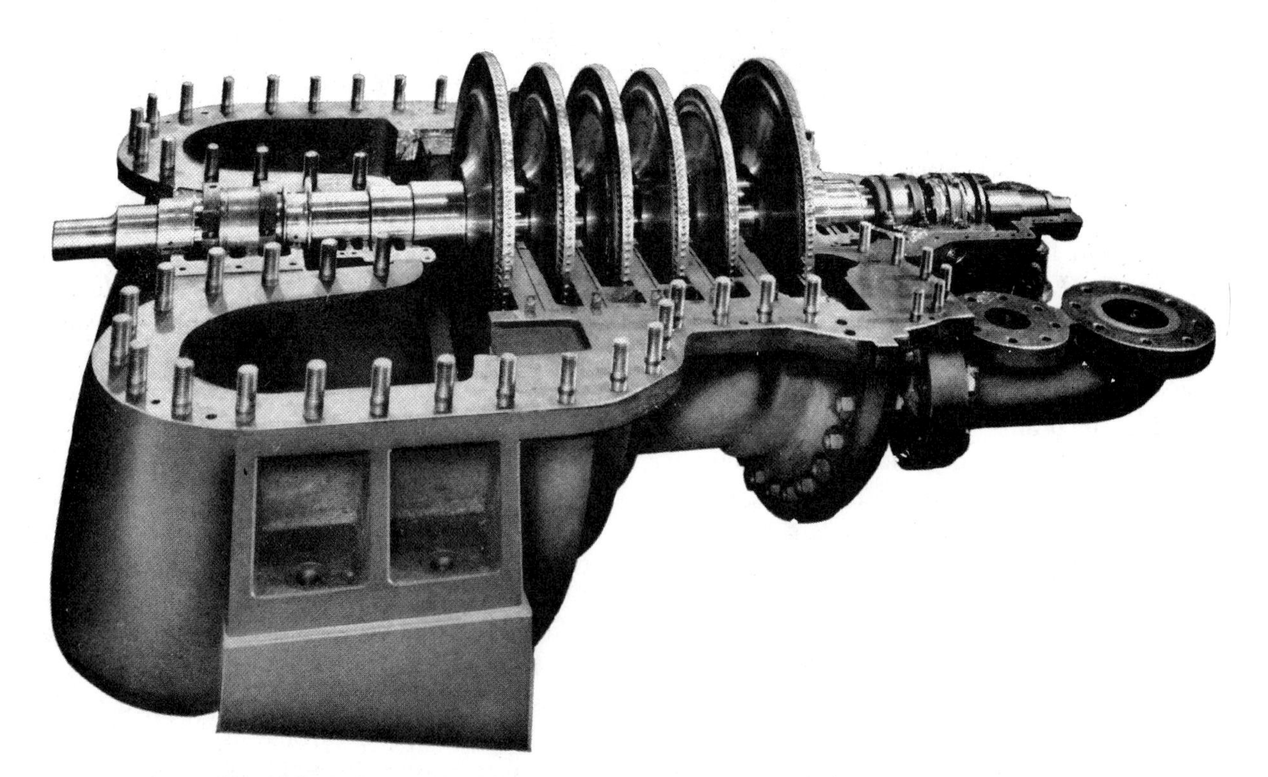

Fig. 14-7 A six-stage impulse turbine with the top half of the casing removed

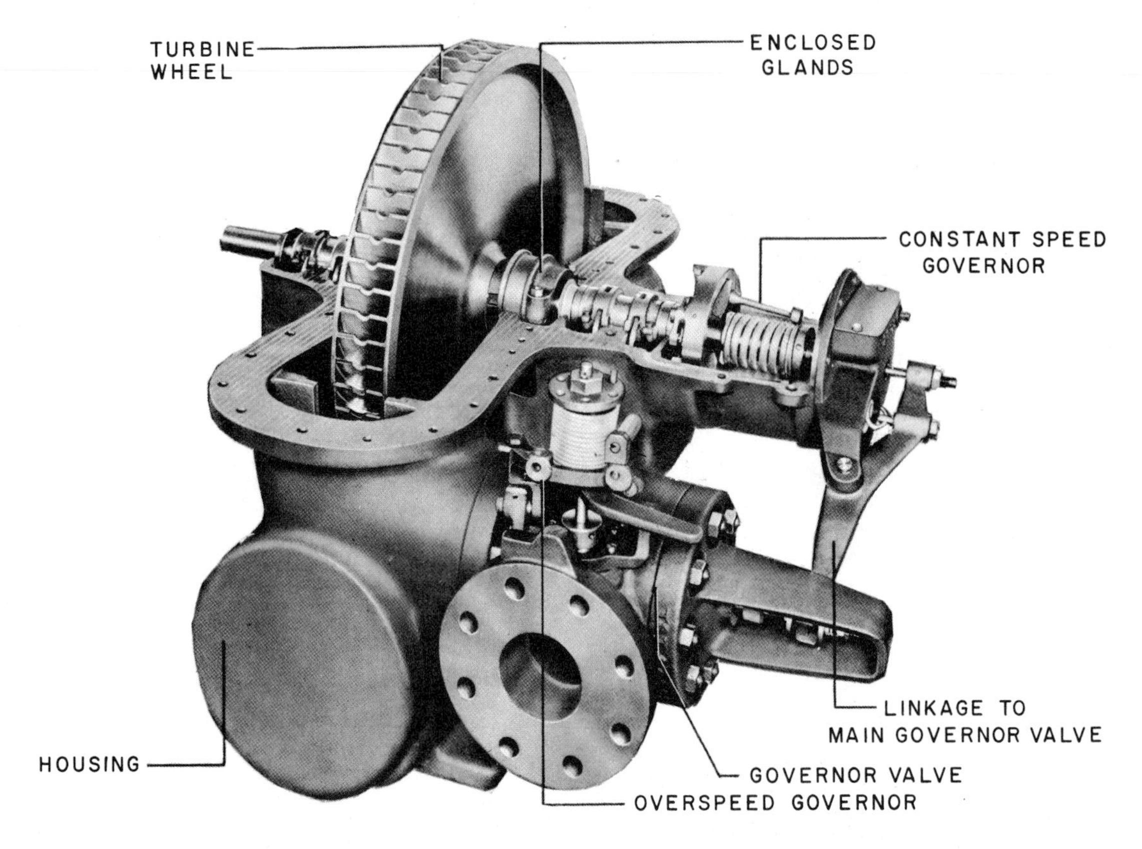

Fig. 14-8 Modern helical flow impulse turbine with the top half of the casing removed

pulse, half from reaction. Most modern steam turbines are a blend of impulse and reaction.

The typical turbine consists of: the rotor, including the main shaft and the turbine wheels and their blading, figure 14-12; the casing, including the nozzle passages, figure 14-13; the stationary blades or diaphragms, figure 14-14; a speed regulator or governor controlling the amount of steam allowed to enter through the nozzle; a lubricating system to satisfy the needs of a continuous high-speed rotor shaft and gearbox, figure 14-15; and, for some applications, a gear-reduction system.

A *non-condensing* turbine allows the spent steam to exit the turbine at atmospheric pressure or higher. In some cases this steam is further put to work in a variety of industrial processes. In the *condensing* turbine the spent steam is sent to a condenser for cooling and reuse. Cooling is accomplished by circulating large amounts of water through pipes in the condenser. This allows a low pressure to be maintained in the condenser, improving the expansion ratio of the steam and increasing the turbine's efficiency.

The general pattern in turbine construction is to reduce the steam pressure in many stages for slow-speed operation, or reduce the steam pressure in few stages for high-speed operation. A high-speed turbine used for driving a generator may reduce the steam pressure in

Fig. 14-9 Action of the steam in a solid wheel machine

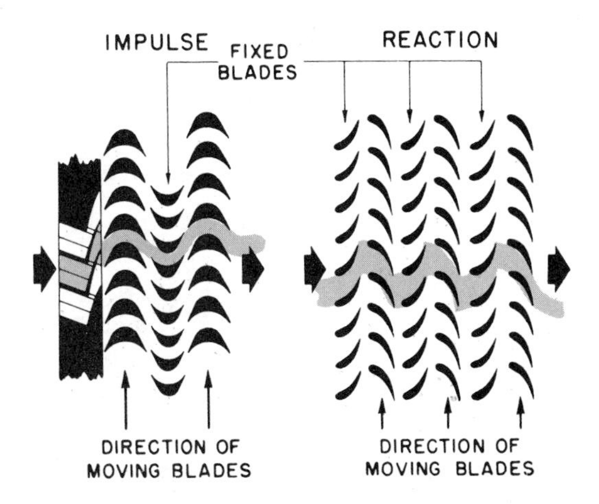

Fig. 14-10 Impulse and reaction blading

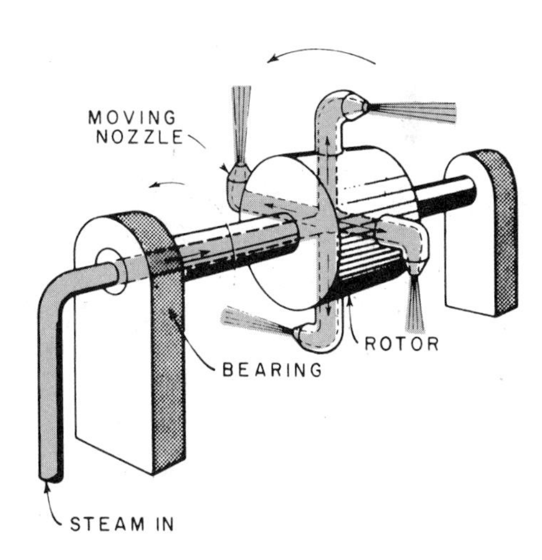

Fig. 14-11 Simplified reaction turbine

50 steps while a low-speed turbine for propelling a ship may reduce the steam pressure in 200 steps.

The primary application of steam turbines is in the field of generating electricity, figure 14-16. The high-speed turbine is well suited to this purpose since generators also operate best at high speed. The steam turbine accounts for two-thirds of the total electrical output in the United States. Often the electrical generating plant is located near a river. This provides a constant source of fresh water and economical barge transportation of the coal or oil that is used as fuel. The coal is pulverized and blown into the fire chamber of the boiler where it literally explodes as its chemical energy is changed into heat energy. Steam is produced in the water tubes in the boiler and sent to the turbines which drive the generators. Steam turbine electrical generating plants have a high efficiency, 20 to 28%.

Modern turbines are designed for a variety of applications. To meet different speed and horsepower requirements, turbine manufacturers develop various turbine styles, although the basic operating principles remain the same. Many modern turbines use steam under as

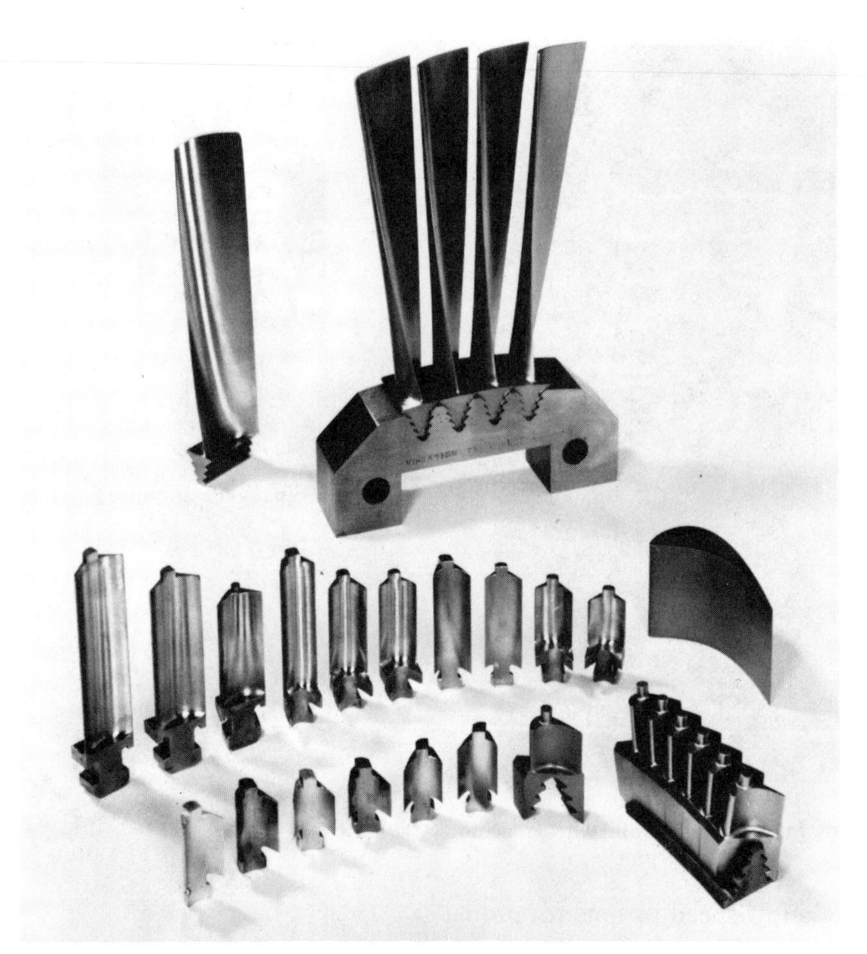

Fig. 14-12 Various styles of turbine blades shown individually and turbine blades partially installed on a turbine wheel

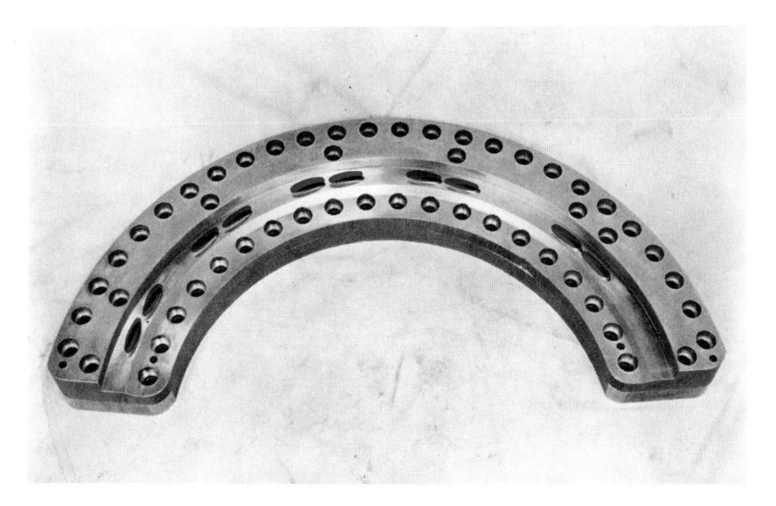

Fig. 14-13 Nozzles for admitting steam to turbine–only one-half shown

Fig. 14-14 Diaphragm of stationary blades–only one-half shown

much as 5000 pounds per square inch (34 megapascals) pressure and 1100°F (593°C) temperature. These extremes are possible because special alloys are now available for use that did not exist when turbines were first developed. Careful machining and workmanship go into every turbine, making it an efficient prime mover of watch-like precision.

Fig. 14-15 High speed gears for pressure lubricating system

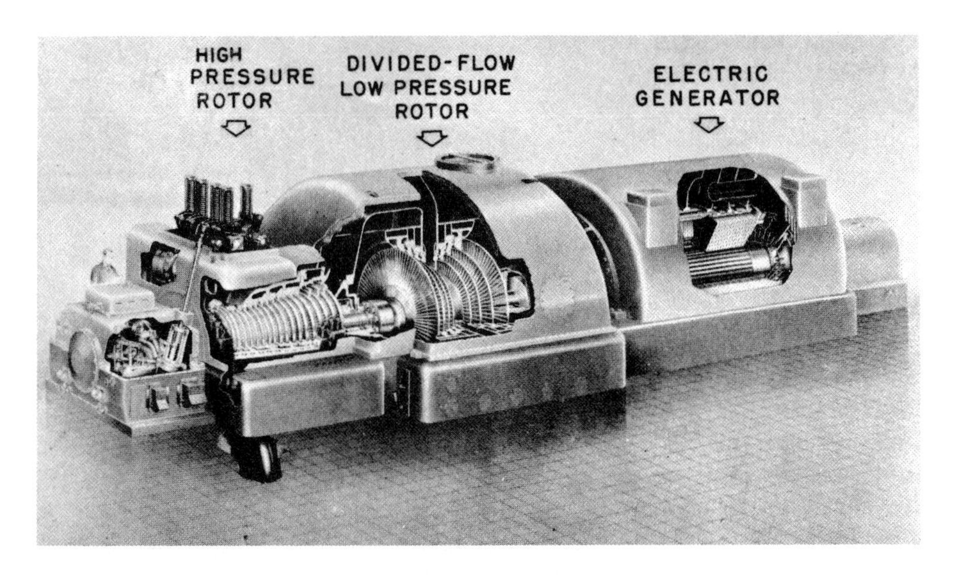

Fig. 14-16 A typical steam turbine generating plant

REVIEW QUESTIONS

1. Explain why a steam engine is classed as an external combustion engine.
2. Briefly trace the energy path within a steam engine or steam turbine.
3. What were the first steam engines used for?
4. Why is Thomas Newcomen's steam engine sometimes referred to as an atmospheric engine?
5. In what way did James Watt improve the steam engine?
6. What improvements made the reciprocating steam engine adaptable to locomotives?
7. Explain why some steam engines need a condenser and some steam engines do not.
8. Explain the action of the steam in an impulse turbine.
9. Trace the path of the steam through a Curtis-type turbine.
10. In what way is a reaction turbine different from an impulse turbine?
11. Are modern turbines pure impulse or pure reaction? Explain.
12. What is the most important use of steam turbines today?
13. What is the efficiency of a modern steam turbine?

SECTION 5
ELECTRICAL ENERGY

Unit 15 Principles of Electricity

OBJECTIVES After completing this unit, the student should be able to:

- Explain the principles of static electricity.
- Recognize series and parallel circuits.
- Discuss magnetism and electromagnetism.
- Identify and explain primary and secondary cells.
- Discuss the charging and discharging cycle of an automobile storage battery.
- Explain how alternators and generators produce electricity.

To appreciate electricity, try to place yourself at a time in history before electricity was commonly used. There were no electric lights, radios, television sets, telegraphs, telephones, or electric motors. There were no power household appliances like washers, driers, vacuum cleaners, toasters, electric coffee pots, or hair dryers. The machines of industry were powered by large, inefficient steam engines. The pace of life was slower before the era of electrical technology. However, both men and women worked harder than their modern counterparts do. Aided by electricity, our work load has been lightened.

Communication devices such as radio, telephone, telegraph, television, and computers have speeded communications and helped to shrink our world, because messages can be transmitted at the speed of light. While the world has, in a sense, been made smaller, our understanding of other people has been broadened as a result of these devices. Modern communication places ideas and pictures before us from all parts of the world. Today, we are made aware of community, state, national, and world affairs and have an opportunity to expand our understanding and knowledge.

Gone is the era of candle light and horse

drawn carriages. Gone too—fortunately—are long, hard working hours and low productivity. Electricity in industry, farm, and home has enabled all persons to enjoy a richer, fuller life.

Electron theory and the characteristics of the atom were discussed in Unit 7. If you remember from that unit, a source of electricity can be mechanical, chemical, or static. In Unit 7 the mechanical source of electricity was discussed.

STATIC ELECTRICITY

It is possible for some substances to build up an excess of electrons and for others to have a deficiency of electrons. A body with an excess of electrons is said to be *negatively charged.* A body with a deficiency of electrons is said to be *positively charged.*

Just as electrons and protons are attracted to each other, unlike charges attract each other; negative and positive charges attract each other. The opposite is also true: like charges repel each other; negative repels negative and positive repels positive. The Greeks observed these electrical properties 2500 years ago, though they did not fully understand them. Their observations pertained to what is called static electricity.

Substances also differ in their ability to lose or gain electrons. For example, if a hard rubber rod is rubbed with fur, the rod becomes negatively charged. Electrons from the fur are brought in contact with the rubber. Some electrons leave the fur, building up an excess of electrons on the hard rubber rod. Another example is a glass rod that is rubbed with silk. In this case, however, electrons on the glass break free and are transferred to the silk. The rod is short of electrons or positively charged.

The action of static electricity can best be seen through the use of an *electroscope,* which is a glass container with a metal rod that passes part way through its stopper. Attached to the lower end of the rod are two metal leaves. When the electroscope is neutral, the metal leaves hang down. When a negatively charged rubber rod touches the electroscope, electrons flow into the rod and leaves. The leaves fly apart (diverge) because they are both now negatively charged, and like charges repel, figure 15-1.

If a positively charged glass rod is brought in contact with a neutral electroscope, electrons flow from the electroscope onto the glass rod, figure 15-2. The metal leaves are both short of electrons. They are positively charged and fly apart because again, like charges repel. Charging by either of these methods is called charging by *contact.*

Charging can also be done by *induction.* Suppose that a negatively charged hard rubber rod is brought close to an electroscope.

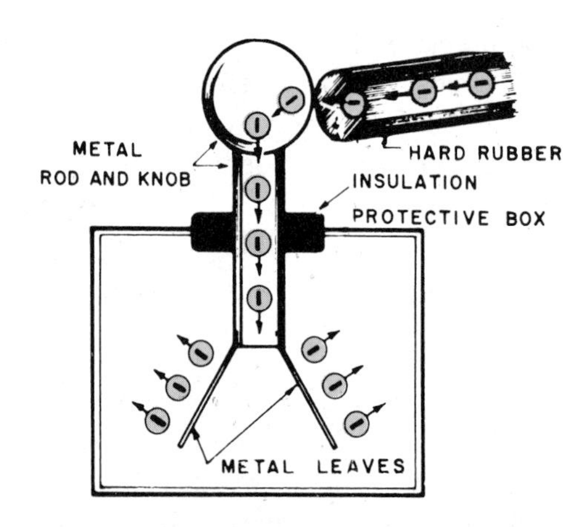

Fig. 15-1 A negatively charged electroscope. Electrons are received from the negatively charged rubber rod.

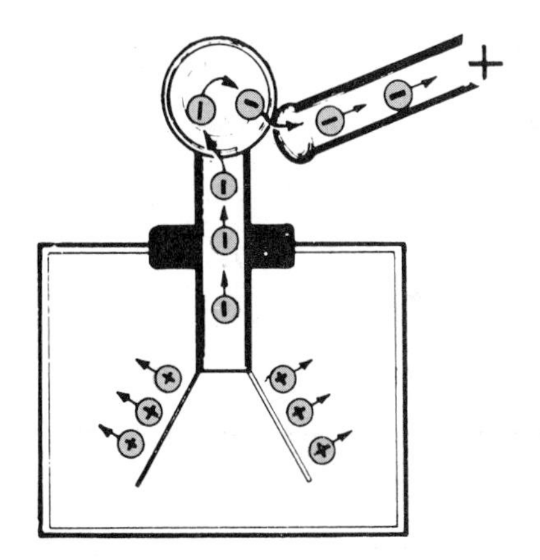

Fig. 15-2 A positively charged electroscope. Electrons are taken from the electroscope by the glass rod.

Electrons in the electroscope are repelled. Now, if a person touches the electroscope providing an escape path for the electrons, some electrons will flow onto the person's finger. When the finger is removed, the electroscope is short of electrons and positively charged. The leaves fly apart and remain apart even after the rubber rod is removed from the area. Positive charges can also be induced onto the electroscope by using a positively charged glass rod.

The electroscope can store an electrical charge, as can another device, the leyden jar. For modern applications, however, a condenser is used to store electrical charges.

Because static charges want to neutralize themselves, electrons will travel from a negative area to a more positive area, since unlike charges attract. They will move until both areas are neutral. They can do this if they are connected to each other by a wire, are in contact with each other, or are brought close enough together for the electrons to jump from the negative body to the positive body. Electrons tend to flow also when their is a difference in charge or electrical potential.

ELECTRONS IN MOTION

Electric current can be defined as a movement of electrons within a conductor toward a more positive body. The basic unit of electron measurement is the *coulomb*. A coulomb is 6,280,000,000,000,000,000 electrons (6.28×10^{18}). Another basic measurement is the *ampere*. It is derived from the fact that if one coulomb of electrons passes a given point in one second, one ampere of electrical current is flowing.

Electrons flow because there is a difference in electrical potential, and this difference is called *voltage*. Electrons flow from the negative to the positive potential.

Another factor, besides voltage, that enters into the flow of electrons is resistance, the opposition to current flow. It is measured in *ohms*. Some materials are good conductors and offer little resistance to electron flow. Large diameter wires offer less resistance to electron flow than small diameter wires. The length of the wire also affects resistance. The longer the wire is, the greater the resistance, assuming the diameter of each is the same.

- *Ampere* is the unit measurement of the rate of electron flow (I).
- *Volt* is the unit measurement of electrical potential or pressure (E).
- *Ohm* is the unit measurement of the resistance a substance offers to current (R).

The three factors, current, voltage, and resistance, are inseparable. Each is related to the other. Their relationship is expressed mathematically in Ohm's Law:

$$I = \frac{E}{R} \qquad R = \frac{E}{I} \qquad E = I\,R$$

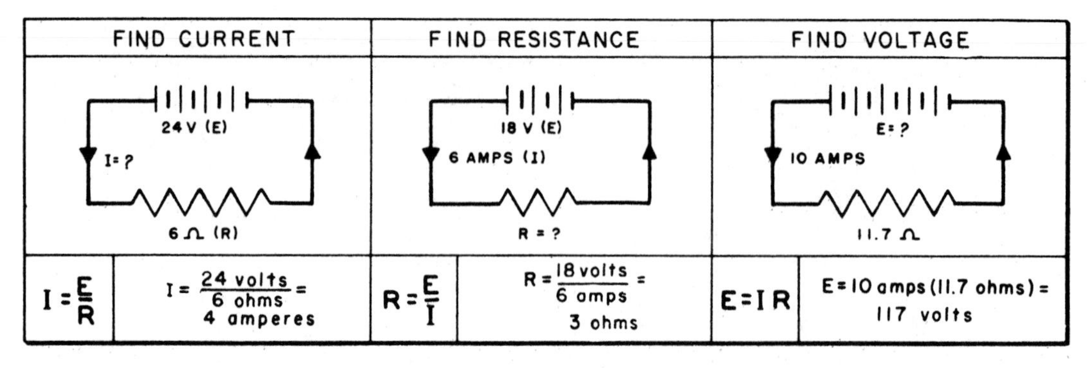

Fig. 15-3 Applications of Ohm's Law

Where:

I = Current intensity or amperes
E = Voltage or electromotive force
R = Resistance stated in ohms

Three simple diagrams, figure 15-3, show how Ohm's Law can be used to calculate electrical problems.

ELECTRICAL CURRENT

In discussing electrical circuits, we shall assume that we have *direct current*–(DC)–a continuous flow of electrons in one direction. Typical producers of direct current are batteries, DC generators and rectifiers. Rectifiers convert AC to DC. Current flows from a battery as long as there is a complete path for it to follow: a complete circuit. A switch is usually placed in the circuit to control electron flow. When the switch is closed, electrons flow and the light is on, for instance. When the switch is opened, current stops. The electrical potential is still present when the switch is open, but there is no way for it to be satisfied; the circuit is not complete. Recall that a complete circuit is necessary for electron flow.

There are two basic types of electrical circuits: series and parallel. If all the resistances in the circuit are in one path, the circuit is a *series circuit.* Current must pass through all of the resistances. Therefore, the total resistance of a series circuit is the sum of the individual resistances. Each resistance will create a voltage drop, so the total voltage is equal to the sum of the separate voltages. In figure 15-4, if one lamp burns out, the circuit is broken and all the lamps go out; this is one disadvantage of series wiring.

Remember: (1) although voltage is divided among the resistances, (2) current is the same

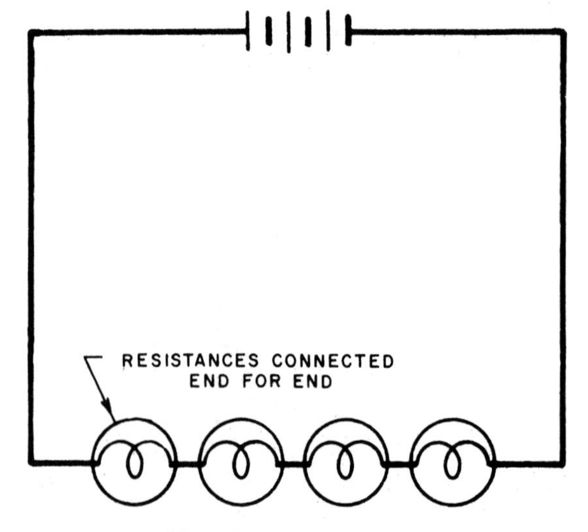

Fig. 15-4 Series circuit

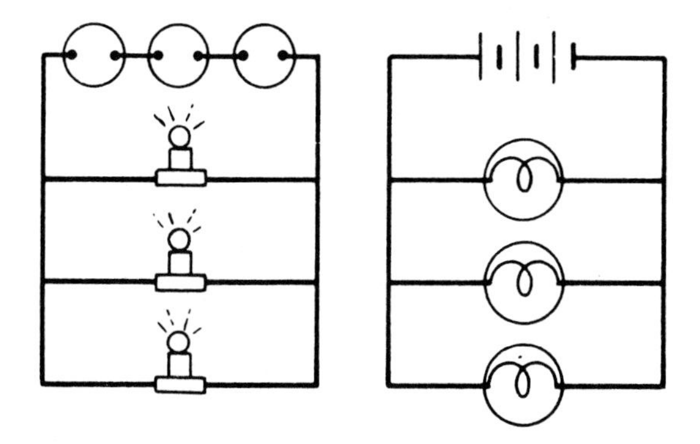

Fig. 15-5 Parallel circuit

through all resistances, and (3) resistance is the sum of the separate resistances.

If the resistances in a circuit are connected side by side, each having one end to the power source, the circuit is a *parallel circuit,* figure 15-5. Current divides itself to flow through the available paths. The amount of current that flows through any part depends on the resistance of that part. The resistance is divided up so that the total parallel resistance is less than the smallest single resistance in the circuit. The voltage across each resistance is the same, the full 117 volts on normal household installations.

Remember: (1) voltage is the same across all resistances, (2) current is divided among the resistances, and (3) total resistance is less than that of the smallest resistance.

MAGNETISM

The ability of certain metals to attract iron or steel is discussed in Unit 7. These metals, called magnets, can also attract nickel, cobalt, chromium, and magnesium. Although the first practical use for the magnet was the compass, this is not the main reason for studying magnetism. Magnetism has achieved greater importance because most of the electricity that is produced originates with the principles of magnetism.

In explaining the phenomenon of magnetism, it is known that atoms of iron and a few other substances have the *special ability* to line themselves up in a pattern whereby the magnetic qualities of each atom complements those of surrounding atoms. These atoms work together, combining their minute magnetic fields to produce a noticeable magnetic field. In nonmagnetic substances, on the other hand, atoms are arranged in a random manner and, therefore, do not exhibit magnetic properties.

Substances that are magnetic have a north-seeking pole and a south-seeking pole, corresponding to the earth's magnetic poles. In effect, each magnet has a magnetic field around it just as the earth does. As a reminder, the like poles of two magnets repel each other while the unlike poles attract each other, figure 15-6.

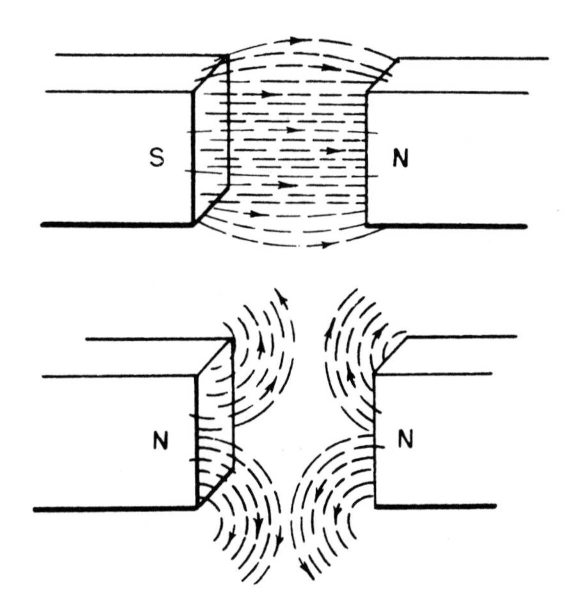

Fig. 15-6 Unlike poles attract each other; like poles repel each other.

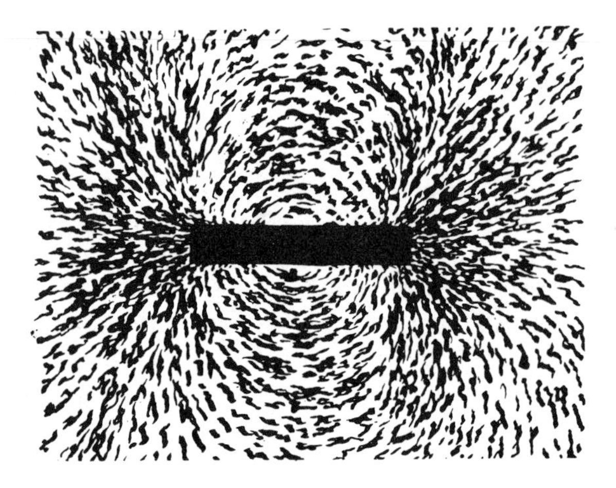

Fig. 15-7 Illustrating a magnetic field by using iron filings

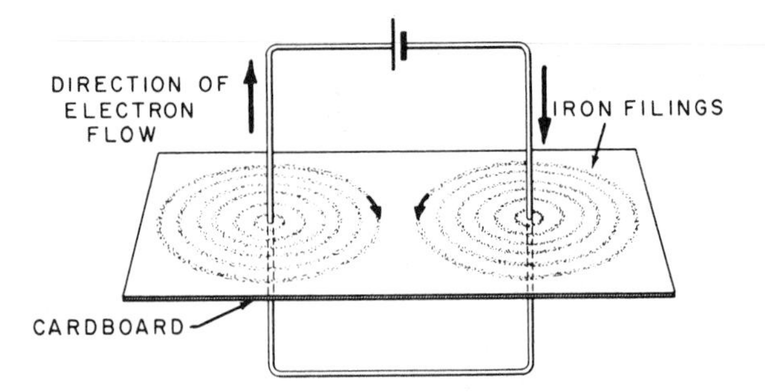

Fig. 15-8 Magnetic field around a wire that is carrying electric current

The magnetic field is strongest at the ends of the magnet; in the center of the magnet the field is weak. A magnetic field can be seen by placing a magnet under a piece of paper and then sprinkling iron filings onto the paper. The iron filings line up to show the magnetic field and its concentration, figure 15-7.

ELECTROMAGNETISM

When electric current flows through a wire, a magnetic field is built up around the wire itself. This can be seen using cardboard, iron filings, and a current-carrying wire. With the wire passed through the cardboard and the current flowing, iron filings are sprinkled onto the cardboard. They can be seen arranging themselves in a magnetic field, figure 15-8. The magnetic field or lines of force are referred to as *flux,* and just as in a natural magnet, the field is strongest near the wire and decreases as the distance from the wire increases.

Flux around a wire also has direction; this is determined by the direction of electron flow within the wire. As shown in figure 15-9, the north pole of the compass needle indicates the direction of flux around the wire. The dot in the center of the wire on the left indicates the point of the current-direction arrow coming toward the observer. The X at the right represents the tail of the current arrow pointing away from the observer. If the direction of electron flow within the wire is reversed, the compass needles will reverse themselves, indicating a change in flux direction.

The left-hand rule enables a person to determine either flux or current direction, providing one of the two is known. If current direction is known, point the left thumb in the direction of flow. When your fingers grasp the wire, they point in the flux direction. If the flux direction is known, wrap the fingers of the left hand around the wire in the direction of the flux. The left thumb will point in the direction of current, figure 15-10.

Unit 7 says that when a current-carrying wire is made into a coil, the whole coil becomes a magnet. A single magnetic field builds up around the coil. The magnetic flux around the turns of wire combines to form one large magnetic field; the coil is said to be electromagnetic.

The magnetic field of the electromagnet has a north pole and a south pole just as a

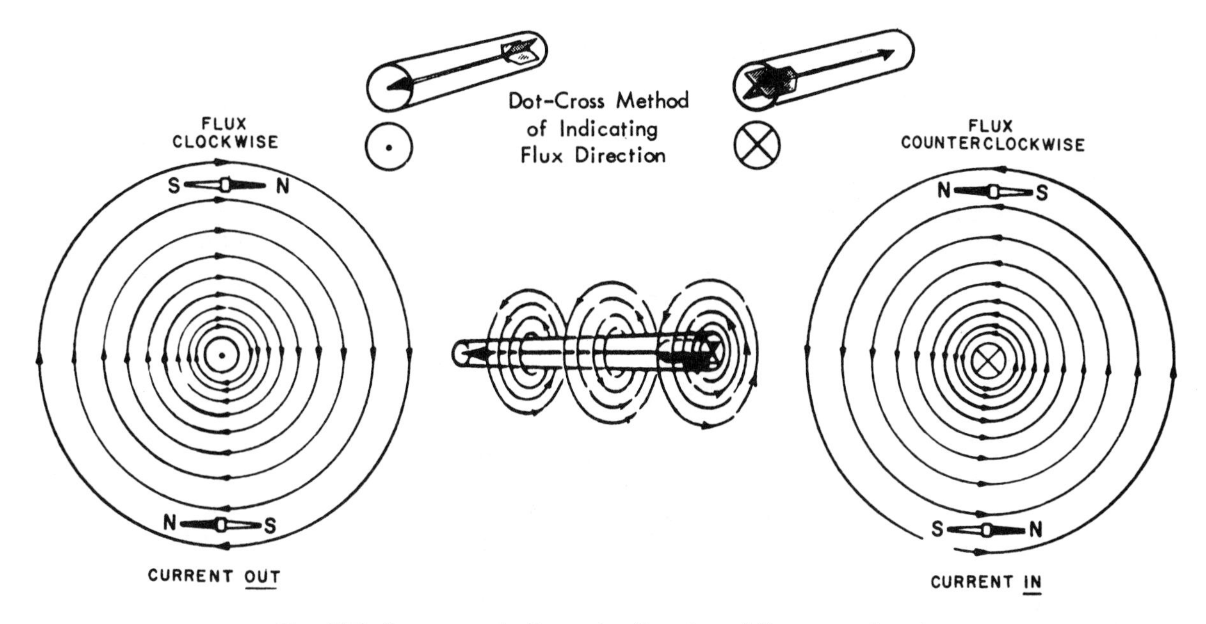

Fig. 15-9 Compasses indicate the direction of flux around a wire.

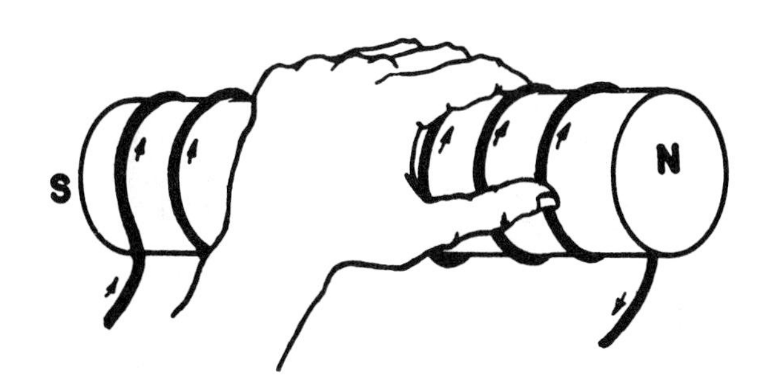

Fig. 15-10 The left-hand rule for determining flux direction or direction of electron flow

natural magnet has. The like poles repel and the unlike poles attract. Electrically speaking, the ends of the coils where the currents are parallel in the same direction are attracted to each other; the ends of the coils where the currents are parallel in the opposite direction repel each other, figure 15-11.

On the electromagnet, a soft iron core is used to strengthen the magnetic field. Electromagnets and other magnetic devices can be used for lifting iron and steel, magnetic separators, relays, bells, solenoids, sound producers, and many other items.

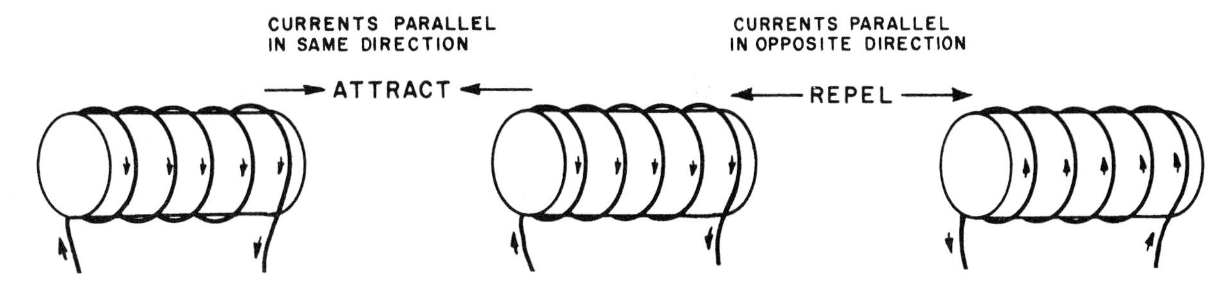

Fig. 15-11 Electromagnets attract or repel each other according to the direction of current in the coil.

ELECTROCHEMICAL SOURCES OF ELECTRICAL ENERGY

Dry cells and automobile storage batteries are perhaps the two most common examples of the conversion of chemical energy into electrical energy. They are separated into two categories, primary cells and secondary cells. The dry cell is a *primary cell*; once its chemical reaction is complete, it cannot be recharged. A storage battery is a *secondary cell*; it can be used and recharged many times during its useful life.

The dry cell is made of two electrodes and an electrolyte. Its operation depends on a difference in the electrical potential of the

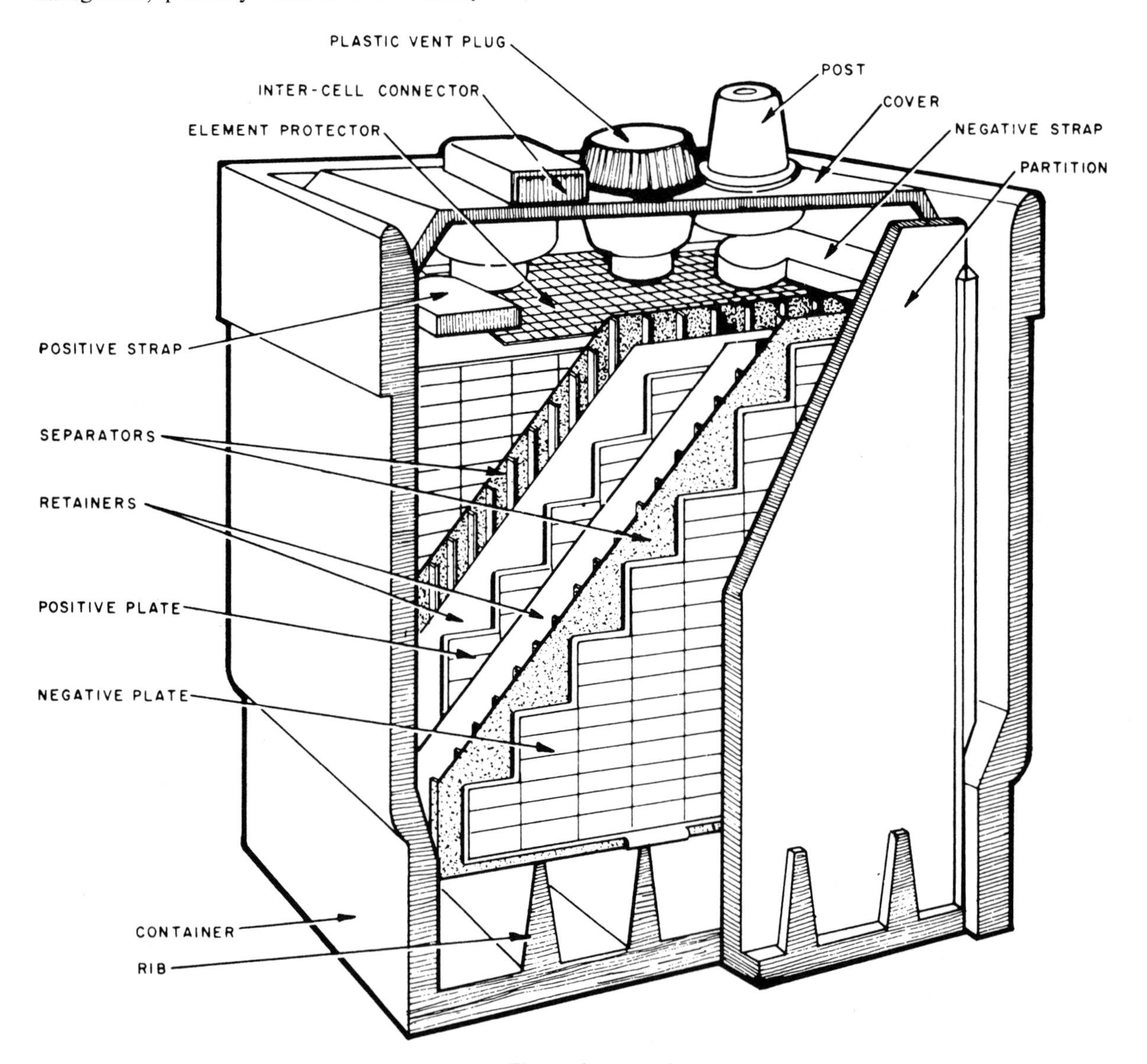

Fig. 15-12 The lead storage battery

two electrodes. Equally important, it depends on ionization for its operation. An *ion* is an electrically-charged atom or group of atoms; they may have lost or gained electrons. Many substances are capable of being ionized.

The dry cell—which really is not dry at all—is made of a zinc outside shell or container; this is one electrode. The container is filled with a thick paste of ammonium chloride dissolved in water; this is the electrolyte. Extended down the center of the cell is a carbon rod; another electrode. The top of the cell is sealed off, preventing evaporation of the electrolyte. In a complete circuit, free electrons in the zinc travel through the wire to the carbon electrode. These electrons are picked off by ammonium ions in the electrolyte. The zinc case is the negative terminal and the carbon rod is the positive terminal. There is a potential difference of 1 1/2 volts between the two electrodes.

Storage batteries such as those commonly seen on automobiles convert chemical energy into electrical energy. The storage battery may be either 6 or 12 volts. They are made of alternate plates of lead peroxide and spongy lead. The plates are prevented from touching each other by separators, figure 15-12. The electrolyte is a solution of sulfuric acid and water. The chemical action of the storage battery is discussed more fully in Unit 7, and additional illustrations show its construction.

It should be noted that many modern storage batteries are almost maintenance free. Little or no attention to the electrolyte level is required since the battery is designed to eliminate water loss. This is done by first moving the plates nearer the bottom of the battery case to allow more electrolyte to cover the plates. The lead grids or plates are designed in a different manner to also reduce the amount of water lost by electrolysis. They are constructed of calcium lead instead of antimony lead, because calcium lead and sulfuric acid do not create the hydrogen gas that is formed in the presence of antimony and which is very explosive.

Cold, subzero weather can reduce the capacity of a fully-charged battery by 45%. In addition, the cranking load due to cold oil and parts might increase 3 1/2 times, figure 15-13. That is the main reason why it is important to have a properly-sized, fully-charged battery in the automobile.

DC AND AC ELECTRICAL GENERATORS

Very simply, generators are used to convert mechanical energy into electrical energy. The energy of free falling water or the energy of hot gases directed against a turbine blade produces mechanical energy in the form of a rotating shaft. This motion is used to turn the generator to produce electricity.

Whenever magnetic lines of force cut a wire, electric current is induced to flow in the wire. There must be a changing magnetic field in order to induce electron flow. If the wire is held stationary in a fixed magnetic field, there is no electron flow. Electron flow occurs only when the wire cuts into the magnetic field, or the magnetic field moves relative to the wire, figure 15-14.

Either an increasing magnetic field or a decreasing magnetic field can cause electron flow. More current can be induced by increasing the speed that magnetic lines of force are cut, or increasing the strength of the magnetic field.

In a generator, the magnetic field is stationary. The coil of wire is moved within the

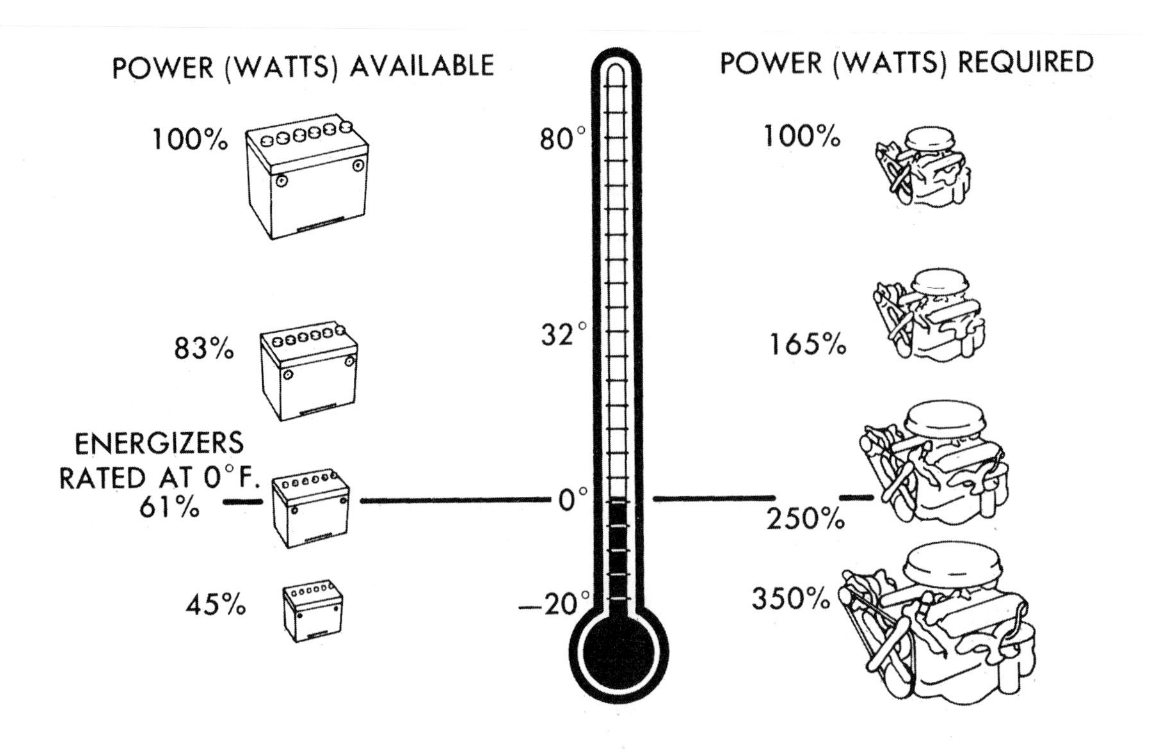

Fig. 15-13 Battery power goes down with temperature while engine starting requirements go up.

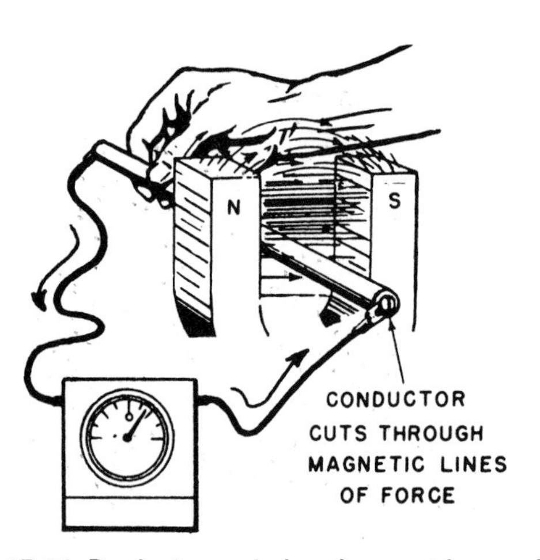

Fig. 15-14 Producing an induced current by moving a wire through a magnetic field

magnetic field, figure 15-15. In small generators, the magnetic field is created with permanent magnets. In large generators, it is normally produced by electromagnets that create a stronger magnetic field. The coils of wire on the armature rotate inside the magnetic field. Actually the *armature* is made of a shaft that has many iron laminations fastened to it that strengthen the magnetic action. Conducting wire is wound around the laminations. The two ends of the wire are connected to a slotted slip ring mounted on the armature. Two brushes, one on either side of the slip ring *(commutator)* provide the necessary path for electron flow.

For the sake of clarity, the drawings and

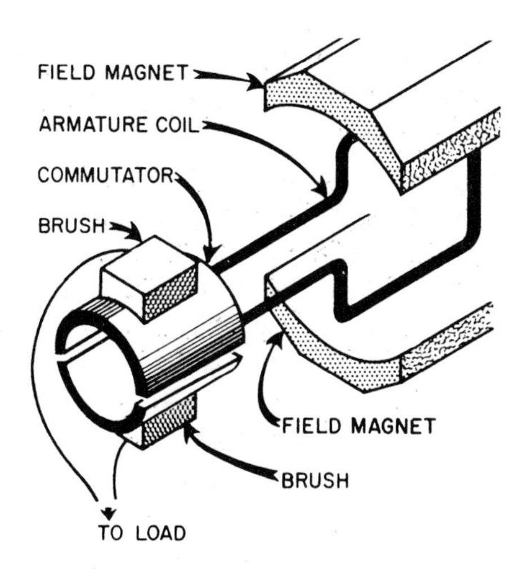

Fig. 15-15 A simplified DC generator

explanation of the DC *(direct-current)* generator have been simplified, figure 15-16. In position #1 there is no electromotive force produced since the wire is moving parallel to the field's magnetic lines of force. However, as soon as the armature begins to move toward position #2 it begins to cut magnetic lines of force. At position #2 the maximum number of magnetic lines of force are being cut; induced current has gradually built up. Electrons flow toward the right-hand brush. As the armature continues on toward position #3 induced current continues to flow in the same direction but at a diminishing rate. At position #3 there is no electromotive force generated. Also, the right-hand half of the ring will now touch the left-hand brush and vice versa. Now the armature moves up to position #4, again cutting lines of force. The direction of armature coil movement has been changed in relation to the magnetic field and the direction of electron flow has been reversed. However, since the commutator segments are touching opposite brushes, the effect is that the current continues in the same direction.

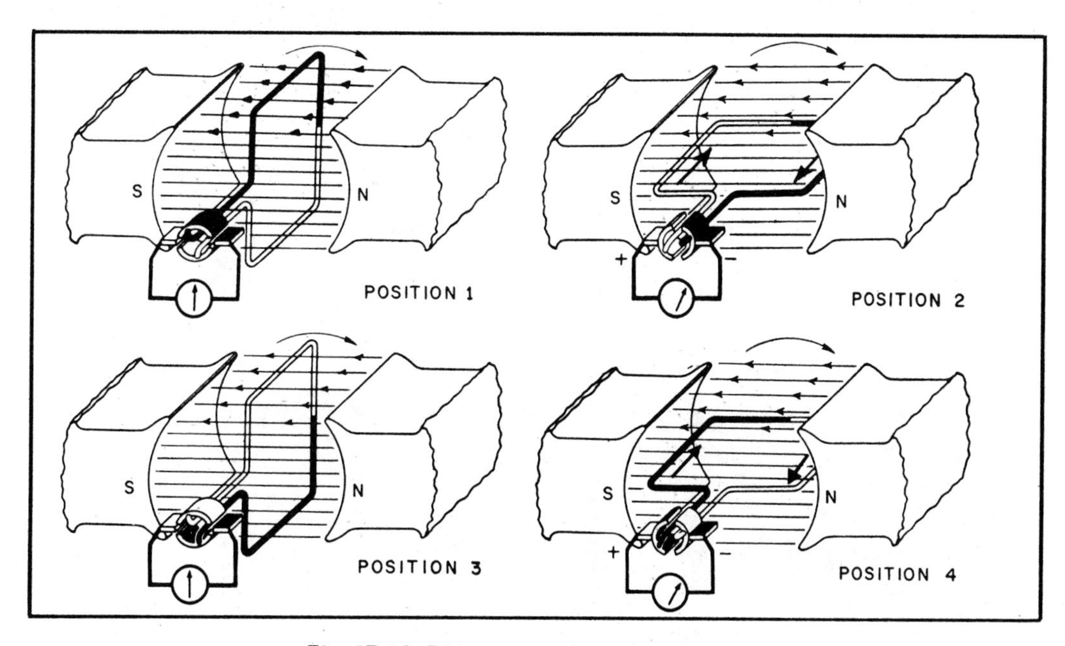

Fig. 15-16 Direct current generator rotation

As the armature continues its rotation, it produces a stream of electrons flowing in one direction; this is said to be a direct current or DC.

This output is not constant in the number of electrons that flow, figure 15-17. Every revolution of the DC generator produces two peaks. This is because, in actual construction, the DC generator is much more complex than these simplified diagrams indicate. In order to smooth out the operation of the generator and to produce a steady electromotive force, several coils of wire are wound around the rotating iron core. This means that the construction of the commutator must also be much more complex.

The AC *(alternating-current)* generator, called an alternator, is very similar in construction and operation to the direct-current generator. However, the current produced by the AC generator changes its direction with every revolution of the armature. Instead of having a split ring commutator, two complete rings are mounted side-by-side on the armature shaft and connected to the armature coil. A brush rides against each ring, one brush always in contact with the same ring. In figure 15-18, the basic action shown is the same as that of a DC generator. At position #1 of the armature there is no electromotive force produced, but as the armature moves on toward position #2 the induced electron flow increases. Current reaches its maximum at position #2, diminishing to zero as it arrives at position #3. The direction of electron flow reverses when the armature starts moving toward position #4. The induced electron flow reaches its maximum at position #4, diminishing to zero as it arrives back at position #1. The slip ring commutator enables the generator to produce a flow of electrons that changes its direction with every armature revolution or cycle, figure 15-19. Ordinary alternating current is supplied at 60 cycles.

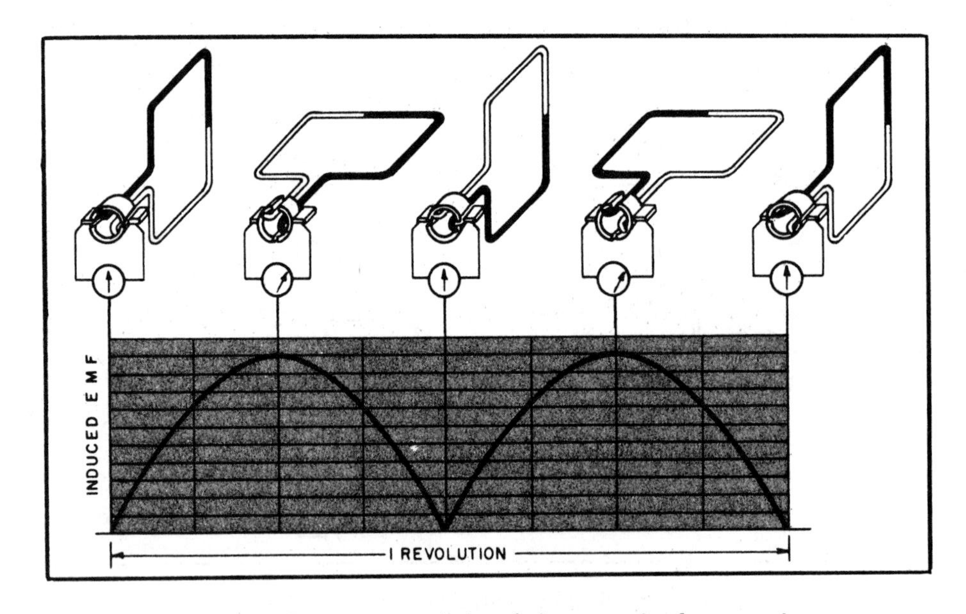

Fig. 15-17 Direct current induced electromotive force graph

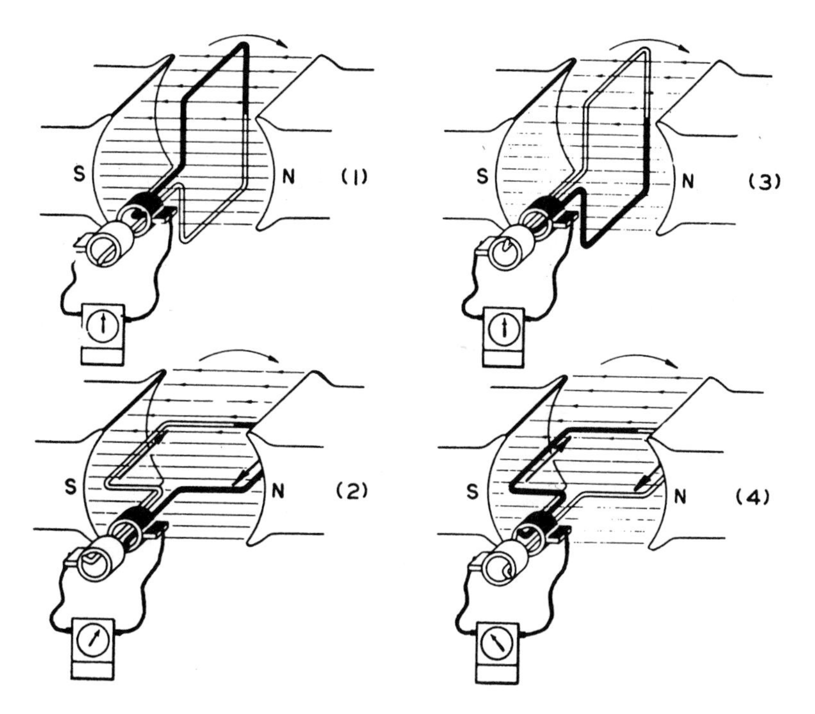

Fig. 15-18 Alternating current generator rotation

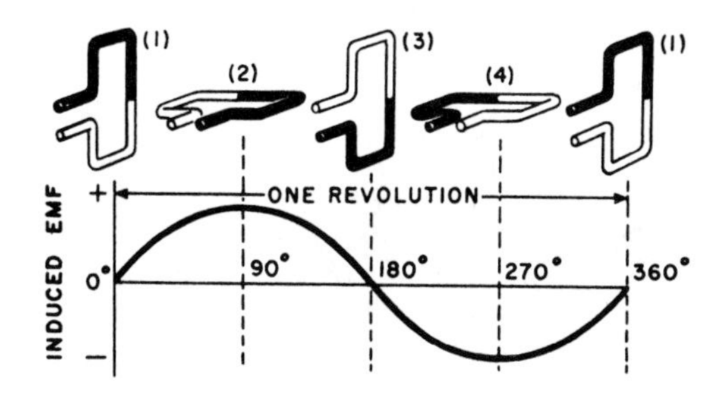

Fig. 15-19 Alternating current induced electromotive force graph

This means that there are 60 complete cycles or revolutions, 120 changes in current direction, every second.

Alternating current and direct current are in common use. Both are good for the operation of electric light, heating units, or universal electric motors. Direct current is needed for electroplating, electrolysis, or the charging of storage batteries. However, the use of alternating current is more widespread because it can be transmitted over long distances more easily than direct current, figure 15-20. The voltage of alternating current can be stepped up or stepped down by the use of transformers. Generally, the voltage from the power station is stepped up to be transmitted over a long distance and then stepped down near the point where it is to be used, figure 15-21. With alternating current, high-voltage transmission lines are made of smaller wires. This means that there is less power loss. Most generators deliver 13,200 volts. This voltage is stepped up as high as 250,000 volts for long distance transmission, and then eventually stepped down to the 120 or 240 volts for use in homes and industry, figure 15-22.

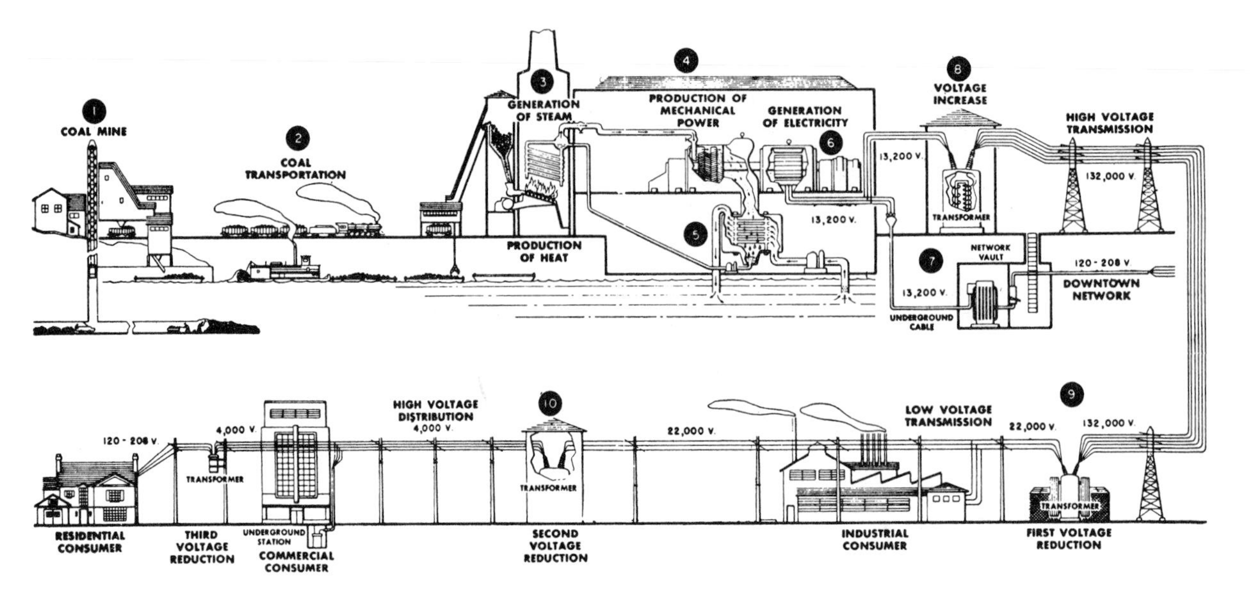

Fig. 15-20 Stages in generating, transmitting, and distributing electricity

Fig. 15-21 Conowingo hydroelectric power plant of the Philadelphia Electric Company (Courtesy Philadelphia Electric Co.)

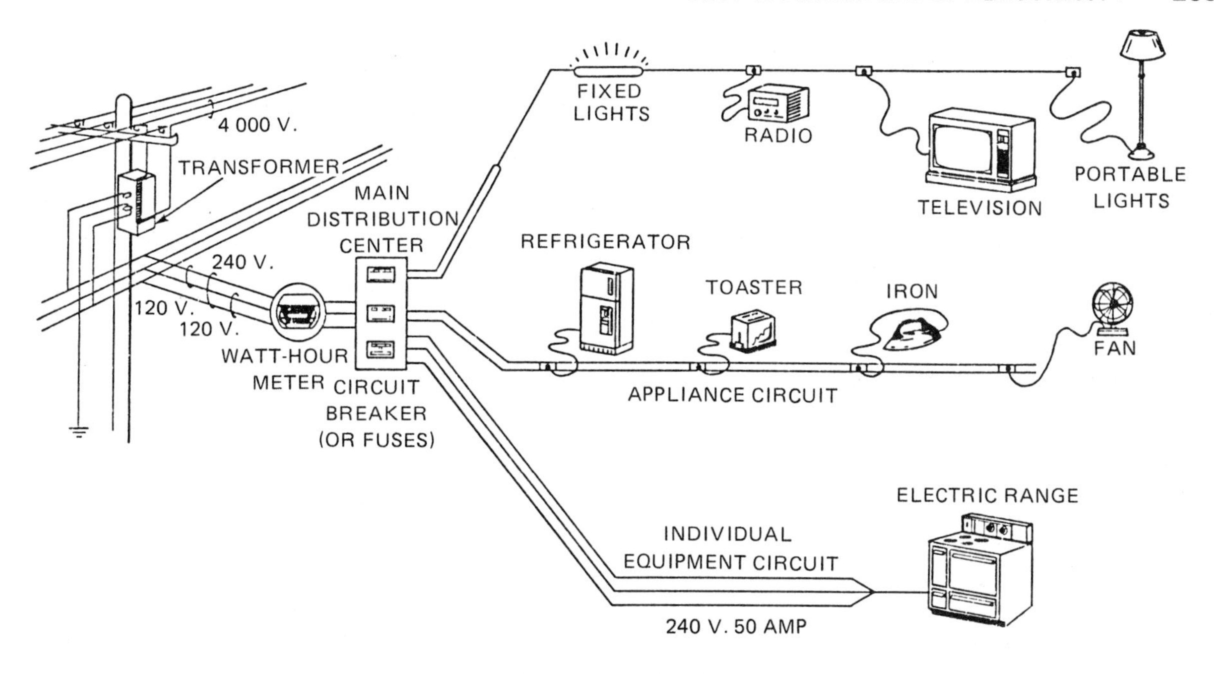

Fig. 15-22 A Typical distribution system

REVIEW QUESTIONS

1. Explain how a pencil or any other object is electrical in nature.
2. List the parts that make up an atom and tell the electrical characteristic of each.
3. What accounts for copper's ability to readily conduct electricity?
4. Why are elements whose atoms have almost complete outer shells poor conductors?
5. Do like charges of static electricity attract or repel each other?
6. Explain how a negatively charged rubber rod can charge an electroscope by contact.
7. Define an electrical current.
8. Define the following terms: amperage, voltage, resistance.
9. Using Ohm's law, calculate the following:
   E – when I is 0.02 ampere and R is 4000 ohms
   I – when E is 110 volts and R is 220 ohms
   R – when E is 120 volts and I is 6 amperes

10. Draw a simple series circuit that contains three resistances.
11. Draw a simple parallel circuit that contains three resistances.
12. How can you account for the magnetic characteristics of iron and steel?
13. State the laws of repulsion and attraction for magnets.
14. How can the left-hand rule be used to determine the direction of magnetic flux when the direction of current is known?
15. Describe how you could build a simple electromagnet.
16. Do electromagnets have any practical use? If so, list some common uses.
17. What is meant by primary and secondary cells?
18. What is electrochemical energy?
19. Explain the underlying principle of a generator.
20. Explain briefly how most generators are powered.
21. What is the basic difference in the construction of a DC and an AC generator?
22. Why is alternating current used in preference to direct current for home and industrial uses?
23. Explain the function of a transformer.

SECTION 6 TRANSMISSION OF POWER

Unit 16 Mechanical Transmission of Power

OBJECTIVES **After completing this unit, the student should be able to:**

- **Explain the five simple machines.**
- **Recognize the three classes of levers.**
- **Calculate mechanical advantages of force, speed, and distance.**
- **Discuss ways of transmitting power to machines.**

DIRECT TRANSMISSION OF POWER

The power of the engine must be *transmitted* to the machines. In some cases, the machine can be fastened directly to the prime mover. This is not only desirable and simple, but few power losses occur this way. When the drive shaft of the prime mover is directly connected to the driven shaft of the machine or device, an ideal situation exists as, for example, a blade fastened to the engine's crankshaft on a rotary lawn mower. Also, direct drive from the shaft can be seen in many machines that are driven by electric engines, such as grinders and polishers. The simple, ideal situation for power transmission can be found, but in most cases the machine application calls for something more complex.

INDIRECT TRANSMISSION OF POWER

Direct transmission of power may not cover all of the operating requirements of the given machine or mechanism. Factors such as speed changes, disengaging the machine from the prime mover, torque requirements, changes in direction of rotation, transmitting energy over long distances, transmitting energy around corners, or converting rotary motion to linear motion, cannot be satisfied by direct, positive linkage with the prime mover. The power transmission system requirements of the automobile point up several of these factors. Appropriate speed changes and direction changes would not be practical if the crankshaft were directly and positively fastened to the wheels with no mechanism in between.

Unit 9 provides a detailed explanation of the automobile drive system and it would be well to review it after studying this unit to see how mechanical transmission principles are applied.

Small engine applications—such as seen on a riding lawnmower—will also use many principles discussed in this unit, and so the power train of a typical riding lawnmower is covered at the end of this unit.

In addition to direct transmission of energy to machine, there are three primary methods of delivering the power of the prime mover to the machine that is to do the work:

- *Mechanical transmission* involves gears, pulleys, chain drive, levers, shafts, and feed screws, as the transmission media.
- *Fluid transmission* involves the use of a liquid or a gas within a dynamic system as the transmission media.
- *Electrical transmission* involves the use of conductive wire to deliver electric current.

Each of the three methods has its advantages and disadvantages. Engineers select the transmission system that fits the requirements; often they use the systems in combination with each other. For example, coal is used to produce heat which converts water into steam. The steam is directed against the blades of the steam turbine. The steam turbine drives a generator producing electrical energy which is transmitted throughout the countryside by the use of wires. The energy arrives at a factory, for instance, where it is used to drive various electric motors. The power produced by the electric motors may be transmitted to the machine through gears, belts, or other mechanical arrangements, or the power of the electric motor may be transmitted to a fluid power system which, in turn, transmits the power to the machine. There is an infinite number of combinations. This section deals with the physical principles and some of the techniques involved in transmitting power.

MECHANICAL TRANSMISSION OF POWER

Mechanical transmission principles enable a person to do *more* with the power that is available. The principles do not *increase* the power coming out of the engine or other force-producing device because the total power being produced cannot be multiplied; the mechanical principles can *adapt* the power to a particular machine.

Several mechanical principles have been applied for centuries, long before engines were a source of power. The principles work well whether the source of power is an engine, a draft animal, a human, or a natural force. The principles discussed in this unit are often referred to as simple machines. These principles are restated by engineers in thousands of variations as seen in transmission systems and the machines themselves. It is sometimes difficult to draw a firm line between the transmission systems and the machine itself, since the two may often blend into an integral unit. Countless applications of the simple machine principles are seen but basically the principles can accomplish these things:

- Change the *size* of the applied force
- Change the *direction* of the applied force
- Change the *speed* that results from the applied force

These simple machines include the lever, wheel and axle, inclined plane, wedge, and screw.

LEVER

Levers allow a person to lift, move, or hold far more than is possible without them. They are commonly seen. A seesaw enables a small child to lift a man off the ground; a man uses a lever to move a boulder; a boy uses a pair of pliers to hold a bolt. Any long shaft or board will do for a lever as long as it has a pivot point *(fulcrum)* to work against. In gaining a mechanical advantage with a lever, less force is used but it must apply over a greater distance to overcome and move the opposing object which resists movement.

Mechanical advantage is the number of times that force is multiplied. What is gained in applied force is lost in having to apply the force over a greater distance. The principle of the lever does not allow both a larger force and smaller distances moved. One is sacrificed for the other. With the lever shown in figure 16-1, an effort or force of 25 pounds (50.5 newtons) can lift 100 pounds (202 newtons), a mechanical advantage of four.

Levers fall into three categories: first class, second class, and third class levers.

First class levers place the fulcrum between the resistance and the effort force. These levers multiply force with a sacrifice in the

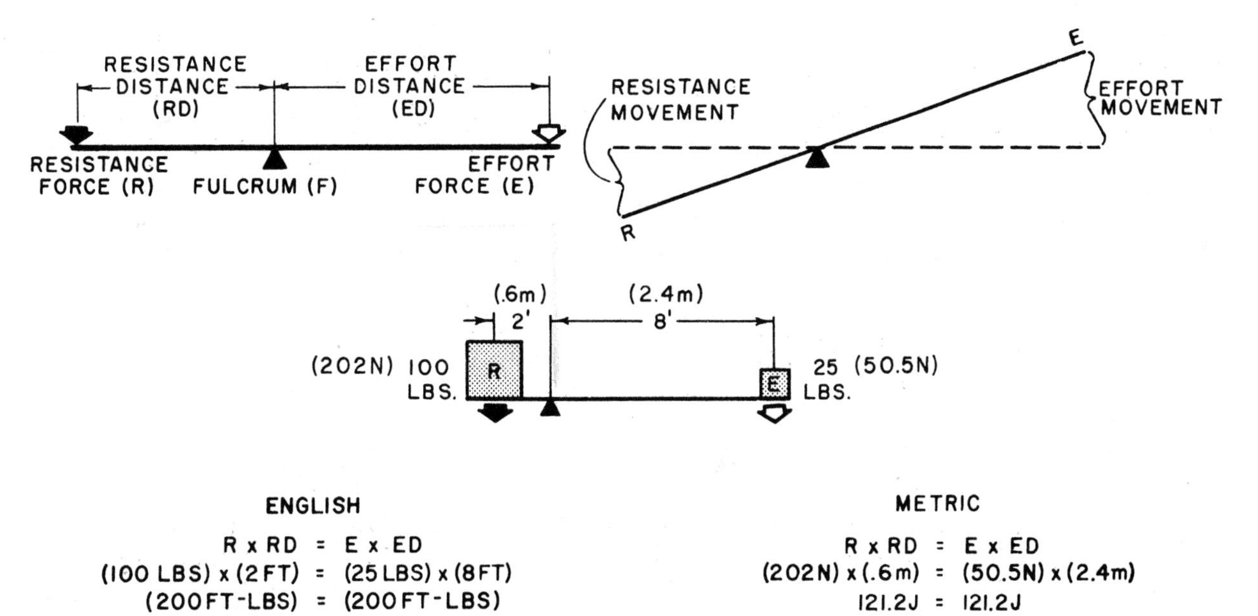

Fig. 16-1 The moments of force around the fulcrum balance each other. Note that for larger force applied (R) or for greater distance applied (RD), the effort distance (ED) must be increased, or increase effort force (E).

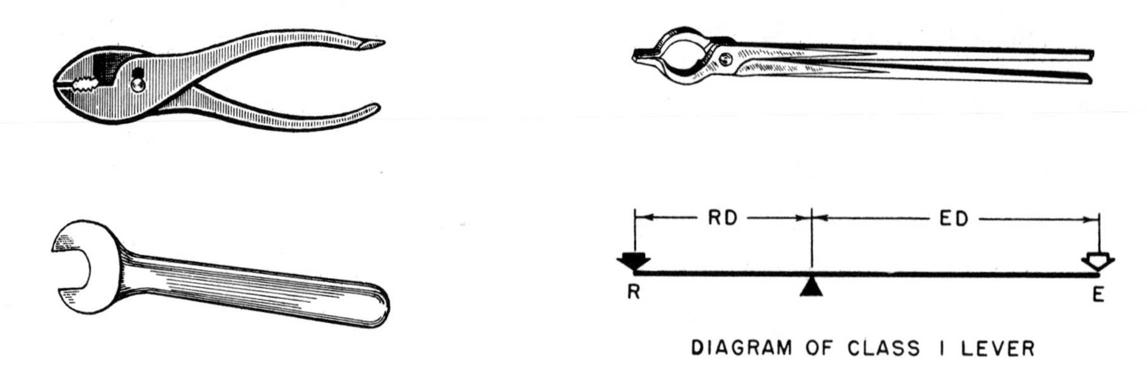

Fig. 16-2 Examples of first class levers

distance the object is moved. The direction of movement is reversed, figure 16-2.

Second class levers place the fulcrum at the end of the lever with the resisting weight at a point along the lever. These levers also multiply force with a sacrifice in the distance the object is moved. The direction of object movement is the same as that of the applied force, figure 16-3.

Third class levers place the fulcrum at the end of the lever with the resisting weight at the opposite end of the lever. The applied force or effort is at a point along the lever. These levers increase effort by multiplying distance moved by the object and, therefore, have a negative mechanical advantage. A fishing rod illustrates this principle. Sometimes the fisherman is mystified at how a small fish can create such a large pull on his rod. The fish has the mechanical advantage, figure 16-4.

WHEEL AND AXLE

Essentially the wheel and axle consists of two wheels firmly held together along the same axis. The axle may be short or it may be a very long shaft. Mechanical advantage is gained when the large wheel is moved through a distance. The axle moves also; its motion is less since its diameter is less, but the force twist, or torque on the axle is multiplied, figure 16-5. A boat trailer winch is an example of this principle. The mechanical advantage of the winch will allow a man of normal size to properly load and secure a heavy boat onto a trailer.

Study figure 16-6; note that the mechan-

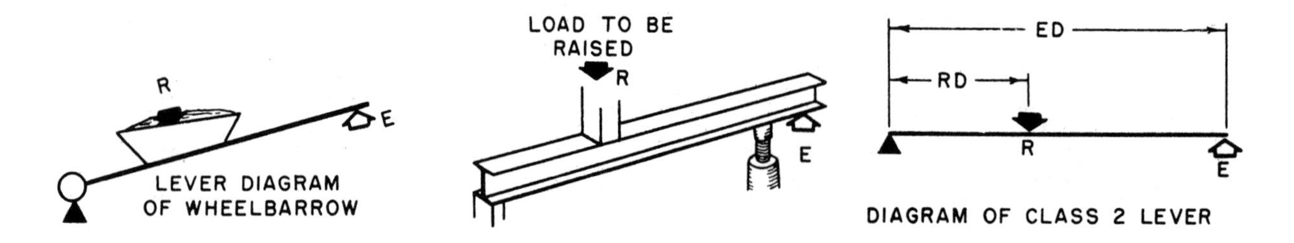

Fig. 16-3 Examples of second class levers

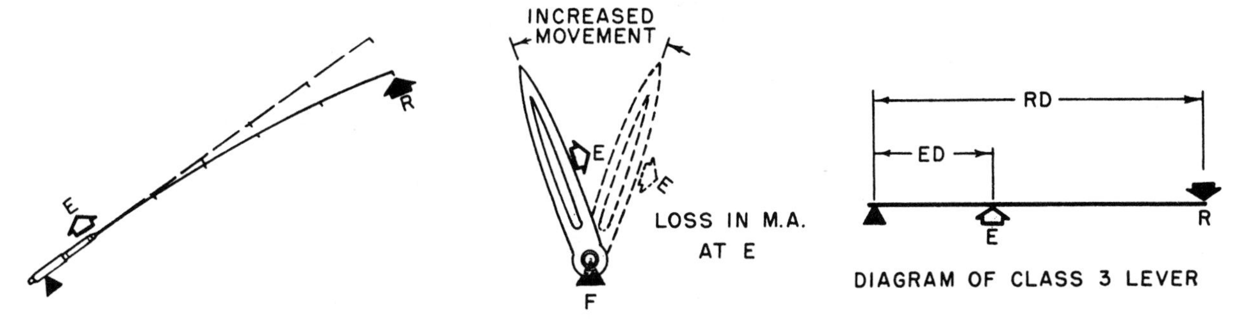

Fig. 16-4 Examples of third class levers

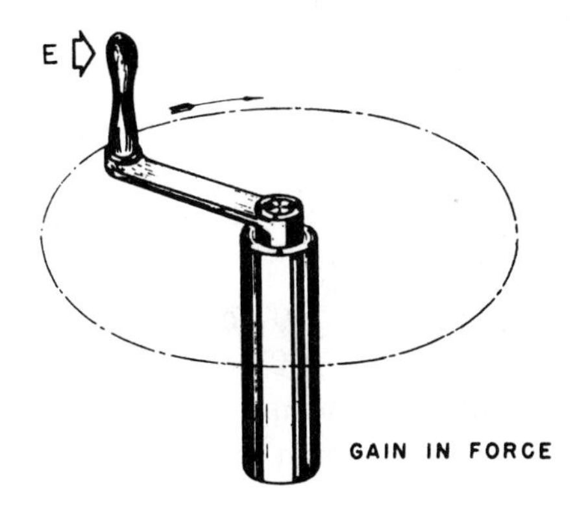

Fig. 16-5 Gaining mechanical advantage of force with the wheel and axle

ical advantage possible with the wheel and axle is essentially that of a Class 1 lever. The fulcrum is at the center of the wheel and the axle. The resistance distance, RD, is measured from the fulcrum to the perimeter of the axle, and the effort distance, ED, is measured from the fulcrum to the perimeter of the wheel.

In the perfect wheel and axle, these moments of force are equal: $E \times ED = R \times RD$.

Figure 16-7 shows an effort force of 10 pounds (44.5 newtons) applied to a 5-inch (127-millimeter) radius wheel. The resistance force acting on the axle of 1-inch (25.4-millimeter) radius which the effort force over-

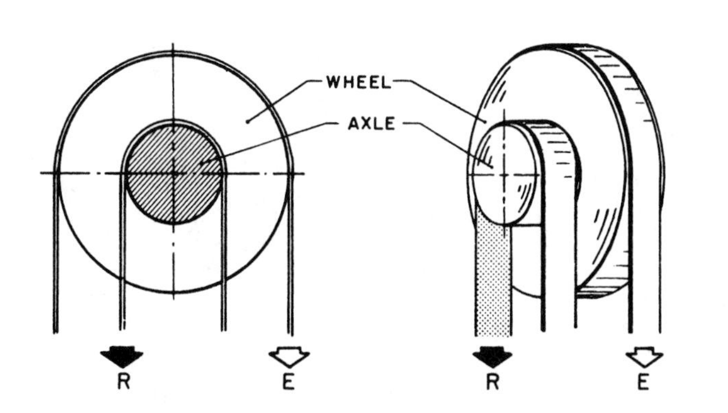

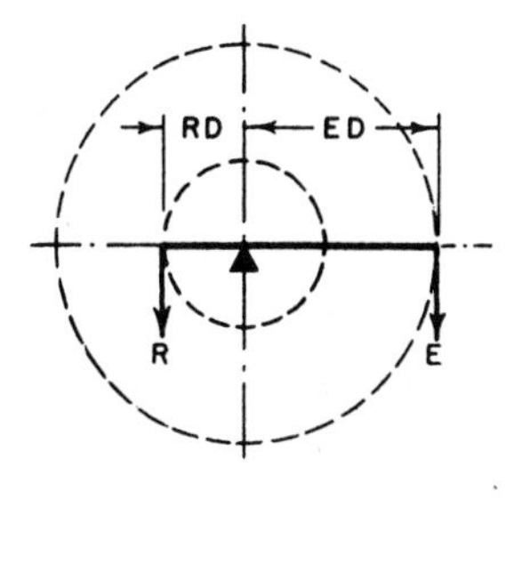

Fig. 16-6 Wheel and axle moments of force

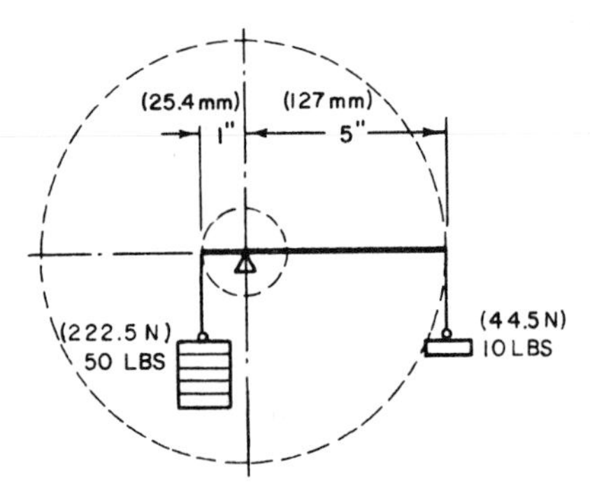

Fig. 16-7

comes is found by the moment formula:

U.S. Customary

E X ED = R X RD

10 pounds X 5" = R X 1"

$\frac{\text{50 pound-inches}}{\text{1 inch}} = R$

50 pounds = R

Metric

E X ED = R X RD

44.5 newtons X 127 millimeters = R X 25.4 millimeters

5651.5 newton-millimeters = R

222.5 newtons = R

With a wheel and axle combination, a smaller force may be used to produce or control a larger force.

INCLINED PLANE

The inclined plane also allows force to be applied over a longer distance in order to handle heavy objects, figure 16-8. This principle—as well as others—was used by the Egyptians in the construction of the pyramids. Stone blocks weighing several tons were rolled up a gently inclined plane and maneuvered into place. Upon completion of the structure, the earthen inclined planes were removed.

Inclined planes or ramps permit the application of a lesser amount of force over a longer distance to ease the job at hand.

Example: a boy lifts his 100-pound (445-newton) body onto a 3 foot (0.9 meter) high loading dock with a single jump.

U.S. Customary

Work = Force X Distance

Work = 100 lb X 3 ft = 300 ft-lb

$\text{Force} = \frac{\text{Work}}{\text{Distance}} = \frac{\text{300 ft-lb}}{\text{3 ft}} = \text{100 lb}$

Metric

Work = Force X Distance

Work = 445 newtons X 0.9 meter = 400.5 joules

$\text{Force} = \frac{\text{Work}}{\text{Distance}} = \frac{\text{400.5 joules}}{\text{0.9 meter}} = \text{445 newtons}$

By walking up a 10 foot (3 meter) long ramp, less than the 100-pound (445-newton) force required in a single jump is applied, but it must be applied through a greater distance.

U.S. Customary

$\text{Force} = \frac{\text{Work}}{\text{Distance}} = \frac{\text{300 ft-lb}}{\text{10 ft}} = $ 30 ft-lb when walking up ramp

Metric

$\text{Force} = \frac{\text{Work}}{\text{Distance}} = \frac{\text{400.5 joules}}{\text{3 meters}} = $ 133.5 newtons when walking up ramp

In this example, the boy actually has a mechanical advantage of force when using the inclined plane. The mechancial advantage of force is 3 1/3.

U.S. Customary

$$MA_f = \frac{\text{R (Resistance Force)}}{\text{E (Effort Force)}}$$

$$MA_f = \frac{\text{100 lb}}{30} = 3\ 1/3$$

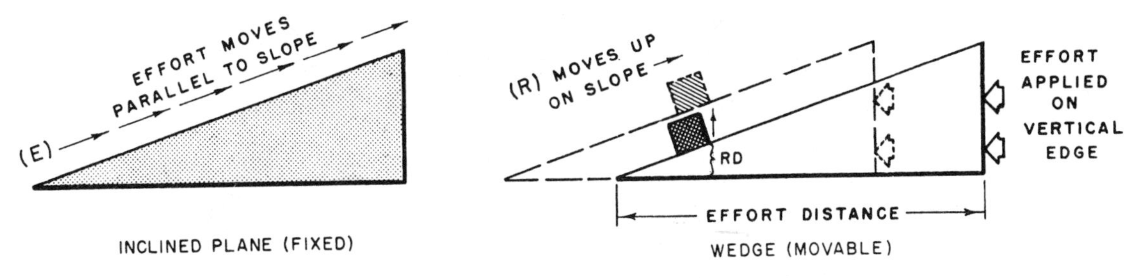

Fig. 16-8 The wedge is a variation of the inclined plane.

Metric

$$MA_f = \frac{\text{R (Resistance Force)}}{\text{E (Effort Force)}}$$

$$MA_f = \frac{\text{445 newtons}}{133.5} = 3\ 1/3$$

The mechanical advantage of distance can also be calculated.

U.S. Customary

$$MA_d = \frac{\text{RD (Resistance Distance)}}{\text{ED (Effort Distance)}}$$

$$MA_d = \frac{\text{3 ft}}{\text{10 ft}} = .3$$

Metric

$$MA_d = \frac{\text{RD (Resistance Distance)}}{\text{ED (Effort Distance)}}$$

$$MA_d = \frac{\text{0.9 meter}}{\text{3 meters}} = .3$$

The mechanical advantage of speed is calculated with the same formula:

$$MA_s = \frac{RD}{ED}$$

Note that these mechanical advantage formulas apply to all simple machines.

WEDGE

The wedge is a variation of the inclined plane. It is driven under or into an object and applies its force through a longer distance. Many cutting tools are actually made up of a series of wedges which are called teeth. Under magnification these wedge teeth get under and break away the material they are cutting, figure 16-9. Axes and chisels are also wedges. In addition, wedges make excellent holding devices; the simple doorstop is one example.

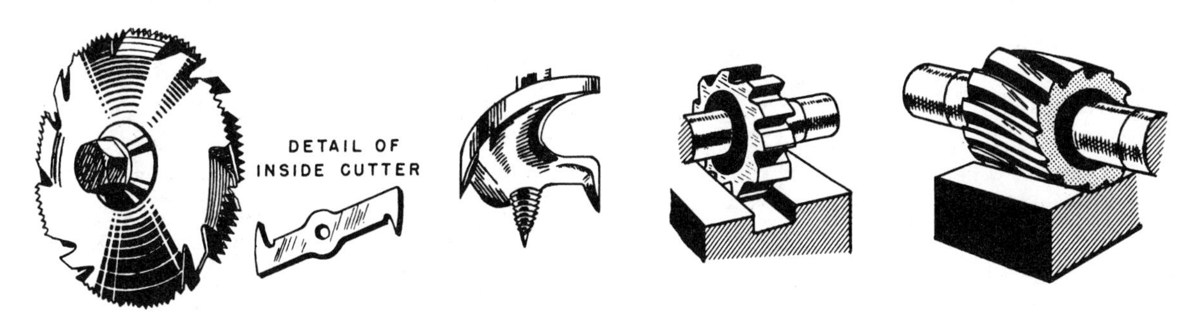

Fig. 16-9 Applications of the wedge as a separating device

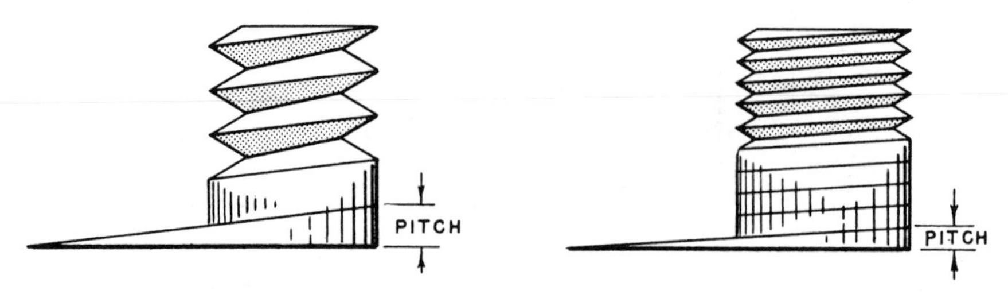

Fig. 16-10 The screw is an inclined plane in spiral form.

SCREW

The screw is also an inclined plane. However, it is one that has been put into a round form and appears as a continuous spiral, figure 16-10. Its most common application is in fasteners such as nuts, bolts, and wood screws. However, screw principles are used in jacks, worm gears, some pumps, and to convert rotary motion to linear motion. The mechanical advantage that a screw applies can be tremendous since the pitch of the screw is usually quite small (RD) and the distance the effort is applied through (ED) is usually quite large by comparison. A jack screw illustrates this very well, figure 16-11.

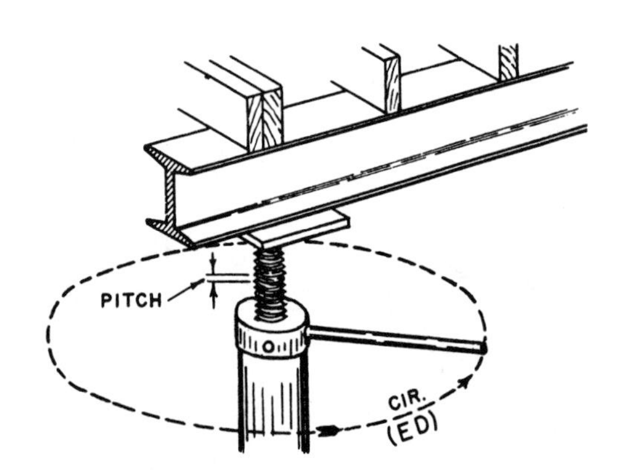

Fig. 16-11 The jack screw applies a tremendous mechanical advantage of force.

PULLEYS

Pulleys in various arrangements create a mechanical advantage which can aid in the movement of heavy objects. In gaining the mechanical advantage, distance moved is sacrificed in the interest of applying less force. The number of ropes supporting the object determines the mechanical advantage. If four ropes support the weight, then the rope is pulled four feet in order to move the object one foot; but the force, disregarding friction, required is one-fourth the weight of the object, figure 16-12.

GEARS

The gear uses the same principles as simple machines. It is merely a wheel with special notches or teeth cut into its perimeter to provide positive contact with another gear. In a sense, two meshing gears are like two levers working against each other.

V-belts and pulleys, or chains and sprockets, have the same mechanical function as do gears, figure 16-13. They are useful if power is transmitted between shafts that are separated by a distance. If construction permits close spacing of the shafts, then gear teeth can mesh against each other, a simpler arrangement.

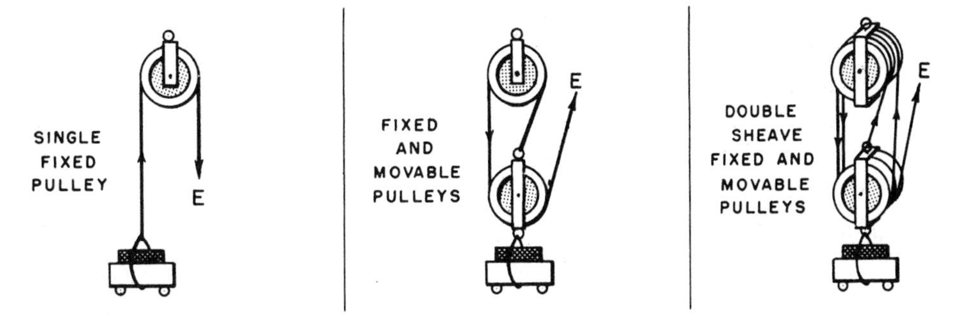

Fig. 16-12 Pulleys increase in mechanical advantage as the number of turns of rope (ED) increases.

Gears of different sizes change speed in proportion to the number of teeth on each gear, figure 16-14. A gear with 24 teeth will drive a gear with 12 teeth at double its speed.

$$MA_S = \frac{\text{Teeth in Driving Gear}}{\text{Teeth in Driven Gear}}$$

$$MA_S = \frac{24}{12} = 2$$

Fig. 16-13 A chain-driven gear train

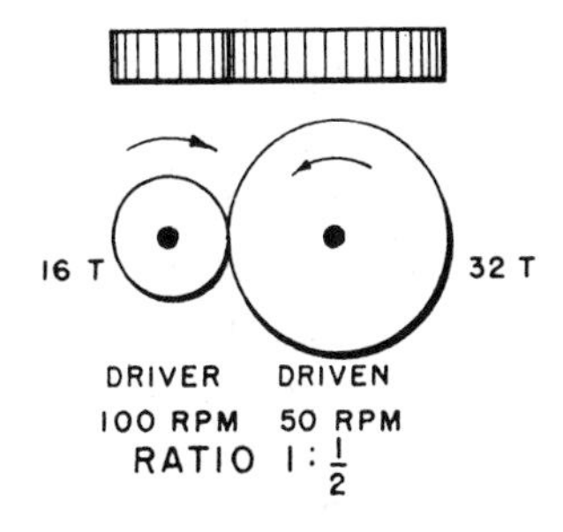

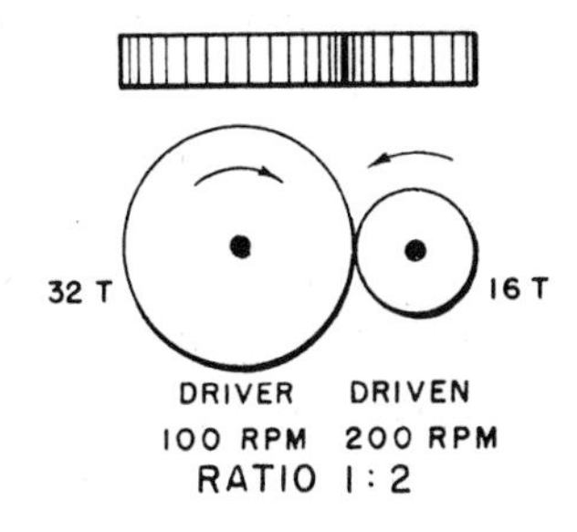

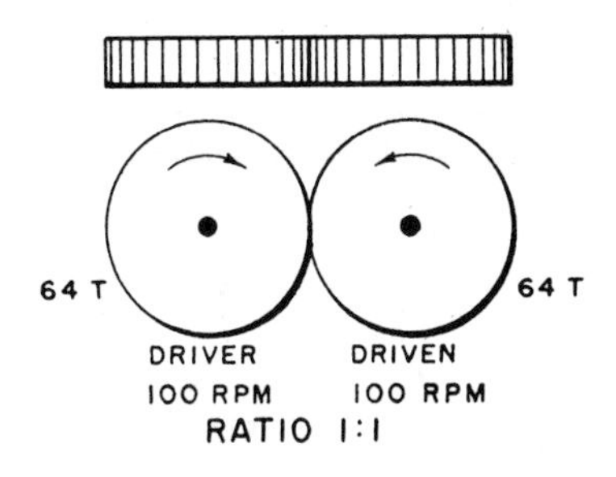

Fig. 16-14 Gears may be used to change speed and force.

When speed changes, the torque or twisting force also changes. Gearing up to increase speed reduces the torque produced. Gearing down to decrease speed increases torque. A gear with 24 teeth will drive a gear with 48 teeth at half its speed, but twice its force.

$$MA_f = \frac{\text{Teeth in Driven Gear}}{\text{Teeth in Driving Gear}}$$

$$MA_f = \frac{48}{24} = 2$$

The gearing in an automobile is an excellent example of this. At low speed it is geared down to gain the torque needed to get the auto moving. At highway speed, the auto is moving nicely and is geared up. Less torque is needed, but more speed is required. The power transmission principles of the automobile are discussed in more detail in Unit 9.

Gears also change the direction of rotation as the power is transmitted from one to another; from clockwise rotation to counterclockwise rotation and so forth, figure 16-15.

Gears are arranged in a gear box in the proper sizes and combinations to satisfy the needs of the machine that is powered, figure 16-16. Sometimes the positions and gear combinations are fixed, but often provision is made to shift the gears around in various combinations as is done in an automobile

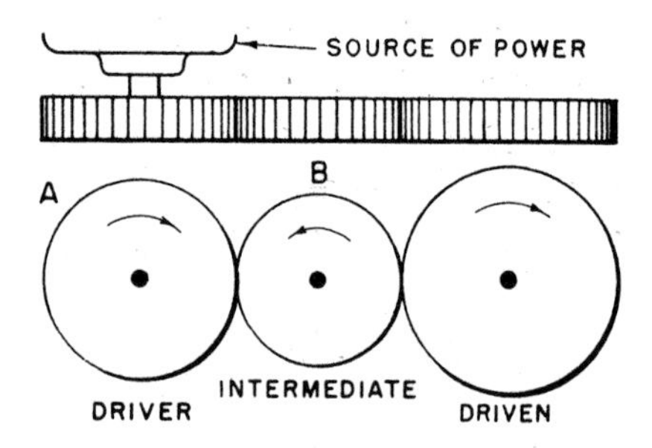

Fig. 16-15 Changing direction of rotation

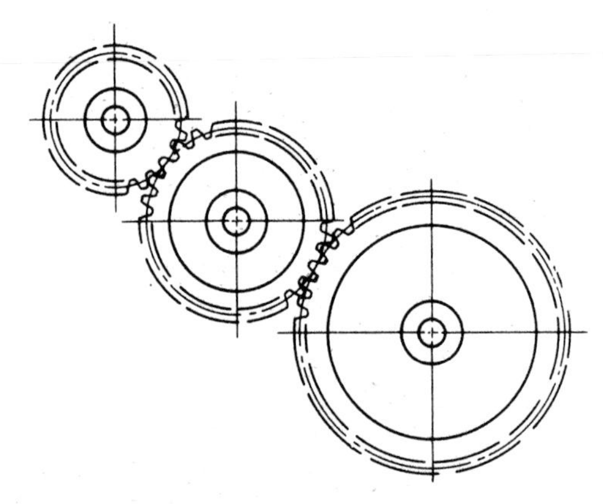

Fig. 16-16 Representation of simple gear train

transmission. Gears in no way change the total power that is developed by the prime mover, they just change the speed or torque.

Mechanical transmission of power is seen in everyday machines and mechanisms. It is important to understand some of the basic principles of machines and power transmission. This takes some of the mystery out of the devices that at first appear too complicated to be understood. Mechanical transmission is the world of gears, wheels, levers, belts, pulleys, chains, shafts, and screws. They are positive linkages between the prime mover and the mechanism. Through years of improvement and research they have become very efficient and reliable in their service.

THE POWER TRAIN

Many uses of engines involve getting the power to the wheels as well as getting power to the machine. For example, a riding lawnmower engine must move the entire machine along as well as turn the blades for mowing grass. A wide variety of applications of levers, gears, pulleys, V-belts, shafts, chains, and sprockets are used to transmit the power of the engine's crankshaft at the correct speed and torque for the machine's requirements. Sometimes it is as simple as the direct drive of a rotary push-type mower. However, many uses are more complex and require a clutch, transmission, universal joint, and differential. These mechanisms are referred to as the power train.

- *Clutch* – connects and disconnects the power without stopping the engine.
- *Transmission* – provides for different speeds and directions (forward and reverse).
- *Universal Joint* – provides a flexible drive shaft.
- *Differential* – allows the wheels to rotate at different speeds for turning corners.

Clutch

The clutch engages or disengages the engine from the rest of the power train. When the engine is started and the machine is standing still, the clutch is disengaged. When the machine is moving, the clutch is engaged. Think of two pie plates, each on the end of a shaft, figure 16-17. One plate and shaft can rotate without the other plate and shaft rotating. However, if the rotating plate is moved into contact with and held against the nonrotating one, both will rotate.

A very simple clutch arrangement is sometimes used on V-belts and pulleys. By having no belt tension the engine is disengaged. Tightening the belt tension with an idler pulley engages the machine, figure 16-18.

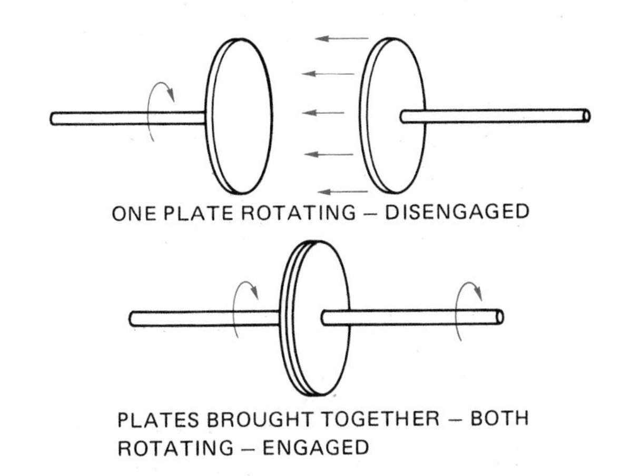

Fig. 16-17 When a nonrotating plate is held against a rotating plate, both plates will rotate.

Transmission

Gears are arranged in a transmission in the proper sizes and combinations to satisfy the operating requirements of the machine that is being powered. Provisions are made to engage or shift the gears into various combinations which provide different output speeds and directions.

A typical transmission for small gasoline engines has three forward speeds and one

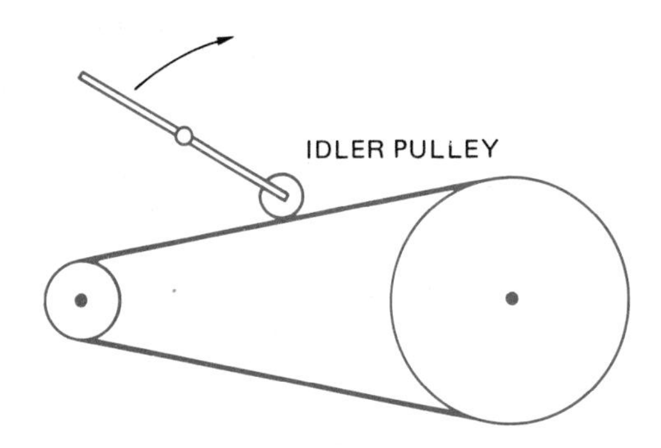

Fig. 16-18 Moving the idler pulley against the belt engages the drive.

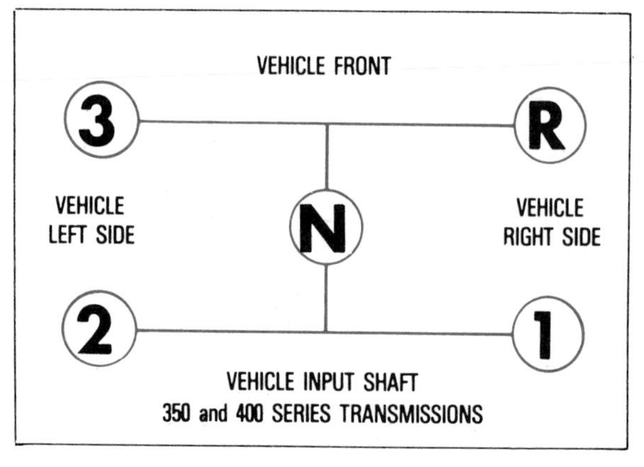

Fig. 16-19 Transmission shifting pattern

reverse speed, figure 16-19. The Tecumseh/Peerless model 350 shown in figure 16-20 is a good example of a typical small gasoline engine transmission. Power input is to the large bevel gear. A shifter fork mechanism shown in figure 16-21 slides the gears into mesh as required and power is transferred to the companion or output shaft. When the transmission is put into reverse, an idler gear is brought into mesh. This "third" gear reverses the direction of rotation on the output shaft.

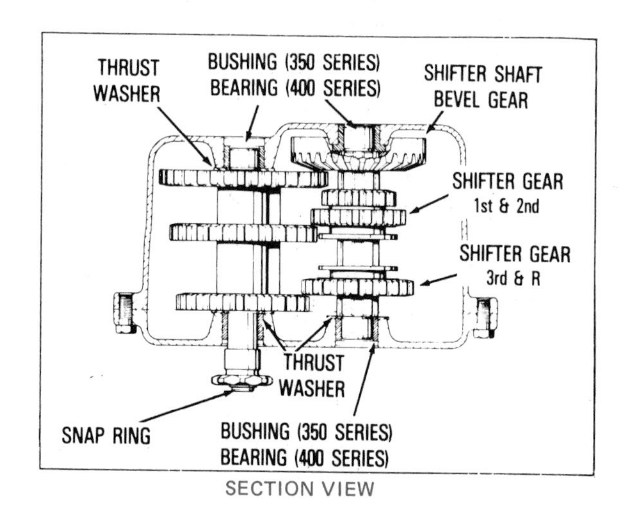

SECTION VIEW

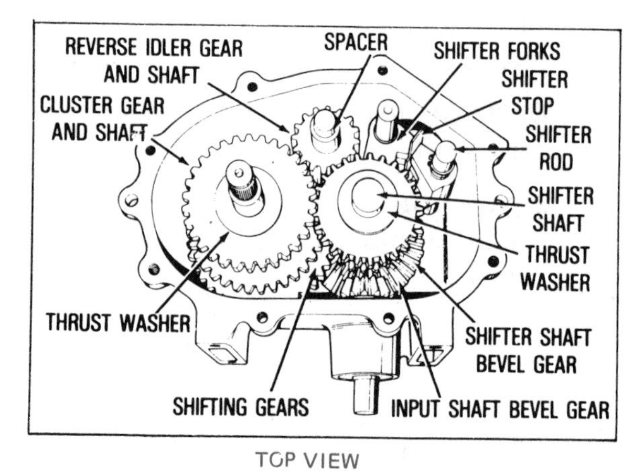

TOP VIEW

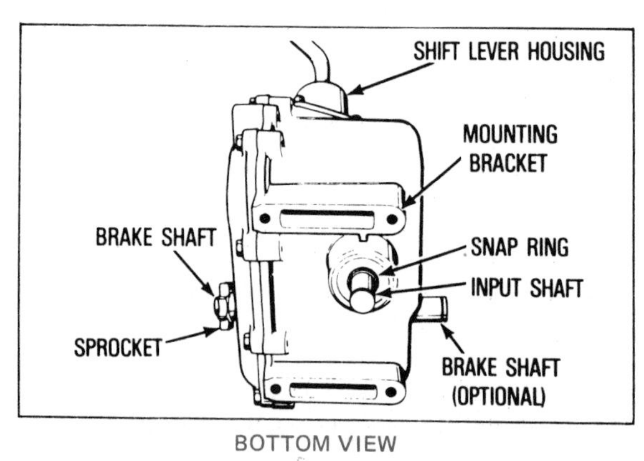

BOTTOM VIEW

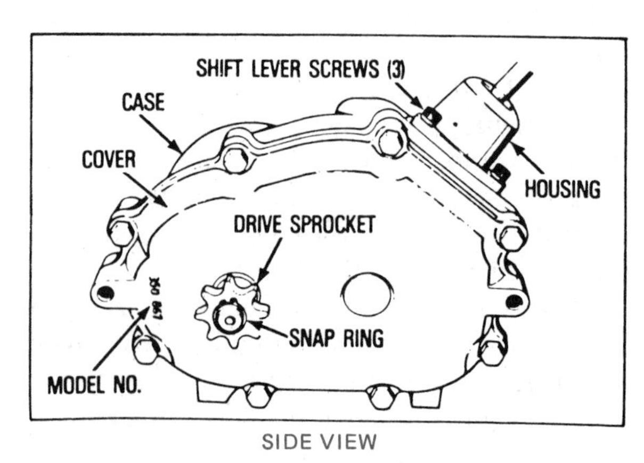

SIDE VIEW

Fig. 16-20 A typical small gasoline engine transmission

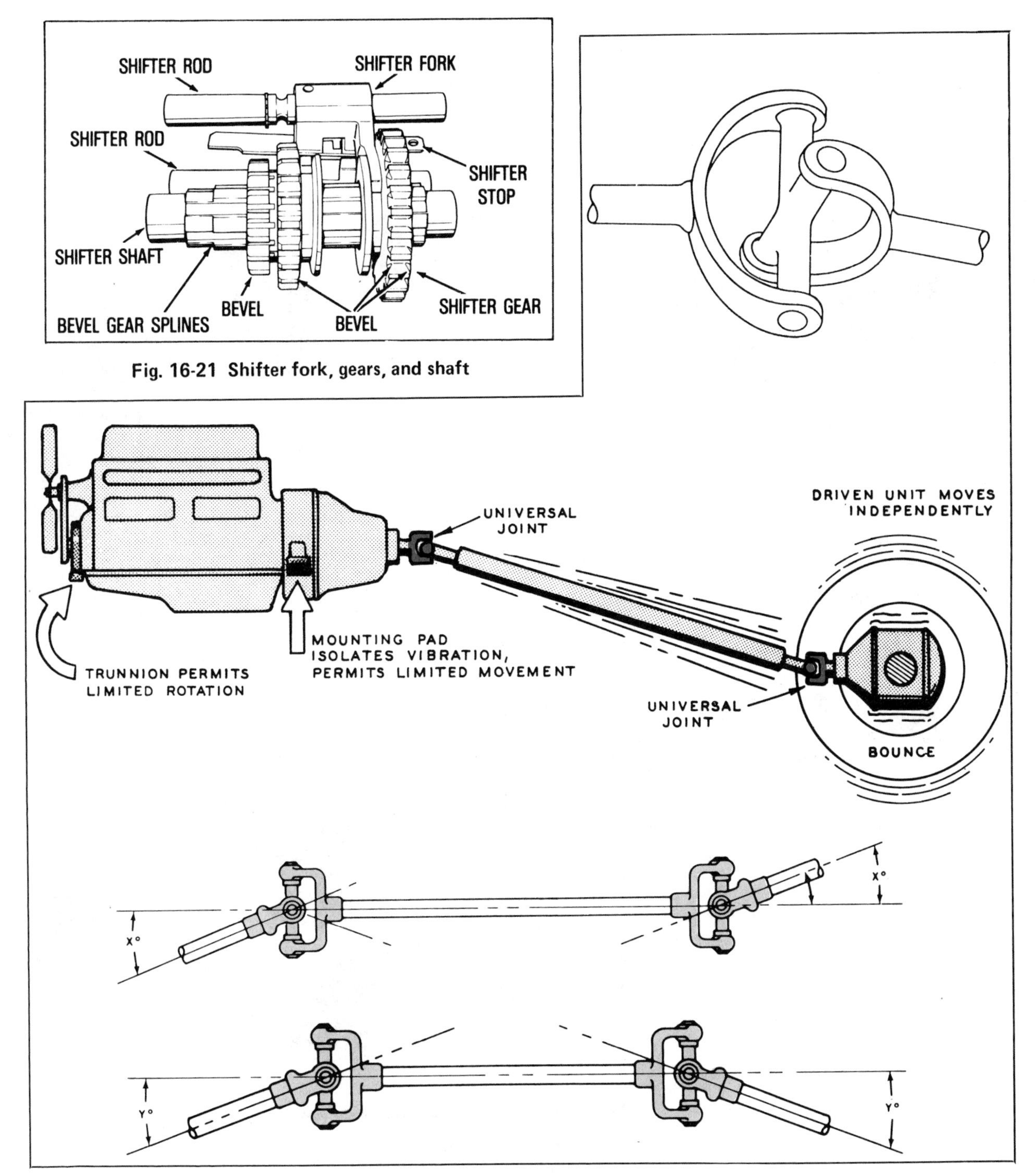

Fig. 16-21 Shifter fork, gears, and shaft

Fig. 16-22 Universal joint

Universal Joint

Universal joints add flexibility to the mechanical transmission of power by connecting shafts that do not share the same axis, figure 16-22. The most common use for universal joints is between the transmission and differ-

Fig. 16-23 Flexible couplings (Courtesy Lovejoy, Incorporated)

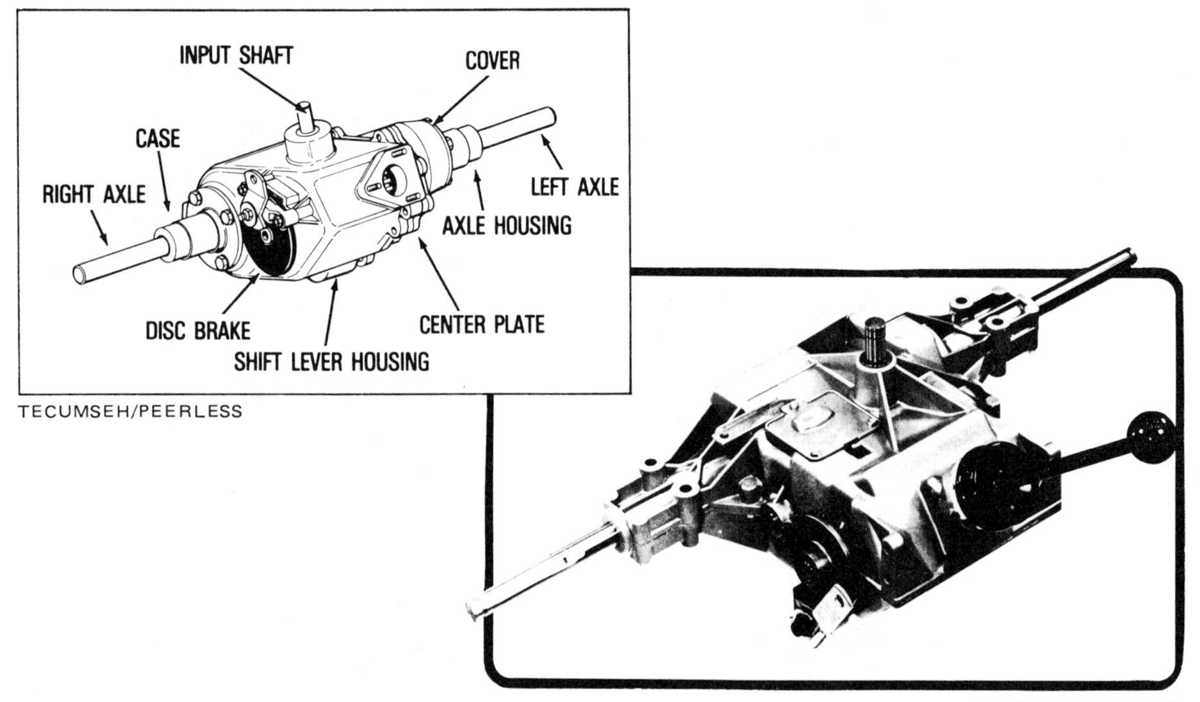

Fig. 16-24 Transaxles

ential. The universal joint compensates for misalignment, bouncing, and movement between the transmission and differntial. The action of the universal joint is similar to the action of the human wrist. In addition to universal joints, flexible couplings can be used to transmit power if there is a small amount of misalignment or motion between two units, figure 16-23.

Differential

The drive shaft and universal joints carry the power to the differential. The differential transmits the power to the axles and also enables the wheels to revolve at different speeds when a corner is turned. The wheel on the inside of the turn needs to travel less distance than the wheel on the outside of the turn, just as a column of marching soldiers do in turning a corner.

The parts of the differential are the case, drive pinion, ring gear, differential side bevel gears on the end of the axles, and pinion gears mounted in the differential case. Details of the differential operation are covered in Unit 9.

Transaxles

Transaxles are a combination of the transmission, differential, and axles within one housing, figure 16-24. This results in a more compact and lighter weight unit. Transaxles are very common for riding lawnmowers, lawn tractors, and similar machines. A transaxle unit with the top housing removed is shown in figure 16-25. The transaxle shown in figure 16-26 has five speeds forward and reverse. Power input is to the bevel gear. The shifter collar moves the key into the shifter spur gear cluster to engage a particular gear. The remaining four shifter spur gears turn freely on the shifter and brake shaft. The spur gear at the end of the shifter and brake shaft turns the output gear; the output pinion engages the differential ring gear. The sprocket and chain drive are engaged for reverse.

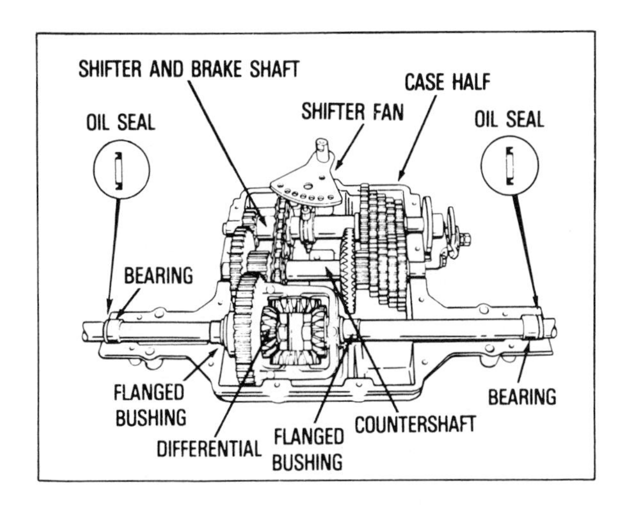

Fig. 16-25 Transaxle with top housing removed

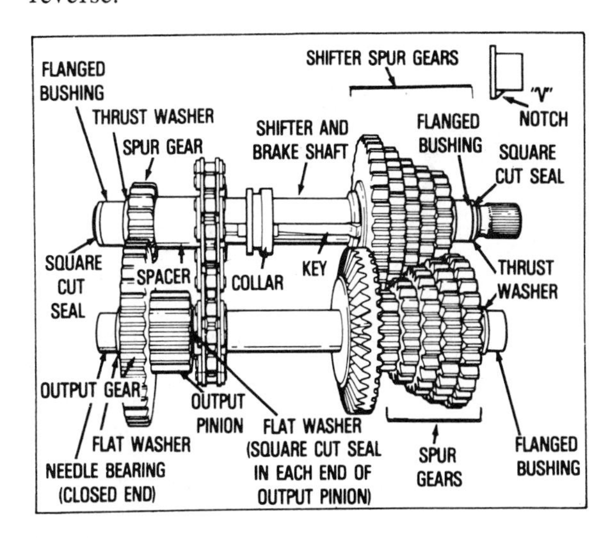

Fig. 16-26 Basic parts of a typical transmission portion of a transaxle

REVIEW QUESTIONS

1. Can direct or fixed transmission of power between the prime mover and the machine satisfy all machine applications? Explain.
2. What are the three basic power transmission systems?
3. List one example of each class of lever: first class, second class, third class.
4. Calculate the force necessary to move a 200-pound (890-newton) weight using a first class lever. The weight is 3 feet (0.9 meters) from the fulcrum. The lever's total length is 13 feet (4 meters).
5. Calculate the mechanical advantage of force found in carrying a 50-pound (222.5 newton) box up a 10-foot (3-meter) ramp that is 2 feet (0.6 meters) from ground level.
6. Compare a conventional wedge to the wedge formed by a cutting tooth.
7. What could be said about the mechanical advantage that is created by the use of the screw as a simple machine?
8. What determines the mechanical advantage of a pulley?
9. When are V-belts and pulleys or chains and sprockets used instead of gears?
10. Explain the relationship that exists between speed and torque.
11. What is the mechanical advantage of speed between a driven gear with 12 teeth and a driving gear with 96 teeth?
12. Explain the term *power train.*
13. What is the purpose of a clutch?
14. Explain why many engine applications need a transmission.
15. When might a universal joint be used?
16. Why do wheeled vehicles need a differential?
17. What is a transaxle?

Unit 17 Fluid Transmission of Power

OBJECTIVES After completing this unit, the student should be able to:

- Explain the difference between hydraulics and pneumatics.
- List and discuss the parts of a typical hydraulic system.
- Discuss the advantages of the hydraulic system.
- List and discuss the parts of a typical pneumatic system.
- Explain the applications of a vacuum.

Fluid power involves the use of a fluid—in either liquid or gas form—to operate a mechanism. The fluid might be air-directed, as wind captured by the sail of a ship, or it might be water flowing in a river, its energy tapped by a water wheel. In the ancient civilizations of the Romans, Cretans, Persians, Indians, and Chinese, water was stored in reservoirs, channeled in aqueducts and pipes, and used for domestic purposes and irrigation. Civilization did not progress beyond these applications of fluid power for many centuries. These, then, give modern fluid power its foundation. To better understand modern uses, two key words must be introduced. These words are the two basic terms that are associated with fluid power: hydraulics and pneumatics.

Hydraulics is that branch of fluid power that deals with the use of liquids, usually oil, to transmit power from one place to another. The hydraulic system receives and controls the power delivered by the prime mover, generates a hydraulic force, transmits the force, and converts the force to power which is delivered to the work load.

Pneumatics is that branch of fluid power that deals with the use of gases, usually air or nitrogen, to transmit power from one place to another. The pneumatic system receives and controls the power delivered by the prime mover, generates a pneumatic force, transmits the force, and converts the force to power which is delivered to the work load.

Hydraulics and pneumatics share many common characteristics:

- Both use a fluid media; a liquid takes the shape of the vessel in which it is contained, and a gas completely fills the vessel in which it is contained.
- Both liquids and gases can flow through a pipe or a hose.
- The rate of flow of both liquids and gases is easily controlled with valves.
- Both liquids and gases can be caused to flow by various types of pumps.

There is one basic difference between liquids and gases which makes the separation of the two areas into different study units easier to follow. Liquids are virtually incompressible, while gases can be compressed. A hammer blow on the piston of a liquid-filled cylinder is like hitting the end of a steel shaft. A hammer blow on the piston of a gas-filled cylinder is like hitting a spring-loaded cylinder; it gives and then bounces back. Although there are many similarities that exist between hydraulic and pneumatic systems, for the sake of clarity and because of their one basic difference, the two are discussed separately.

HYDRAULICS

The key principle for modern hydraulics was formulated by Blaise Pascal in the seventeenth century. Pascal's Law states that pressure set up in a confined liquid acts equally in all directions and *always* at right angles to the containing surfaces. As an example, consider a flat fire hose which becomes cylindrical when it is filled with water. Pascal theorized, "There is a vessel

1. that is filled with water.
2. The vessel is closed on all sides.
3. It has two openings.
4. One opening is one hundred times as large as the other.
5. Each opening is supplied with a piston that *fits exactly*.
6. A person is pushing on the small piston.
7. The force exerted *equals* that of 100 people pushing on the large piston.
8. If *one* of the 100 people leaves, the force exerted by the person on the small piston will *overcome* the remaining 99 people pushing on the large piston.

This law even more simply stated means the pressure at any point in a static liquid is the same in every direction and exerts equal force on equal areas, figure 17–1.

Pascal's Law is an application of mechanical advantage to a liquid. Just as a lever can multiply force at the sacrifice of distance, so can hydraulics.

In the illustration of the lever and the weight, figure 17–2, a force of 2 pounds (8.9 newtons) moves through 100 inches (2540 millimeters) to produce 200 inch-pounds (22.6 joules) of work. This work is capable of producing equal work on the other end of the lever and can move 100 pounds (445 newtons) through a distance of 2 inches (50.8 millimeters), or 200 inch-pounds (22.6 joules).

The hydraulic creation of mechanical advantage is similar to that of the lever, figure 17-3. Two pounds (8.9 newtons) of force pushing the piston down the cylinder 50 inches (1270 millimeters) is equal to 100 inch-pounds (11.3 joules) of work. The two pounds (8.9 newtons) of force is transmitted

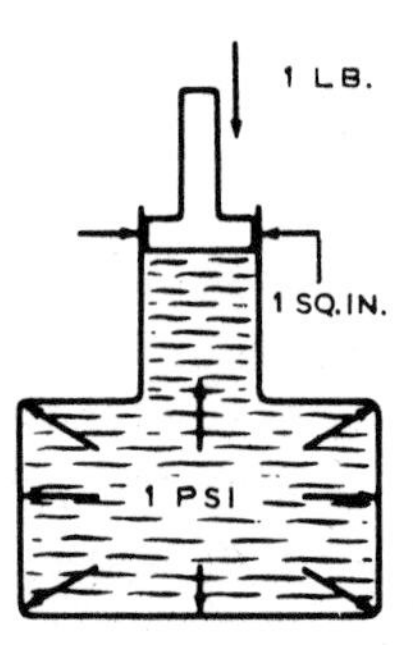

Fig. 17-1 Pressure is transmitted equally in all directions.

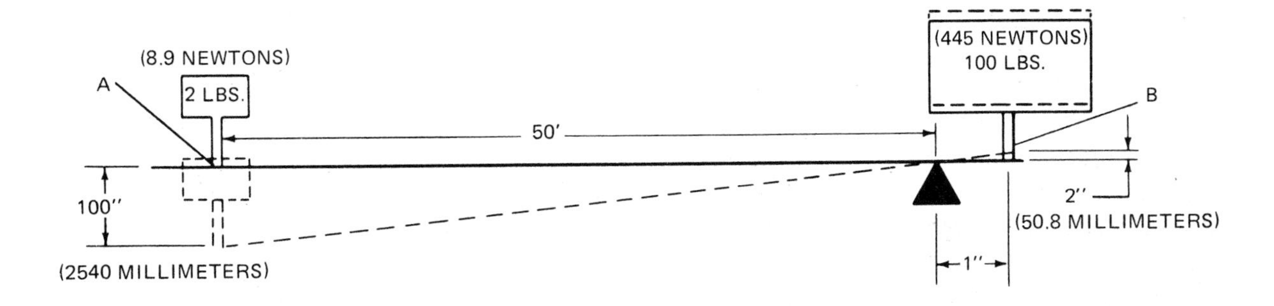

Fig. 17-2 The mechanical advantage of a lever also applies to hydraulics.

undiminished in all directions and pushes against the 50 square inches (32 258 square millimeters) of the large piston with a total force of 100 pounds (445 newtons). Work = Force × Distance:

U.S. Customary

$$\text{Distance} = \frac{\text{Work}}{\text{Force}} = \frac{100 \text{ inch-pounds}}{100 \text{ pounds}} = 1 \text{ inch}$$

The large piston moves upward one inch.

Metric

$$\text{Distance} = \frac{\text{Work}}{\text{Force}}$$

$$= \frac{11.3 \text{ newton-meter}}{445 \text{ newtons}} = 25.4 \text{ millimeters}$$

The large piston moves upward 25.4 millimeters.

Pascal's Law deals with *static liquids,* that is, liquids that are not in motion; the *dynamics of liquids,* or liquids in motion, is quite another thing. As indicated earlier, friction is not taken into account, but friction is a factor in fluid transmission of power just as it is in the mechanical transmission of power. Friction in dynamic fluids is always present but it can be minimized if (1) the pipes or hoses are large enough to prevent excessive velocity and the resulting turbulence, (2) the lengths of the pipes or hoses are not excessive, (3) the number of bends in the pipe or hoses is not excessive, and (4) the corners are not too sharp.

The principle of how friction affects liquid

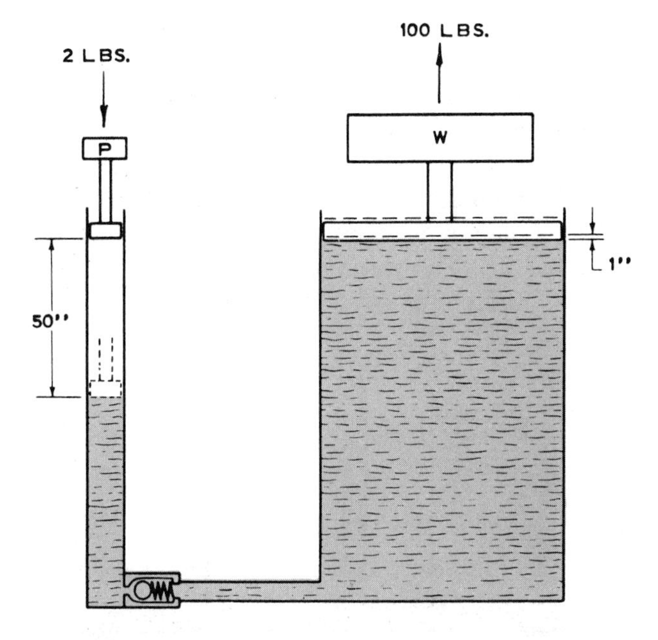

Fig. 17-3 Creating a mechanical advantage in hydraulics

flow is seen in figure 17–4. The main body of the liquid between A and C creates what is called *static head pressure.* When point B is opened up, the water is free to flow and dissipate the entire pressure. Notice that the water does not rise to the level of A in pipes D, E, and F. It is noticeably less the further the pipes are from the source. The drag of friction prevents the water from rising in these pipes, because friction has robbed some of the static head pressure.

Turbulence is another factor in fluid dynamics. A smooth laminar (in layers) flow of liquid is the most efficient. The layers of moving fluid are parallel to each other, but layers near the center of the pipe are moving more rapidly than those near the surface of the pipe where friction creates some drag. Therefore, if the velocity becomes too great due to a pipe that is too small, the laminar flow is disturbed and then efficiency-destroying turbulence sets in. Laminar flow might be compared to the smooth flow of traffic on a superhighway under moderate traffic load. Rush hour traffic—on the other hand—slows, it is more turbulent, and therefore slower and less efficient.

Bernoulli's principle concerns itself with the dynamic behavior of fluids. This principle states that the static pressure of a moving liquid varies inversely to its velocity. The static pressure on the surface of the pipe or hose decreases as the velocity of the fluid increases. This principle is applied in carburetors when the fluid flowing is air rushing through the venturi section. Low pressures are created in the venturi section. Figures 17–5 and 17–6 show how static pressure is reduced in constricted areas where the velocity is increased.

These are some of the principles that provide guidelines for engineers to follow as they design various hydraulic units that make up hydraulic systems of power transmission. Hydraulic systems are commonplace today, and, in terms of history, is a young industry. Modern hydraulics is not water flowing in aqueducts; it is liquid under high pressure in high-pressure pipes and hoses and in close tolerance machines with leak-proof pistons and cylinders. The development of reliable fluid power transmission systems had to wait for

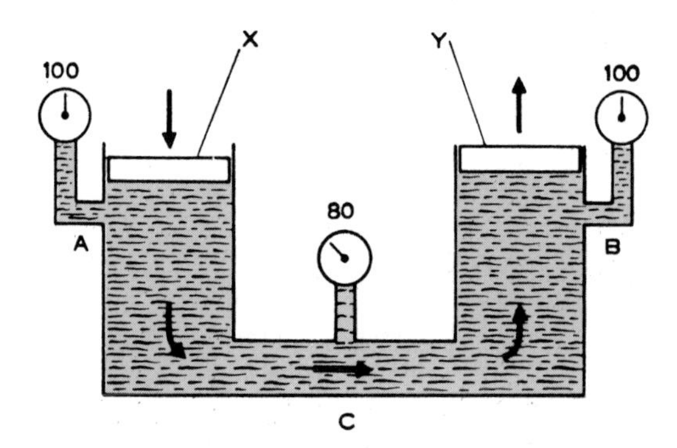

Fig. 17-5 Static pressure decreases as velocity increases.

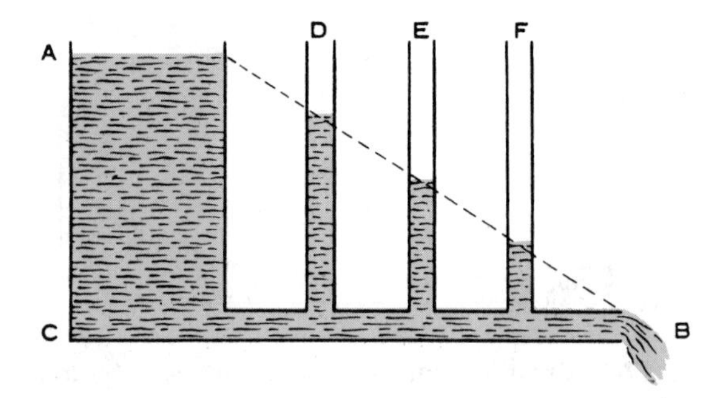

Fig. 17-4 The drag of friction also applied to liquids.

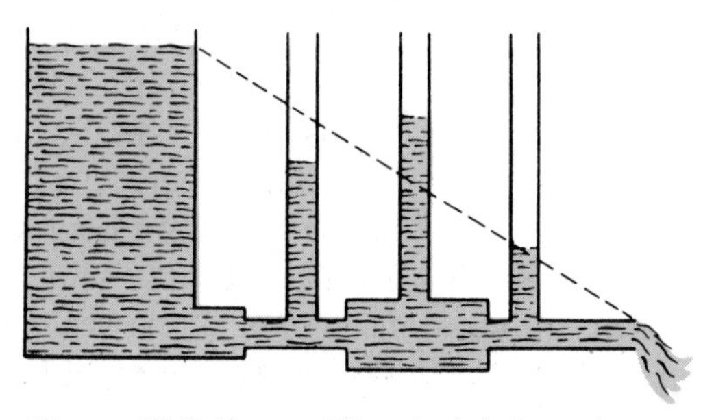

Figure. 17-6 Bernoulli's principle in action

improved materials and manufacturing techniques before it became a reality.

HYDRAULIC COMPONENTS

The average person is probably unaware of the many hydraulic systems in common use. Such uses include the automobile brake system, power steering, a garage hoist, or a dump truck, figure 17-7. There is also the wide application of hydraulics to the manufacturing industries such as machine tools, food and chemical processors, die casting, plastics,

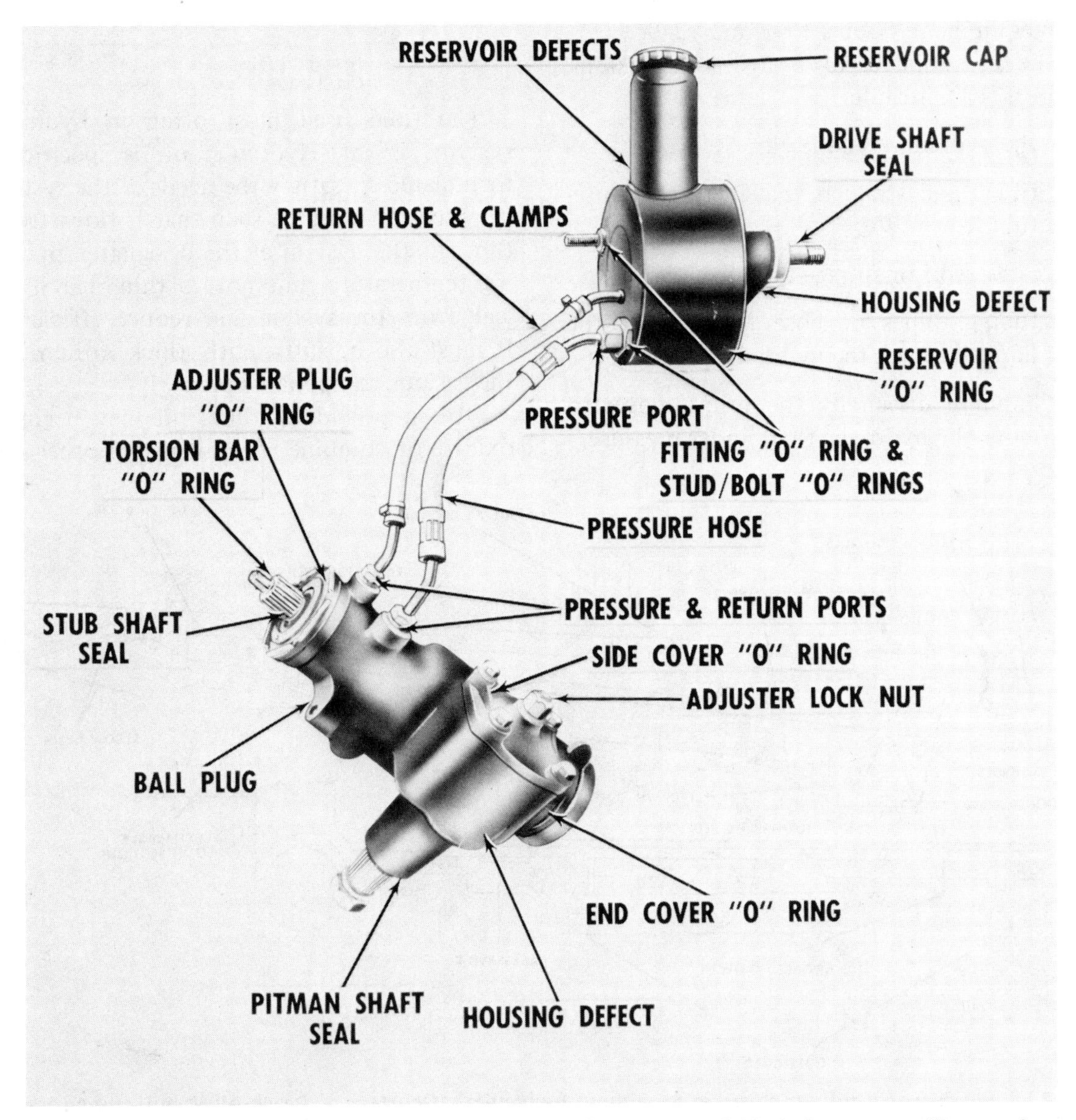

Fig. 17-7 Automobile power steering uses hydraulics. Points to check if a leak occurs are illustrated.

presses, glass machinery, textile mills, paper products, ordnance, and deck machinery. Applications in the aerospace industries include commercial and military aircraft, rockets, satellites, and helicopters. Applications for mobile equipment include lift trucks, earth movers, farm machinery, log skidders, trucks and buses, utility vehicles, mining, and railroad equipment, figures 17–8 and 17–9.

The components of a hydraulic system, figure 17–10, include:

- the hydraulic fluid itself.
- filters and strainers that increase its useful life.
- a reservoir or storage tank for liquid.
- tubing, pipes or hose to transmit the liquid through the hydraulic circuit.
- a pump driven by the prime mover to move the hydraulic fluid under pressure.
- valves to change direction, pressure, or amount of flow.
- motor or cylinder which is acted on by the hydraulic fluid, converting hydraulic power back into mechanical power for use by the mechanism.

Let us now cover these one by one.

HYDRAULIC FLUID

The fluid used most often in hydraulic systems is oil. Hydraulic oil is specifically formulated to satisfy the needs of the system. Its viscosity must be such that it flows freely through the system at the designated operating temperature, but not so thin that it can leak from the system and reduce efficiency. It must also be sufficiently thick to lubricate the various components properly.

Over a period of time, oil may begin to oxidize or combine with air just as other sub-

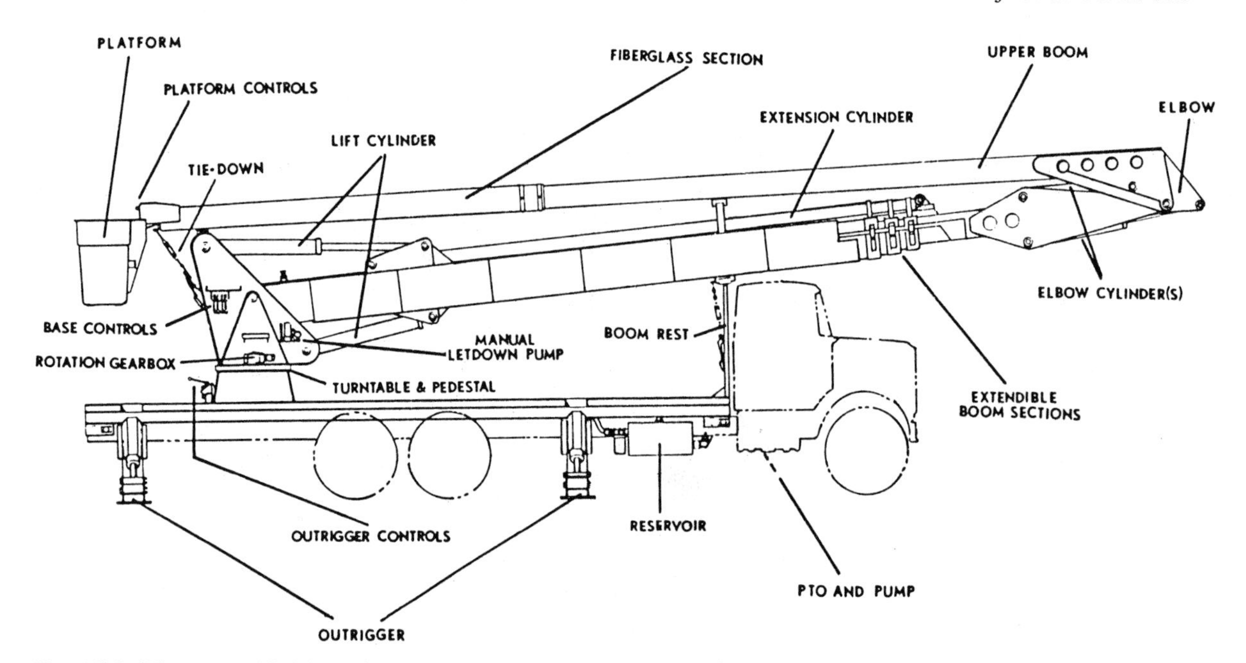

Fig. 17-8 Where would this aerial basket be without hydraulics? (Courtesy of Reach-All Manufacturing and Engineering Company)

Fig. 17-9 Typical applications of hydraulics (Courtesy Massey-Ferguson)

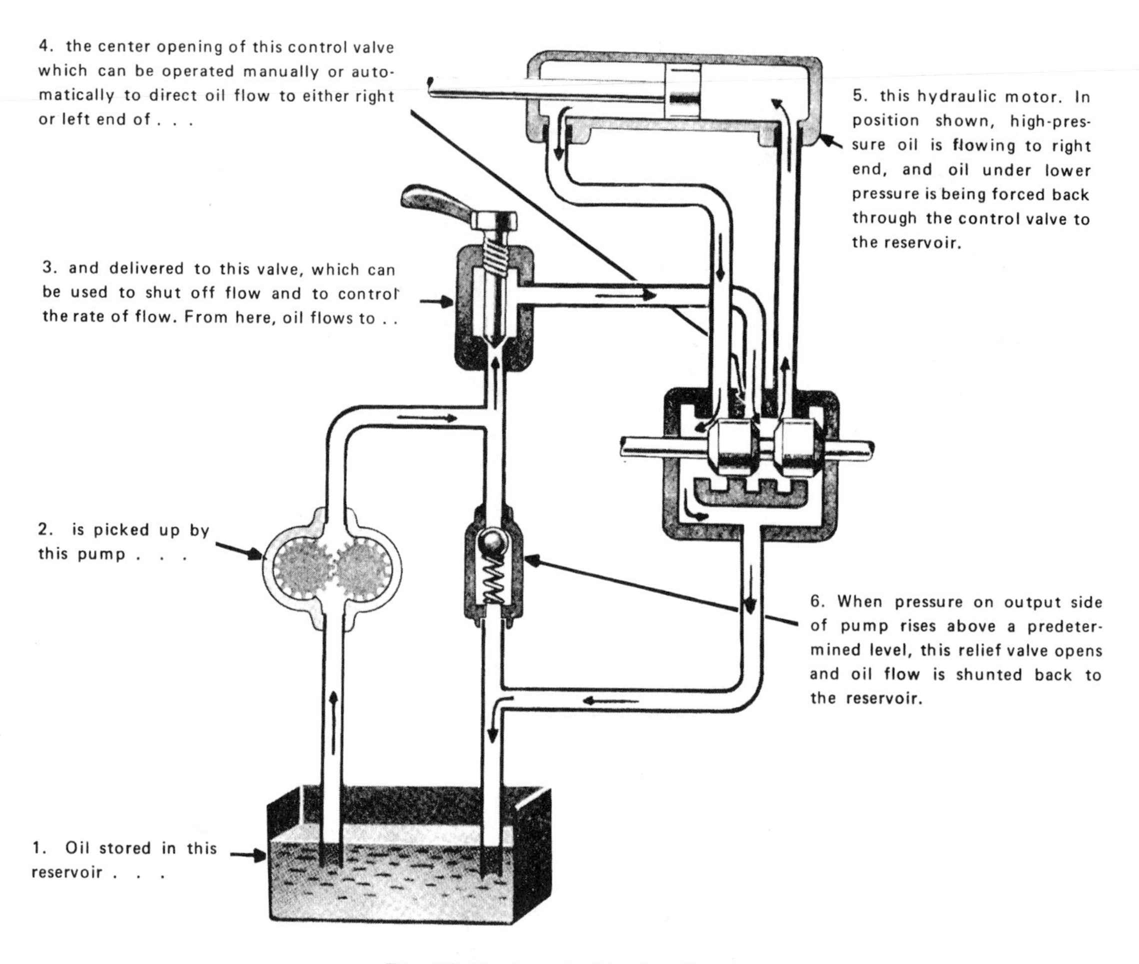

Fig. 17-10 A typical hydraulic system

stances do. This must be controlled because it causes the development of resins, varnishes, lacquers, and sludges. High operating temperatures increase the rate of oxidization in oil, and the contaminants caused by oxidization are corrosive acids. Therefore, corrosive effects on the precision hydraulic components can be minimized by using equipment that works at moderate pressure and temperature, and by careful preventive maintenance.

FILTERS AND STRAINERS

Filters and strainers are important in every system to prevent excessive wear, figures 17-11 and 17-12. Since small particles of dirt in the oil cause an abrasive action in hydraulic parts just as they do in engine parts, strainers are usually made of fine wire mesh capable of removing coarse foreign matter. Filters—by comparison—remove much smaller impurities, such as combustion byproducts and water.

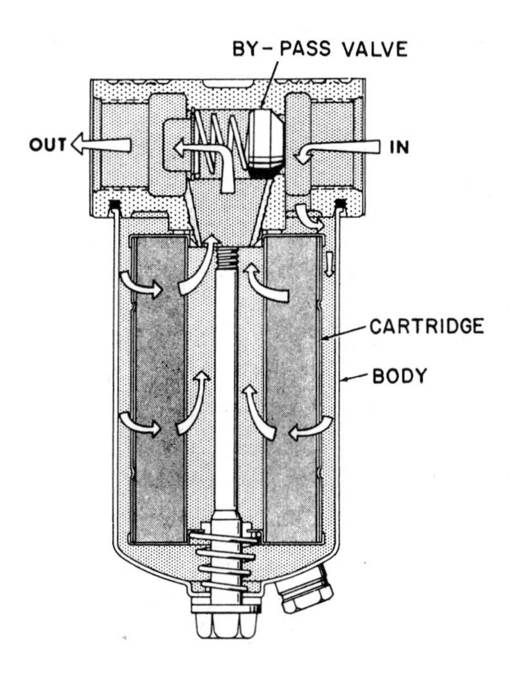

Fig. 17-11 Full flow oil filter

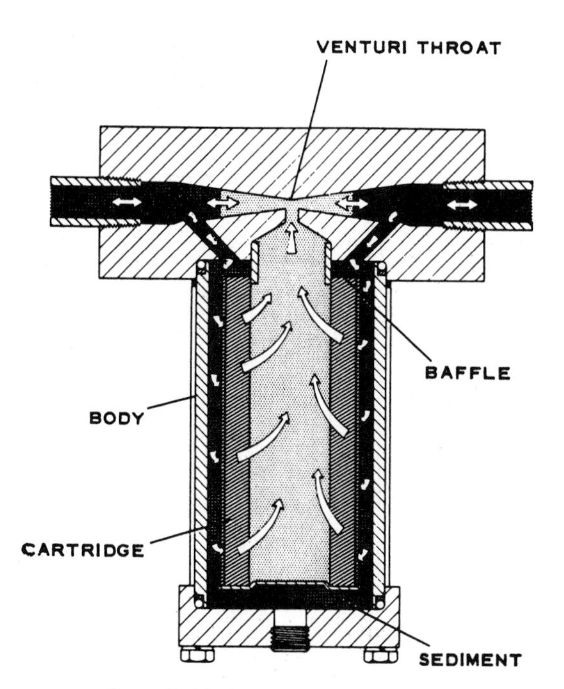

Fig. 17-12 Proportional flow oil filter

Most filter elements can be removed and replaced much the same as automobile filters. Today's filters remove particles as small as 78 millionths of an inch (2 nanometers).

RESERVOIR

The reservoir is a tank large enough to hold a sufficient reserve supply of hydraulic oil, figure 17-13. Most reservoirs are vented to the atmosphere to prevent a vacuum from forming as the oil level varies. The vent is fitted with an air breather or filter to prevent foreign materials from entering the system. As the oil returns to the reservoir, it enters with a good deal of force and turbulence. Baffle plates slow down the oil and allow time for air bubbles to rise to the top and the oil to cool before it is recirculated.

PUMPS

The hydraulic pump is driven by the prime mover or electric motor, and converts mechanical energy into hydraulic energy. It picks up the hydraulic fluid and forces it along its way. People often think of these pumps as *building up* an oil pressure, but since oil is virtually incompressible, it is better to think of the pump as delivering a driving rod of solid oil through the piping of the system. The pump generates both the flow and the resulting force. Pressure or resistance to the flow occurs further down the line as the machine places a load against the hydraulic force.

Pumps are rated or sized by the volume of hydraulic fluid they can deliver within a given time. This output varies with the speed of the

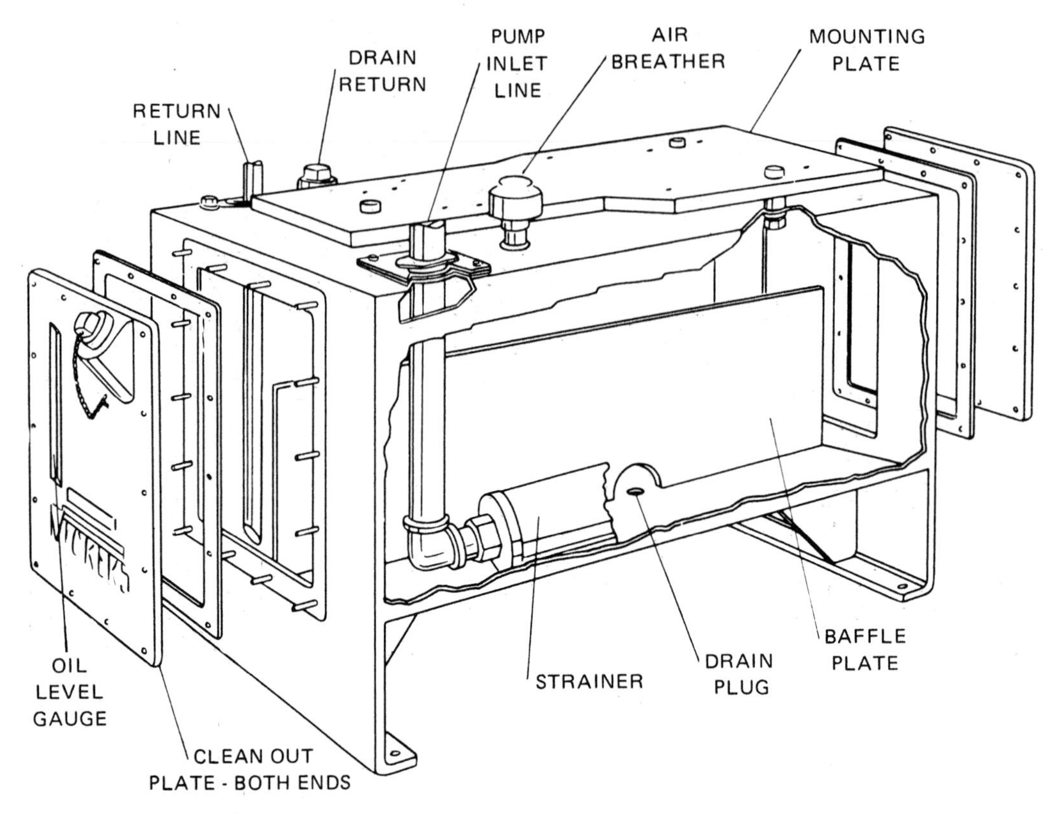

Fig. 17-13 A typical reservoir

pump, so the pump speed must be established as a factor in a pump rating. Some pumps are rated by their displacement or the amount of oil they will move during one cycle.

Pumps operate on either of two basic principles: nonpositive displacement or positive displacement.

Nonpositive Displacement Pumps

The nonpositive displacement pump is characterized by the centrifugal pump, figure 17-14. The speed of the pump and the resistance to flow determine how much fluid is delivered. The principle is simply that centrifugal force throws the fluid from the eye of the pump—where it first enters—to the outside perimeter of the spinning impeller, and through the pump outlet. If the pump discharge was completely

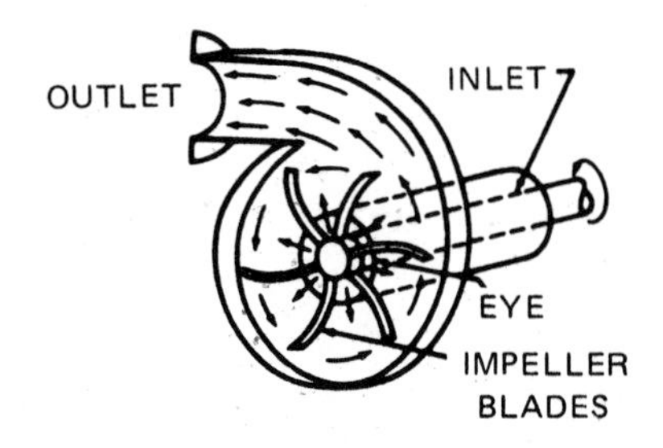

Fig. 17-14 Operating principle of a centrifugal pump which is a nonpositive displacement pump

shut, the pump would churn through the liquid, producing only heat. With the centrifugal pump 100% slippage is possible. These pumps are used in applications that require a large volume of flow at relatively low pressures.

Positive Displacement Pumps

Positive displacement pumps deliver a fixed volume of fluid at any given speed. If resistance to flow is increased, the force of the liquid and its velocity is increased. If the resistance becomes absolute or the flow is cut off, then pressure builds up to the stall or breaking point. A relief valve which opens at a predetermined pressure setting prevents this possible damage. There is little if any slippage in a positive displacement pump.

The pump speed is usually not varied to meet differing operating requirements. If the application calls for flexibility in the volume of the fluid flow, then variable displacement pumps are used. In these pumps the physical relationship between the pump components can be changed to provide for different volume requirements. Flow control valves can also change the volume of fluid flow.

Rotary pumps are of the positive displacement type and include gear, lobe, vane, and piston pumps. In any of these pumps the hydraulic fluid is allowed to enter the pump chamber, trapped by the closing pump chamber, and then moved to the discharge side of the pump where it is forced out. The key parts may be gears or lobes rotating against each other, movable vanes that work against the walls of the pump housing, radial pistons in a rotating off-center cylinder block, or axial pistons in an offset or tilted cylinder block. Several simplified drawings of these pumps are shown in figure 17-15A through E.

ACCUMULATORS

An accumulator is a device sometimes used to store oil under pressure for use when faster-than-normal delivery is required. Also, the accumulator absorbs and cushions shocks or surges in the system. Accumulators are spring or weight loaded, or gas charged.

TUBING

Fluid from the pump is transmitted through pipes, tubing, or flexible hose. These transmitting lines are strong enough to contain the hydraulic fluid at its working pressure and accept higher surges in pressure should they occur. Often hydraulic lines contain a pressure as high as 5000 pounds per square inch (34 megapascals).

VALVES

Valves are located in the hydraulic circuit to control the rate of fluid flow, the pressure of the system, or the direction of the fluid flow.

Flow Control Valves

Rate valves, commonly known as flow control valves, operate on valve principles that are rather well known and commonly used in handling many different fluids. Rate valves include gate, plug, globe, and needle valves, figure 17-16. Gate and plug valves are normally ON–OFF valves. They are not suited for use as a throttle since in a partially-opened position they cause excessive turbulence. The globe valve is also normally ON or OFF, but it can be used as a throttle since the liquid flows on all sides of the globe and

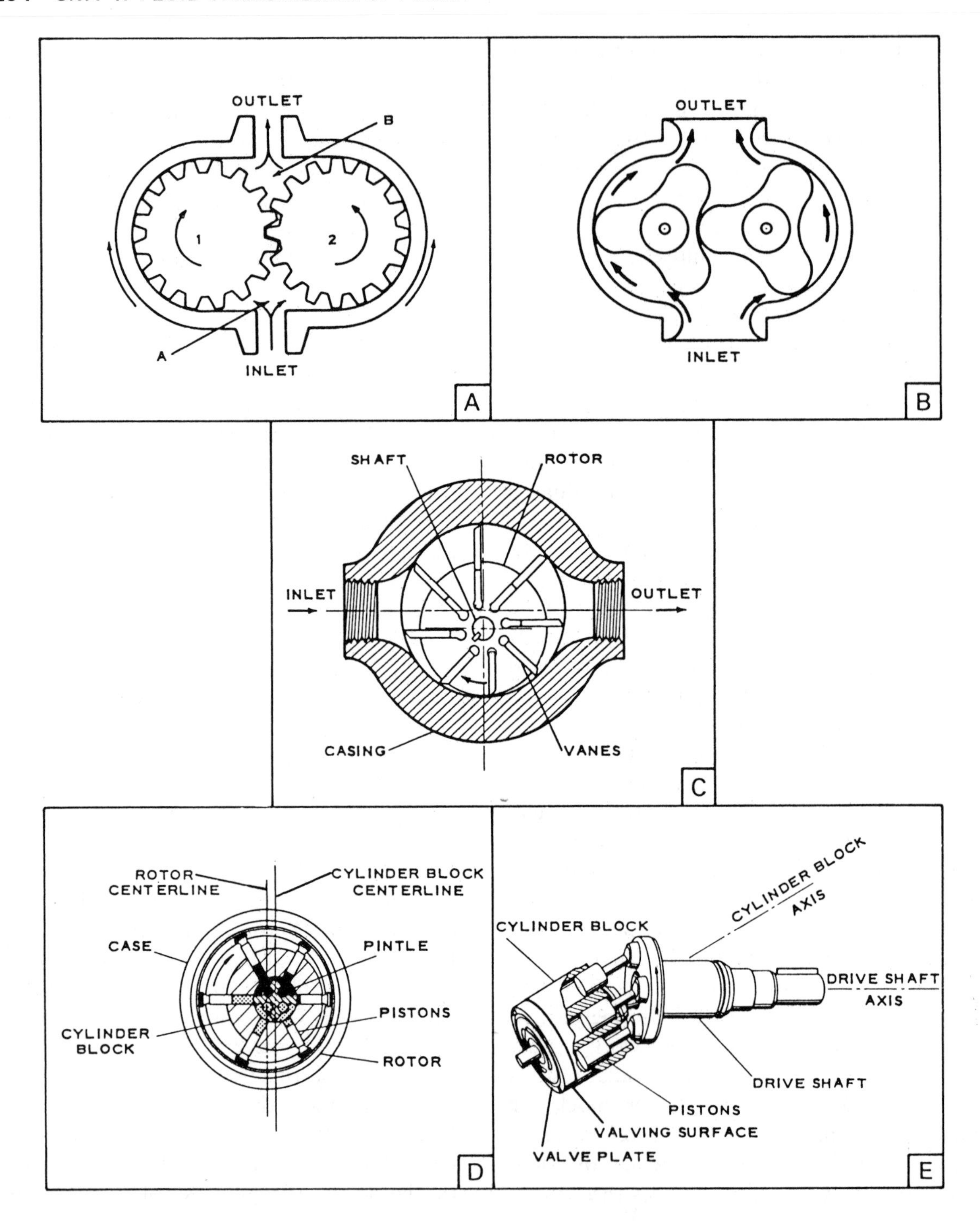

Fig. 17-15 Positive displacement hydraulic pumps: (A) gear pump, (B) lobe pump, (C) vane pump, (D) radial piston pump, and (E) axial piston pump

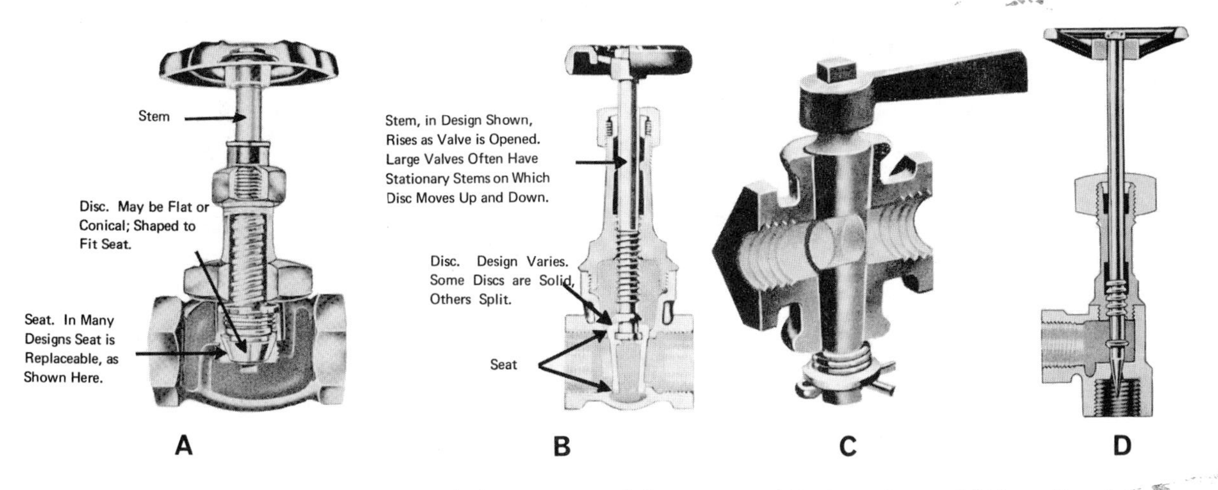

Fig. 17-16 Typical flow control valves: (A) globe valve, (B) gate valve, (C) plug valve, and (D) needle valve

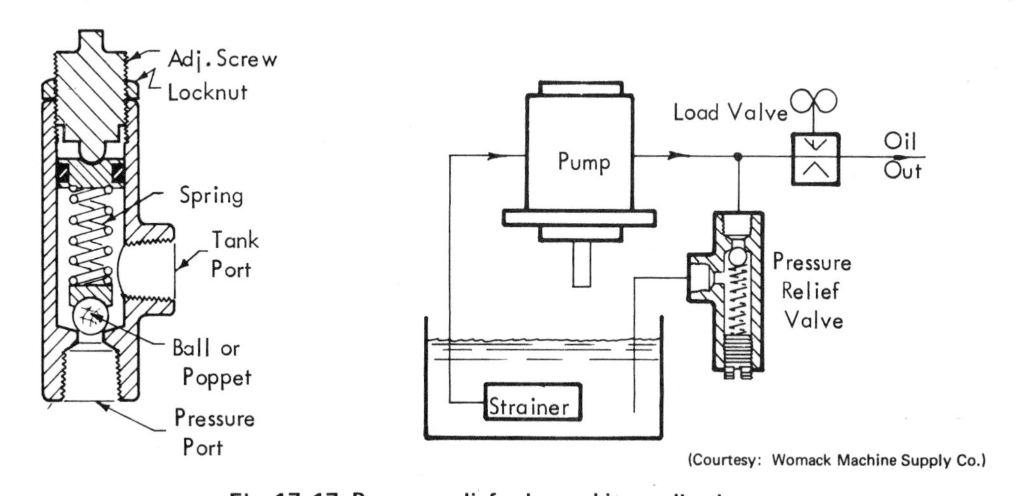

(Courtesy: Womack Machine Supply Co.)

Fig. 17-17 Pressure relief valve and its application

it is more balanced in operation. Needle valves are well suited to vary the amount of flow through small openings.

Pressure-compensated flow control valves accomplish the same basic task as a governor does on an engine. These valves provide a constant pressure and hence a constant speed at the working machine, even though the load on the machine may vary. A spring-loaded piston moves to increase the valve opening when the pressure begins to drop at the discharge side of the valve. The pressure balances the spring to provide a uniform rate of fluid flow.

Relief Valves

Pressure relief valves, figure 17-17, are set to open at a predetermined pressure. They are a safety feature which prevents excessive pressure from building up. They are spring-loaded; the valve is closed when the spring pressure is equal to or greater than the pressure of the hydraulic fluid. At fluid pressures higher than the spring pressure, however, the valve is pushed open and a portion of the fluid is bled off, thereby reducing the force or pressure in the system.

Another type of relief valve is the pressure-reducing valve which allows a branch of the main hydraulic circuit to operate at a lower pressure than the main circuit. These are spring-loaded valves which bleed off fluid to balance the valve at a preset pressure.

Directional Valves

The check valve is perhaps the most common directional valve. This valve permits flow in one direction only. As soon as pressure in the opposite direction is applied, the valve mechanism is forced tightly against its seat. In figure 17-18, the check valve is a spring-loaded piston. If the force of fluid flowing through the valve is stopped, the spring (A) pushes the piston (B) against the valve seat preventing any fluid flow through the valve in the opposite direction.

Spool Valves. Spool valves are designed to control the direction of flow, figures 17–19 and 17–20. The liquid flows around the spool shaft when and where a path is provided. Shifting the spool shaft back and forth opens and closes the various paths. These valves, like many others, can be located in remote positions and be actuated by solenoids.

Motors

The hydraulic fluid now arrives at the motor where hydraulic energy is reconverted into mechanical energy. If linear motion is the desired output of the system, a hydraulic cylinder is used as the motor, figure 17–21. If rotary motion is the desired output, a rotary motor is used. Regardless of which motor is used, its output is converted into power that moves against the work load at the machine. Rotary motors perform the opposite

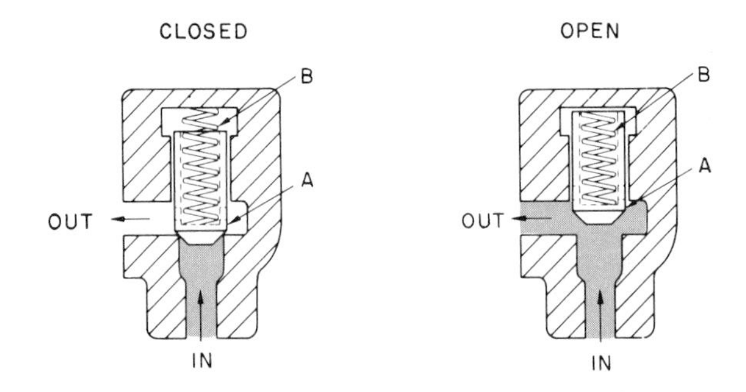

Fig. 17-18 Check valve allows flow in one direction only.

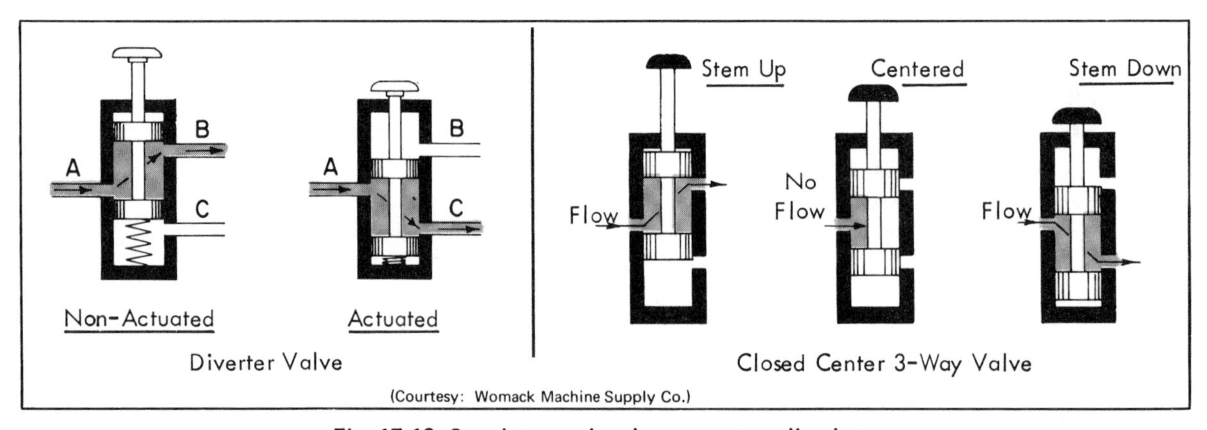

Fig. 17-19 Spool-type valves have many applications.

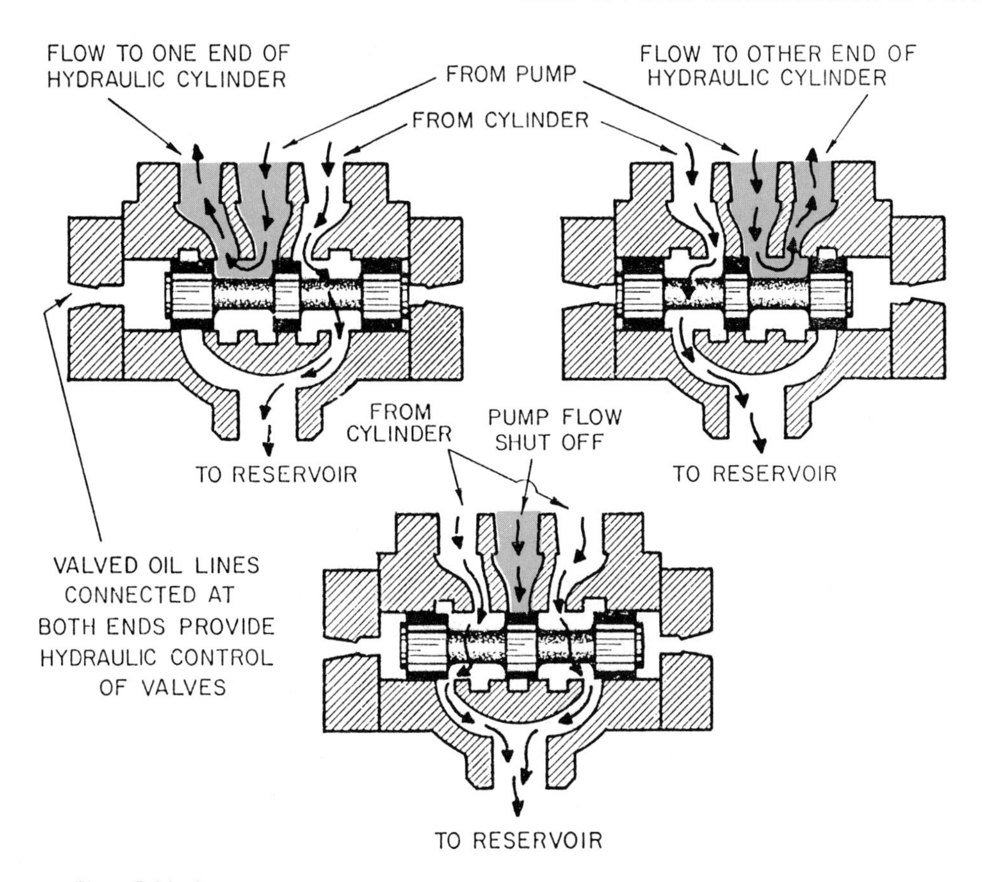

Fig. 17-20 Spool valves are directional valves that provide quick, positive action.

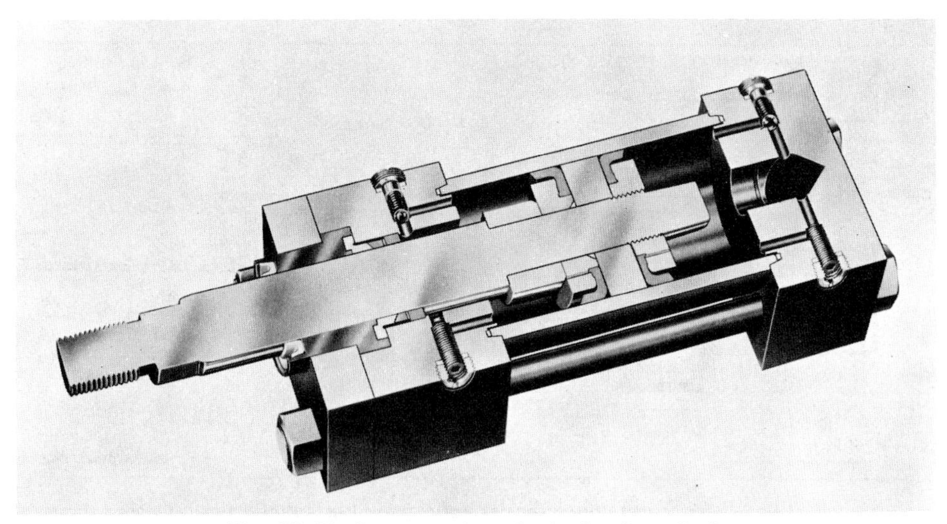

Fig. 17-21 Cutaway view of a hydraulic cylinder

function of rotary pumps, and, in many cases, pumps can be used as motors with little or no modification. Motors are selected on the basis of the speed and torque that are required.

HYDRAULIC CYLINDER ACTION

In a hydraulic cylinder, figure 17–22, the fluid applies a force in all directions inside the cylinder, but only the piston is free to respond to the force, which it does by sliding down the cylinder. This linear motion of the piston produces work at the machine. When the cylinder has completed its stroke, the fluid direction is changed and hydraulic force on the opposite end of the piston moves the piston back to its original position. The area of the piston face has a lot to do with the speed the piston travels. The choice is speed and long distance, or larger force. As the size of the piston increases, the speed decreases and the force is applied through a shorter distance.

If the efficiency of a cylinder is to be high, it must be well sealed at both the rod and the piston. The development of superior packing and sealing materials was a problem in the hydraulics industry for many years, but modern materials prevent leakage within the cylinder. There are a surprising number of applications for cylinders, figure 17–23.

SUMMARY OF HYDRAULICS

The complete hydraulic system consists of the reservoir, pump, transmission lines, pressure, flow, and directional valves, and motor. Figure 17–24 shows a typical hydraulic system.

Hydraulic circuits are usually drawn with symbols rather than with picture or cutaway-type drawings. Fortunately, hydraulic symbols are standardized throughout the industry. The graphical diagram, figure 17–24, is the usual way to show the hydraulic components and their relationships.

Hydraulics is a large and growing field. It can offer advantages which make it an attractive power transmission system.

- Hydraulics produces linear or rotary motion without the use of gears, chains, or belts which require more maintenance.

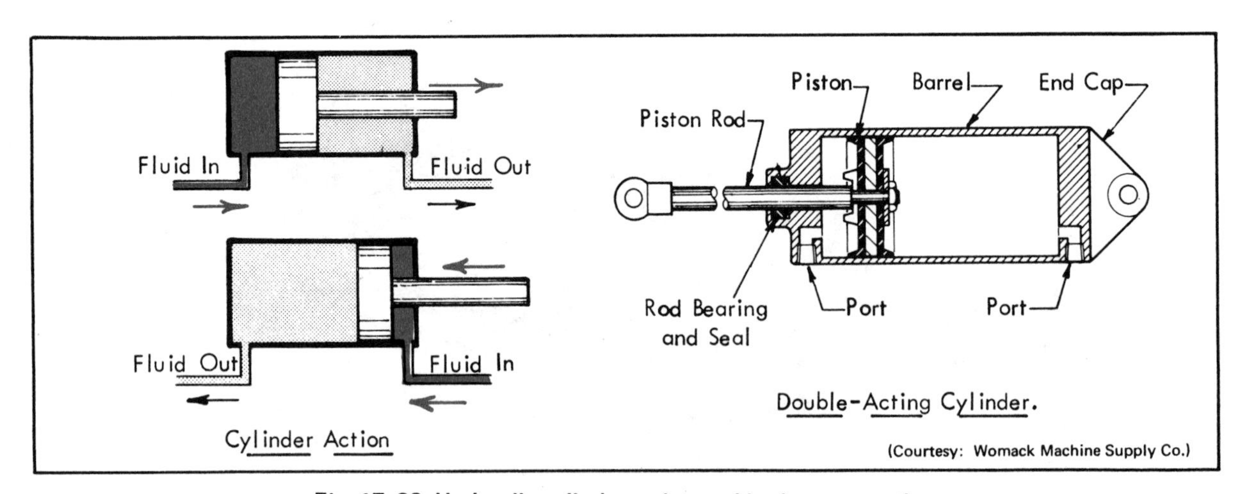

Fig. 17-22 Hydraulic cylinder action and basic construction

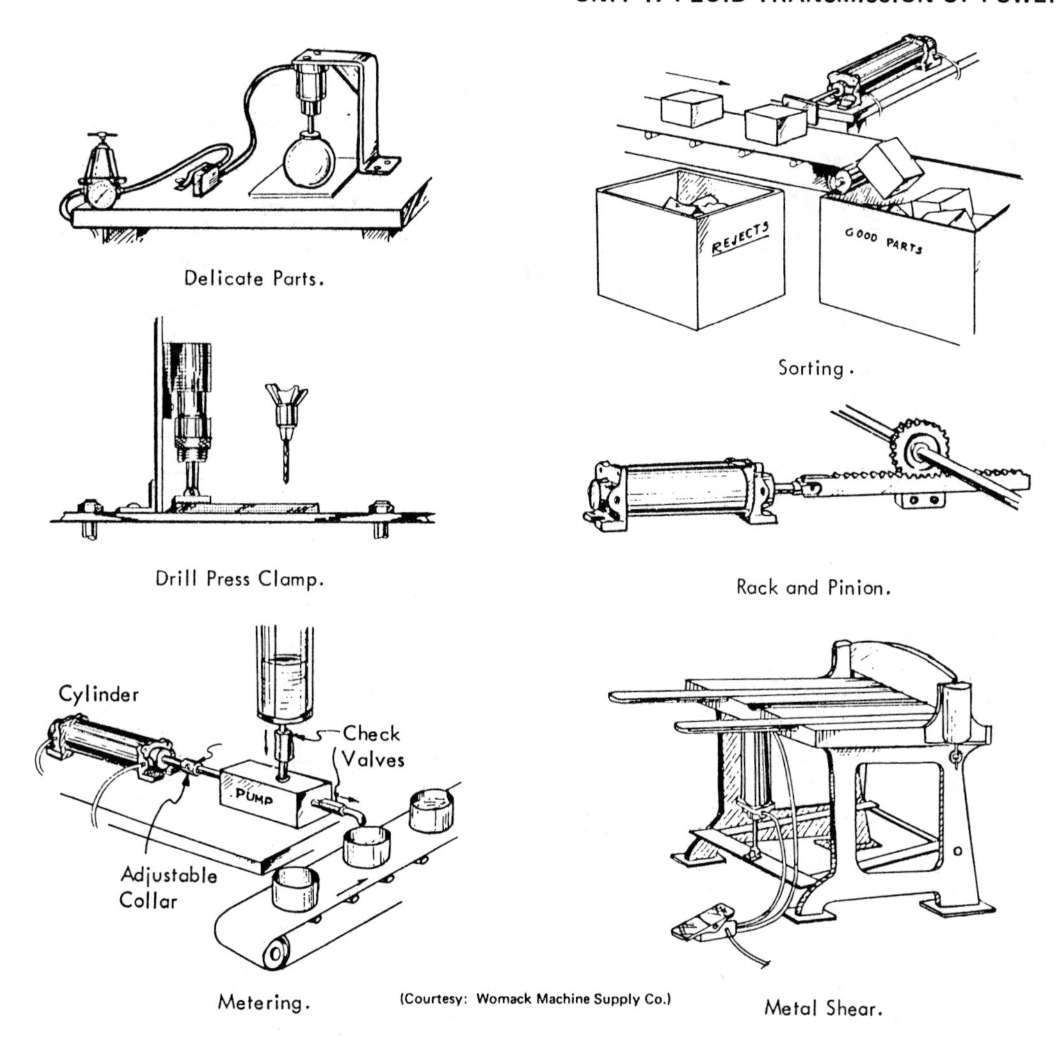

Fig. 17-23 A few of the many applications for hydraulic cylinders

- Hydraulic components lubricate themselves with hydraulic oil.
- Hydraulic power is flexible. A hydraulic hose can go around corners, flex with the machine motion, or go across the room.
- Hydraulics are capable of exerting forces to fit the need, either tremendous forces or very delicate forces.
- Hydraulic power is smooth. Since hydraulic oil is virtually incompressible, instant power response is possible.

PNEUMATICS

Pneumatics and hydraulics—as mentioned before—are both fluid power systems. Much that has been said about hydraulics also applies to pneumatics. Pneumatics does with gases what hydraulics does with oil to transmit

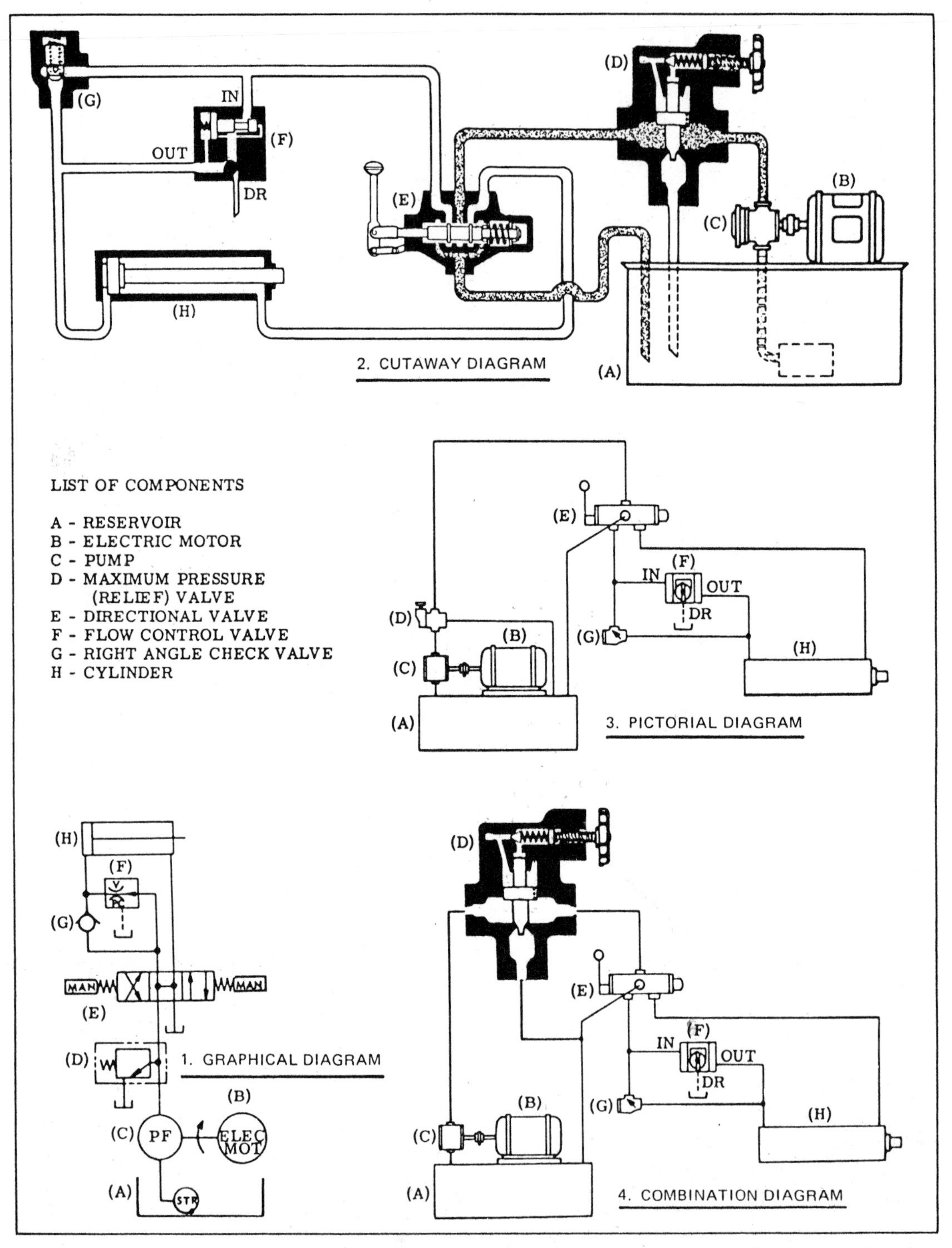

Fig. 17-24 Four types of hydraulic drawings: 1. graphical diagram, 2. cutaway diagram, 3. pictorial diagram, and 4. combination diagram

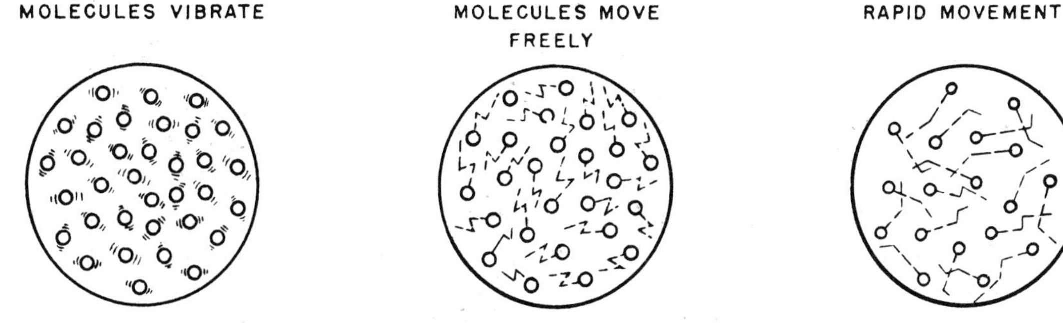

Fig. 17-25 Effect of heat on gas molecules

power from one point to another. The pneumatic system generates the pneumatic force, transmits, and controls the power it receives from the prime mover. Modern pneumatics goes beyond many of the historical applications of gases in motion such as sailboats and windmills because it involves generating and transmitting the gaseous flow within the pneumatic system itself.

LAWS OF PNEUMATICS

Pneumatics is controlled by physical principles just like hydraulics. Pascal's Law, the pressure at any point in a static fluid is the same in every direction and exerts equal force on equal areas, applies to both liquids and gases under pressure. Pneumatics can gain a mechanical advantage just like hydraulics.

The molecules of a gas are far apart and very light. When a gas is heated, molecular activity increases, figure 17-25. Since there is more motion of the molecules, they tend to strike or bump into each other more often. If the gas is in a container, the molecules strike the sides of the container more often and this is seen as an increase in pressure. Temperature has more effect on the expansion of gases than it does on either liquids or solids.

The temperature behavior of gases was observed by Jacques Charles, and his findings resulted in the physical law that bears his name. Charles' Law states that all gases expand and contract in direct proportion to the change in the absolute temperature, provided the pressure is held constant. This means: if the volume is held constant, then the pressure varies in accordance with the temperature, figure 17-26.

Boyle's Law is one of the most common gas laws applied to pneumatics. This law states that "the absolute pressure of a confined body of gas varies inversely as the volume, provided the temperature remains constant." That is to say, double the pressure on a gas and its volume is reduced in half. Quadruple the pressure and the resulting volume is one-fourth

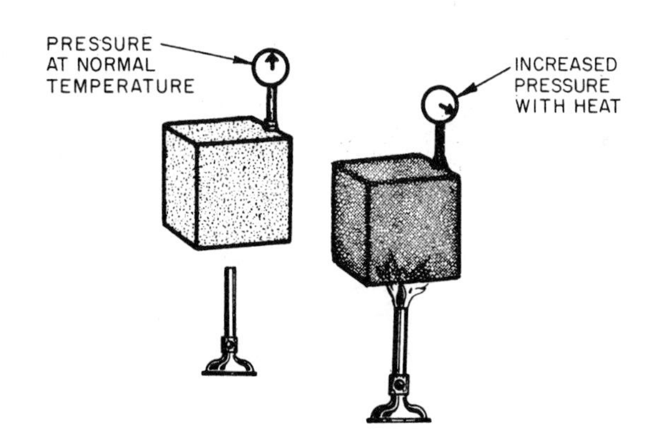

Fig. 17-26 Temperature rise affects movement and pressure.

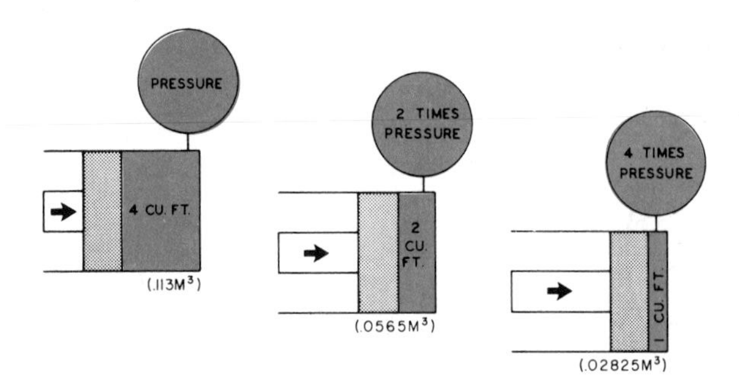

Fig. 17-27 Boyle's Law on the volume of gases

the original volume, figure 17-27. P_1 and V_1 are the original pressure and volume. P_2 and V_2 are the pressure and volume after expansion or compression.

$$P_1 \times V_1 = P_2 \times V_2$$

The formula can also be written as:

$$P_2 = \frac{P_1 \times V_1}{V_2} \quad \text{or} \quad V_2 = \frac{P_1 \times V_1}{P_2}$$

When these formulas are used in actual problems the pressures on the gauges that are in the system must be converted to absolute pressure. Gauges are calibrated at sea level pressure which is 14.7 pounds per square inch (100 kilopascals). The gauge starting point or zero is therefore 14.7 (100) less than absolute pressure of zero. Therefore, 14.7 pounds per square inch (100 kilopascals) is added to the original pressure, P_1, converting it to absolute pressure. For example, find P_2 in figure 17-28.

U.S. Customary

$$P_1 \times V_1 = P_2 \times V_2$$

$$P_2 = \frac{P_1 \times V_1}{V_2}$$

$$P_2 = \frac{(15 \text{ psi} = 14.7 \text{ psi}) \times 5 \text{ cu ft}}{1 \text{ cu ft}}$$

$$P_2 = \frac{29.7 \times 5}{1} = 148.5 \text{ psi}$$

$$\begin{aligned} \text{Absolute} &= 148.5 \text{ psi} \\ &- \ \ 14.7 \\ &\overline{\ 133.8 \text{ psi}} \end{aligned}$$

$$P_2 = 133.8 \text{ psi gauge pressure}$$

Metric

$$P_1 \times V_1 = P_2 \times V_2$$

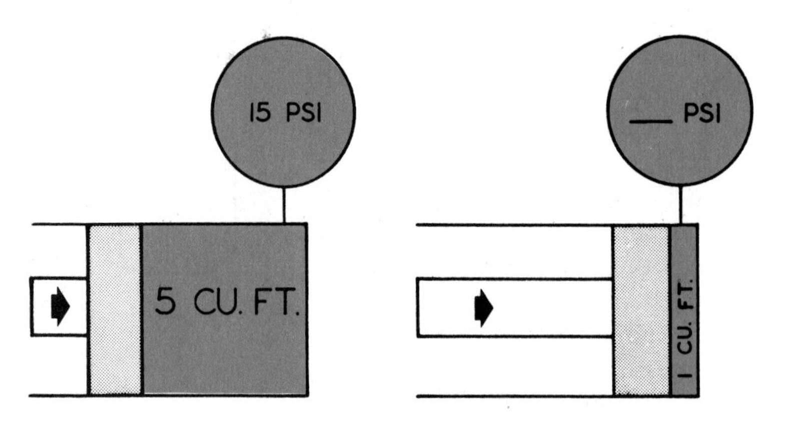

Fig. 17-28 A problem in volume and pressure

$$P_2 = \frac{P_1 \times V_1}{V_2}$$

$$P_2 = \frac{(102 \text{ kPa} = 100 \text{ kPa}) \times 0.14 \text{ c}^3}{0.028 \text{ c}^3}$$

$$P_2 = \frac{28.28}{0.028} = 1010 \text{ kPa}$$

$$\begin{aligned} \text{Absolute} = 1010 \text{ kPa} \\ -\ 100 \\ \hline 910 \text{ kPa} \end{aligned}$$

P_2 = 910 kPa gauge pressure

The same parts are required in a pneumatic system that are required in a hydraulic system. The main difference is that the transmitting media, a gas, is not rigid like a column of confined oil. However, under pressure and in motion, gas can exert a force and produce work.

PNEUMATIC COMPONENTS

Physical laws govern the design of the parts that make up a pneumatic system which must have these components:

- pump or compressor to supply pressurized gas.
- receiver, storage cylinder, air cylinder, or air bottle to act as a reservoir for the pressurized gas.
- actuators.
- dryers.
- hoses or pipes to channel the air.
- valves to control pressure and direction of flow.
- motor which is acted on by the pneumatic fluid, converting pneumatic power back into mechanical power for use by the mechanism or machine.

Some of the components of a pneumatic system are identical to hydraulic components. Nearly all are similar in their construction and operation. They are just designed to handle a different type of fluid, figures 17–29 and 17–30.

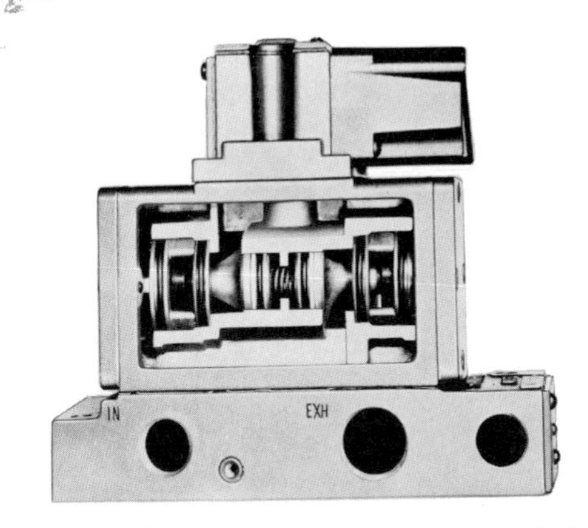

Fig. 17–29 A piston poppet pressure-actuated air valve

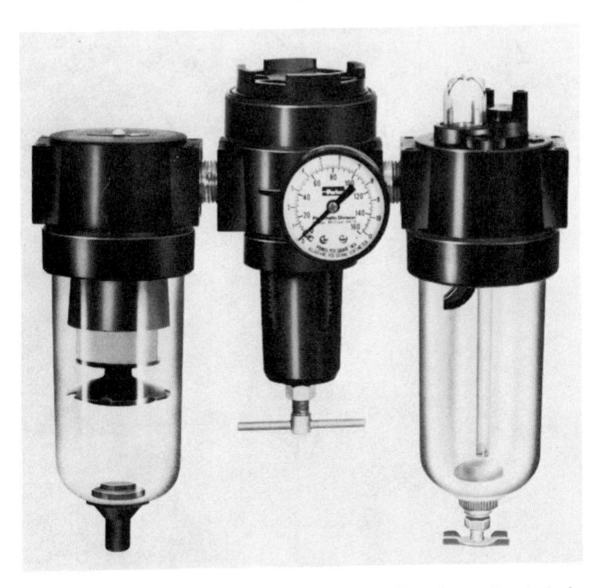

Fig. 17–30 A three-way unit for filtering air, lubricating air, and regulating the air pressure

Compressors

Air compressors are driven by an electric motor or by a prime mover. Compressors are centrifugal, rotary, or reciprocating types, figure 17-31. Air compressors often build up their final pressure over a series of stages, figure 17-32. For example, in a three-stage reciprocating air compressor, the air travels successively through three cylinders before it arrives at the air receiver or reservoir. The size of the cylinders and the piston displacement is progressively reduced as the volume decreases and the pressure increases. Special attention is given to the lubrication of the parts of all air compressors. Many of the same techniques used in reciprocating internal combustion engines are used, such as splash oil systems and oil pumps.

Compressing air creates heat, and a certain portion of the heat is carried away to prevent damage to the compressor. Air cooling, water cooling, or oil cooling systems are used.

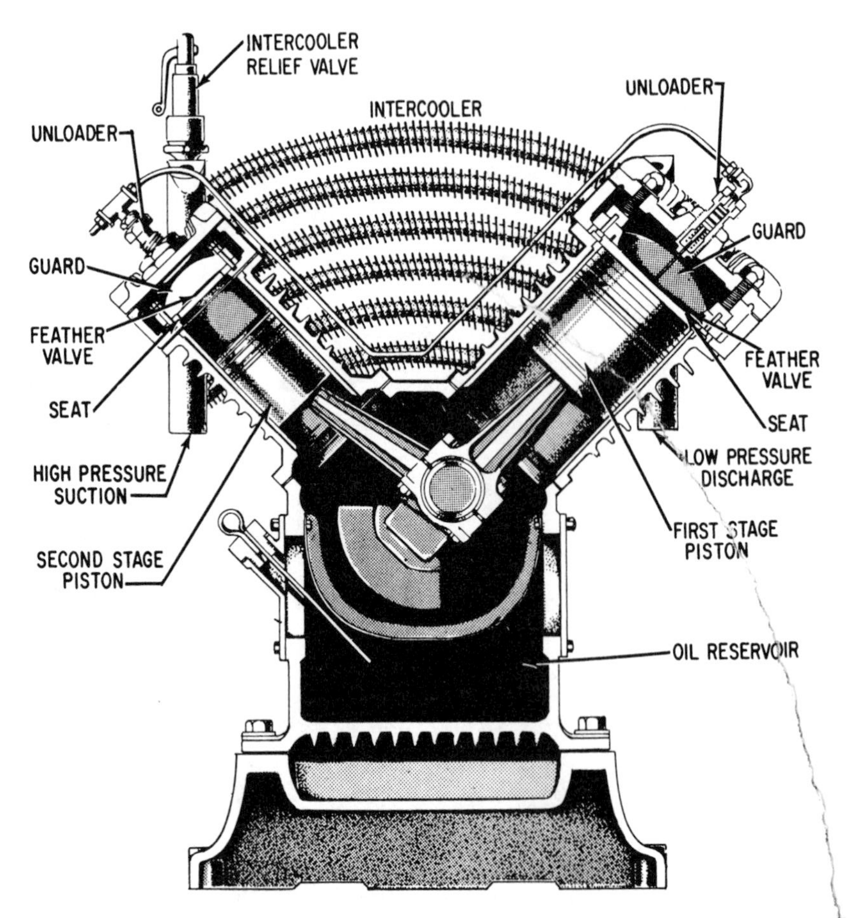

Fig. 17-31 A simple two-stage reciprocating low-pressure air compressor

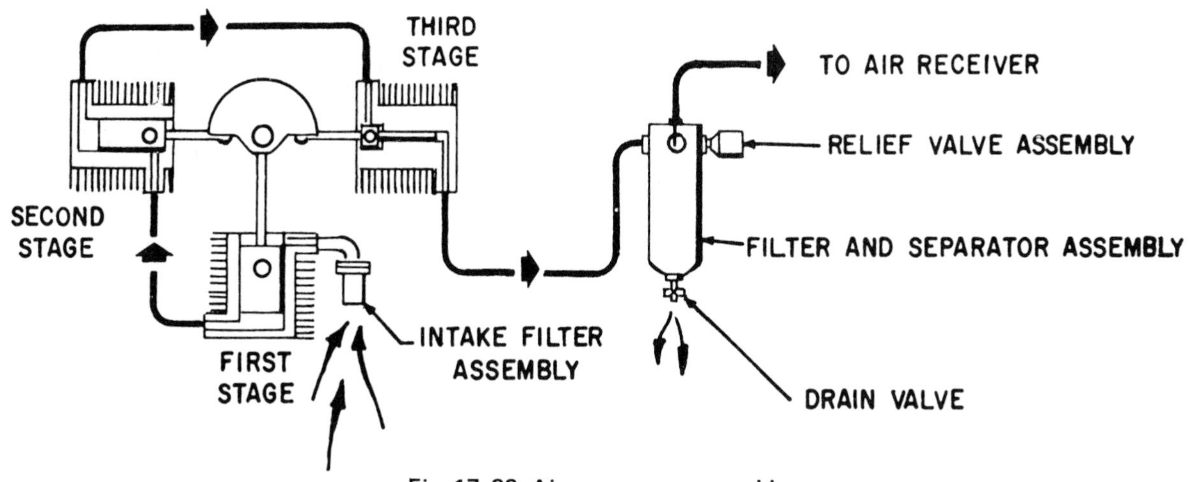

Fig. 17-32 Air compressor assembly

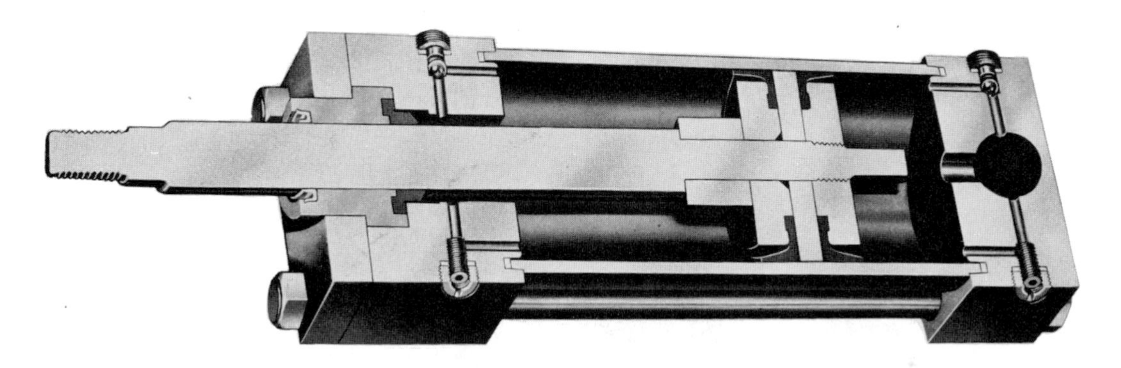

Fig. 17-33 Air cylinders are very similar to hydraulic cylinders.

Air Cylinders

Motors use the compressed gas to apply force to the machine and its work load. Air cylinders are very similar to hydraulic cylinders, figure 17-33. They are single-acting, double-acting, or telescoping cylinders. The double-acting cylinder is the most common. Air is directed to either side of the piston, giving power alternately in each direction.

Turbine Motors

If rotary motion is the desired output of the pneumatic system, a turbine motor is usually employed. Jets of air are directed against the turbine blades. The construction principles are much the same as those found in steam turbines or gas turbines, but usually the application is much smaller and lighter. Both single stage and multiple stage turbines

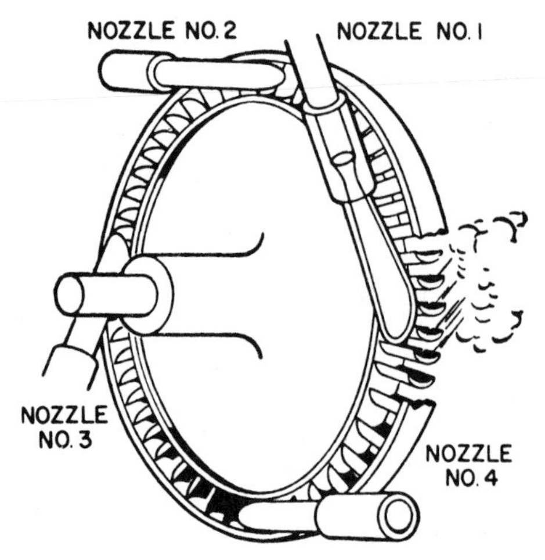

Fig. 17-34 Single-stage turbine

are used, figures 17-34 and 17-35.

Another turbine is the pinwheel turbine which is simply jet action applied to pneumatics. Its action is like that of a lawn sprinkler, figure 17-36.

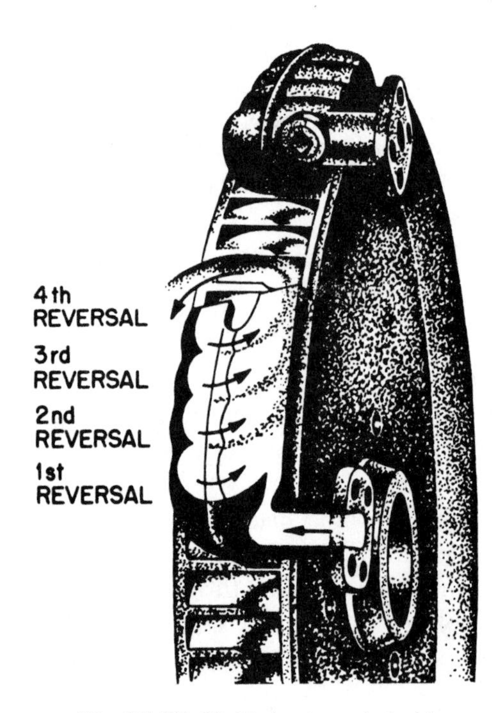

Fig. 17-35 Multiple-stage air turbine

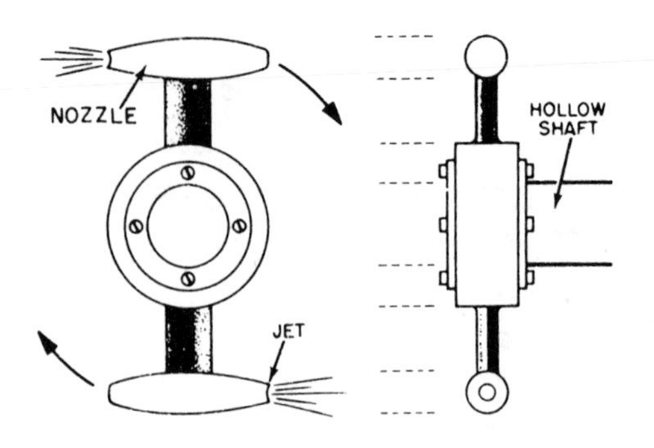

Fig. 17-36 Pinwheel air turbine

SUMMARY

Pneumatics has many applications. Most persons can identify some of the many pneumatic applications, such as air hammers or compressed air tools, used in manufacturing operations, but these are only a few of the many modern uses. Others include:

- Tools such as grinders, buffers, sanders, drills, screwdrivers, nut setters, and wrenches
- Air hoists
- Rivet, chipping, and sand hammers
- Clamping devices
- Blast cleaning
- Spraying
- Air control circuits
- Pneumatic systems as backup or emergency systems for hydraulic systems
- Air gauging
- Pneumatic conveying of materials

Vacuum Applications

Another phase of fluid power involves vacuum applications. Vacuum is just the opposite of fluid under pressure. It is fluid, usually air, removed from a space. Vacuum applications depend on atmospheric pressure. Simple vacuum principles can be seen in a collapsing soda straw. Reducing pressure on the inside of the straw allows the atmospheric pressure of 14.7 pounds per square inch (100 kilopascals) to push in and collapse the straw.

Atmospheric pressure working against areas of very low pressure or partial vacuum can be used to clamp and lift materials, figure 17-37. Vacuum cups can lock on to any smooth surface.

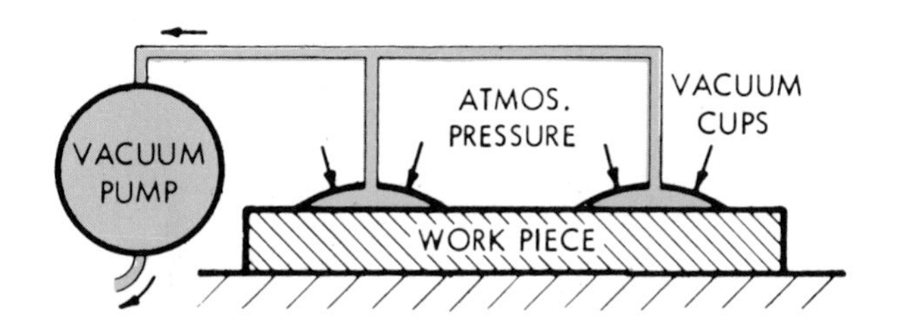

Fig. 17-37 Lifting concrete by vacuum clamping

The vacuum forming of pliable plastic sheets is another application. A vacuum pump produces the low pressures that are necessary to put this principle to work, figure 17-38.

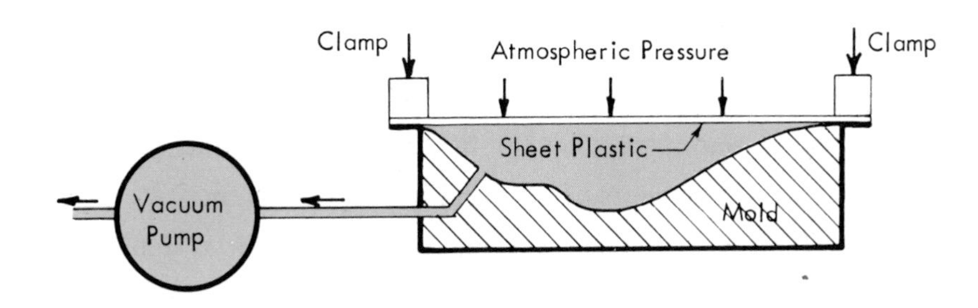

Fig. 17-38 Vacuum molding of plastic

REVIEW QUESTIONS

1. Explain several ways in which fluid power has been used throughout history.
2. Define the term hydraulics.
3. Define the term pneumatics.
4. What can be said about the compressibility of liquids?
5. The hydrostatic principle tells us that $P_1 = P_2 = F/A = Y/B$, assuming no pressure deviation. If A = 1 square inch (6.45 square centimeter), F = 1 pound (.45 kilogram), and B = 10 square inches (64.5 square centimeter), what is "Y", in figure 17-39?

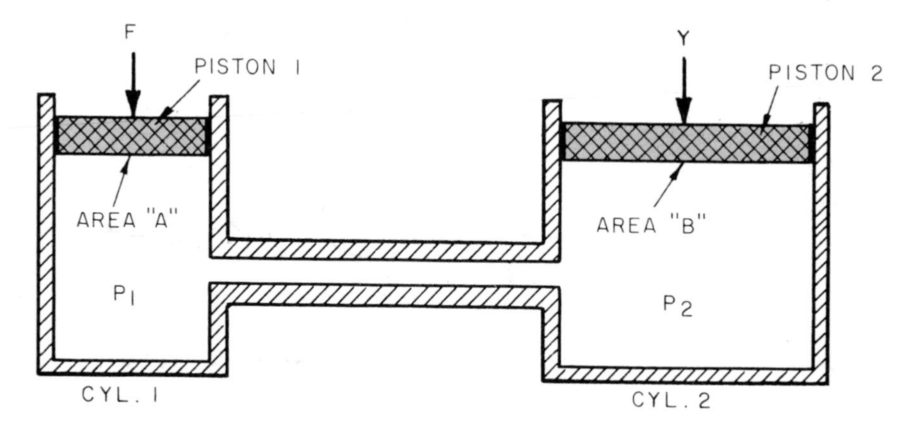

Fig. 17-39

6. How can the mechanical advantage possibilities of hydraulics be compared to that of a lever?
7. What can cause excessive friction on fluids flowing in hoses or pipes?
8. What effect does turbulence have on the efficiency of fluid dynamics?
9. What does Bernoulli's principle deal with?
10. List several common, everyday applications for hydraulic systems.
11. List the basic components of a typical hydraulic system.
12. Why are filters of particular importance to a hydraulic system?
13. What is the basic difference between a positive displacement pump and a non-positive displacement pump?
14. What three purposes can valves be designed to satisfy?
15. What hydraulic device is used to provide rotary hydraulic power at the load?
16. List several applications for hydraulic cylinders.
17. List several advantages that the field of hydraulics can offer.
18. Why does an increase in temperature have more of an effect on the pressure within a pneumatic system than it does in a hydraulic system?
19. State Boyle's Law in a brief, basic form.
20. What basic components are necessary for a pneumatic system?
21. Prepare a list of several modern uses for a pneumatic device.

SECTION 7 ALTERNATIVE ENERGY SOURCES

Unit 18 Solar Energy

OBJECTIVES After completing this unit, the student should be able to:

- Discuss the role of government in developing alternative energy sources.
- Briefly explain the operation of a solar collector.
- Explain how solar heat can be stored.
- Explain the term passive solar heating.

INTRODUCTION

Until the 1880s, the main energy source was wood. Coal took over during the early 1900s. In the 1950s the United States was using more oil than coal. Today, oil and natural gas are used in about equal amounts. At present 95% of the energy the United States uses comes from fossil fuels: coal, oil, and natural gas. For millions of years the deposits of fossil fuels have been built up by the photosynthesis in plants as the sun's energy has been converted into plant growth.

For generations fossil fuels have been used at an ever-increasing rate as our culture became more industrialized. People have surrounded themselves with countless energy demanding devices and machines. The energy has been cheap and abundant and, because of this, has been used wastefully. Recent history, however, including alarming political developments in the Middle East and oil shortages have convinced many that, unless solutions and alternative energy sources are developed, increasing fuel shortages and higher prices are certain.

Development of other energy sources provides alternatives to fossil fuels. Some alternative energy sources are new; some are quite old and time tested, while some have been discarded in favor of cheap, convenient fossil fuels. This section discusses the most significant of these energy sources.

- Solar (Unit 18)
- Wind (Unit 19)
- Hydroelectric (Unit 20)
- Biomass (Unit 21)
- Geothermal, Wood, Tidal (Unit 22)
- Nuclear (Unit 24)
- Direct Conversion (Unit 25)

At present it is not expected that all of the alternative energy sources can solve the energy crisis. The solution will come from (1) the intelligent conservation and use of present fossil fuel reserves, (2) increasing the efficiency of engines and other prime movers, (3) changing life-styles to reduce consumption, and (4) increasing the use of alternative energy sources. All four of these elements are needed to bring energy demands in line with the energy available.

Government research activities are being directed by the Energy Research and Development Administration (ERDA). In 1975 the Atomic Energy Commission was abolished and replaced by two new agencies, the Nuclear Regulatory Commission and the Energy Research and Development Administration. ERDA's mission is to do research and development in nuclear and other fields of energy. Its goal is to make the nation self-sufficient in energy while protecting the environment and the public welfare. Currently ERDA is giving its highest priority to energy conservation.

SOLAR ENERGY

The sun has poured its energy upon the earth day in and day out for millions of years. The existence of fossil fuels is traced to the sun. The wind, tides, running water, wood, and plant life are also created by the sun's energy. It is predicted that the sun will continue to send its radiant energy to the earth for another four billion years. For all practical purposes, the sun is an ever-present, renewable, free source of energy.

This massive amount of energy can be used for heating and cooling as seen in figure 18–1. It has been estimated that an average of 1400 Btus (British thermal units) per square foot (16 megajoules per square meter) falls on the United States every day. Recall that a Btu is the quantity of heat required to raise the temperature of one pound of water one degree Fahrenheit. Take that 1400 Btus per square foot, look at the area of the solar collectors in the figure, and it's easy to see that there is a lot of energy that is being trapped by those collectors instead of being wasted. It also makes it easy to understand why solar heat is an important alternative energy source because it conserves fossil fuels. About 20% of the total energy used in the United States is for residential heating, cooling, and hot water. A 20% energy savings is an enormous amount to consider.

The use of solar heating does present problems, however. The problems of cloudy days, night, and seasonal changes are basically solved by storing the extra heat during the day for use at night and cloudy periods.

The sun's radiant energy is collected and transferred in the form of heat through air or water. The water is stored in large water tanks, figure 18–2. The heated air is

Fig. 18-1 In 1973 the University of Delaware built this house to experiment with active and passive solar systems. Hot air collectors, some containing experimental cadmium solar cells, were integrated into the roof and south walls; phase change storage for both heating and cooling was located in the basement. Research was continued in the house for about a decade.

stored in a container filled with fist-sized rocks. The stored heat energy is returned to the living spaces when needed by hot-water heating or through the rocks by a forced-air heating system. For very cold or extended cloudy periods, a backup heating system is provided.

A basic solar collector consists of window glass, a black metal panel, and insulation, figure 18-3. The principle of operation is commonly called the greenhouse effect. Ordinary glass allows the short wavelengths of visible sunlight to pass through. When they are absorbed by the collector panel, the sunlight changes to longer wavelength heat waves which do not pass back through the glass. The collector panel is metallic and coated with black paint to help gather the heat. If a water system is used, tubes or pipes are bonded to the collector panel where water flowing through the system picks up the heat. On air systems the collector is corrugated to increase the collecting area. Insulation under the

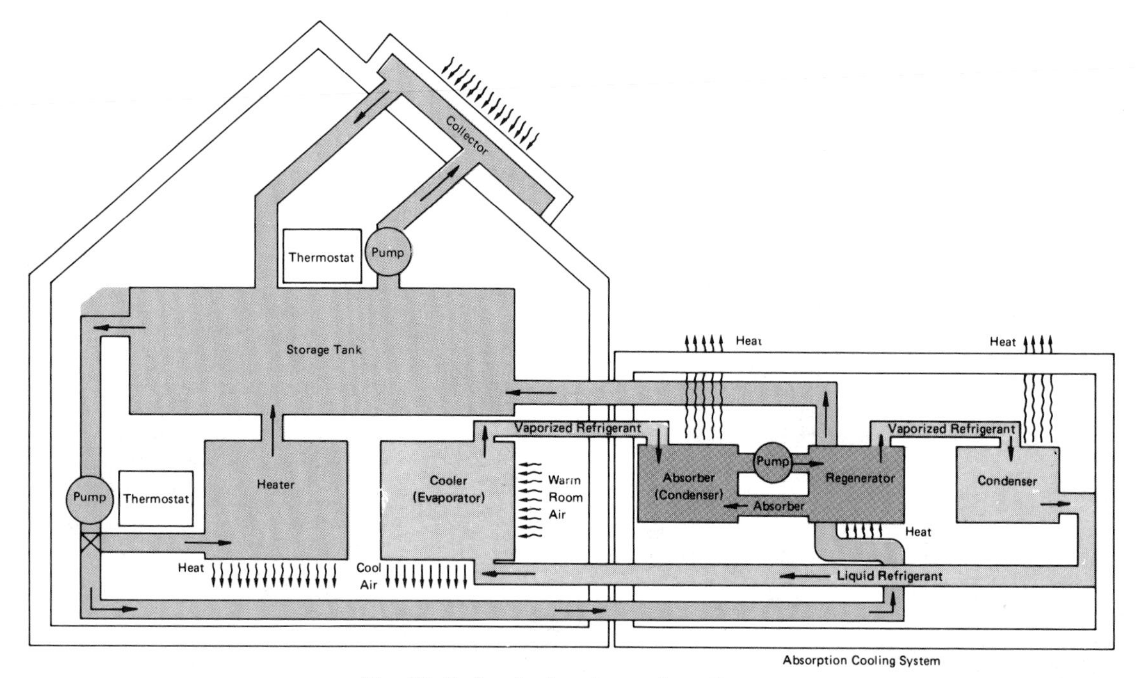

Fig. 18-2 A solar heating and cooling system

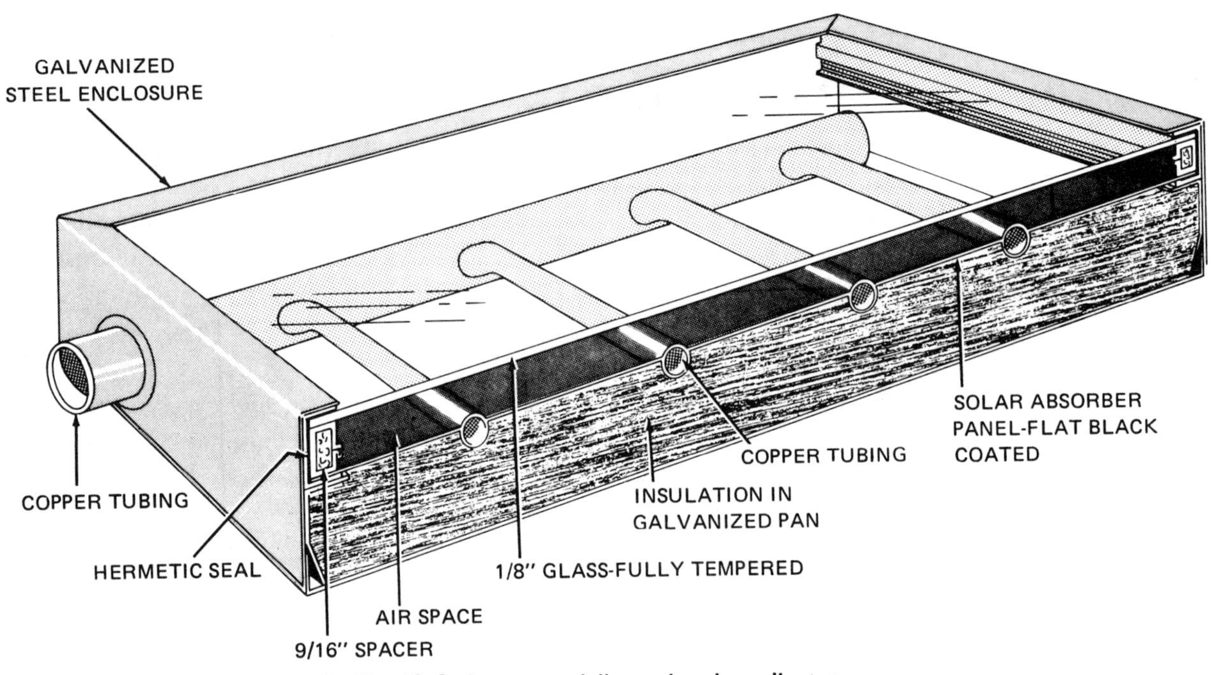

Fig. 18-3 A commercially-made solar collector

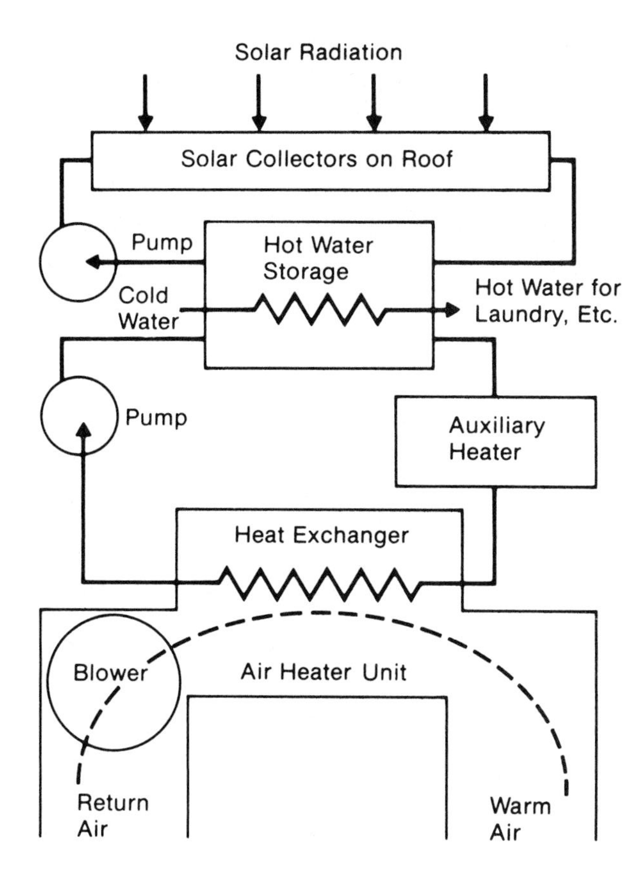

Fig. 18-4 Major elements of solar heating concept

collecting panel prevents heat loss. The collecting panel can reach a temperature of 300°F (149°C) with this simple but effective system.

The collected heat can be stored in a large water tank, perhaps 2000 gallons (7570 liters) or more. One cubic foot (0.028 cubic meter) of water stores 62.4 Btu's of heat for each degree Fahrenheit (0.47° Celsius) above the desired room temperature. This heat can be returned to the living area as needed, figure 18-4. If rock storage is used, the rock size must allow air to be blown through the system of rocks, but rocks are not as effective as water in absorbing heat. Therefore a large amount must be used, 25 to 50 tons (22½ to 45 metric tons). This is often stored in the basement of the building to be heated. Of course, using air and rock storage does have the advantage of being free of plumbing and freezing problems. Certain salts are even better for holding heat. Glaubers salt, sodium sulfate decohydrate, is being carefully tested. Five tons can store heat equal to 25 tons (22½ metric tons) of water or 125 tons (112½ metric tons) of rock.

Many solar heated homes are operating today, although they are far from being common. It is difficult, but not impossible, to convert a conventional older home into a solar home; a group of solar collectors as large as a billboard must face the south and be elevated at the angle of latitude plus approximately 10°.

In the future it is possible that the largest users of solar heating will be public, industrial, and commercial buildings. The school shown in figure 18-5 is a large scale solar heating and cooling installation. It is designed to provide 60 percent of the energy requirements and 80 percent of the hot water.

Solar cooling uses the absorption-refrigeration method, figure 18-6. This is the same principle that refrigerators use. Ammonia is the refrigerant and water is the absorber. Evaporating ammonia picks up heat and the water easily picks up the ammonia. The heat indoors is thereby absorbed. Solar heated water is used to boil off or vaporize the ammonia from the heated water absorber. The ammonia is then condensed and returned to the system to again evaporate and pick up the heat from inside the house.

Fig. 18-5 A large scale solar heating and cooling installation at the George A. Towns Elementary School in Atlanta, Georgia (Courtesy Westinghouse Electric)

PASSIVE SOLAR SYSTEMS

Passive solar heating systems are the most simple. They consist of a home designed to make the most of the solar energy striking the building. They are not mechanical in nature; pumps and fans are not used. Simply having a large amount of window area facing the winter sun will help to heat the living area through the greenhouse effect. When the amount of overhang of the roof is correct the low angle of the winter sun will permit the sun's rays to strike the windows while the overhang will shade the windows during the summer when the sun is higher in the sky. Passive solar heating usually involves the extensive use of

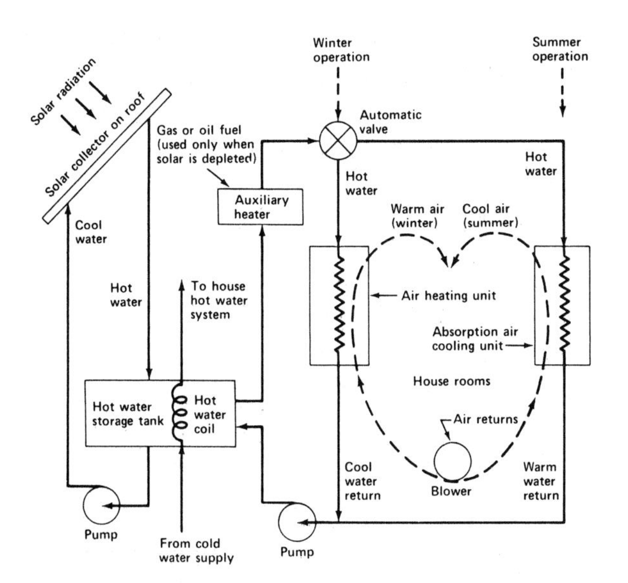

Fig. 18-6 A system that uses solar energy for heating and cooling

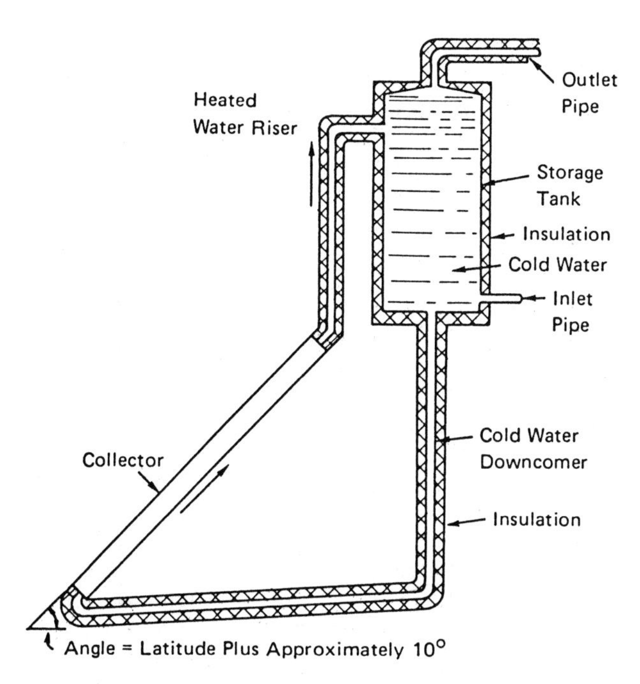

Fig. 18-7 Thermosyphon solar water heater

Fig. 18-8 The "Sun Shower" for campers is made of clear plastic on one side and black plastic on the other. When placed in the sun with the clear side up for a few hours, water will be heated enough for a comfortable shower. (Courtesy Basic Designs Inc.)

brick and stone inside the home. The materials will absorb a good deal of heat during the day and then slowly release the heat during the evening. Large drums of water can also be used to store heat during the day. Most passive solar heating techniques are built into the home at the time of construction.

A passive solar water heater is small and simple. The most common heater operates on the thermosyphon principle, figure 18–7. It consists of a solar collector, an insulated storage tank mounted above the collector, a heated water riser pipe, and a cold-water downcommer pipe. As the solar rays heat the water in the collector, it rises up the system to the storage tank. Cool water, being heavier, moves from the bottom of the storage tank, through the downcommer, and into the bottom of the collector. Using this method, about 40 to 50 gallons (150 to 190 liters) of water a day are heated with a 4 by 8 foot (1.2 by 2.4 meter) collector. The temperature of the water reaches 165°F (74°C) on a hot summer day or 115°F (46°C) on a cold winter day. This type of solar water heater is ideal for mild climates where freezing weather is not a

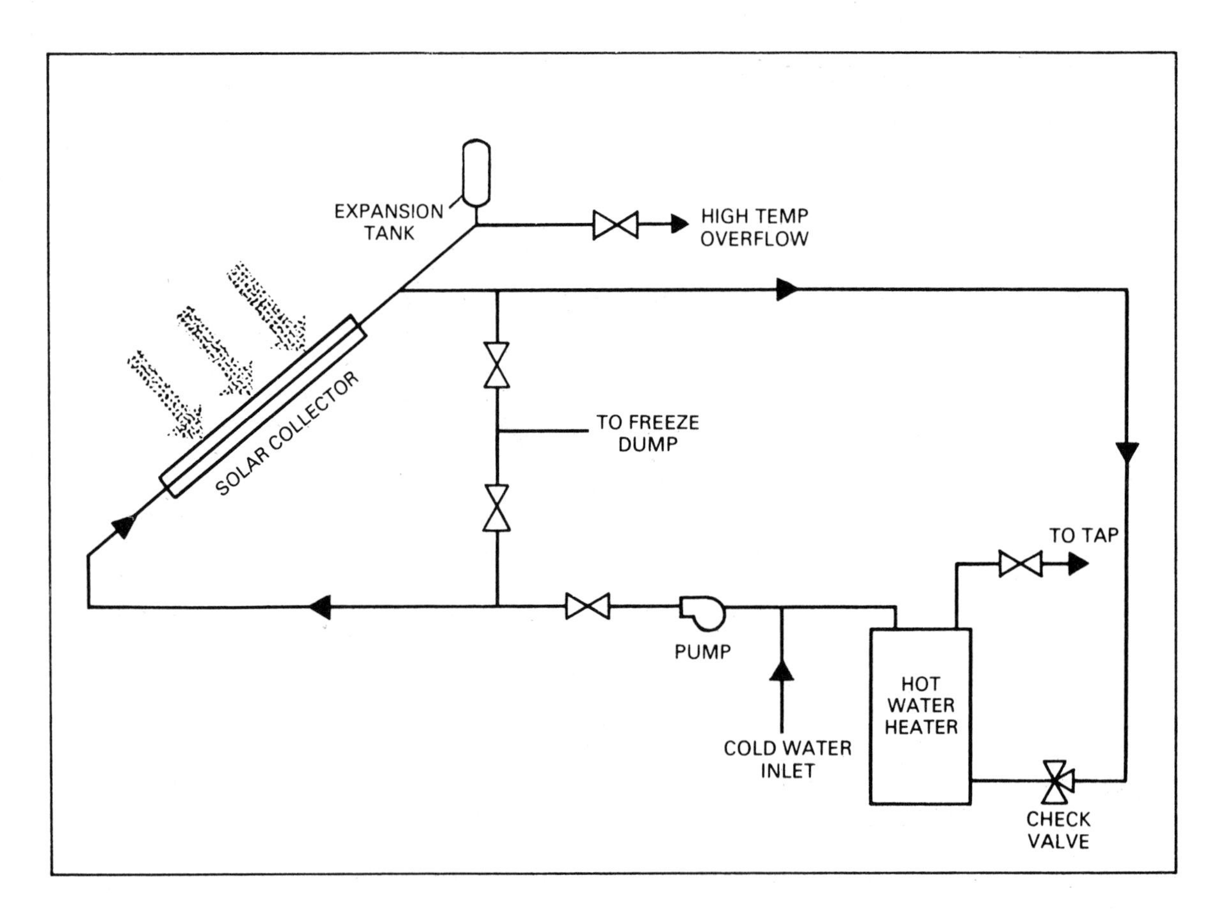

Fig. 18–9 Solar hot water heater using a pump to circulate the water

problem. The system can be simplified, as in figure 18–8 or made more complex for better control, figure 18–9.

Solar cooking is an old idea that has been used for years in areas where the sun is dependable and fuel is scarce. A solar oven is little more than an insulated box with a hinged double glass top. The inside of the box is lined with aluminum foil to reduce heat loss. Cooking is done in a blackened pot to help absorb the heat. A solar oven can reach a temperature of 400°F (204°C).

Reflector-type solar cookers use a parabolic reflector to concentrate the sun's rays at one point, usually the grill that supports a pot. The parabolic reflector must face the sun; some solar cookers can automatically turn to follow the direction of the sun as it moves across the sky. Concentrating collectors represent a safety hazard to the eyes and also a burning hazard.

On a large scale, the sun's rays are reflected by mirrors to a central receiving point, figures 18–10 and 18–11. The heat received boils the water, producing steam for mechanical work or the generation of electricity. In the sunny southwest it is estimated that 10 square miles (26 square kilometers) of solar reflectors could operate a 1,340,000 horsepower (1000 megawatt) power plant. If such a plant were working at 60 percent capacity, it could supply the needs of a city of one million people. Building costs would be high, but the fuel would be free. A five-megawatt solar thermal installation is now under construction at ERDA's Sandia laboratories in New Mexico. A megawatt is one million watts, or the equal of 10,000, one-hundred watt light bulbs.

Solar cells convert solar energy directly into electricity when light falls onto their silicon surface. These solar cells are used in nearly all of our space vehicles. They are very dependenable, but at this time too expensive to compete with other sources of electricity. Solar cells and batteries are discussed in Unit 24.

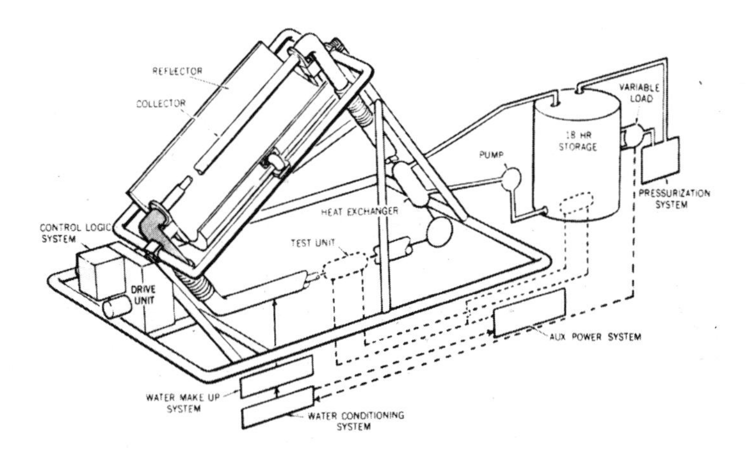

Fig. 18-10 Curved metal plates on the collector cause the sun's rays to focus on a glass collecting tube in which water or other fluid is flowing.

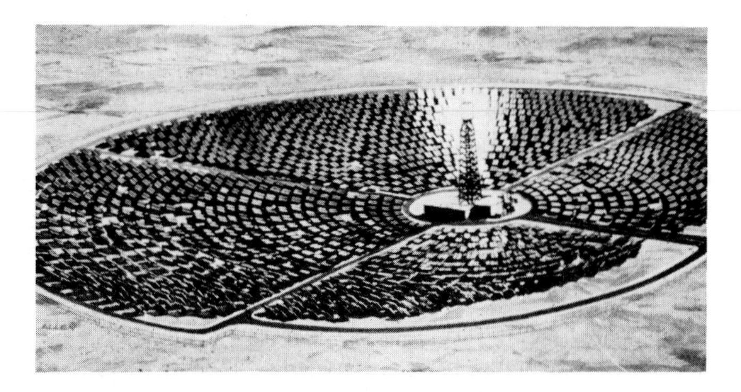

Fig. 18-11 Large solar farm collector system (Courtesy Olympic Resource Management)

REVIEW QUESTIONS

1. How is the sun used as the source of 95 percent of our energy?
2. Discuss how people have been wasteful with energy.
3. Will fossil fuel ever run out? Explain.
4. Discuss how changing a person's life style can conserve energy.
5. What government agency is responsible for developing alternative energy sources?
6. Solar energy is most likely to be used for what purposes?
7. What are the three main parts of a solar collector?
8. Discuss the greenhouse effect.
9. How can heat be stored in a solar home heating system?
10. Explain the principle of the thermosyphon water heater.
11. Is it possible to cook with solar energy? Explain.
12. What is a solar cell?

Unit 19 Wind Energy

OBJECTIVES After completing this unit, the student should be able to:

- Explain the problems associated with wind energy.
- Discuss the potential for using wind energy.
- Describe the conditions necessary for the successful operation of a wind turbine.

The topic of wind energy brings to mind the days of sailing ships and windmills. The vision of old Dutch windmills for pumping water and grinding grain usually comes to mind–part of the lifestyle of another country–but windmills were once on virtually every farm in the United States. Six million windmills were in operation, and even today it is estimated that there are still 150,000 working.

These windmills had metal fan blades of 12 to 16 feet (3.6 to 4.9 meters) in diameter and could develop about 1/6th horsepower (124 watts) in a 15 miles per hour (24 kilometers per hour) wind. Such windmills were commonly used for pumping water and for generating electricity. Their DC output was sometimes used to charge batteries. Special wind generators were also developed. At one time more than 300 companies were producing generators to power radios and electrical devices.

The Rural Electrification Administration strung wires and brought electricity to most United States farms in the 1930s. This convenient energy from a central power plant caused the windmill to temporarily lose its importance as a source of power.

Today the energy of the wind is an almost unlimited, untapped resource. It is calculated that the energy of the winds blowing across the continental United States is twenty-eight times greater than the estimated total power needs of 1986. But the trememdous potential of wind energy is not easy to use. Storage of the wind generator output is often necessary, figure 19-1. Sometimes there is too little wind blowing to be productive, and, on occasion, the wind blows with destructive force; contrary to much popular belief, with wind energy, more is *not* better.

Winds come from the uneven heating of the earth's surface, figure 19-2. Usually the land heats more rapidly than the water. As the warm air rises, the cool air from the water moves in under it as an on-shore breeze. In the evening, the land cools more quickly than the water, and breezes move from the land to the water. Mountain and valley breezes occur when hot air travels up the mountain slopes during the day, and cool air travels back down the mountainside at night. The prevailing winds are created by the cool polar air moving toward the equator, and the heated equatorial air rising to the upper atmosphere and moving toward the poles.

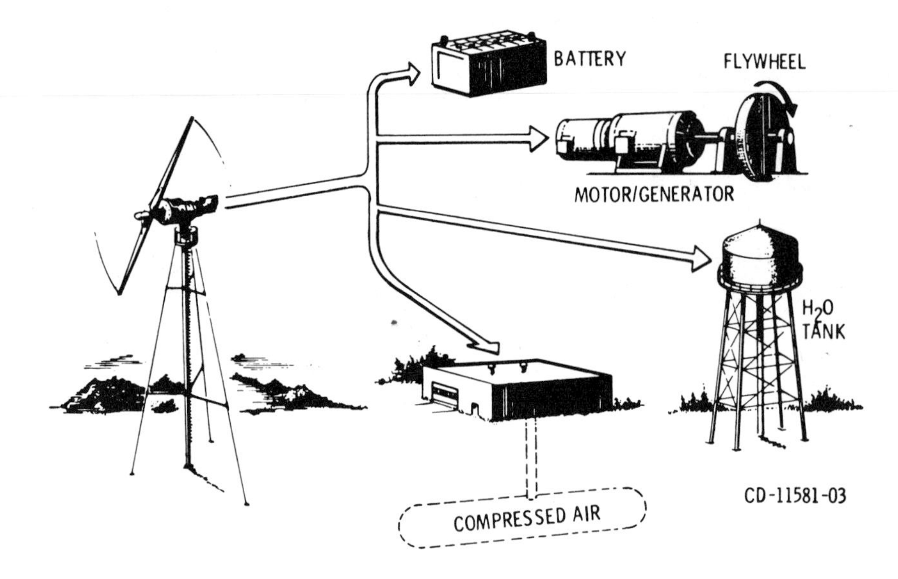

Fig. 19-1 How wind energy might be stored

To operate a wind generator, a minimum wind of 6 miles per hour (10 kilometers per hour) is needed. Twelve to 25 miles per hour (19 to 40 kilometers per hour) winds provide the best wind energy. The energy of the wind is proportional to the cube of the wind speed. This means if the wind speed is doubled, the power output will be eight times greater. Short periods of high wind speed are more effective than long periods of low or moderate wind speed.

Small wind generators are on the market

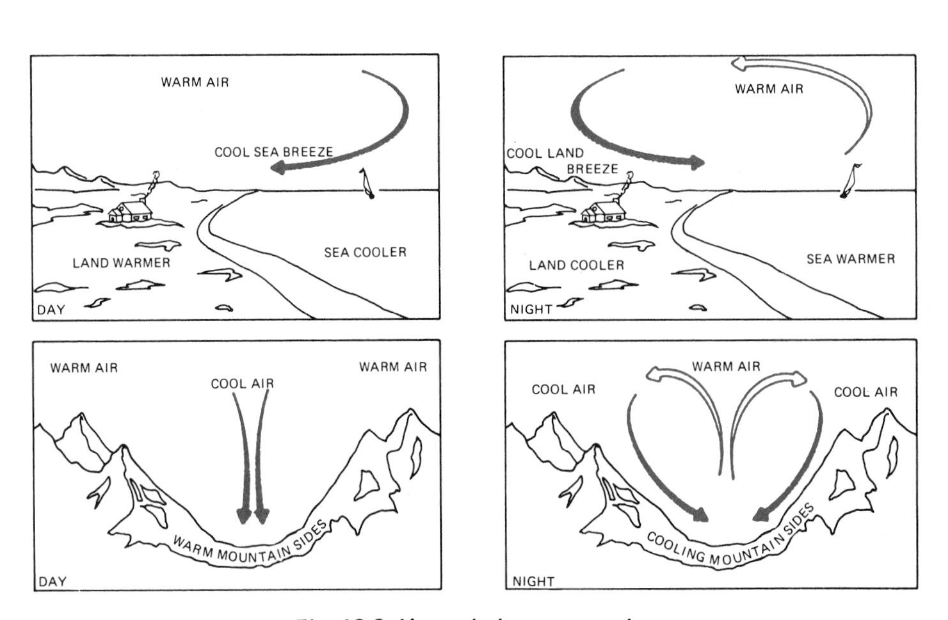

Fig. 19-2 How winds are created

for farms, ranches, cabins, and remote locations. Often these can compete economically with the cost of bringing wires to a distant location. Such generators are usually 12- or 24-volt systems, delivering their energy to storage batteries, figure 19-3. These systems are large enough for lighting, light refrigeration, and small appliances. However, electric heating, cooking, or air conditioning will overload all but the most expensive systems.

Homes designed to use wind generators are usually wired for both AC and DC. Incandescent lights, some electric motors, and some hand-held electric tools can use DC power directly from the batteries. For certain appliances that cannot operate directly from batteries, a solid-state inverter converts the DC into 110-volt AC.

The wind system must have (1) a tower that is at least 30 feet (9 meters) above the ground in a clear area, (2) special batteries,

0-watt generator for battery operation
cbaugh Wind Electric)

referred to as stationary or houselighting, which have a long life and can be charged and discharged up to 2000 times, (3) a slow speed generator or alternator and (4) blades and a hub.

All modern wind systems use two or three propeller blades. Propeller wind generator will rotate five to ten times faster than the older style, multiblade windmill. Propeller speeds are in the range of 300 to 400 revolutions per minute. If high winds threaten to increase the propeller speed beyond its designed limit, the pitch of the blades is automatically changed. Reducing the pitch changes the angle of the blade into the wind, reducing the speed of the generator.

The largest wind generator ever constructed was the Smith-Putnam machine located on a 2000-foot (600-meter) hill known as Grandpa's Knob, Rutland, Vermont. This machine, designed by Palmar Putnam, was to feed energy into the Central Vermont Public Service Corporation electrical system. This huge machine had a 110-foot (34-meter) tower and a 175-foot (53-meter) diameter blade. Rotor speed was 28.7 revolutions per minute with an adjustable pitch blade. At wind speeds of 30 miles per hour (50 kilometers per hour) or greater, the generator would produce 1250 kilowatts. While technically successful, wartime shortages, a bad main bearing, and an overstressed blade plagued the project. It was halted in 1945 after only 1100 hours of logged time.

Serious efforts and experiments are again underway with large scale wind generators, figure 19-4. ERDA and the National Aeronautics and Space Administration, NASA, have constructed a 100-kilowatt wind turbine at the Lewis Research Center's Plum Brook facility in Sandusky, Ohio.

Two 62.5 foot blades are designed to rotate at 40 revolutions per minute. In an 18

Fig. 19-4 The Mod-2 wind turbine generates 2,500 kW of power at its Goodnoe Hills, Washington location. The rotor is 300 feet in diameter with a center of rotation 200 feet above ground.

mile per hour (29 kilometers per hour) wind, it produces 133 kilowatts. The slow speed and high torque of the hub are changed into low torque and high speed for the generator with a 45:1 ratio gearbox, figure 19-5. The assembly rests on a 100-foot (30-meter) tower, figure 19-6.

Wind energy can be tied into existing power systems. Figure 19-7 shows wind turbines supplementing an existing hydroelectric plant. Similarily, a home owner with a wind generator can feed any unused electrical power into the local utilities systems–selling his extra power to the utility.

Wind energy is clean and free and when well designed and constructed the generators are not even visually polluting, figure 19-8. Wind turbines are no longer a novelty–they are successfully operating in many areas, figure 19-9. They are especially practical and competitive in rural or remote areas where electrical power is expensive, figure 19-10.

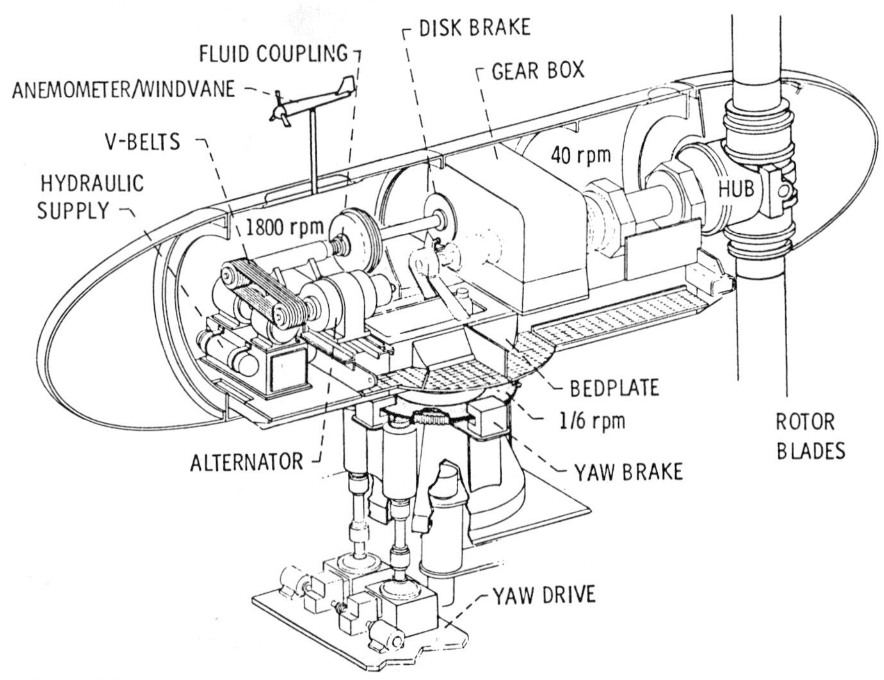

Fig. 19-5 Cutaway view of the DOE/NASA 100-kW experimental wind turbine

Fig. 19-7 Wind turbines working with a hydroelectric plant

Fig. 19-6 DOE/NASA 100-kW experimental wind turbine (Courtesy NASA, Lewis Research Center)

Fig. 19-8 The Darrius style of wind generator has foils that react to wind from any direction

Fig. 19-9 World's largest vertical axis wind turbine. It can produce 250 kilowatts in a 35-kilometer per hour wind, enough for lighting and appliances for 50 homes. (Courtesy of National Research Institute of Canada)

Fig. 19-10 Wind-powered turbine is used to power a stripper oil well in the Texas panhandle. (Courtesy Phillips Petroleum Co., by the American Petroleum Institute)

REVIEW QUESTIONS

1. Why were windmills used on farms in the United States? What happened to them?
2. Why is it difficult to harness wind energy?
3. What wind speeds are best for wind turbines?
4. How is the electricity from small wind generators usually stored?
5. How can wind turbine blades be protected from high winds?
6. Explain what ERDA/NASA is doing in the field of wind turbines.

Unit 20 Hydroelectric Energy

OBJECTIVES After completing this unit, the student should be able to:

- Explain the difference between impulse and reaction turbines.
- Discuss the importance of hydroelectric energy to our nation.
- Explain the uses of several types of turbines.
- Explain the use of the hydraulic ram.

Water does not qualify as a new and little used alternative energy source. Rather, water energy is the oldest source of motive power. Today the technology of hydroelectric plants is highly developed, figure 20-1. However, although water power is an alternative to fossile fuels, it is currently contributing a great deal to fill the electrical generation needs of our country. About 15 percent of the electric energy used in our country is from hydroelectric plants, figure 20-2.

The United States uses about 1/3 of the electricity generated in the world, and the demand for electricity has doubled about every ten years. However, the limits of hydroelectric energy are being approached since most of the best sites have already been developed with power plants. That means that hydroelectric expansion in the future will be modest.

The water turbines that drive electric generators operate on either the *impulse* or *reaction* principle. Impulse can best be demonstrated by directing the force of a garden hose against a cupped hand. Hit by the force of the water, the hand moves in the direction of the stream of water. Impulse turbines usually operate with high water pressures. The reaction principle can be seen from the force a person needs to use to hold the hose. If the hose could be mounted on the rim of a wheel, its motion would be opposite that of the water stream.

The most common impulse turbine designs are the Pelton and turbo. The Pelton wheel, figure 20-3, has buckets cast or bolted on its perimeter. One or more jets of water are directed against the buckets. The buckets are shaped somewhat like an open walnut shell. Water striking the buckets splits into two streams which fall clear of the wheel or runner. This type of turbine must be run in air. If surrounded by water, its motion would be slowed. More than any other turbine, the Pelton wheel resembles the old fashioned water wheel on the grist mill. While used very little today, this type of turbine does produce a tremendous amount of power. For example, a large Pelton wheel using water jets 12 inches (0.3 meter) in diameter produces 20,000 horsepower (15 megawatts). A turbo-type impulse turbine directs the water at the runner from the side.

Common types of reaction turbines are the Francis, Kaplan, and Deriaz; reaction turbines are more widely used than impulse

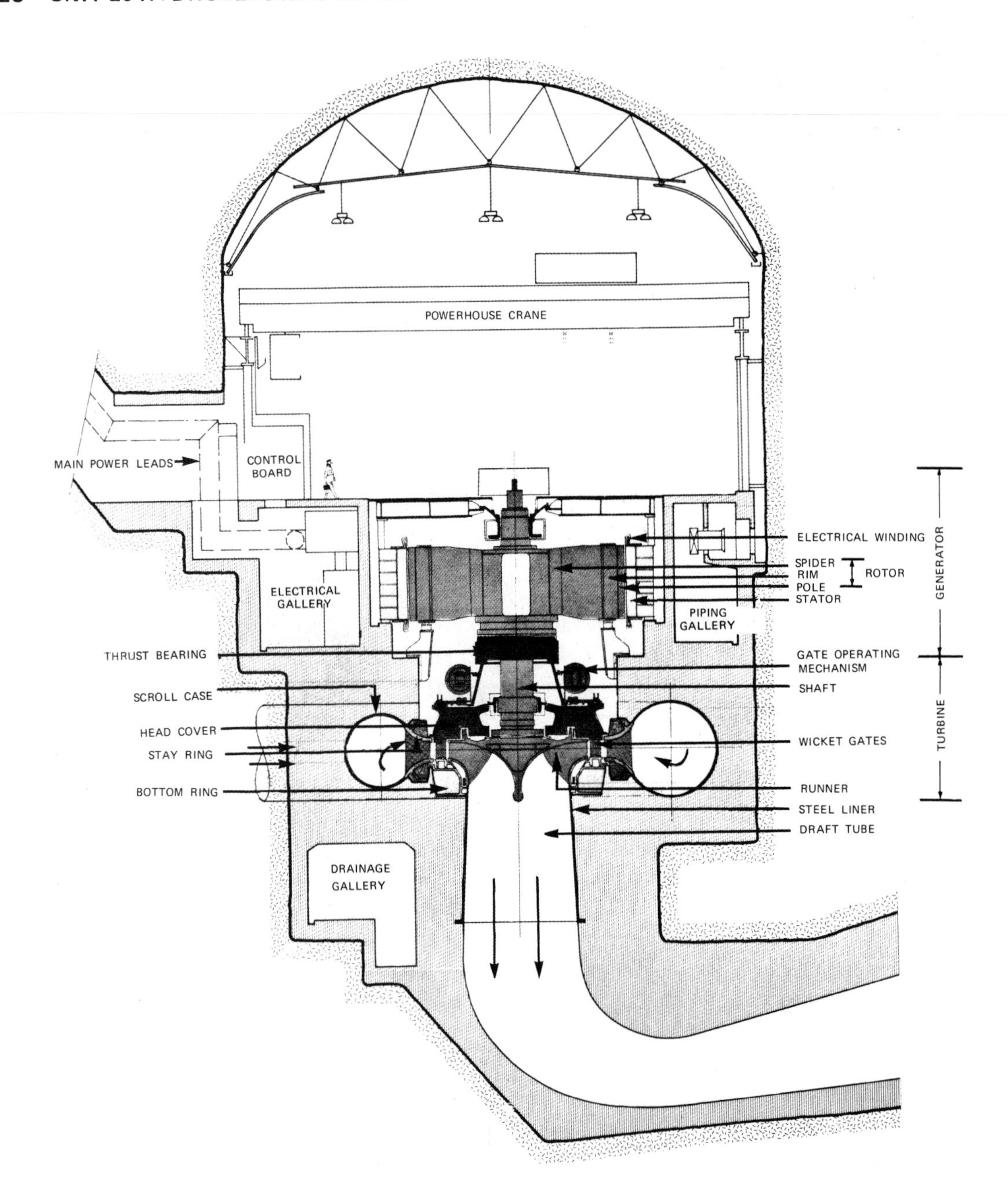

Fig. 20-1 Cutaway view shows the turbine and generator section. Note the size of the man by the control board.

Fig. 20-2 Hydroelectric projects contribute 15% of our electricity (Courtesy Dominion Engineering Works, Montreal, Canada)

turbines. The Francis turbine brings the water to the runner in a spiral casing, figure 20-4, that has a snail shell appearance. The runner, figure 20-5, is usually operated horizontally, rotating on a large vertical shaft. The flow through the runner is radial; water goes from the outer edge of the turbine towards the middle. The water exit is axial, flowing from the center of the turbine downward. Moving gates or wickets, figure 20-6, in the casing can regulate the quantity of water flow. When water strikes the curved vanes of the runner, it changes direction. The acceleration of the water by this direction change moves the wheel. Francis turbines are well suited for power sites that have low water pressure, but still have a large quantity of water flow.

Larger turbines are more economical than several smaller turbines at one site, and three of the largest turbines ever installed are located at the new third power house at Grand Coulee Dam, and are of the Francis type. Each of these turbines is capable of delivering over one million horsepower (750 mega-

Fig. 20-3 A variety of impulse turbine runners ranging in output from 3500 horsepower (2600 kilowatt) in the foreground to 150,000 horsepower (110 megawatt) in the background.

Fig. 20-4 Casing brings water to the turbine.

Fig. 20-5 Runner for a Francis turbine

watts). The assembled runner is 32 feel (10 meters) in diameter and weighs 1,100,000 pounds (500 000 kilograms). Each of the 32 wicket gates weighs 14,000 pounds (6400 kilograms).

The Kaplan style turbines, figure 20-7, have the appearance of a ship's propeller. With low water pressure, the Kaplan turbine has the advantage of higher speeds than the Francis turbine. The Kaplan turbine blades are movable to correctly match the blade angle to the amount of water flow, figure 20-8. Another type turbine, the Deriaz, also has movable blades to adjust the pitch.

An interesting water pumping device is the *hydraulic ram,* figure 20-9. It uses the fall or head of water, 20 inches (0.5 meter) or more, with a flow rate of 1.5 gallons (6 liters) or more per minute. It will pump water to a height of 25 feet (7.6 meters) for each foot (0.3 meter) of fall. The ram does this pump-

Fig. 20-6 Wicket gates control the quantity of water entering the turbine.

Fig. 20-7 Cutaway showing the construction of a hydroelectric power unit

Fig. 20-8 Runner and shaft assembly of a Kaplan turbine (Courtesy Dominion Engineering Works, Montreal, Canada)

Fig. 20-9 Rife No. 40 heavy-duty universal ram, 4-inch inlet, 2-inch outlet (Courtesy Rife Hydraulic Engine Manufacturing Co.)

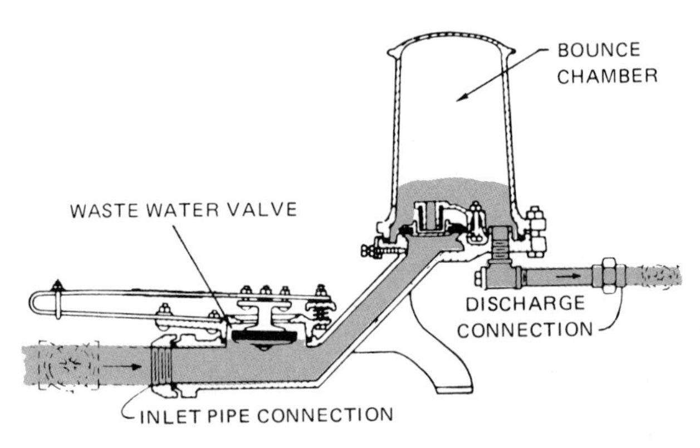

Fig. 20-10 Section drawing of series B ram

ing action on its own; no outside power is required. Water flows to the ram through an inlet pipe, figure 20-10, and out a waste water valve. At a certain speed the waste water valve automatically closes. The downward movement of the water column opens a second valve leading to the air-bounce chamber. When air pressure and water pressure equalize, the second valve closes. This action is repeated and, after several strokes, pressure in the bounce chamber will be pushing water to a considerable height. This type of pump is ideally suited for farm operations and can be used at springs, wells, and river sites, figure 20-11. Although not hydroelectric in any way, the ram can substitute its action for electric or gasoline engine pumps.

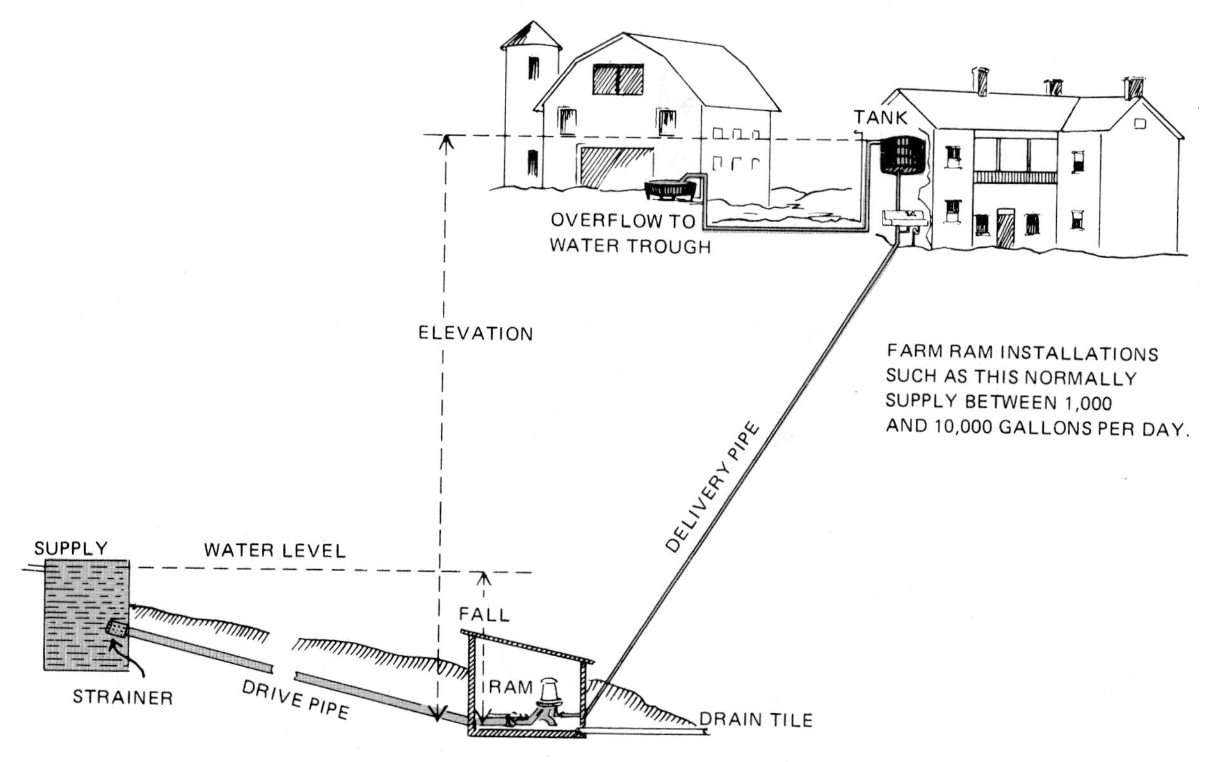

Fig. 20-11 Installation of a hydraulic ram on a farm

REVIEW QUESTIONS

1. What percentage of our electricity is generated by hydroelectric power?
2. Explain the impulse turbine principle.
3. Explain the reaction turbine principle.
4. What type of turbines are installed at the Grand Coulee Dam?
5. What is the main advantage of a Kaplan style turbine?
6. What is the source of energy to operate a hydraulic ram?

Unit 21 Energy From Biomass Sources

OBJECTIVES

After completing this unit, the student should be able to:

- Define biomass.
- Explain what photosynthesis has to do with biomass.
- Discuss biomass burning.
- Explain how biomass can be distilled into alcohol.
- Understand the difference between ethanol and methanol.

Obtaining energy from biomass has been accomplished for many years. Its primary importance today is to provide a fuel source to supplement our diminishing fossil fuels. The Department of Energy estimates that biomass is now supplying about 1% of our energy. There is a great possiblity that this percentage will be raised.

Simply stated—biomass is organic material. All vegetation or plant material that is grown can be used as biomass. For energy usage biomass is usually converted in one of three ways: (1) biomass may be directly burned, (2) biomass may be converted into alcohol, and (3) biomass may be converted into a gas called methane which is also the principle ingredient in natural gas.

DIRECT BURNING

All trees and plants grow by the chemical process called photosynthesis. Plants capture light energy and use it to create carbohydrates within the chlorophyl-containing portions of the plant. A direct result of this process is the formation of oxygen, which the plant gives off as a waste product. Carbohydrates—if you remember—are molecules containing carbon and hydrogen. Of course, there are also hydrocarbon molecules in fossil fuels which causes them to burn. Dry plant material will also burn; we do not have to wait millions of years for them to be transformed into fossil fuels. Studies show that a pound of dried plant material can produce up to 7,500 Btus of heat energy. This is more than half the amount of heat energy obtained from burning one pound of coal.

Collecting a significant amount of organic material for large scale burning for energy is possible but not entirely practical. For example, an electric power plant would require several hundred square miles of a forest "tree farm" to operate. Remember that before the widespread use of fossil fuels many countries had almost exhausted their supply of trees that they had been using for energy sources. Also, such a use competes directly with the lumber and wood products industry; really quite a major consideration.

Research is directed toward producing the maximum amount of biomass for the least cost. Eucalyptus trees are very productive with yields of 8 to 25 tons of biomass per acre.

Burning the solid wastes in garbage and trash to produce energy is very practical but, as of this date, there are fewer than 30 processing plants in the country. This means that less than 1% of the 150 million tons of garbage and trash we throw away each year is burned to produce heat energy for electrical generation. In Sweden, Denmark, and the Netherlands—on the other hand—some 40% of their municipal waste is burned for energy.

The production of energy by burning biomass is too rich a resource to neglect. It is a renewable source and if it is soundly managed it will do no harm to the environment. Wood as a *home* heating fuel is making a serious comeback, however, and is discussed in Unit 22.

ALCOHOL FROM BIOMASS

Alcohol, as just mentioned, can be produced from all organic material. Two important types of alcohol can be distilled: ethyl alcohol (ethanol or grain alcohol), and methyl alcohol (methanol or wood alcohol). Both processes also produce useful by-products other than the alcohol.

Ethanol is produced from the distillation of plant sugars; many fruit and grain crops are commonly used. The biomass is ground and mixed with water to form a mash. The mash contains starch which must be converted to sugar. This is done by adding malt which contains an enzyme that changes the starch to sugar. Another enzyme which is contained in yeast is added; this enzyme ferments the sugar into ethyl alcohol.

To remove the alcohol from the mash it is distilled; heat is applied and the alcohol, which boils at 171°F, vaporizes and can be condensed on a cool surface and separated from the mash. A bushel of corn can yield up to 2.4 gallons of ethanol. The byproducts from the process can be used as feed for livestock.

Ethanol has many industrial uses in plastics, solvents, drugs, and food processing. Today ethanol is also much in the news as a motor fuel. When combined in the proportions of 90% gasoline and 10% ethanol it is commonly called gasohol, figure 21-1. Adding ethanol not only stretches the gasoline supply, it also helps to raise the octaine rating of the fuel. Gasohol sales are on the increase; more than 270,000 barrels are sold each day.

Methanol is accomplished by the *destructive* distillation of the biomass. Wood, sawdust, or other material is dried, placed in an oven, and heated to between 400 and 500°F. Gases are driven from the material and are condensed into a liquid. Chemical processes are then used to separate out the methanol. Charcoal remains in the ovens after all the liquids and gases are driven off by the heat. Methanol has much the same industrial uses as ethanol, however it does contain more heat energy.

Ethanol and methanol are the principle types of the many types of alcohol that can be produced from biomass. Both of the al-

Fig. 21-1 A service station attendant filling a tank with gasohol (Courtesy of the American Petroleum Institute)

cohols can also be produced directly from petroleum. Today 30% of the alcohol used by industry comes from biomass. None of the alcohols discussed in this unit are beverage alcohols, since they are treated with toxic materials to make it unfit—to say nothing of downright deadly—for human consumption.

METHANE GAS FROM BIOMASS

The production of methane gas has received considerable attention as an alternative energy source. Many experiments involving the production of methane are underway with several successful installations already at work in the U.S., figure 21-2. In fuel-short India, some 2500 units are operating in farms and villages. During World War II, because of massive gasoline shortages, intensive research on methane was carried out by the Germans as an alternate source of fuel. Methane plants are in operation today at locations throughout the world.

The production of methane offers other advantages besides the production of a combustible gas for heating and engines. It also yields high quality organic fertilizer and helps to reduce the problems of waste disposal.

The source of methane gas is organic matter that is decomposed (broken down) by bacteria. Manure, organic garbage, or leaves are mixed into a watery paste called a slurry and fed into a digester tank. The digester tank might be as small as a 55-gallon (190-liter) oil drum or it might take on the dimensions of a 3000-gallon (11 000 liter) pit.

When *aerobic* (with air) decomposition of plants, animals, and manure take place in nature, the byproducts are ammonia, carbon dioxide, and the solid we call humus. However, methane is produced by a biological disgestion in which two type of *anaerobic* (without air) bacteria break down organic substances. Anaerobic bacteria can live and grow without oxygen, while aerobic microorganisms require oxygen to live and grow. Acid-producing bacteria are capable of breaking the matter into materials such as acetic acids. The other anaerobic microorganisms produce methane. The digester must have the proper balance of these organisms. The bacterial action works best at 95°F (35°C).

The gas produced is sometimes referred to as bio gas, marsh gas, or sewer gas. It contains about 70% methane, carbon dioxide, and traces of sulfurated hydrogen. The trace of sulfurated hydrogen gives the gas an odor of rotten eggs.

The slurry in the digester will begin to produce methane within several days; the entire process taking about 40 days. In that amount of time, 95 percent of the methane will be given off. If the digester is to be operated continuously, 1/40 of the digester capacity should be renewed each day, with an equal amount of sludge removed.

On a large scale, the production of methane is economically sound. Sewage treatment plants can produce energy and power, and fertilizer as a byproduct. Increasingly, large livestock feeding operations are faced with the dilemma of waste disposal. Here methane digesters have a real potential. At the individual home site, the practicality is in question since there is the initial large cost involved. Also regular feeding and monitoring of the digester is required. Unless an individual has the time and an interest in methane, he may not wish to start up a home unit.

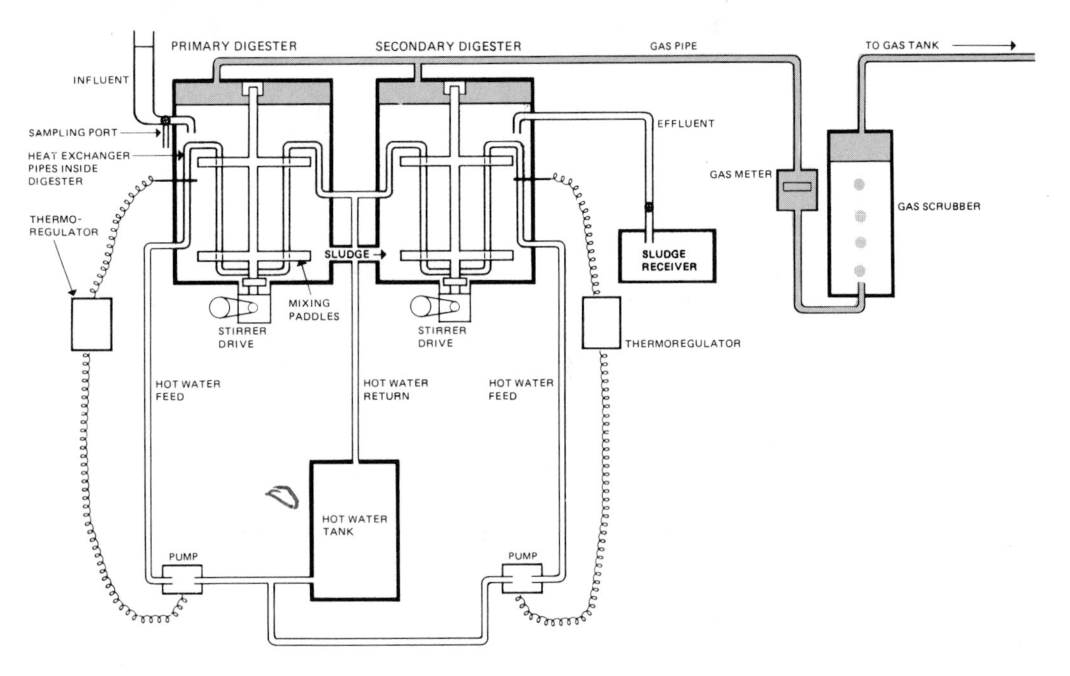

Fig. 21-2 Diagram of experimental livestock waste digester

REVIEW QUESTIONS

1. Explain how biomass and fossil fuels are related.
2. Compare the heating value of biomass with that of coal.
3. What biomass materials can be burned in addition to wood?
4. Explain why burning biomass has its limitations.
5. Briefly outline the distillation of ethanol.
6. What is gasohol made of?
7. What are the advantages of gasohol?
8. What types of biomass are often used in methane production?
9. What is the difference between aerobic and anaerobic decomposition?

Unit 22 Geothermal, Wood, and Tidal Energy

OBJECTIVES After completing this unit, the student should be able to:

- Discuss the term geothermal.
- Locate several major geothermal areas.
- Discuss the problems of using wood for home heating.
- Explain the advantages and disadvantages of several wood burning methods.
- Explain how the energy of the ocean tides can be harnessed.

GEOTHERMAL ENERGY

The internal heat of the earth is referred to as *geothermal energy*. The thickness of the earth's crust varies from 6 to 30 miles (10 to 50 kilometers). Below this crust lies the magma at a temperature of 6000° to 8000°F (3300° to 4400°C). This intense heat radiates very slowly to the earth's surface. As a mine shaft goes down, a temperature rise of 100°F per mile (24°C per kilometer) can be noted. Drilling technology has not reached the magma, and scientists are not sure how to harness this energy even if they could reach it.

At the base of the continental crust, a layer just above the magma, the temperature is 400° to 1800°F (200° to 980°C). It is at this level that interest in geothermal energy lies. Geysers, such as Old Faithful, figure 22-1, hot springs; and bubbling mud pots are found in the United States and around the world. Water seeps into the underground cracks in the rocks or through porous rocks, picks up heat, and returns to the surface. It

Fig. 22-1 Old Faithful displays geothermal energy.

is possible for the water to return as dry steam or as wet steam, a combination of steam and water.

Dry steam is the most useful because, on a commercial basis, it can be fed directly into standard steam turbines to generate electricity. Dry steam is not found in many places throughout the world. The geyser region near San Francisco, California, figure 22-2, is one such dry steam field. Steam is collected from over 100 wells and sent to the turbines where about 500 megawatts of power are produced. While it already provides San Francisco with about one-half of its electricity, the capacity of this site is currently being increased.

In Larderello, Italy, there is a dry steam geothermal region that has been producing electricity since 1913. Today the area is working at capacity with thirteen generating plants producing 365 megawatts.

Superheated underground water is under sufficient pressure to keep it liquid. When it is brought to the surface through a well, some of it will boil into steam as the pressure is reduced. Geothermal fields that have this structure are called wet steam fields, figure 22-3. They are far more common than dry

Fig. 22-2 The Geysers, a geothermal steam field in California

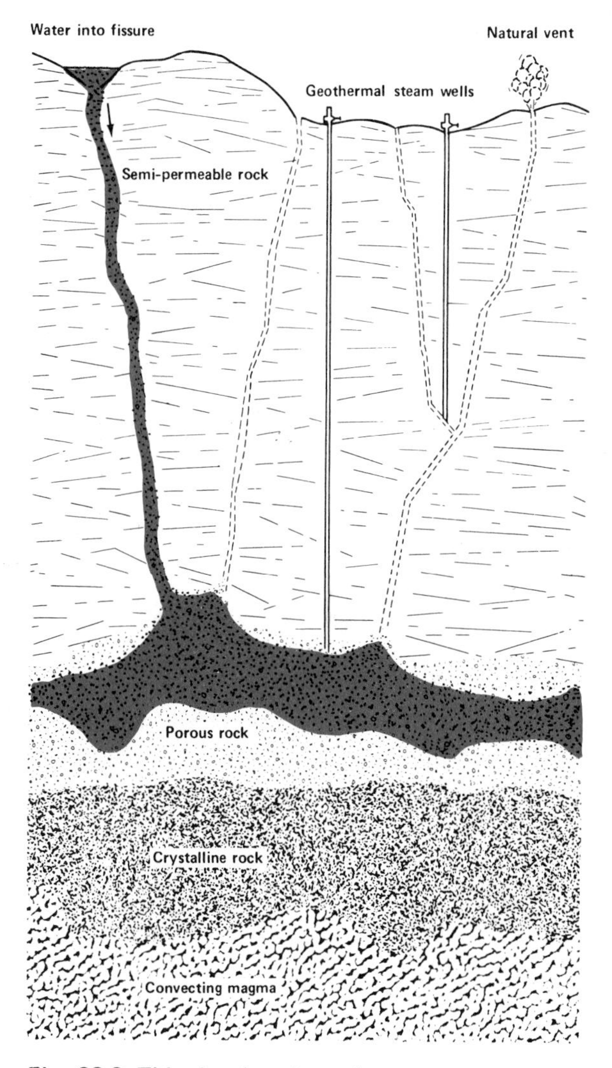

Fig. 22-3 This drawing shows how water seeps into the earth to pick up the heat in the porous rocks above the crystalline rock and magma.

steam fields; this steam is used primarily for heating. Interestingly enough, Iceland was developing wet steam fields in the 1930s. Ninety percent of the homes in the capital city of Reykjavik are heated with geothermal power. In the United States, wet steam heating is used in Klamath, Oregon and Boise, Idaho.

The secondary fluid cycle, figure 22-4, brings superheated water up, but keeps it under pressure and does not allow it to boil. This high-pressure secondary fluid is directed against the turbine and blades. The spent fluid from the turbine is condensed and recycled through the heat exchanger. Most of its heat is given up in the heat exchanger where a fluid like isobutane or freon, which has a lower boiling point, is vaporized. The hot water, having lost a good deal of its heat, is reinjected into the ground to conserve the total heat of the well area.

Wet and dry steam systems are in operation today. They are successful and economical, but limited to those hot spots on the

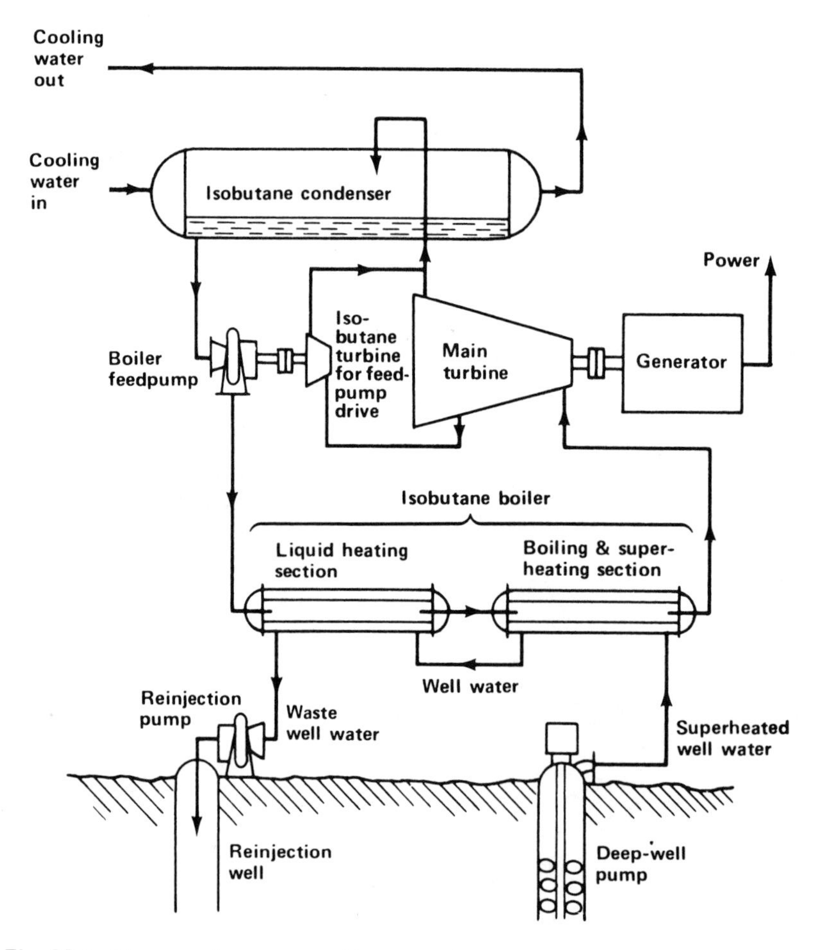

Fig. 22-4 Diagram of a vapor-turbine cycle for geothermal power generation

earth's surface where the heat is easy to reach. Dry hot rock deposits also exist, representing a potential for energy probably ten times that of known wet and dry steam fields. When such dry fields are found, a well is drilled down to the hot rock area. Cold water—pumped under high pressure down the well—opens a vertical crack in the granite above the hottest area. A second well is drilled to intersect with the cracked area. Cold water poured down into this hole continues to crack the hot rocks and allow the water to pick up heat. This water returns to the surface through the first hole in the form of steam and water. Here it is harnessed with a turbine. Hot rock technology is still in the experimental stages, but it does hold much promise. Government agencies hope that by 1986, 2% of the electric heating needs of the nation will be from geothermal sources.

WOOD—ENERGY FOR THE HOME

Can it be true that in our space age society some people are seriously returning to wood to heat the family home? Yes, more than just a few are renewing their interest in wood heating. The reasons include fear of oil and natural gas shortages, the rising costs of fossil fuels, and a nostalgic return to some of the old values and ways.

This renewable energy source can have an impact by conserving other fossil fuels. No one will pretend that heating homes with wood is right for everyone. Some want wood heat, using their conventional system as a backup. Others want conventional heat, with wood heat as a backup. Most people living in metropolitan areas will probably not even consider wood heat, because, ideally, a person interested in wood heat has a wood lot or lives on acreage that has an abundance of trees. Clearing fallen trees, limbs, and diseased trees can also yield considerable wood, as can thinning trees. However, it would take the yield from several acres of woods to heat a home in a cold climate. Therefore, wood could not be a total alternative fuel source for a nation.

In the 1600s, before coal came into use in England, there was an energy crisis because most of the forests had been cleared for firewood. Likewise, in the Mediterranean countries, many of the forests had been harvested clean with serious environmental results in the form of erosion and ecological damage.

Since many persons purchase wood for fireplace and recreational burning, it should be mentioned that wood is sold by the cord, figure 22-5. A stack 8 feet (2.4 meters) long, by 4 feet (1.2 meters) wide, and 4 feet (1.2 meters) high is a cord. A face cord is 8 feet (4.2 meters) long, by 4 feet (1.2 meters) high, and cut to any desired length. A rick is a face cord of 16-inch (0.4 meter) logs.

Buying wood by the cord does not consider the heating value of the type of wood. Among the hardwoods: ash, red oak, white oak, beech, birch, hickory, hard maple, pecan, and dogwood are high in heat output. Aspen, basswood, cottonwood, chestnut, and yellow

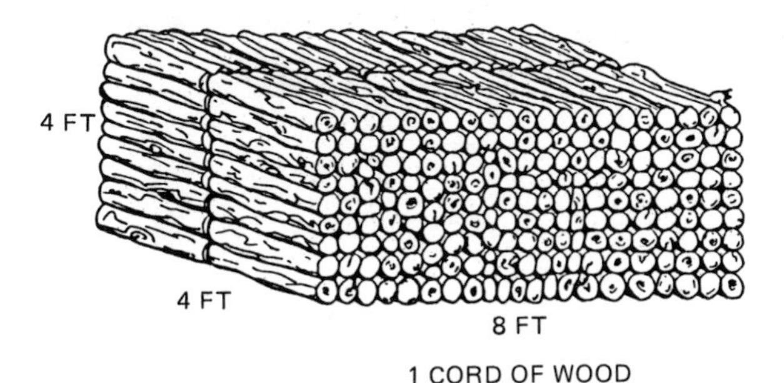

Fig. 22-5 Diagram of one cord of wood

poplar are low in heat output.

In the burning of wood, the cellulose and lignin of the wood are heated to the point where they break down into a gas. This gas, combined with oxygen, is heated to ignition temperature. Upon ignition, heat is given off. The heat supports the fire and warms the room.

The way the wood is burned is important. Generally, open fireplaces are the least efficient. It is pleasant to watch the burning logs, but much of the heat goes up the chimney. A poorly designed fireplace an even result in a loss of heat because it draws warm air in the house up the chimney, especially after the fire has died down and the damper is left open.

Fig. 22-6 There is renewed interest in home heating stoves. (Courtesy Portland Stove Foundry)

A Franklin stove–with its flue fitted into an existing fireplace–has an advantage over a fireplace because it has doors. When the doors are closed, the rate of combustion can be controlled for longer, more even burning of the wood. The cast-iron construction of the Franklin stove helps hold and spread the heat throughout a room. Many other heating stoves of interesting designs, in addition to the Franklin, can be found on the market, figure 22-6. There are, for instance, fireplace *inserts,* made of a heavy cast iron that fit snugly inside the existing fireplace. The only part that shows is an attractively designed heavy cast door. They feature a damper for controlling the rate of burn and operate similar to a complete combustion heater.

Complete combustion heaters, figure 22-7, are the most efficient. They are airtight, with the cold air being preheated as it comes into the combustion area. Smoke is forced back through the combustion area before it goes up the chimney. Wood can also be burned in complete combustion heaters set up as central heating units, much the same as any other furnace. Furnaces are on the market that can use several different fuels: wood, oil, or natural gas. Thermostatically controlled dampers regulate the rate of burning by allowing the correct amount of air to enter the fire box. This type of heater can be loaded in the evening and burn through the night unattended until morning, or even longer,

Fig. 22-7 Complete combustion wood heater (Courtesy Riteway-Dominion Manufacturing Inc.)

figure 22-8. In the basement, this furnace can be installed into the central hot air duct system. While they are efficient, they are not usually as pleasant to view as a fireplace or heater of old fashioned design.

Wood smoke in small amounts is usually not a problem and, in fact, is low in irritating pollutants. For example, wood smoke contains no sulfur dioxide. Wood as an alternative energy source does have a place in the total energy picture, if used wisely.

Fig. 22-8 Airflow through a complete combustion heater

TIDAL ENERGY

Anyone who lives near or has visited the ocean is familiar with the tides. Tides are caused by the gravitational pull of the moon on the earth's large bodies of water. What is known as the cycle of the tides occurs twice a day. This means that water beginning at high tide will go down to low tide and then back up to high. This also means that the interval between low tide and high tide is about six hours. Tides are generally higher the further the location is from the equator. The difference between low tide and high tide is sometimes as much as forty feet. If water flows into an estuary, bay, or up a river, and if a dam is built and equipped with water turbines, electricity can be produced, figure 22-9.

Interest in tidal power continues but, to date, only one large scale plant is in operation. The plant is located at the junction of the Rance River and the Gulf of St. Malo, Brittany, France. This plant consists of a dam and 24 reversible Kaplan turbines that drive 10,000-kilowatt alternators. Turbines can produce electrical energy when the tide is rising, and also when the tide is going out. In order to best match the production of electricity to the community demand, the turbines can also be used as pumps for filling the river behind the dam during slack electric demand, thereby providing additional water for the turbines when it is needed. The tidal rise and fall at this location averages 26 feet (8 meters).

To be considered for a tidal power plant, an area must have a great difference between high tide and low tide, and be a tidal river or bay that contains considerable water volume.

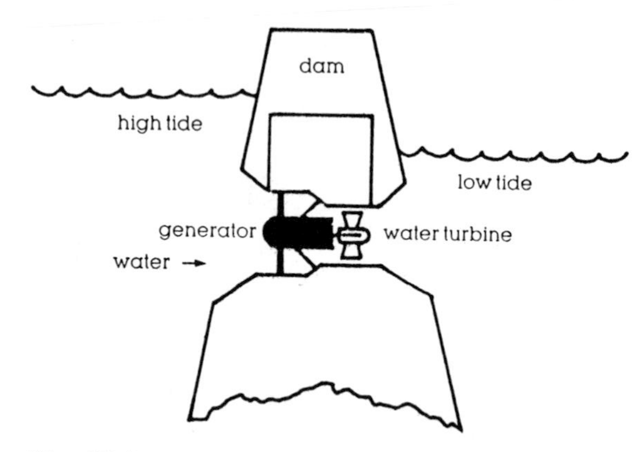

Fig. 22-9 Water turbine uses the tides to generate electricity. (Courtesy of Standard Oil Company of Indiana)

REVIEW QUESTIONS

1. Is geothermal energy practical? Are any geothermal systems in operation today?
2. How thick is the earth's crust?
3. What is the temperature of the magma?
4. Who is the ideal person to use wood as a home heating fuel?
5. What are the measurements of a cord of wood?
6. What types of woods have a high heat output?
7. Explain how an open fireplace can create a net loss of heat.
8. Where will tidal dams be most likely located?

Unit 23 Nuclear Energy

OBJECTIVES After completing this unit, the student should be able to:

- Discuss matter, atomic structure, and radiation.
- Explain nuclear fission.
- Compare a fossil fuel power plant to a nuclear power plant.
- Discuss the workings of a nuclear reactor.
- Explain nuclear fusion.

Nuclear energy as an energy source is by far the most promising, the most awe-inspiring, and the most fearsome that civilization has yet uncovered. The tremendous energy reserve of nuclear power promises to be an economical power source for all nations for centuries to come. Nuclear power can be relied on long after conventional power reserves of coal and oil are exhausted. When nuclear energy is used, natural resources of coal and oil can be conserved.

It is difficult to comprehend the energy locked within the atom. It is almost inconceivable that enormous energy exists within something that small. Yet the potential energy that lies within a pound (.45 kilogram) of uranium is nearly equal to 3,000,000 pounds (1364 metric tons) of coal, about twenty-five railroad cars full of coal, figure 23-1.

An appreciation of nuclear or atomic en-

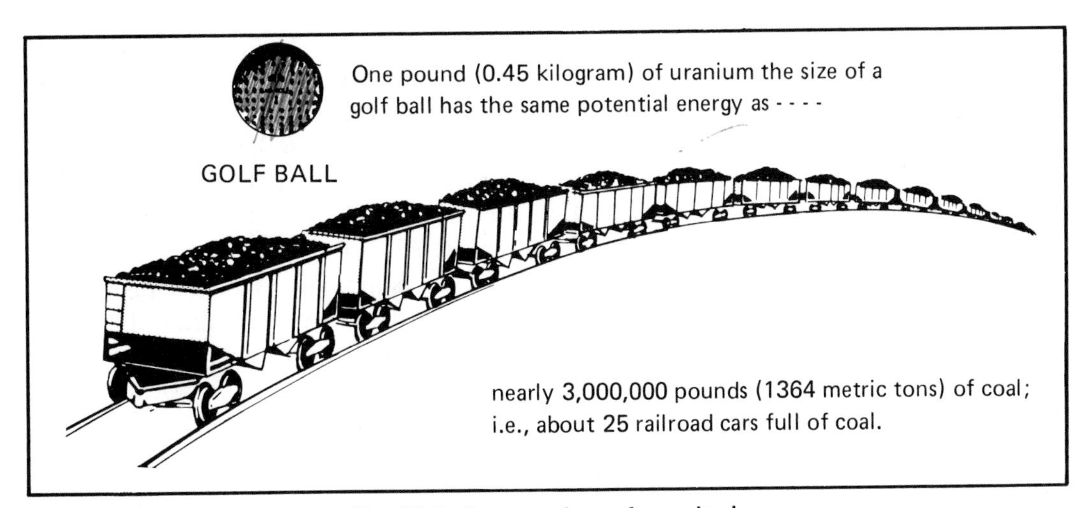

Fig. 23-1 A comparison of magnitude

ergy and an understanding of how it is produced and controlled is important. Basic knowledge of the structure of matter should be gained, however, before a more detailed investigation of nuclear energy is begun.

NATURE OF MATTER

The smallest complete unit of matter is called the *atom;* they are the building blocks of all substances. If it were possible to magnify any substance an infinite number of times, it would be seen as being made up of atoms. Some substances, in turn, are made up of all the same type of atoms. These substances are therefore *pure* and are known as the *elements.* There are ninety-two different elements that occur naturally; additional elements have been man-made.

Other substances could be magnified and found to be made up of *different kinds* of atoms that are bound together in a cluster—known as a molecule, figure 23-2. A *molecule* is the smallest group of different atoms that retain the properties of the substance. When

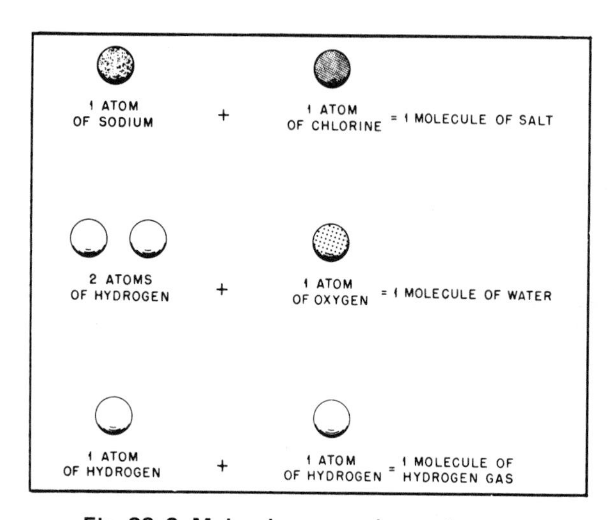

Fig. 23-2 Molecules are made up of atoms.

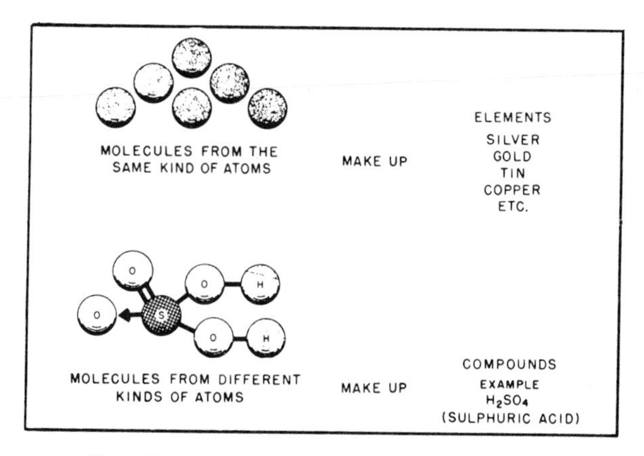

Fig. 23-3 Matter is made up of molecules.

two or more different atoms make up a molecule, it is referred to as a *compound,* figure 23-3. There are many ways that atoms can arrange themselves to form many different compounds. Over 400,000 different combinations are known, among these: wood, cloth, water, coal, and rock.

The Atom: Structure and Characteristics

The atom itself is made up of protons, electrons, and, with the exception of hydrogen, neutrons. The *protons* are heavy particles at the center of the atom; they have a positive electrical charge. The *neutrons* are also heavy particles at the center of the atom; they have no electrical charge. The protons and neutrons, bound tightly together, form the center or *nucleus* of the atom. The *electrons* are in orbit around the nucleus; they are extremely light and carry a negative electrical charge. With these facts in mind, the atom might be thought of as a solar system in miniature, figure 23-4. Electrons are held in their orbits by the pull of the protons in the nucleus. The number of electrons in orbit around the nucleus is equal to the number of protons in the atom.

The electrons are arranged in neat and predictable patterns called shells around the nucleus of the atom. Some elements are chemically *satisfied,* and since all of their shells are complete, they do not react with other elements; they are *inert.* Most elements have atoms with incomplete outer shells, however. The electrons in these outer shells are loosely held and they are capable of combining with other atoms. As these atoms share the electrons of their incomplete outer shell with those of other atoms, they form a molecule. This sharing of electrons produces chemical compounds. A good example is sodium chloride—common table salt—figure 23-5. As you can see, the chlorine atom has only seven electrons in its outer shell; it is one short. The sodium atom has only one electron in its outer shell. These two atoms share their outer shell electrons. Each outer shell is now complete, and a stable compound is formed.

To explain further, chlorine normally has only seven electrons in its outermost range or shell. For an atom to be chemically inert (satisfied) it should have eight electrons in this shell. Sodium has only one electron in its outermost shell. Now, if one atom of sodium combines with one atom of chlorine in such a way that the single atom in the sodium shell is shared with the outer shell of chlorine, then each atom is considered to have eight electrons in its outermost shell. In effect, then, the sodium loaned its one electron to the

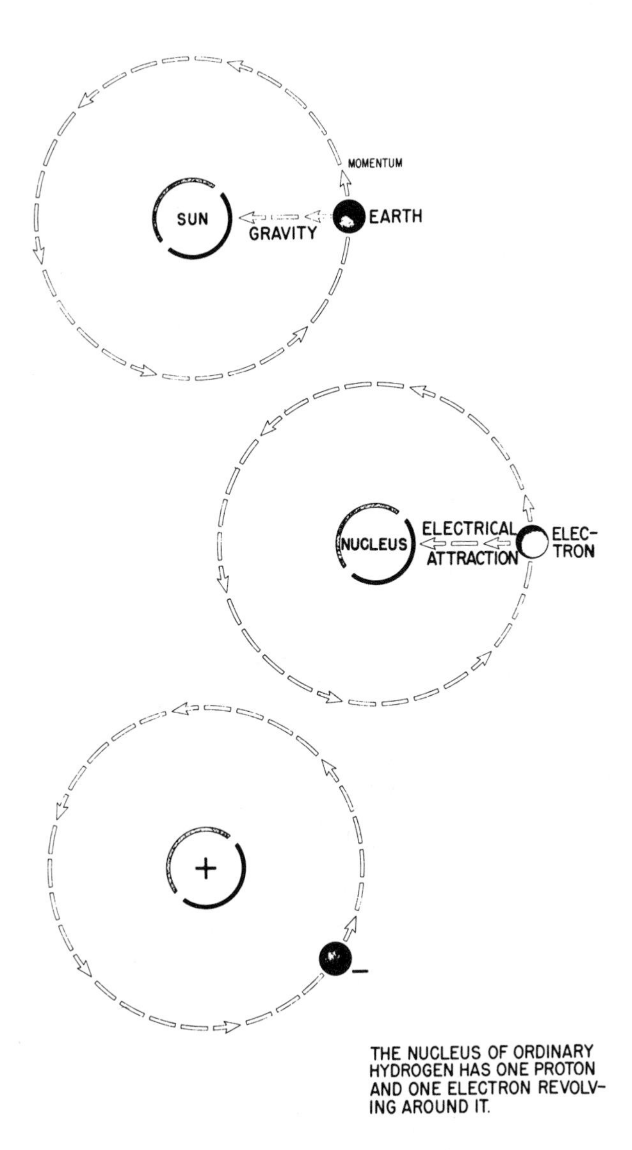

Fig. 23-4 An atom of hydrogen

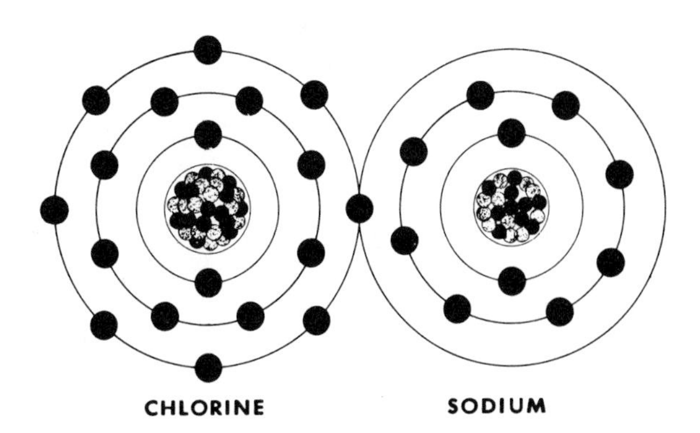

Fig. 23-5 Molecule of sodium chloride

chlorine to complete the eight electron outer shell and the chlorine loaned its seven electrons to do the same thing for the sodium. Under these conditions, a stable chemical compound results from the chemical reaction between sodium and chlorine and a molecule of sodium chloride is formed.

There is space between the molecules of any substance. Some molecules are very close together; some are very far apart. Gases have the most space between molecules, liquids have less space between their molecules, and molecules of solids have the least space between them; they are the most dense.

Molecules do not remain still; they are in constant motion. In gases there is tremendous movement of the molecules, while in solids the movement is reduced to a slight vibration. Also, the molecule's speed increases as it is heated. This can be seen on even a simple—supposedly solid—concrete road surface. The road surface is not laid in one piece, rather there are spaces between sections—expansion joints—so that the surface, heated and expanding under the summer sun, has room to move without creating stress cracks or *heaving.* This type of road construction is seen more frequently in southern states where heat can be more of a problem to a road surface. It can also be seen when water boils; some molecules get so hot and move so fast that they are thrown off the surface and appear as steam.

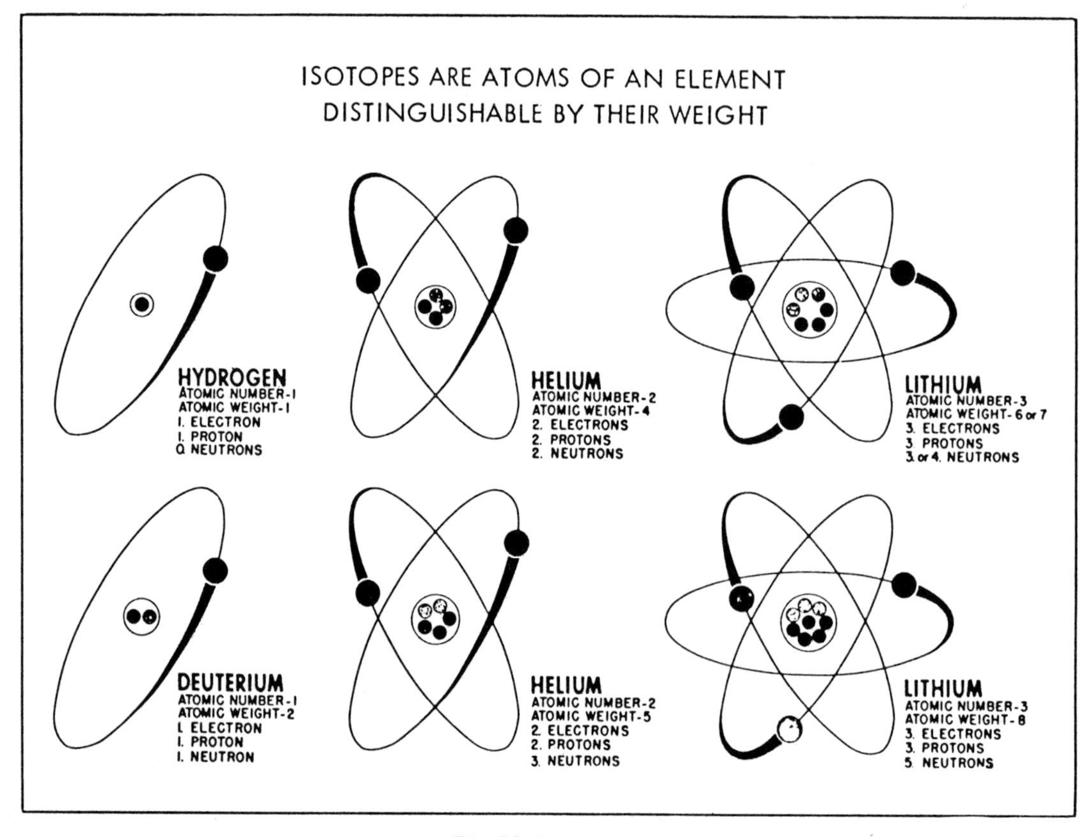

Fig. 23-6 Isotopes

The atom of each element has an *atomic weight*. This weight is the weight of the nucleus, or the combined number of protons and neutrons. Hydrogen, the lightest element, has one proton; its atomic weight is 1. Zinc has 30 protons and 35 neutrons; its atomic weight is 65. Uranium, a very heavy element, has 92 protons and 146 neutrons; its atomic weight is 238. The electrons are infinitely light and are not considered in atomic weight. Elements are also assigned *atomic numbers*, the number of protons in the nucleus. The atomic numbers for these elements then would be: hydrogen, 1; zinc, 30; and uranium, 92.

Another point in the understanding of atomic energy is isotopes. *Isotopes* are forms of the same element that have different atomic weights. The isotopes of an element are chemically alike and have the same atomic number, yet their total weight is different. The difference is in the number of neutrons in the nucleus. Isotopes are common; all elements have them. There are more than 1000 isotopes known today, figure 23-6.

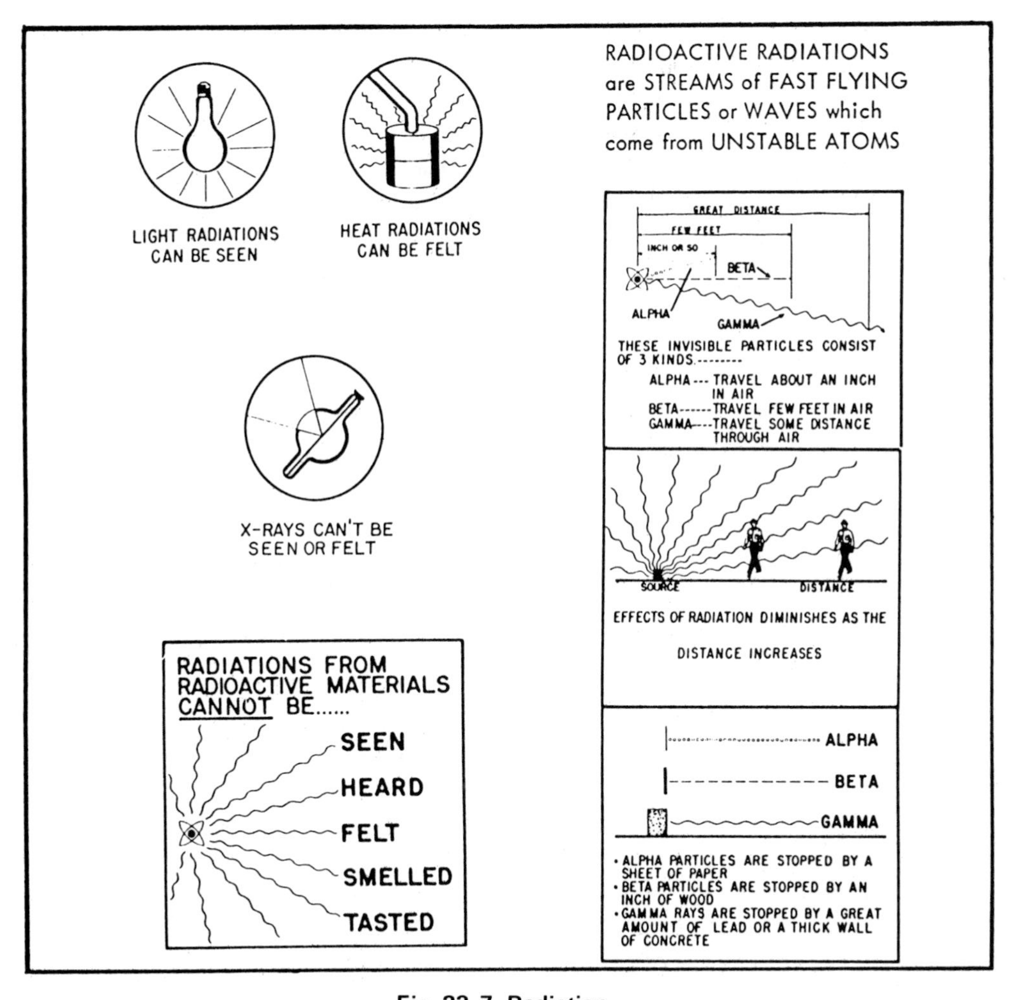

Fig. 23-7 Radiation

Radiation

Radiation is not new; we have always been exposed to it. It comes from the stars that are intensely radioactive and from the earth which is slightly radioactive. There are many *types* of radiation: light radiation which can be seen, heat radiation which can be felt, X-rays which can neither be seen nor felt, and radiation from radioactive material, figure 23–7. Our senses cannot detect radiation from radioactive material. Until recently, this last type of radiation occurred only in nature; now there are many man-made materials.

Radiation is a stream of fast-flying particles (waves) coming from atoms. There are three types: alpha, beta, and gamma. The *alpha* particle is heavy, travels a short distance, and can be stopped by human skin or a piece of paper. The *beta* particle is the same as an electron; it too can be easily shielded. However, alpha and beta particles are dangerous if taken into the body. The *gamma* ray is the most penetrating and dangerous. It is the radiant energy from the excess energy in the atom's mass. Great care must be taken to protect living tissue from gamma rays. Radiation, like fire or electricity, can be dangerous. It is a matter of the degree of exposure.

Uranium is a naturally radioactive element. The uranium atom loses protons in the form of radiation as it ages. The uranium slowly becomes radium and then finally becomes lead, a stable element. This process of one element changing into another is called *natural transmutation.*

NUCLEAR FISSION

As scientists studied the atom, they found that the atomic weight of each atom was less than the weight of the total mass of protons, neutrons, and electrons that it is composed of. Some of the mass appeared to have disappeared, contrary to the law of the conservation of matter which states that matter can be neither created nor destroyed, but only altered in form. As early as 1905, Albert Einstein reasoned that energy and mass were two forms of the same thing. He thought that the loss in mass was converted into energy. When free protons and neutrons were formed into a nucleus, they lost some mass. This loss in mass was converted into the energy that binds the nucleus together. If the atom could be broken or split, this binding energy could be released. Einstein calculated that this energy release would be tremendous.

Uranium-235 has been found to be the only natural element that can be split. An atom of uranium-235 is split when its nucleus is struck by a free neutron that is traveling at the correct speed. This neutron *bullet* can strike the nucleus of the atom and split it, figure 23–8.

Suppose that free neutrons are bombarding a mass of uranium atoms. Many of the neutrons will miss their target completely. However, when a neutron does hit the nucleus of the uranium atom, the atom will split. The result of this split is two smaller atoms, krypton and barium, figure 23–9. Also two more neutrons are freed. The process of splitting the uranium-235 atom is called *fission.* Just as Einstein had predicted, tremendous energy is released during the process.

Other isotopes of uranium, U-234 and U-238, will not undergo fission. That makes it unfortunate then, that 99.3% of uranium occurs as U-238. Only 0.7% is the valuable U-235, and the amount of U-234 is too small to be significant, even if it were fissionable.

A chain reaction is begun with the initial fission of the first uranium-235 atom, figure

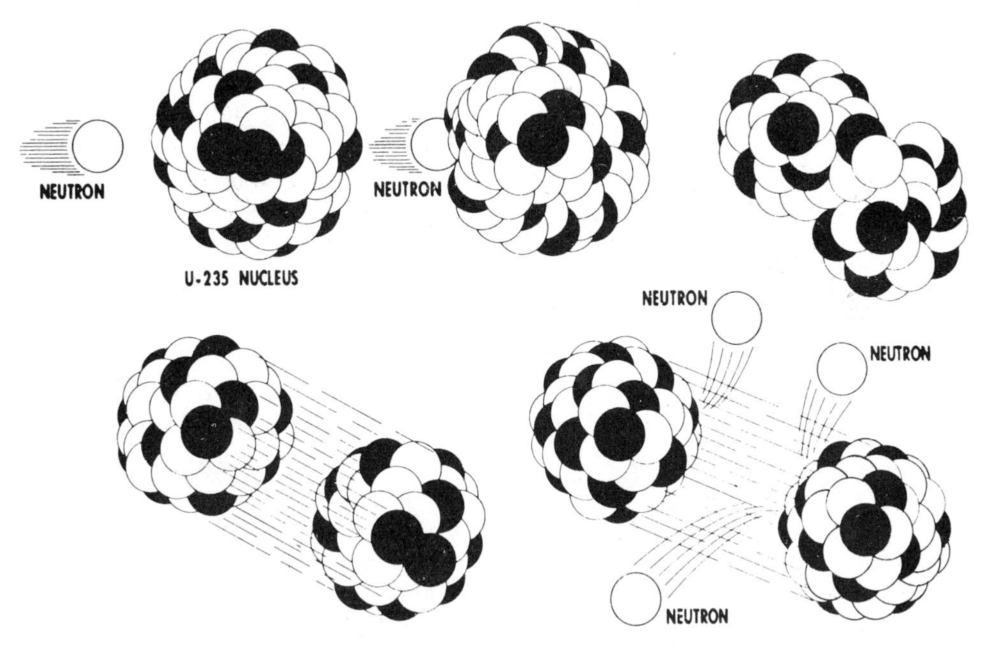

Fig. 23-8 Fissioning of uranium-235

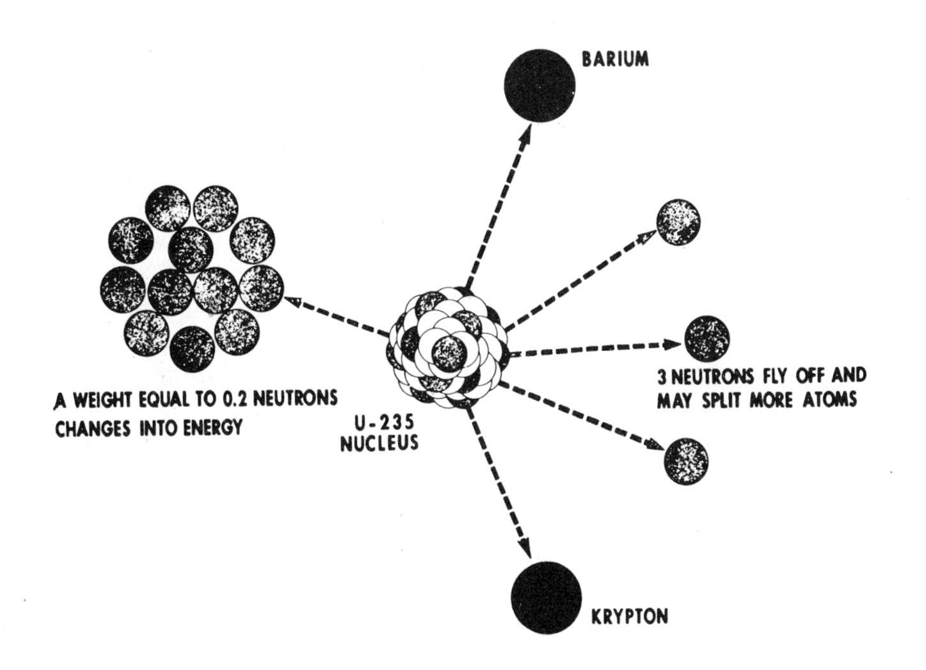

Fig. 23-9 Weight into energy

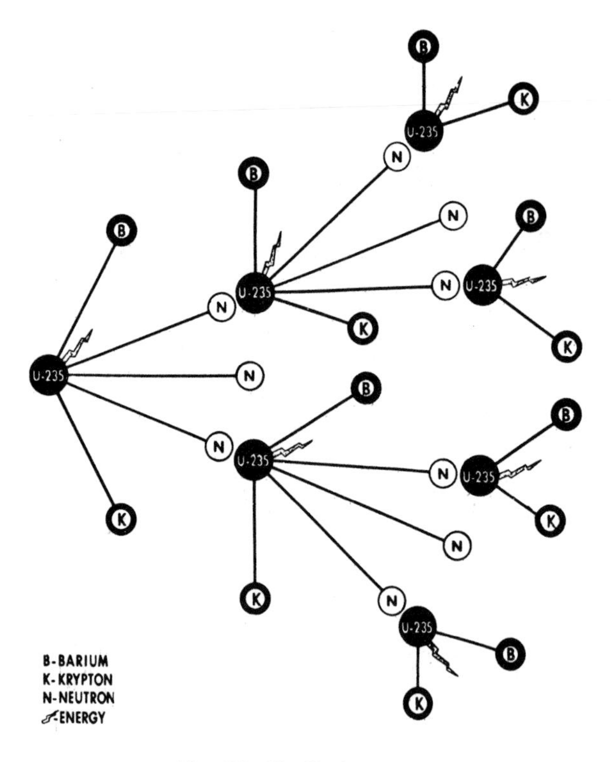

Fig. 23–10 Chain reaction

23–10. The free neutrons released in the fission travel until they strike and split other uranium atoms. With each fission in the chain reaction, energy is released and more neutrons are set free to strike and split still more atoms. An almost unlimited number of free neutrons can be released in an incredibly short amount of time. The result is an enormous release of energy. If the chain reaction is left uncontrolled, we have an atomic bomb, a destructive release of energy. The chain reaction can be controlled, however, with the use of an atomic reactor.

A chain reaction cannot be supported by a single atom. Neither can the reaction be supported by only two, three, or four atoms. There must be a large enough mass of uranium atoms present to support and propagate a chain reaction. If the mass is too small, the reaction will die out because too many neutrons are escaping from the surfaces. A mass where neutrons escape from the surfaces rapidly is said to be *sub-critical.* In order to maintain a productive chain reaction, there must be more than one additional atom split for each atom previously split. When a productive chain reaction is maintained, the mass is said to be *super-critical.* When the process of nuclear fission is proceeding at a controlled rate, as in a nuclear reactor, the mass is said to be *critical.*

NUCLEAR REACTOR

The nuclear reactor is designed to provide a controlled fission of the uranium atoms, releasing energy at a constant rate for a long period of time. The chain reaction is slowed down by diluting the uranium with another substance such as graphite. As fission occurs, the free neutrons are slowed down by the graphite *moderator.* This combination of uranium and graphite is referred to as a *pile.* The first atomic pile was built at the University of Chicago in 1942.

In figure 23–11, the initiating neutron works it way through the graphite atoms, striking a uranium nucleus at A, and causing its fission. Two more neutrons are emitted. The neutron following path #1 collides with graphite nuclei. These collisions reduce its speed until it causes the disintegration of another uranium nucleus at B. The disintegrating nucleus gives two more neutrons. The neutron following path #3 passes through the uranium mass C without causing disintegration, and proceeds to D, where it causes further reaction. The neutron proceeding along path #2 is captured by the uranium nucleus at E, since it is moving at a relatively high speed. The neutron following path #4 escapes through the surface of the pile.

The neutrons released by the fission of the U-235 may do one of three things. First, if the chain reaction is to be maintained, at least one of the neutrons must strike the nucleus of another U-235 atom causing its fission. Second, some of the neutrons that have been slowed down may be captured by the U-238 isotope. There are 140 of this heavier uranium isotope for every one U-235 isotope. When U-238 captures a neutron it becomes U-239. Neptunium U-239, which is radioactive, soon decays into a more stable form of U-239, plutonium. The formation of plutonium is important because it, too, is fissionable. Plutonium then can be used to help sustain the chain reaction at a later date. Third, the neutron can escape the pile. This is undesirable and should be kept to a minimum since it represents a loss.

Reactor: Parts and Function

The nuclear reactor itself is made up of several main parts. Atomic fuel, uranium U-235, is distributed within tons of U-238. The moderator, usually carbon in the form of graphite, or heavy water, has the effect of slowing the neutrons and also slowing the chain reaction. The control and safety rod mechanism regulates the chain reaction properly. Boron and cadmium rods are moved in

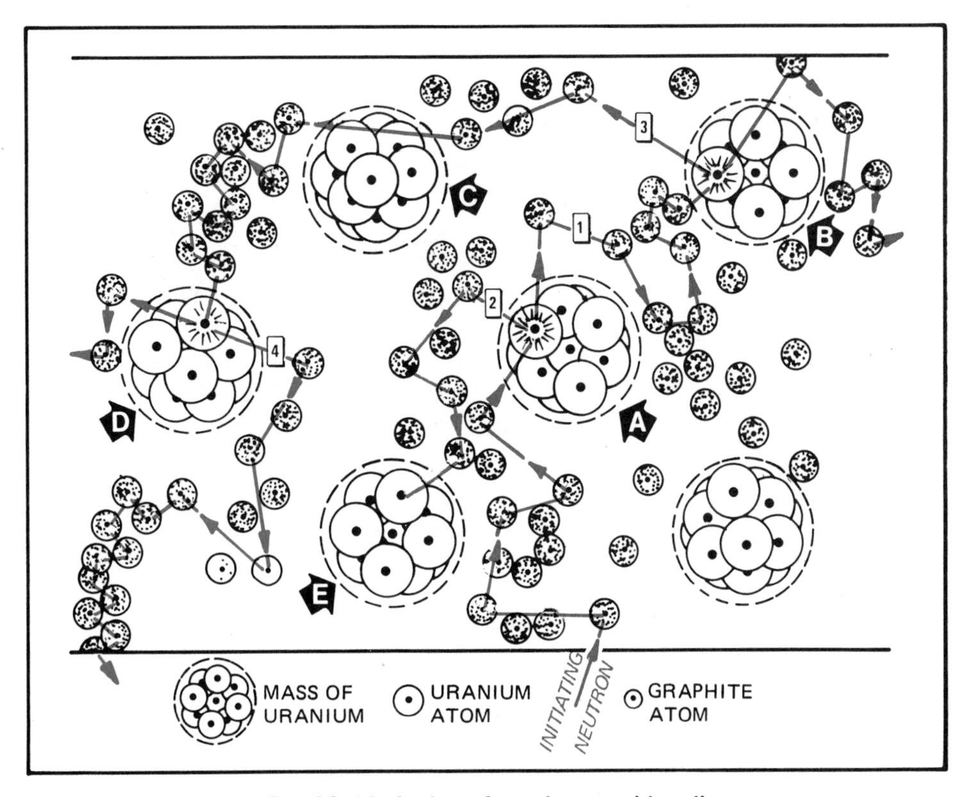

Fig. 23-11 Action of uranium-graphite pile

and out of the atomic pile to control fission, figure 23–12. Both of these elements have a strong ability to capture neutrons. When they are well into the pile they capture so many neutrons that there are not enough left to sustain the chain reaction. If the rods are all the way out of the pile, the chain reaction can reach sufficient intensity to burn up or melt down the materials. The rods are positioned to allow the desired level of reaction. The shield of a reactor is important for the protection of operating personnel. Concrete, several feet thick, and iron are used to completely enclose the reactor. The fission process emits neutrons and gamma rays that are very dangerous. The shield traps these danger-

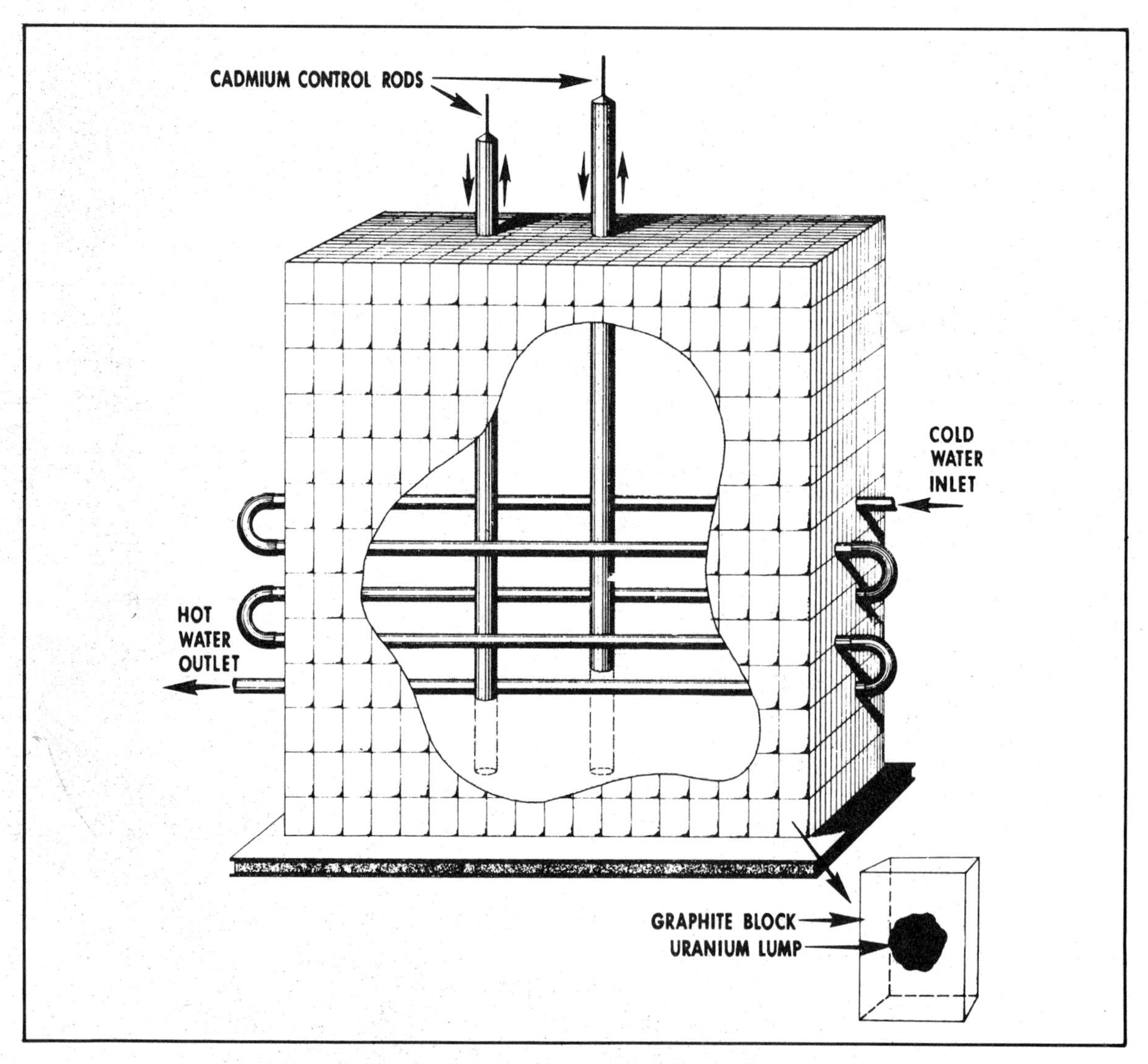

Fig. 23–12 Uranium-graphite pile

ous rays and particles. The coolant is necessary to keep temperatures in the reactor low enough to prevent heat damage. The coolant is water or other fluids.

In power reactors, the coolant picks up this heat energy and uses it to produce power, which is perhaps the greatest potential use for the nuclear reactor, figure 23-13. Further, the

Fig. 23-13 Nuclear power plant—Pilgrim Station Unit 1 (Courtesy Boston Edison Co.)

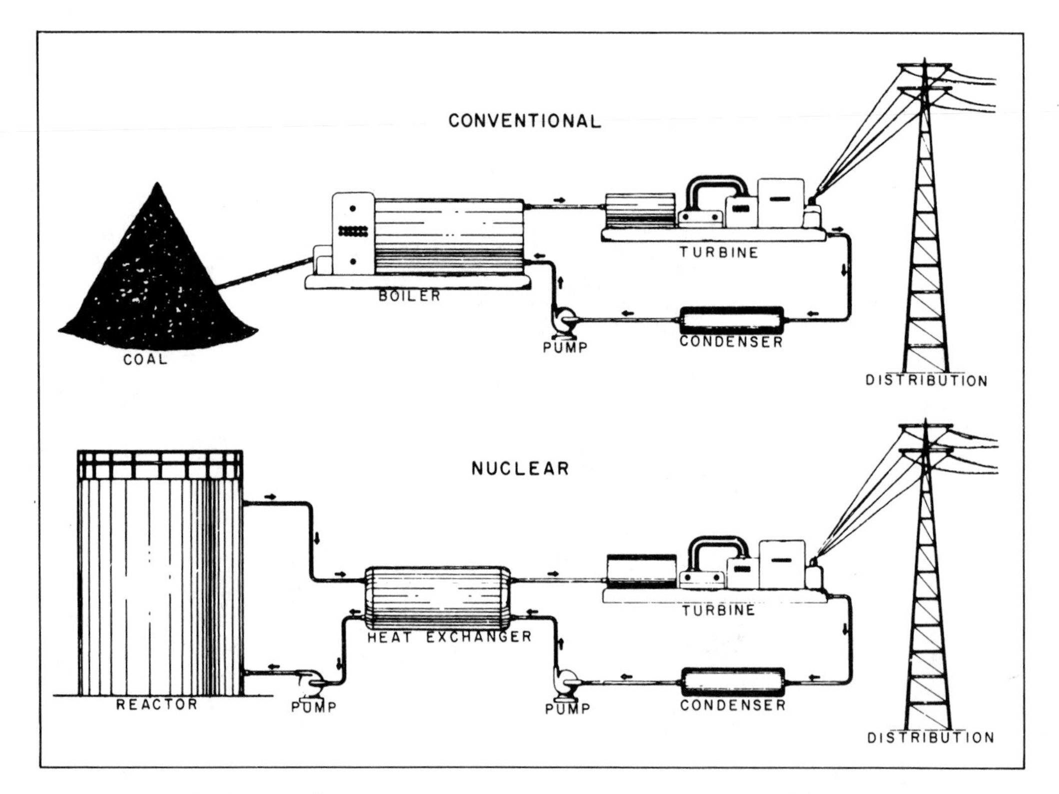

Fig. 23-14 Electrical power plants

intense heat energy of the reactor can be converted into electrical power because the action of a nuclear reactor used to produce electricity is very similar to that of a coal or oil-fired electrical generating station, figure 23-14.

As you can see in the figure, the intense heat energy inside the reactor is picked up by the coolant or heat exchange medium that flows through pipes within the pile. When heat is transferred to ordinary water in a heat exchanger or boiler, high-pressure steam is produced. The high-pressure steam is directed against the turbine blade, revolving the turbine which drives the electrical generator; power is produced. The coolant then returns

Fig. 23-15 Nuclear powered aircraft carrier Dwight D. Eisenhower (Courtesy Newport News Ship Building by American Petroleum Institute)

Fig. 23-16 Uranium mine and mill in northern Saskatchewan, Canada (Courtesy Gulf Oil Corporation by American Petroleum Institute)

to the reactor by way of a condenser, through the heat exchanger, and then to the reactor, where it is ready to begin a new cycle.

Eighty-five nuclear reactors are in operation throughout the nation producing almost 13% of our electrical energy. Fifty-seven more nuclear power plants are under construction. There are many other uses for nuclear reactors, such as nuclear-powered submarines and ships, figure 23-15. Nuclear locomotives and airplanes are also a possibility of the future. There is only a limited supply of uranium-235 available, but the known reserves of U-235 are expected to be adequate well into the next century, figure 23-16.

FAST BREEDER REACTOR

The liquid-metal fast breeder reactor is one of the answers to the long-term energy problem, hopefully stretching the uranium supply of several decades into several hundred years. To date, the breeder reactor is the only method that can extend our use of uranium, but at present the United States is lagging behind other countries in developing breeder reactor plants. Our Clinch River Breeder Reactor Project near Oak Ridge, Tennessee was scheduled to be in operation by 1980. Although 1.4 billion dollars has been spent on the project and plant design is almost 90% complete, political decisions have stalled the project.

The fast breeder reactor, figure 23-17, can use U-238, a non-fuel and convert it into U-239, a fuel. The breeder reactor does this by using fast, high-energy neutrons created by fission. They are captured, in turn, by U-238 (uranium) to create a new element, U-239 (plutonium) which is fissionable.

The liquid metal used in the fast breeder reactor is sodium. It allows neutrons to travel in the reactor with less interference. Liquid sodium does not degrade the neutrons, and it is the medium that cools the fuel and transfers the heat energy for electrical generation. The breeder reactor will not only replenish its own fuel supply but will also produce plutonium fuel for other nuclear plants.

NUCLEAR FUSION

If the goal of nuclear fusion is reached, the world may never face an energy crisis again. Fusion means coming together, while fission means splitting apart. The fusion process is the method the sun and the H-bomb use to create heat and light. Under controlled fusion, our own miniature suns can be created right on earth.

Fusion takes place when the nuclei of light elements collide and merge. The mass of the new nuclear particle is less than the total of the two separate masses. For example, if two

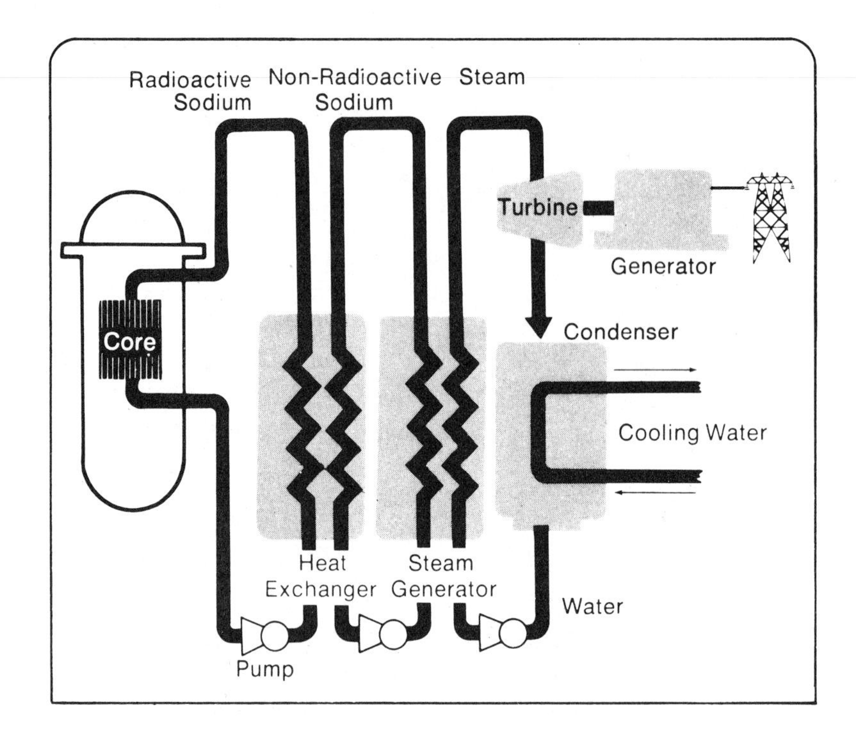

Fig. 23-17 Schematic of liquid metal fast breeder reactor

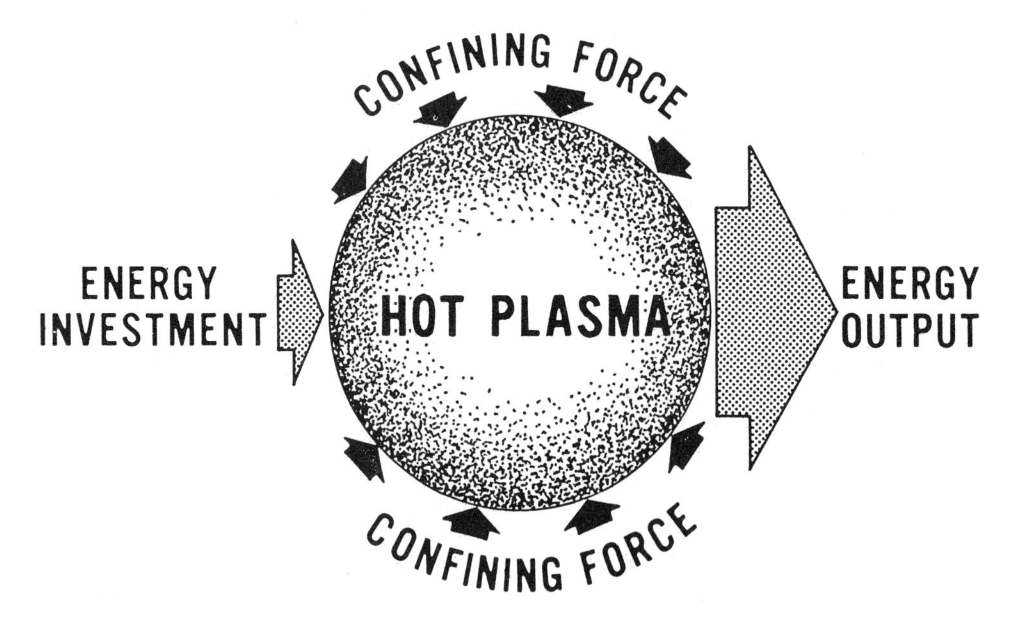

Fig. 23-18 Fusion requirements

Fig. 23-19 Tokamak fusion reactor at Princeton University Plasma Laboratory (Courtesy of Electric Power Research Institute)

isotopes of hydrogen, deuterium and tritium, collide and merge, the result is a nucleus of helium and a free neutron. Just as in fission, mass has been changed to energy in the fusion process.

However, fusion does not take place easily because the positive nuclei will repel each other. In order to get the nuclei to fuse, very high temperatures are needed to speed up their movement. Also, the nuclei are densely packed to increase the chances for collision. The high temperature and high density is maintained long enough for the fusion to become self-sustaining, figure 23–18.

Several methods are being developed to produce the high temperatures necessary for fusion, figure 23–19. In magnetic confinement, a gas is heated inside a magnetic field. The gas looses its electrons, which form an envelope around the nuclei mass. Temperatures are fantastically high, 50 000 000°K. The container which is shaped like a hollow

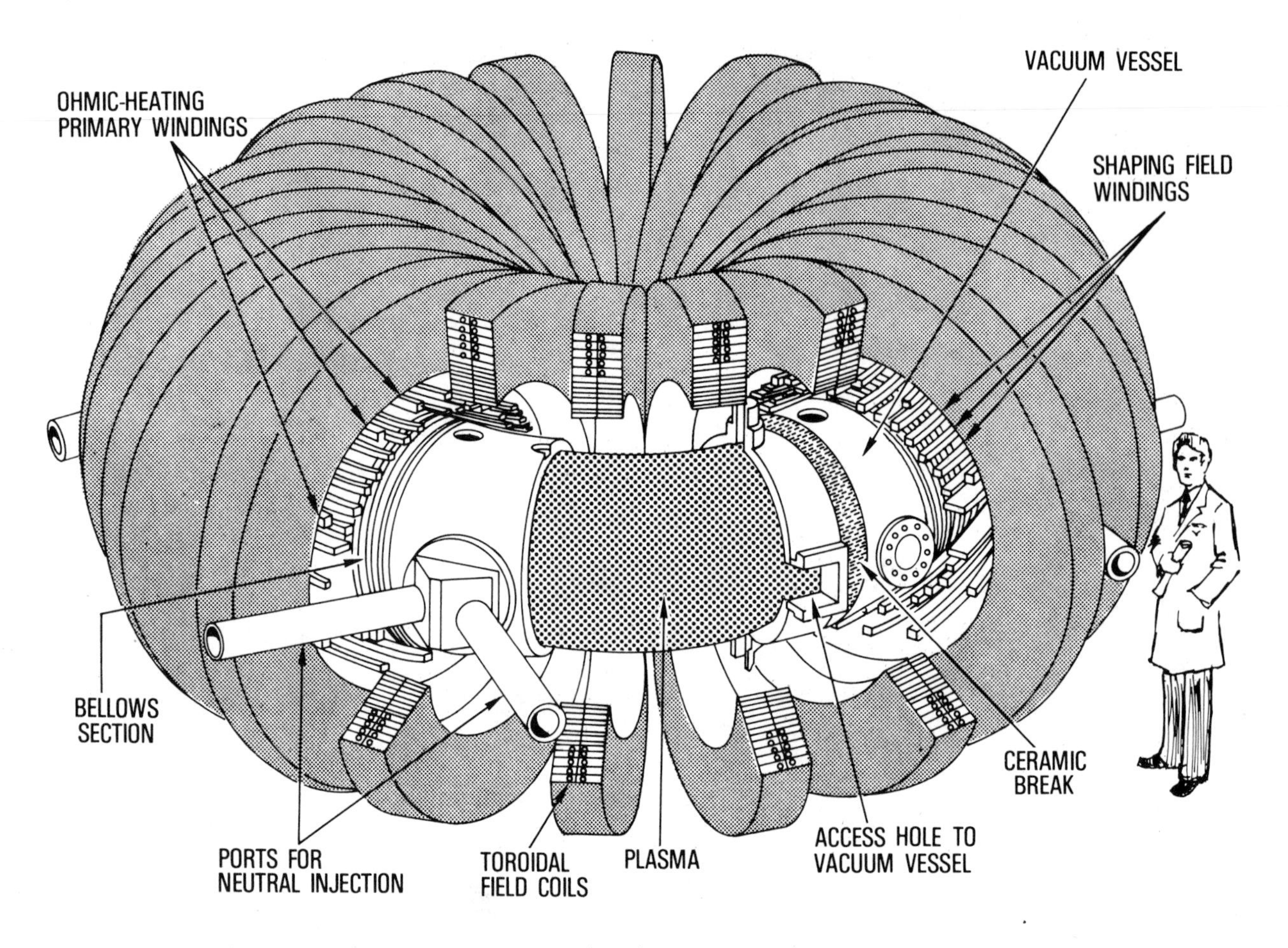

Fig. 23-20 Princeton Large Torus (PLT) is a toroid utilizing the tokamak concept

doughnut, is called a *toroid,* figure 23–20. Laser devices are also being developed to produce and contain the high temperature required for fusion.

It is possible that fusion power plants could become a reality early in the next century, figure 23–21. The fuel sources of deuterium and tritium can be found in sea water and, for all practical purposes, can be considered an inexhaustable fuel. When the energy of the sun is produced on earth through fusion, our energy needs can be met without relying on fossil fuels.

Fig. 23-21 Drawing of a nuclear fision facility (Courtesy Standard Oil Company of California)

REVIEW QUESTIONS

1. Define the following terms: (a) atom, (b) molecule, and (c) compound.
2. What parts make up an atom? Explain their location in the atom.
3. What happens when atoms share their electrons?
4. Define atomic weight.
5. Define atomic number.
6. Define isotopes.
7. What is radiation? What are the three types of radiation?
8. What is fission?
9. List the products of uranium-235 fission.
10. Explain how a chain reaction takes place.
11. What is a nuclear reactor designed to accomplish?
12. What is plutonium?
13. List the main parts of a nuclear reactor and briefly explain the function of each part.
14. Explain how the heat energy within a nuclear reactor can be converted into electrical energy.

Unit 24 The Direct Conversion of Energy

OBJECTIVES After completing this unit, the student should be able to:

- Explain why direct conversion of energy is more efficient than other systems.
- List and discuss the methods used to produce electricity directly from primary energy sources.

Energy should be put to work in as direct a manner as possible. The long route of changing energy through several different states is costly and wasteful. Under usual circumstances energy cannot be destroyed, but it can be lost to friction, or be converted into other forms that are not involved in producing the desired power. For example,

TO ↓ FROM →	ELECTROMAGNETIC	CHEMICAL	NUCLEAR	THERMAL	KINETIC (MECHANICAL)	ELECTRICAL	GRAVITATIONAL
ELECTRO-MAGNETIC		Chemiluminescence (fireflies)	Gamma reactions (Co^{60} source) A-bomb	Thermal radiation (hot iron)	Accelerating charge (cyclotron) Phosphor	Electromagnetic radiation (TV transmitter) Electrolumines-cence	Unknown
CHEMICAL	Photosynthesis (plants) Photochemistry (photographic film)		Radiation catalysis (hydrazine plant) Ionization (cloud chamber)	Boiling (water/steam) Dissociation	Dissociation by radiolysis	Electrolysis (production of aluminum)	Unknown
NUCLEAR	Gamma-neutron reactions ($Be^9 + \gamma \rightarrow Be^8 + n$)	Unknown		Unknown	Unknown	Unknown	Unknown
THERMAL	Solar absorber (hot sidewalk)	Combustion (fire)	Fission (fuel element) Fusion		Friction (brake shoes)	Resistance-heating (electric stove)	Unknown
KINETIC (MECHANICAL)	Radiometer Solar cell	Muscle	Radioactivity (alpha particles) A-bomb	Thermal expansion (turbines) Internal combus-tion (engines)		Motors Electrostriction (sonar transmitter)	Falling objects
ELECTRICAL	Photoelectricity (light meter) Radio antenna Solar cell	Fuel cell Batteries	Nuclear battery	Thermoelectricity Thermionics Thermomagnetism Ferroelectricity	MHD † Conventional generator		Unknown
GRAVITATIONAL	Unknown	Unknown	Unknown	Unknown	Rising objects (rockets)	Unknown	

†Magnetohydrodynamics

Fig. 24-1

notice the waste of energy involved in producing electricity, which is the most universally used type of power for home and industry.

1. The chemical energy of coal is converted into thermal energy as it burns.
2. This thermal energy is absorbed by water.
3. Steam is produced and directed against the blades of a turbine, converting the thermal energy into mechanical energy.
4. The mechanical energy of the turbine

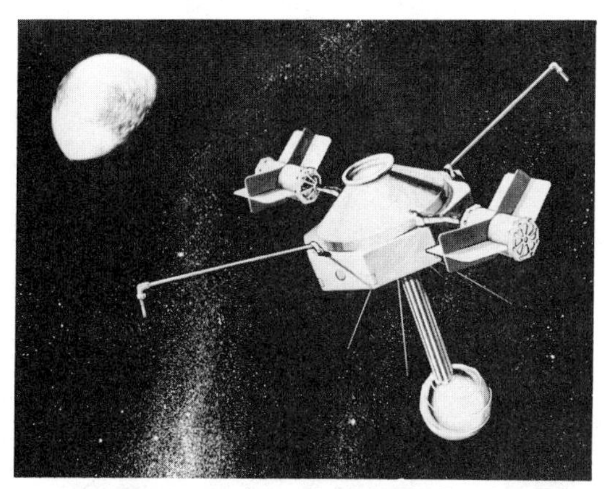

A. Atomic generators convert heat from radioactive fuel into electricity to power radio transmitters of satellites.

B. A nuclear-powered device, center, supplied electrical power for explorations on the moon.

C. Atomic fuel provides the power for instrumentation of this weather satellite.

D. A nuclear-powered laboratory for soft landing on the lunar surface.

Fig. 24-2 U.S. atomic energy SNAP program (Systems for Nuclear Auxiliary Power) utilize direct energy converters. (Courtesy Atomic Energy Commission)

rotates the generator and the energy is at last converted into electrical energy for transmission and use.

5. The electrical energy may again take many forms as it is used by the consumer.

The described method is good; it is reliable; it produces the vast portion of all electrical energy, but it does represent the long route of energy conversion.

Imagine how much more efficient it would be if an energy form could be converted directly into electrical energy. The idea is not impossible, and it is certainly not new. Several direct conversion techniques have been demonstrated by scientists in the laboratory for many years, figure 24-1. The conventional battery has also been converting chemical energy directly into electrical energy for a long time, but it does have limitations of weight and length of service.

In our space age technology, many of the laboratory phenomena of direct energy conversion are under constant, intensive development. The result has been a whole new family of direct energy conversion devices. A good deal of the development of these energy conversion devices has come from the nation's space program, figures 24-2 and 24-3.

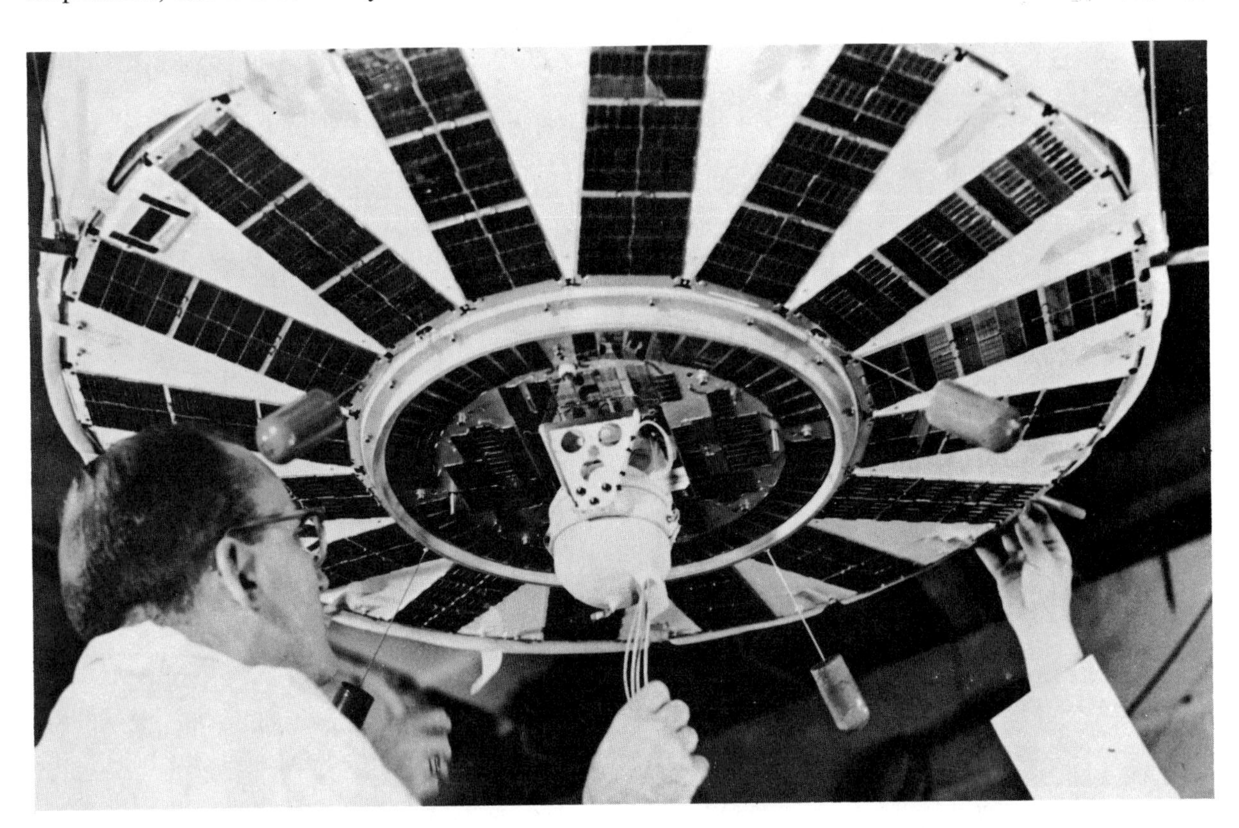

Fig. 24-3 Tiny nuclear generator is fastened to base of a TRANSIT satellite. In the orbiting generator, nuclear fuel generates heat which is converted through thermocouples to electrical energy. (Courtesy Atomic Energy Commission)

Space systems need rockets for basic propulsion, but, in addition, they need electrical systems that fit the requirements of long service life, low weight, and high reliability. Direct energy converters can meet these requirements. They play a part in meeting the electrical, electronic, and life support functions involved in outer space projects.

The direct energy conversion devices also have applications beyond their uses in space exploration. They can be used wherever long life, low weight, and high reliability are of great importance. These energy converters are now the heart of several rapidly maturing industries. In many cases, these products are out of the laboratory, off the drawing board, and are being manufactured and put into space, military and commercial use.

When direct energy converters are discussed, the topic concerns various devices that are capable of producing electricity directly from one of the primary energy sources. Several direct energy converters are discussed in this unit:

- *Solar cell,* electrical energy from the sun; electromagnetic energy
- *Thermophotovoltaic (TPV) converter,* electrical energy from thermal photons
- *Fuel cell,* electrical energy from chemical energy
- *Nuclear battery,* electrical energy from nuclear energy
- *Thermoelectricity and Thermionics,* electrical energy from thermal energy
- *Magnetohydrodynamic (MHD) generator,* electrical energy from kinetic energy

These energy converters do not rely on the dynamic conversion of energy through reciprocating or rotary machines. Quietness of operation is inherent in these energy converters since there are no moving parts. These techniques are highly sophisticated in their manufacture, but still very reliable and safe under normal operating conditions.

While heat engines are basically the domain of the mechanical engineer, direct energy converters are more the domain of the electrical engineer, chemist, physicist, and nuclear scientist.

The sun gives the earth heat and light. This huge, intensely hot body of gases is responsible for most of the energy we use, such as fossil fuels, wind, and water power. Section 7, Alternative Energy Sources, discusses solar heating, hydroelectric, and wind energy. In this unit, the aspect of solar energy discussed is the direct conversion of solar energy into electricity.

Devices that directly convert the sun's radiant energy into electrical power have been in existence for many years. The photogalvanic cell had its beginning in 1839. The barrier-layer photovoltaic cell was developed around 1876. These devices still have many applications today, but they do not produce much electricity. They are, therefore, suitable for measuring instruments and controlling devices that require small voltages. The devices mentioned are generally unsuitable for producing electricity in quantity or in competition with conventional electrical generators. At best, these units are only one percent efficient when converting solar energy into electrical power.

The *photogalvanic cell* is made of two electrodes suspended in an electrolyte. When light falls on only one of the electrodes, an electromotive force is produced.

The *photovoltaic cell* produces an electromotive force when it is exposed to light. A light-sensitive semiconductor, such as selenium, is used in this cell. The photocell is often used in photographic exposure meters, photoelectric eyes, and photoswitches.

No significant breakthroughs in the search for ways to directly convert more of the sun's radiant energy into electricity were recorded until 1954. In that year, three scientists from the Bell Telephone Laboratories reported that they had improved the photovoltaic cell and increased its efficiency to six percent. The scientists, D. M. Chapin, C. S. Fuller, and G. L. Pearson, later succeeded in raising the efficiency of their silicon photovoltaic cell to eleven percent. A device with an efficiency of eleven percent does have possibilities of becoming a practical electrical power source. For example, on a clear day in Phoenix, Arizona, if 1000 watts of solar energy falls on a square meter of the silicon photovoltaic cell, 110 watts of electrical power can be provided.

THE SOLAR CELL

The photovoltaic cell, commonly called the solar cell or battery (and made of semiconductive materials) is now a reality and has found important uses in providing electricity for more than 80 communications satellites that can relay telephone conversations, radio, and television around the world, figure 24-4. Space communications of all types rely heavily on the solar cell. For the consumer, the solar cell is seen in the millions of economical solar-powered calculators sold almost every-

Fig. 24-4 Solar cell array on a communications satellite (Courtesy NASA)

where. Their solar cell needs just ordinary daylight or artifical light to produce sufficient electrical energy to operate the device.

Conductors, Insulators, and Semiconductors

The previous paragraph stated that solar cells were made up of semiconductive materials. At this point, it will be wise to review the various qualities of this and other materials to determine why semiconductor construction is the best choice.

Conductors, as you remember, are materials with loosely bound electrons that can easily move about. Insulators are materials with tightly bound electrons that cannot move about easily. Ninety percent of the substances we know are either conductors or insulators. However, there are a few substances that cannot be classed as either conductors or insulators. They are called semiconductors; silicon and germanium are examples, figure 24-5.

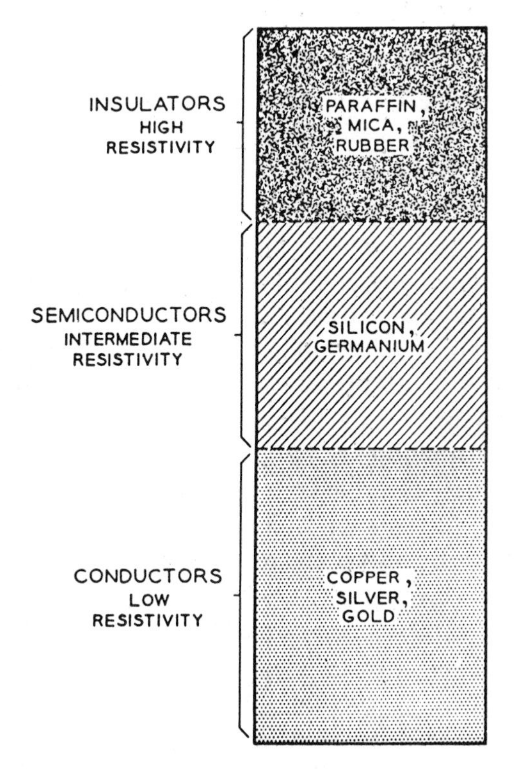

Fig. 24-5 Resistivity of insulators, conductors, and semiconductors

Insulators have a high resistance to the flow of electrons while conductors have a low resistance to the flow of electrons. Semiconductors, because they are artificially made in the laboratory, can have any amount of resistance designed into them.

The conducting qualities of semiconductors change with the temperature of the material. As a semiconductor is heated, it becomes a better conductor of electricity. More electrons are free to leave their atoms and move about. The reverse is also true. Cooling the semiconductor makes it a better insulator since fewer electrons break out of their orbits and become free electrons.

Figure 24-6 shows what takes place in a semiconductor. Though oversimplified, the drawings do illustrate the point. Drawing (A) shows a tray with ball bearings covering its bottom. The ball bearings represent the semiconducting material. Drawing (B) illustrates the semiconductor at absolute zero. The tray is tipped (electrical field), but the electrons will not flow and so the material will not conduct. In (C) the semiconductor is heated, the tray is shaken (thermal energy) and an electron hops out of orbit. The material becomes conductive. In (D) a potential is again applied. The free electrons can now move. The ball bearing moves to the lower end, leaving a hole at the upper end.

Semiconductors are heat sensitive and photosensitive. Light falling on a semiconducting material causes more electrons to leave their orbits, creating free electrons and electron holes which lack electrons. Photons are light energy and their effect on a semi-

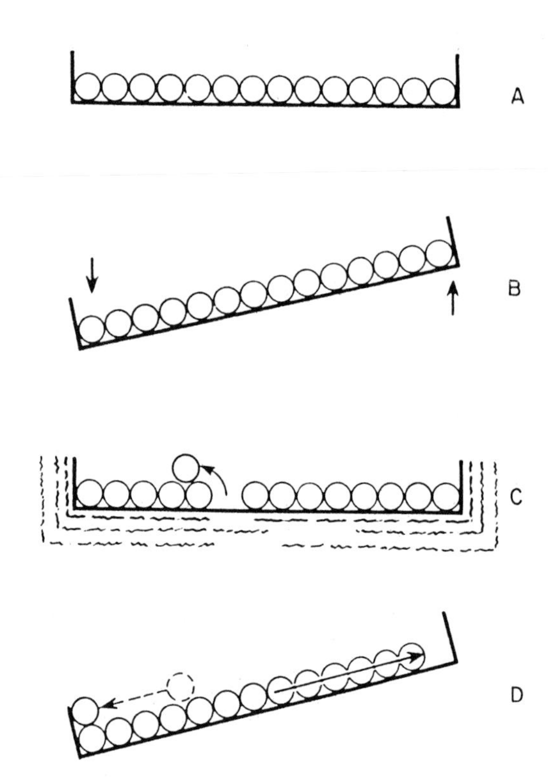

Fig. 24-6 Ball bearings in a tray illustrate the action of a semiconductor.

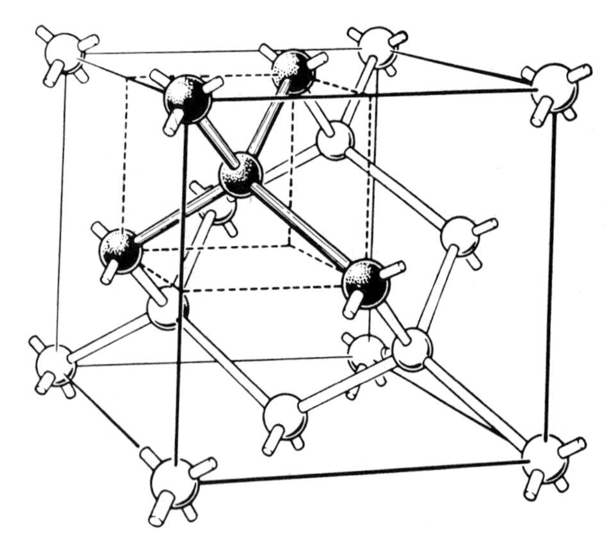

Fig. 24-7 Crystal lattice structure of silicon or germanium

conductor could be compared to pellets striking a tray of ball bearings.

Silicon and germanium are good semiconducting materials. They can be made very pure and both have a very desirable crystalline structure, figure 24-7. Electrons can be set free in both of these materials by the action of thermal energy or by striking the materials with photons of light energy.

Construction of the Solar Cell

A solar cell is made by combining pure silicon with a minute amount of arsenic by a special process. The combination of silicon and arsenic creates a new lattice pattern containing a free (extra) electron and is called an N-type (negative) semiconductor. The addition of boron in a later stage of the battery's construction produces yet another lattice pattern with a missing electron and is called a P-type (positive) semiconductor, figure 24-8. The area of the battery where the two semiconductors meet is referred to as the PN-junction.

Ultra pure silicon is *doped* (mixed with an impurity) with a minute amount of arsenic. The two materials are melted together and cast in the form of a small ingot, a single crystal of silicon. The addition of the arsenic makes the silicon rich in electrons. Silicon treated in this manner is called an N-type semiconductor.

The single silicon crystal is then sliced into very small, thin wafers. Rectangular in shape, these wafers are cut about 1/25th of an inch (1 millimeter) thick by a special crystal cutting machine. The silicon wafers are then polished to a high degree.

Boron is added to the surface of the silicon wafer. The wafer is placed in a quartz tube with a vapor containing boron. The boron is diffused into the surface of the silicon. Boron

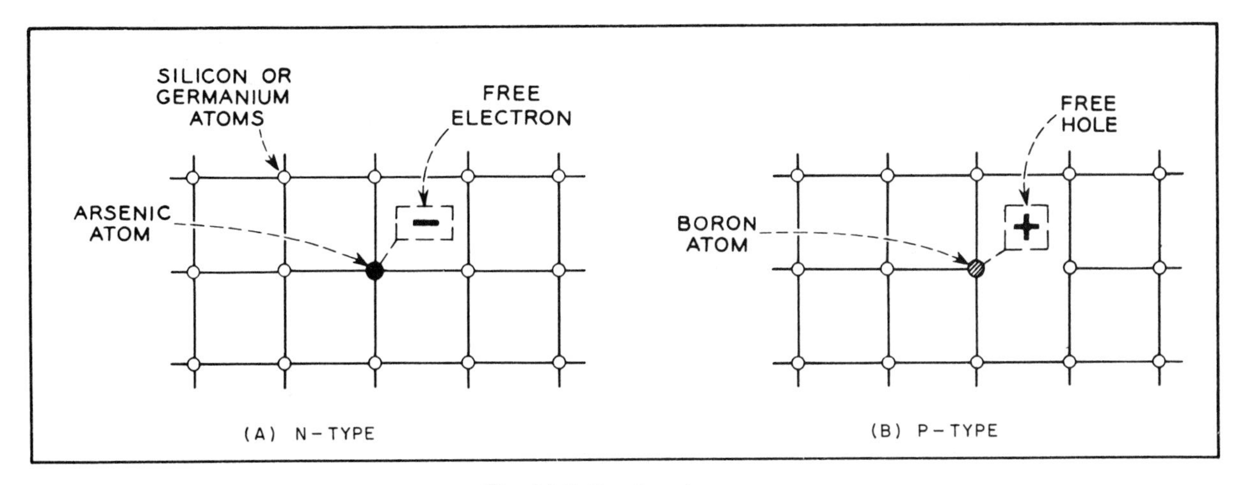

Fig. 24-8 Semiconductors

has the effect of reducing the total number of electrons that are potentially free. The boron-treated area of the wafer is said to be a P-type semiconductor. The boron penetrates only a small distance into the wafer, about 1/10,000 of an inch (.0025 millimeter). The area where the N-type semiconductor and the P-type semiconductor meet is called the PN-junction. The PN-junction is the heart of the solar cell.

Finally, the wafers are cleaned and then plated so that conducting wires can be attached to the wafer. Connected in series, several wafers are combined to form a solar-energized power source of electrical energy, figure 24-9.

Operation

The PN-junction is important to consider. The N-type silicon is, by nature, electron rich while the P-type silicon is hole rich. When light strikes the solar cell, and particularly the PN-junction, photons break electrons away from their atoms. Electrons and holes are

Fig. 24-9 Steps in constructing a solar battery (Courtesy Bell Telephone Lab.)

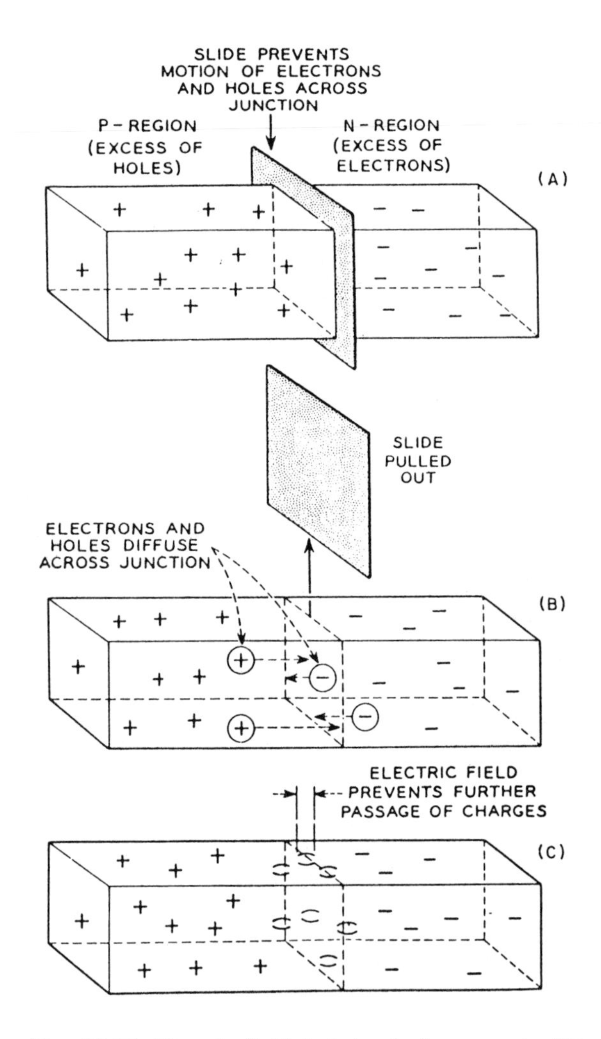

Fig. 24-10 Electric field is being built up at the PN junction.

created, electron hole pairs. The electric field at the PN-junction forces the electrons into the N-region and the holes into the P-region. The N-region is negative; the P-region is positive. There is a potential difference of about 0.6 volt between the two regions when the battery is exposed to direct sunlight. If a load is connected between the two terminals, the potential will cause electrical current to flow. Light energy has been converted direct-

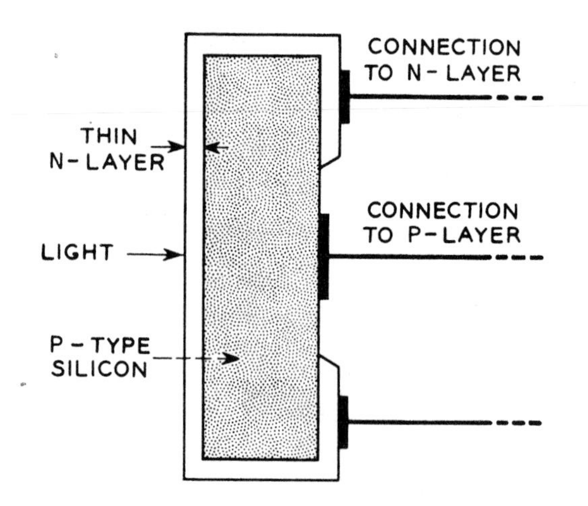

Fig. 24-11 Cross-section shows construction of the solar cell.

ly into electrical energy by the solar cell.

Figure 24-10 illustrates how an electric field is built up at the PN-junction. Drawing (A) shows the P-region and N-region separated by an imaginary slide. In (B), the slide is pulled out and electrons and holes diffuse across the junction. In (C), an electric field quickly builds up at the PN junction to prevent further flow.

If wires are connected to the N-region and the P-region, and a load is placed across the terminals, electrons will flow from the N-region to the P-region, figure 24-11.

Cost

Costs of producing electricity from solar cells started at $1000 per watt of power in the 1950s. Ten years ago the cost had been slashed to $20. Today's cost is $5-9, still a high price considering that electricity from conventional sources is 6¢ per kilowatt hour. However, a technique for producing solar cells in a ribbon form is reducing prices even further—down to 40¢ to $1 per kilowatt hour, figure 24-12.

The thin film solar cells are just one ten-

Fig. 24-12 A commercial solar cell array currently priced at $200 a square foot. The smaller model (left hand) portrays a comparable thin film cell which researchers hope can be made to sell for $2-3 per square foot. (Courtesy Standard Oil of California)

thousandth (0.0001″) of an inch thick. The cells are silicon but their atomic structure is not in crystalline form. Their amorphic, random form allows them to be deposited on a thin film. These solar cells have an efficiency of 5% compared with 12% for a crystalline solar cell but the cost of production far outweighs the loss in efficiency.

THERMOPHOTOVOLTAIC CONVERTERS

Thermophotovoltaic converters use heat (thermo) to develop an incandescent light source, producing thermal photons (photo) which are directed to a semiconducting material capable of inducing voltages across its terminals (voltaic). This new concept was introduced in 1961 by Pierre Aigrain during his lecture series at the Massachusetts Institute of Technology.

Figure 24-13 illustrates a thermophotovoltaic (TPV) converter. In this converter, the photons that excite the semiconducting material do not come from the sun as they do in the case of the solar battery. They come from a hot incandescent light source. The photons that are generated are directed toward the surface of the semiconducting converter by a parabolic mirror.

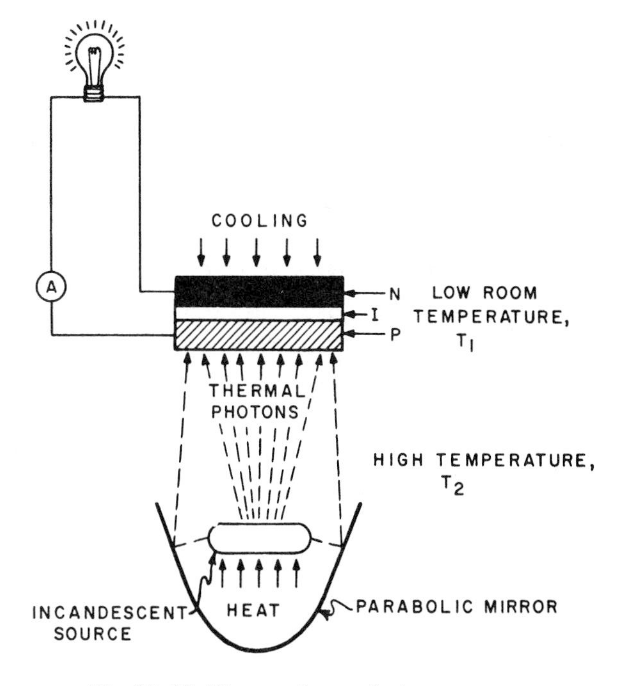

Fig. 24-13 Thermophotovoltaic converter

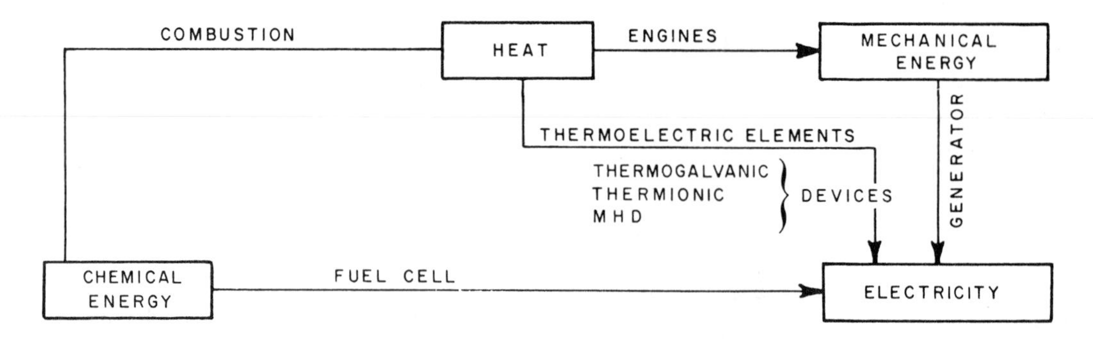

Fig. 24-14 Paths for conversion of chemical energy into electrical energy. Note that fuel cells bypass the heat conversion step.

Much exploration remains to be done with the TPV process, but it has several advantages. A practical efficiency of 30 to 35% is anticipated. Heat is converted to light through incandescence and, in turn, light is converted to electrical energy by a photocell at room temperature.

FUEL CELL

The fuel cell produces electrical energy directly from chemical energy, figure 24-14. To date, these cells are the most efficient nonnuclear devices for generating electrical power. Theoretically, they are 100% efficient, but actual gross efficiences of 80% have been recorded, figure 24-15. Electrical generation by heat engines can seldom top 40%. The Gemini spacecraft used fuel cells to produce the electrical power needed for operation of the space capsule's electronic and life-support systems. The nation's moon shot project, Apollo, also used fuel cells for these purposes.

The applications of fuel cells are not limited to the space program. In 1962 Allis Chalmers announced plans to market a forklift driven by fuel cells. In 1966 the Esso Research and Engineering Company demon-

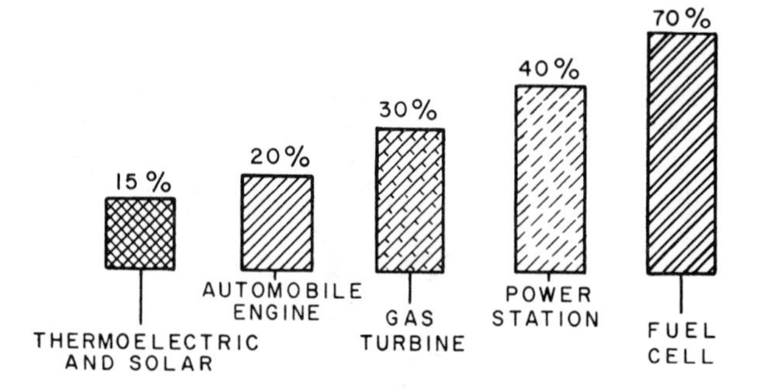

Fig. 24-15 Comparison of efficiencies of energy conversion systems. Note superiority of the fuel cell.

strated an experimental automated fuel cell battery. If inexpensive materials can be found to replace the expensive platinum catalysts used, such systems may eventually have wide applications as a portable power supply, figure 24-16.

The fuel cell is an electrochemical device, like a storage battery, which converts chemical energy into electrical energy. But it is different in important ways. Where the storage battery has its fuel built in as expendable or rechargeable electrodes, the fuel cell has its fuel fed in from outside the cell. The electrodes of the fuel cell are not consumed as are the battery's.

Fig. 24-16 Fuel cell, delivering current for a power drill, operates on the low-cost liquid fuel, methanol. (Courtesy Esso Research and Engineering)

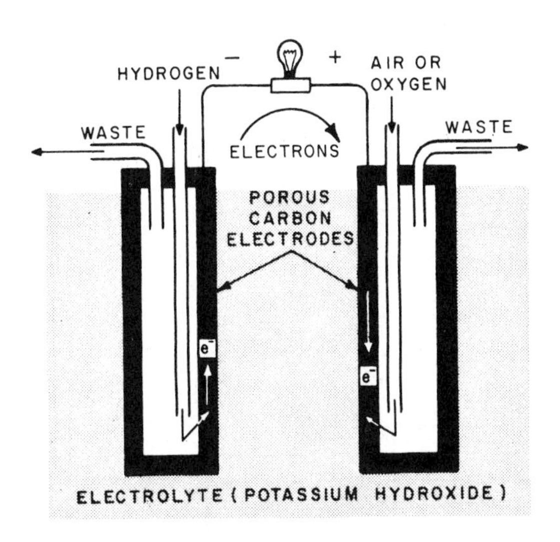

Fig. 24-17 Basic hydrogen-oxygen fuel cell

Fuel Cell Components

In addition to the fuel, the cell contains two electrodes, one positive and one negative, and an electrolyte which serves as the electrochemical connection between the electrodes. Catalysts promote the reaction.

The electrodes bring the reactants in contact with the electrolyte in a controlled way. They also serve as catalysts and carry the current generated by the reaction. Common electrode materials include silver, nickel, palladium, mercury, platinum, and carbon, figure 24-17. The *anode* is the electrode at which the fuel gives up electrodes for delivery to the external circuit. The *cathode* is the electrode that gives up electrons to the oxidizer.

The electrolyte provides the means for ionic conduction. Some common electrolytes include potassium hydroxide, sea water, phosphates and zirconates.

No less than two fuels are needed for operation of the fuel cell. One is the oxidant or electron donor, and the other is the reductant, or electron acceptor. A few common oxidants are oxygen, air, chlorine and brine. Reductants include hydrogen, carbon monoxide, natural gas, methanol, and zinc.

Operation

The hydrogen-oxygen (hydrox) fuel cell is one of several types, and is used for space applications. Others are oxygen-concentration cells, redox cells, hydrocarbon cells, and ion-exchange cells. The basic principle on which the hydrox fuel cell operates is simple. When hydrogen and oxygen are combined to form water, they release electrical energy.

Figure 24-18 shows the cell and the basic chemical action. Two hollow porous carbon electrodes are immersed in a potassium hydroxide electrolyte. Hydrogen is pumped into one electrode and oxygen into the other. The hydrogen molecules containing two hydrogen atoms flow into the pores of the negative electrode. A catalyst splits them into two separate hydrogen atoms. The atoms drift to the reaction zone while ions from the potas-

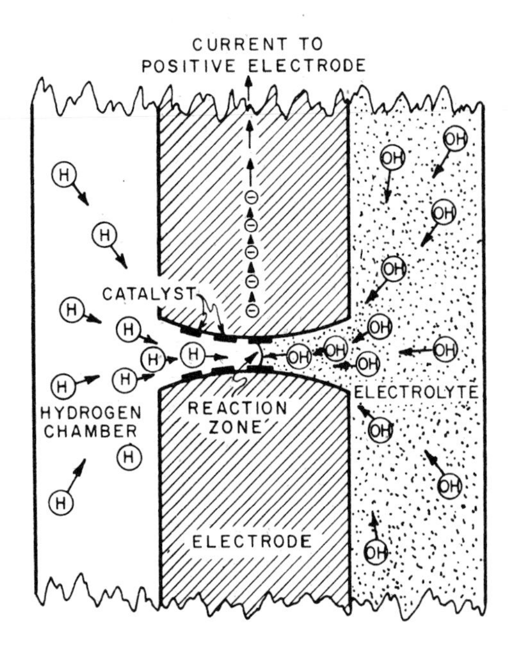

Fig. 24-18 Reaction in pores of electrode between hydrogen and hydroxide produces electrons that flow toward positive electrode.

sium hydroxide electrolyte drift toward the same zone. Here the hydroxyl ions combine to form water and release electrons.

The surplus electrons flow toward the positive electrode. Here oxygen atoms are flowing into the pores of the electrode and combined with water molecules from the electrolyte. The incoming electrons combine with this combination to form ions which replace those being used on the negative electrode. The cycle is continuous.

Advantages and Uses

The advantages of fuel cells are numerous. They have high efficiencies of 70 to 90% because the thermal cycle necessary in the conventional conversion of chemical energy to electrical energy has been eliminated. Fuel cells have no moving parts, do not produce noxious fumes, and can be built in any size and capacity. They are rugged, require little maintenance, and are efficient in all sizes. An offshoot of the process in the hydrogen-oxygen cell is drinking water, which is utilized in such applications as manned space vehicles. Another important advantage is that when no load is connected to the cell, no energy is consumed.

For spacecraft use the two fuels, hydrogen and oxygen, are contained in high-pressure, low-temperature tanks which liquefy these gases, conserving valuable space and weight, figure 24-19. The gases are put through a heat exchanger which brings their temperature to about that of the cell. A fuel cell using methanol and air presents fewer fuel storage problems for land use and has greater promise of approaching a competitive energy cost. Methane is an inexpensive and readily available fuel.

The development of inexpensive materials and less complex systems are the goals for further research; there are also many problems which require such research. One problem is elimination of the water which results from the reaction and weakens the electrolyte in the hydrox-type cell. This very drawback on land is the main reason it is used in the space program. In the manned space application the astronauts will make good use of this water. Another problem is that gases of high purity are required. One attempt to avoid this is the redox cell, *red*uction and *ox*idation, figure 24-20. Although these and other problems remain to be overcome, the fuel cell holds promise as a significant power source.

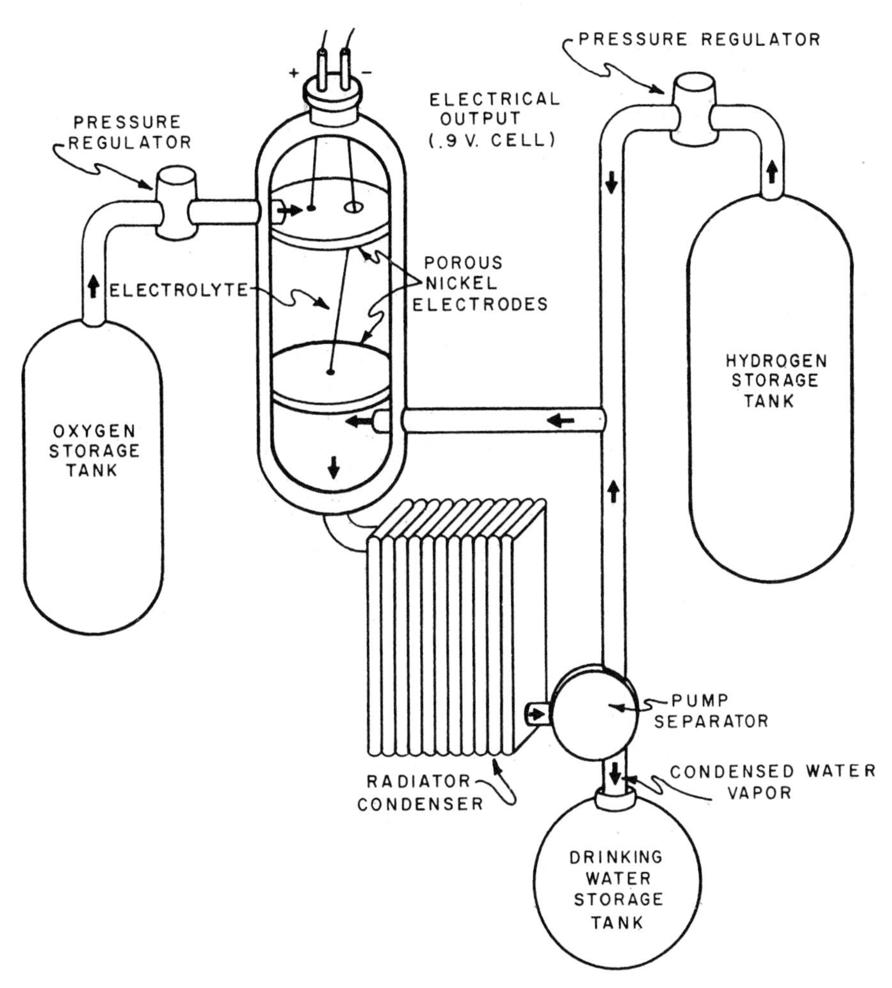

Fig. 24-19 Simplified representation of a fuel cell for manned space vehicles

THERMOELECTRICITY

The thermoelectric principle has been known for some 140 years. This principle, one that converts heat directly into electricity, was observed by T. J. Seebeck in 1821. Two dissimilar materials, such as two different wires, were joined together and heated. At the unheated terminals a small voltage was recorded. The voltages produced were so small that little practical use could be seen for the thermoelectric principle until recently when semiconducting materials were developed. Thermocouples made of semiconductive materials are vastly more efficient than earlier thermocouples.

A review of semiconductor principles will be helpful in understanding thermoelectricity.

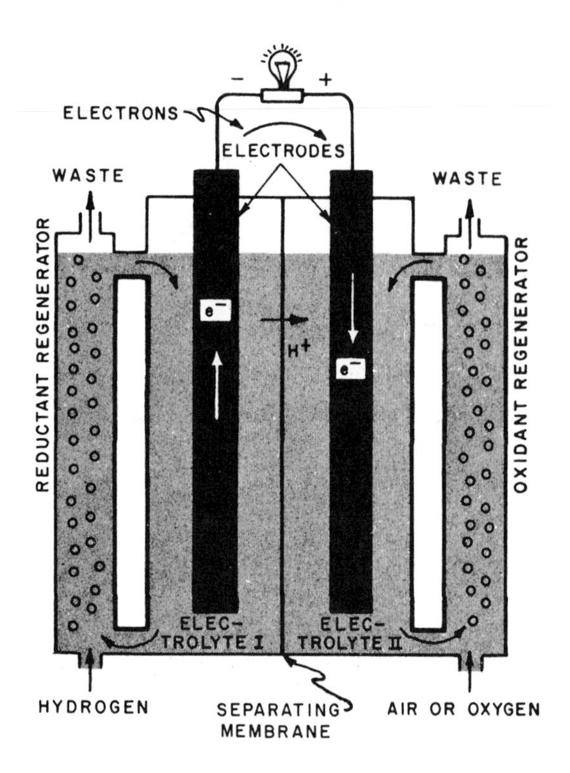

Fig. 24-20 Redox cell with membrane separating the two electrodes

The semiconductor does not have many loose electrons that can become current as do metallic materials. The semiconducting material would, in fact, be an insulator if impurity atoms were not introduced into the material. If the impurity atom that is introduced has more than enough electrons to satisfy the valence-bond needs of the surrounding atoms, there is an electron surplus. A semiconductor of this type is called an N-type semiconductor.

If the impurity atoms introduced on the semiconducting material are basically short of electrons themselves and therefore do not have sufficient electrons to satisfy the valence-bond needs of the surrounding atoms, the material is full of positive holes. The holes can move within the lattice framework of the material, wandering around much the same as electrons do. But the hole is positive; hence these semiconductors are referred to as P-type semiconductors.

Figure 24-21 shows a simplified version of heat-to-electricity conversion in N-type and P-type semiconductors. When heat is applied to one end of the semiconductor (A), the electrons in that end increase their velocities

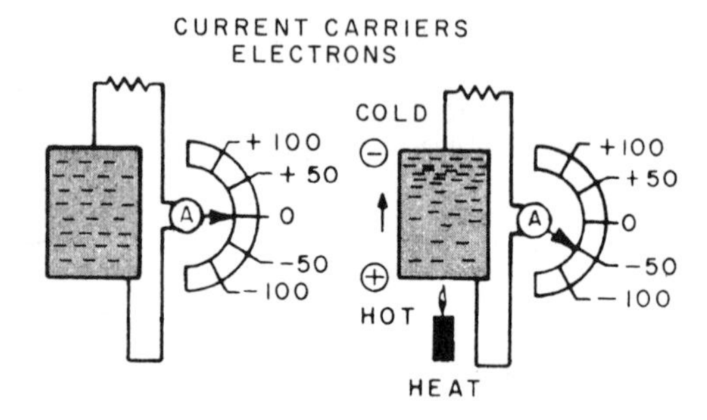

(A) N-Type Semiconductor

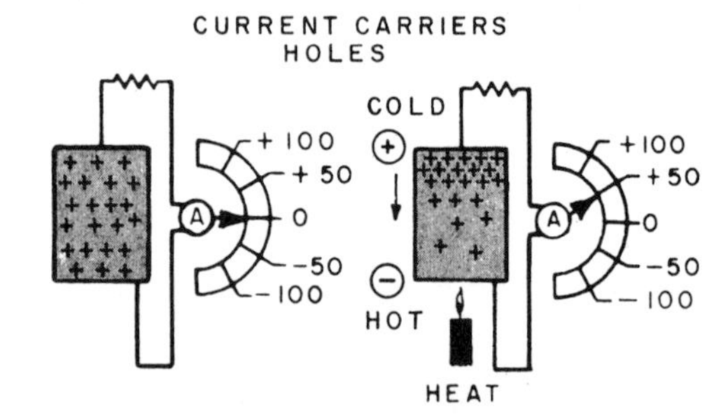

(B) P-Type Semiconductor

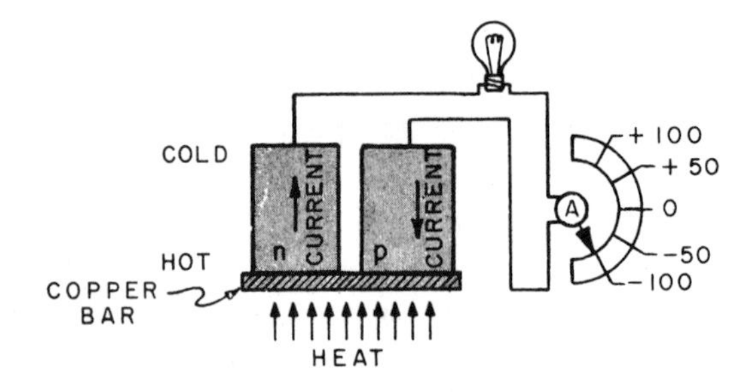

(C) Thermoelectric Generator

Fig. 24-21 Thermoelectric converter

and their kinetic energy. These electrons travel toward the cold end of the semiconductor and pile up. When the circuit is closed, these electrons will flow through the semiconductor and through the wire from negative to positive.

When a P-type semiconductor is used (B), the positive holes pile up at the cold end, and electrons flow from the cold toward the hot end of the semiconductor, from negative to positive through the wire.

A thermoelectric generator (C) connects both the types of semiconductors into an effective producer of electric power. It transforms heat directly into electrical energy.

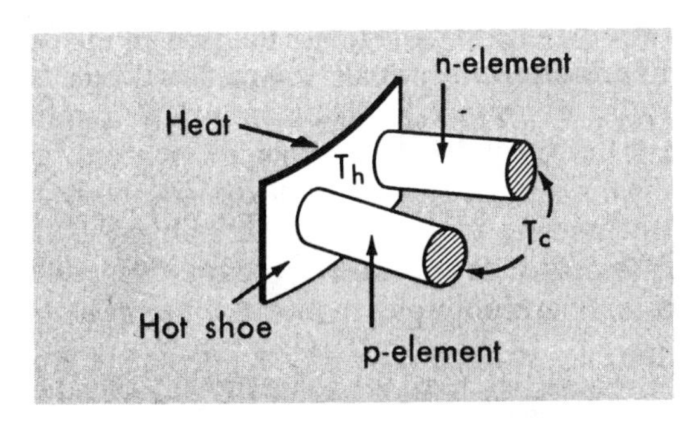

Fig. 24-22 Thermoelectric couple

Advantages

The whole operation of producing electricity from a heated thermocouple is still inefficient, in the range of one percent to five percent, whereas a steam power plant might have an efficiency of 35 to 40%. There are still distinct advantages to this direct conversion principle, however. There are no moving parts, so maintenance and lubrication are not a problem, figure 24-22. The device is silent. Any source of heat can be used to stimulate the flow of electrons and holes: gas and radioisotopes, for example. Thermoelectric devices can, therefore, operate for long periods of time unattended. It is on this point they find their greatest application; powering navigational buoys, remote weather stations in the polar regions, powering floating weather stations, and powering undersea navigational beacons, figure 24-23.

Fig. 24-23 A SNAP generator being installed in an unattended weather station 700 miles from the south pole (Courtesy Atomic Energy Commission)

In the space program, a thermoelectric generator of electricity has several advantages over solar batteries in that it can provide power when the satellite or space capsule is in a shadow and cannot use the sun. Also, when solar cells pass through radioactive belts, they tend to deteriorate. Compared to traditional batteries, the thermoelectric device has the advantage in both length of service and weight.

The thermoelectric converters that are used in the space programs of today have been largely developed under the SNAP program which was initiated by the Atomic Energy Commission in 1956. SNAP stands for Systems for Nuclear Auxiliary Power.

One of the basic concepts of the program is to apply the heat from decaying radioisotopes to the junctions of a thermoelectric mechanism to produce electricity. Another branch of the SNAP program is at work producing electricity by using small nuclear space reactors as a power source.

Radioisotopes and Thermocouples

An isotope of an element has the same number of electrons and protons as the element, but it has a different number of neutrons. Therefore, the chemistry of the isotope and the element is the same but their atomic weights are different. Of the many isotopes, some are unstable and decay, emitting three types of radiation: alpha, beta, and gamma rays. These unstable isotopes are called *radioisotopes.*

Radioisotopes can be used in several different ways. They may be directed at a material with the purpose of altering the material, such as destroying cancerous tissue with radiation treatment. They may be used by directing a fixed source at the object to detect the amount of radiation that is reflected or penetrates the object, such as using radioactive material to gauge a moving sheet of steel. They may be used in medical and agricultural research where small amounts of radioactive material can be introduced into and followed

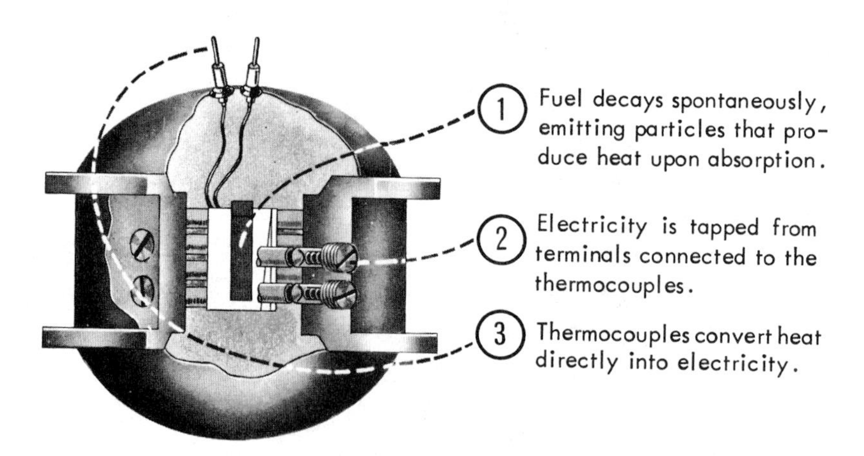

Fig. 24-24 Operation of thermoelectric generator powered by radio isotopes

DESIGNATION	USE	POWER (watts)	GENERATOR LIFE	ISOTOPE
SNAP-3	Demonstration Device	2.5	90 days	Polonium-210
SNAP-3A	Satellite Power	2.7	5 years	Plutonium-238
	Axel Heilberg Weather Station	5	2 years min.	Strontium-90
SNAP-7A	Navigational Buoy	10	2 years min.	Strontium-90
SNAP-7B	Fixed Navigational Light	60	2 years min.	Strontium-90
SNAP-7C	Weather Station	10	2 years min.	Strontium-90
SNAP-7D	Floating Weather Station	60	2 years min.	Strontium-90
SNAP-7E	Ocean-bottom Beacon	7.5	2 years min.	Strontium-90
SNAP-7F	Offshore Oil Rig	60	2 years min.	Strontium-90
SNAP-9A	Satellite Power	25	5 years	Plutonium-238
SNAP-11	Moon Probe	21-25	90 days	Curium-242
SNAP-13	Demonstration Device	12	90 days	Curium-242
SNAP-15A	Military Use	0.001	5 years	Plutonium-238
SNAP-17	Communication Satellite	25	5 years	Strontium-90
SNAP-19B	Nimbus-B Weather Satellite	30	5 years	Plutonium-238
SNAP-21	Deep Sea Use	10	5 years	Strontium-90
SNAP-23	Terrestrial Uses	60	5 years	Strontium-90
SNAP-27	Lunar Landings	60	5 years	Plutonium-238
SNAP-29	Various Missions	500	90 days	Polonium-210

Fig. 24-25

or traced through a system. The energy released in radioisotope decay can be used to produce thermal energy. SNAP devices produce thermal energy as the radiation is absorbed.

Scientists have successfully combined the semiconducting thermocouple and the radioisotope. The first of these devices was completed in 1959. It was the size of a grapefruit and capable of producing 2.5 watts of electricity for a 90-day period, figure 24-24. It used polonium-210 as the radioisotope fuel. Other SNAP devices of various power outputs and longevity have followed, figure 24-25.

The SNAP device is made up of an outer shell that protects the internal components. A radioactive shield is under the shell. If alpha-emitting fuels are used, there is little need for the shield, but if gamma radiation is present, a massive shield may be needed.

Thermoelectric converters or thermocouples that convert a portion of the decay into electricity, and the fuel capsule of radioactive material are located at the very center of the generator.

The radiation particles themselves do not escape from the device. All are trapped inside the fuel capsule where their vast kinetic energy is converted into heat energy. Of the heat energy that is produced, only 5 to 10% is used to produce electricity, the rest eventually escapes to the outer shell and to the media the device is operating in, figure 24-26.

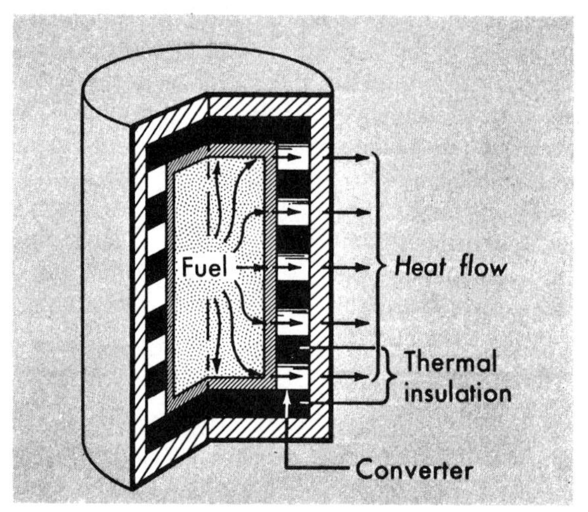

Fig. 24-26 Radioisotope generator showing unavoidable heat loss

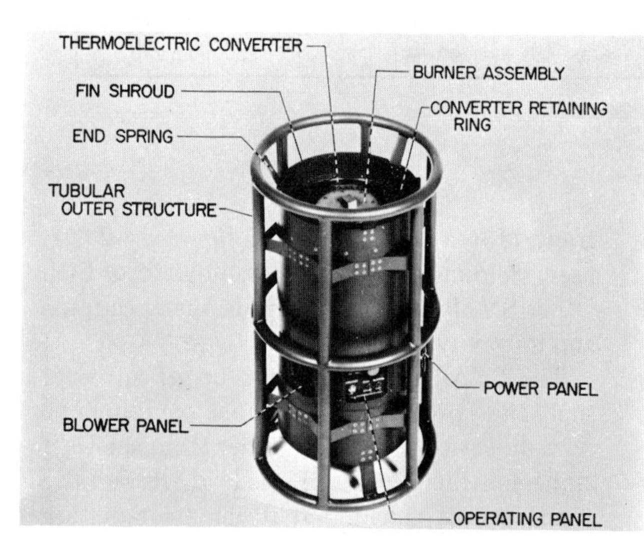

Fig. 24-27 A thermoelectric device that can operate on leaded gasoline, jet fuel, or diesel fuel

Other Heat Sources. Thermoelectric devices that employ more conventional heat sources have also recently come out of the laboratory and into production. These units employ propane or natural gas fuels fired through an enclosed burner. The primary advantage of hydrocarbon fuels versus radioisotope fuels is that of cost; radioisotope fuels are expensive, and hydrocarbon fuels can even compete with conventional batteries. Remote site power can be provided for $2 to $3 per kilowatt-hour using a gas-fired thermoelectric generator, while primary batteries may cost $10 to $18 per kilowatt-hour.

Like the SNAP devices, the gas-fired thermoelectric devices are excellent for remote, unattended service. Of course, periodic replenishment of the fuel is necessary. The Minnesota Mining and Manufacturing Company (3M) is currently producing commercial units that range in output from less than one watt to fifty watts, with larger wattage units being developed, figure 24-27.

THERMIONIC CONVERTERS

Thermionic converters are low-voltage, high-current generators. Their main uses have been in space vehicles. The idea of thermionic conversion, (the generation of electricity from heat) is based on Thomas Edison's discovery that electrons are emitted from a heated cathode. It is called the *Edison effect.* Boiling off electrons from the hot cathode to the cold

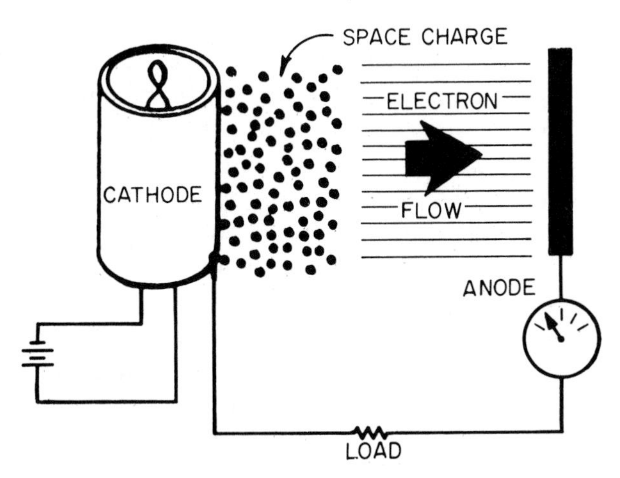

Fig. 24-28 Space-charge effect

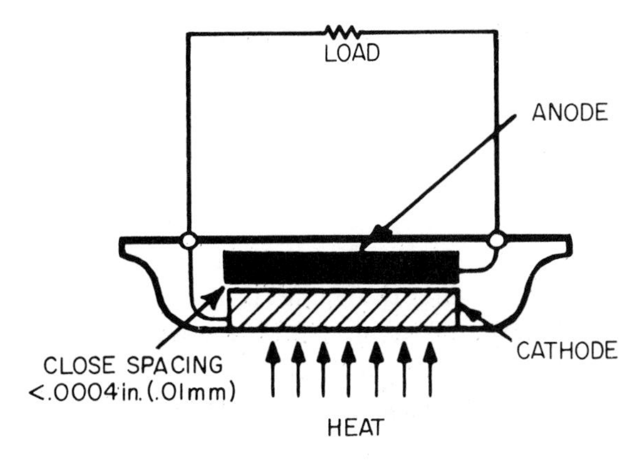

Fig. 24-29 Vacuum close-spaced diode

anode results in an electrostatic charge of negative electrons forming near the cathode. This *space charge* repels other electrons and blocks their flow, figure 24-28. Depending on the methods used to overcome the space charge barrier, three main types of thermionic converters have been developed:

- Vacuum close-spaced diode
- Magnetic triode
- Cesium plasma diode

Vacuum Close-Spaced Diode

This vacuum tube places the cathode and anode very close together to avoid the detrimental effect of the space charge, figure 24-29. Close-spaced diodes were first developed at MIT in 1956 and were called thermoelectron machines. They are heat engines which use the electron gas to deliver useful work as a result of passing through the electrostatic field.

The efficiency of these devices is from 12 to 14% and they weigh the least of all energy converters. However, the cathodes must operate at high temperatures, which shorten their life considerably and reduce efficiencies.

Magnetic Triode

This device reduces the space charge effect by using crossed electric and magnetic fields. A hot cathode and a cold anode are placed in the same plane and close together. Above them an accelerating anode is placed. A battery produces an electric field between the cathode and anode. An external magnetic field is superimposed by operating the entire device between the poles of a large magnet, figure 24-30.

The electron emitted from the hot cathode

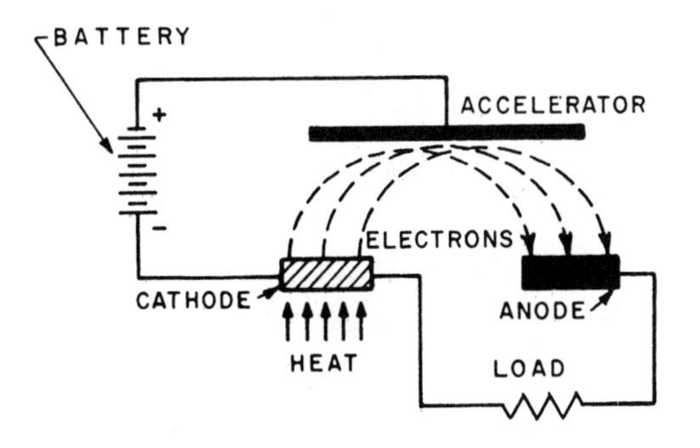

Fig. 24-30 Magnetic triode

speeds towards the accelerator. But the magnetic field deflects it toward the surface of the anode. During its acceleration, the electron gains kinetic energy from the electric field. When it falls back to the anode, it returns that energy to the field. The result is that the energies of the electrons arriving at the anode are the same as they were when they left the cathode. In theory then, the magnetic triode should produce more power than the close-spaced diode. However, due to a number of electron losses, this is not the case.

Cesium Plasma Diode

Use of an ionized gas is another way of counteracting the space charge. In diodes filled with cesium plasma, the positive ions neutralize the negative space charge and allow the electrons to flow toward the anode, figure 24-31. A single ion can effectively cancel several hundred electrons.

As in the vacuum close-spaced diode, the cathode must operate at exceedingly high temperatures which shorten its life. Other fabrication difficulties are involved with the corrosiveness of the cesium vapor. Efficiencies of 10 to 25% have been achieved.

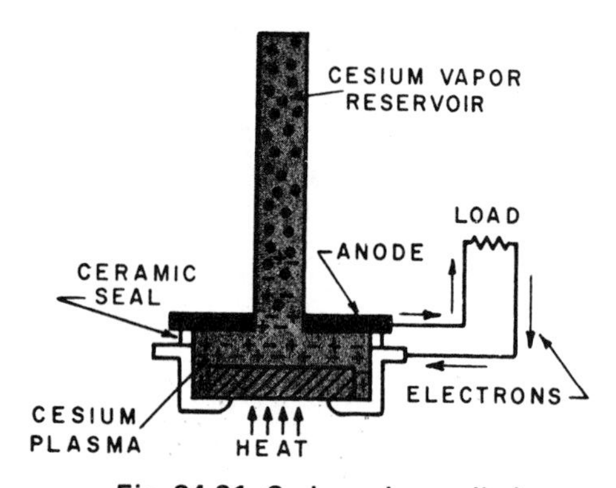

Fig. 24-31 Cesium plasma diode

Further Research

Intensive thermionic research is underway with the goal being the development of reliable converters for space use. These thermionic devices can be heated by the sun, by decaying radioisotopes, or by the heat of a fission reactor. If scientists could simply wrap the thermionic converter emitter around a reactor fuel element, the result would be an ideal combination. However, many problems are involved in developing materials that can withstand the extremely high operating temperatures, 3092°F (1700°C), in the physical construction of the device, and in the nuclear physics factors that must be considered. Much work remains to be done in the area of thermionic conversion devices before practical applications can be made.

THE NUCLEAR BATTERY

The nuclear battery is capable of directly converting the nuclear energy of a strong beta emitter, like strontium 90, directly into electricity. The device is simply a central rod that is coated with a beta-emitting radioisotope, the source of electrons. The rod is in a vacuum-sealed container and the beta particles (electrons) cross the vacuum gap to a metal sleeve that acts as a collector. The sleeve is connected to the load. Current flows from the sleeve, through the load, and back to the central rod, figure 24-32.

Space charge effects do not bother the nuclear electrons. They are leaving the radioisotopes with a million times more kinetic energy than electrons leaving a thermionic surface.

While nuclear batteries apply voltages of 10,000 to 100,000 volts, the electron flow is only a few millionths of an ampere.

A NUCLEAR BATTERY

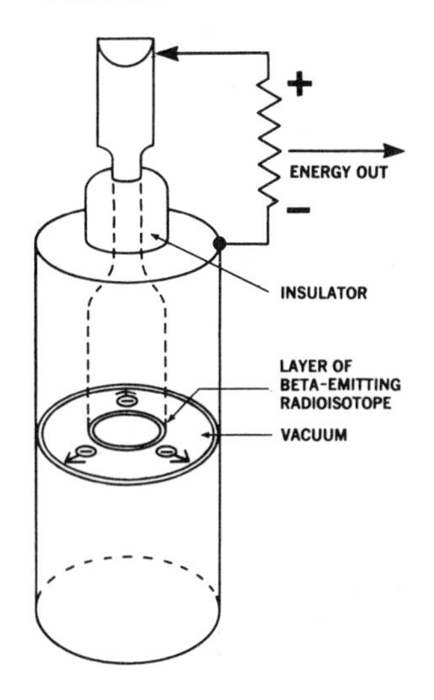

Fig. 24-32 A nuclear battery

MAGNETOHYDRODYNAMIC GENERATOR

The magnetohydrodynamic power generator makes use of the basic principles of electrical generation that were discovered by Faraday in the early nineteenth century. He discovered that moving an electrical conductor through a magnetic field, or vice versa, induces a voltage. Steam, water, and gas turbines are often used in this way, applying their energy to rotate a conductor through a magnetic field. In the magnetohydrodynamic (MHD) generator, a pressure-driven conductive gas replaces the copper wire of the armature that moves through the magnetic field of the normal generator, figure 24-33.

The idea of a gaseous conductor sounds

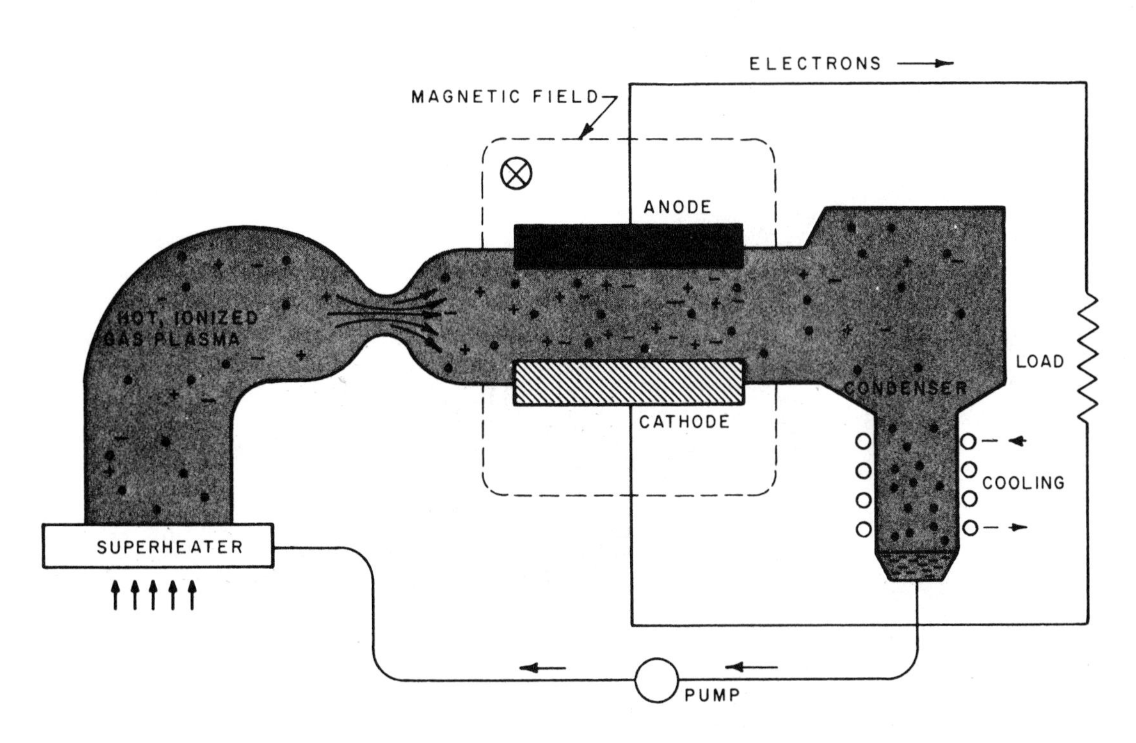

Fig. 24-33 Magnetohydrodynamic (MHD) generator

simple at first, but in order for a gas to be an effective conductor of electricity it must be ionized, and this can only be accomplished by intense heat. In the ionization process, electrons boil off and become free electrons, leaving the gas atom with a positive electrical charge; it is ionized. These free electrons drift in the ionized atoms, but many are free at any given time. Many free electrons are drawn to and picked up by the anode and delivered to the load.

The hot gaseous ionized fluid is referred to as *plasma.* Simply burning chemical fuel such as powdered coal, oil, or natural gas with air does not produce temperatures sufficient enough to promote the slight ionization that is necessary. To solve this problem a gas is seeded with a material that will ionize easily, such as potassium. When seeded with potassium, ionization will result at ordinary oxygen combustion temperatures of 2500–3000°K (2227–2727°C). To insure efficiency and to obtain the high temperatures needed, the air is often oxygen-enriched and preheated.

Temperatures of the magnitude required in an MHD generator are destructive to ordinary turbine generators, but in the MHD generator there are no moving parts that the gaseous stream is directed toward. In the actual production of electricity, there are no moving parts at all.

The combustor itself is very similar to a rocket engine, figure 24-34. The hot plasma exits from the nozzle into a duct of increasing diameter. A magnetic field surrounds this duct. The motion of a conductive plasma gas through this field induces voltage across the electrodes. In actual practice the electrodes are segmented, each segment being separately attached to the load to minimize the Hall

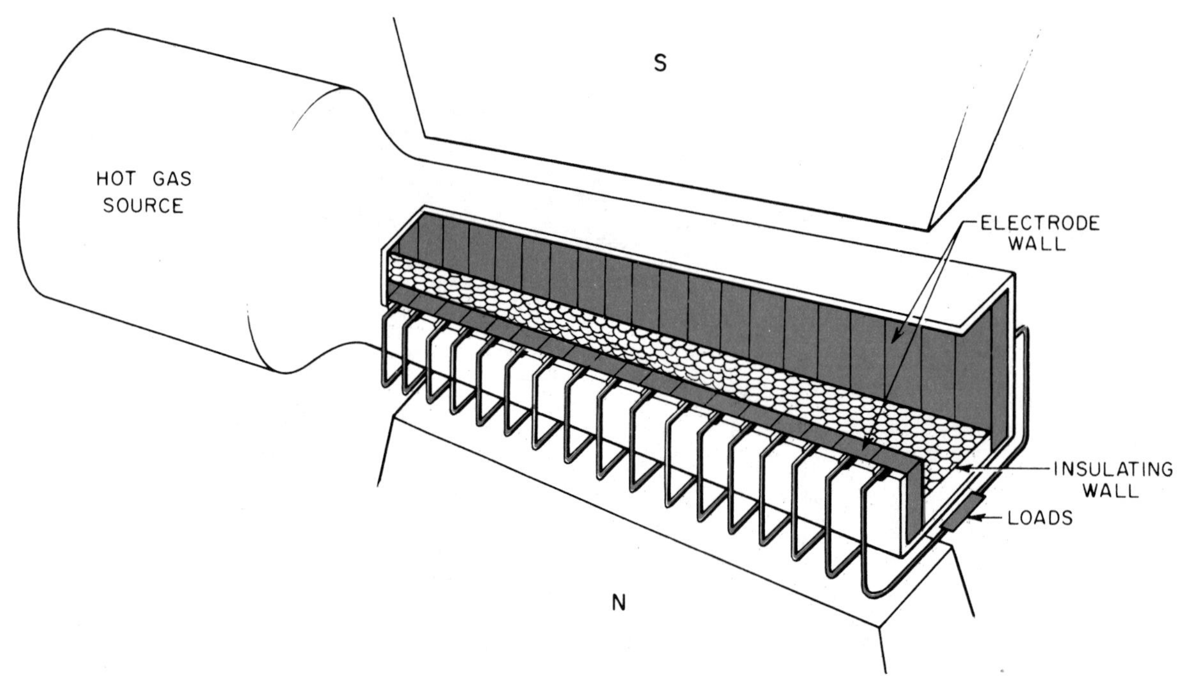

Fig. 24-34 Drawing of MHD generator, normal segmented hookup

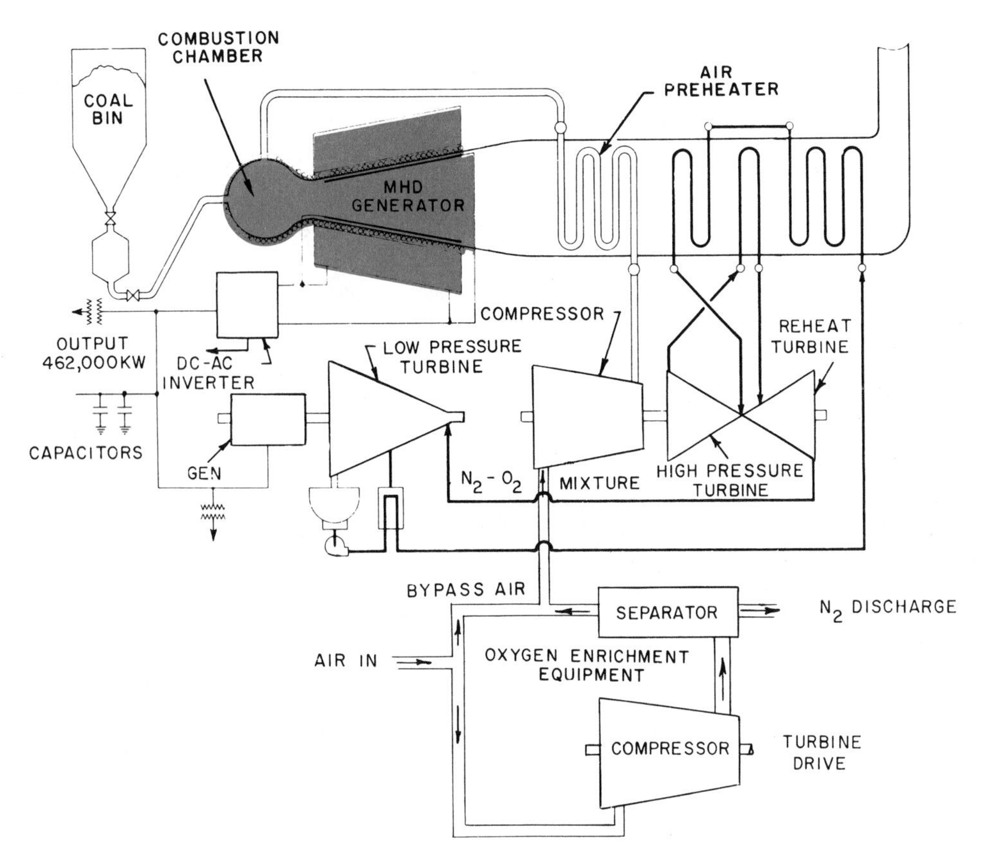

Fig. 24-35 Design for a coal-burning, open-cycle MHD power plant

effect. This is the tendency of the conductor to set up a magnetic field around itself, which cancels out the force of the field coil.

The MHD generator is ideally suited for producing large quantities of electricity at central power stations. When these units are installed they will incorporate several other principles and machines to attain efficiencies of an expected 50%, ten percent higher than conventional plants are now capable of achieving. To attain high efficiency, hot gases cannot simply be exhausted. If the hot plasma is exhausted, it would represent a tremendous loss. A portion of the heat is used in a regenerator to preheat the incoming air. Another portion of the hot gases is used in a steam cycle whose turbine turns both a compressor and a conventional generator. Even the seed material is reclaimed from the ash of the exhaust gases. This type of operation is referred to as an open cycle since the plasma gas is eventually exhausted from the mechanisms when as much energy as practical has been derived from it, figure 24-35.

The possible use of nuclear reactor power for MHD generators presents another area of promise. However, the temperatures attained are too low to cause ionization. Either of two routes can be taken to solve this limitation; reduce the temperature at which an MHD generator operates, or build nuclear reactors that produce greater heat intensities.

Current finds indicate that future nuclear-powered MHD generators will be a closed-cycle design. The plasma, probably argon or helium, is heated by the reactor but does not interact with it. The ionized gas passes through the duct but is not exhausted from the system. It is drawn off at the discharge end of the duct, reheated and expanded through the nozzle again. Safety factors of radioactive contamination present problems for the design of these generators.

At present the prime area of developement of MHD generators is directed toward the generation of electrical power at the central power station. Large MHD generators can produce huge amounts of power at high efficiency. The intense heat required to ionize gases always presents the problem of developing materials and components that can take the heat for long periods of time. Concentrated work on MHD generators has shown much progress in only a few years.

REVIEW QUESTIONS

1. What common direct energy conversion device has served man for many years?
2. What energy form is produced by most all direct energy converters?
3. What are the general requirements for space exploration electronic power supplies?

Solar Cell

4. Explain how a power source such as coal can be traced to the sun.
5. Name three early devices for converting the sun's energy directly into electricity.
6. What group of men is responsible for the latest improvements in the solar battery?
7. What is the efficiency of present day solar batteries?
8. What applications have been found for the solar battery?
9. What is a semiconductor?
10. How can a semiconductor be made conductive?
11. What type of semiconducting material is used in the solar battery?
12. What is the arsenic doped region called?
13. What is the boron doped region called?

14. What is the PN-junction?

Thermophotovoltaic Converter

15. What is the source of the photons that excite the semiconducting material of the thermophotovoltaic converter?
16. Where and when was this converter introduced?

Fuel Cell

17. How does the efficiency of the fuel cell compare with all other non-nuclear energy conversion systems?
18. Does a fuel cell store the energy of a fuel or does it furnish power as it consumes a fuel?
19. What potential uses are seen for the fuel cell?
20. Has the fuel cell seen any practical application to date? Explain.

Nuclear Battery

21. Explain the basic operating principle of the nuclear battery.
22. Why are space charge effects of little consequence to a nuclear battery?

Thermoelectricity

23. Explain the basic principle of a thermocouple.
24. Why are semiconducting materials used in modern thermoelectric devices?
25. What characteristics of a thermoelectric device makes it suitable for use at remote, unattended locations?
26. How can radioisotope materials be used for thermoelectric devices?
27. What is the SNAP program?

Thermionic Converter

28. Explain the term space charge. What has it to do with the problems of thermionic converters?
29. Describe the basic principle upon which the vacuum close-spaced diode operates.
30. What is the major disadvantage of the vacuum close-spaced diode?

31. How does the magnetic triode solve the problem of space charge?
32. On what principle does the cesium plasma diode operate?

Magnetohydrodynamic Generator

33. Explain the principle on which the MHD generator operates.
34. What is plasma?
35. What is the difference between the open cycle principle and the closed cycle principle?
36. What difficulties are encountered in MHD generator design?

SECTION 8 ENERGY FROM THE EARTH—FOSSIL FUELS

Unit 25 Coal

OBJECTIVES After completing this unit, the student should be able to:

- Discuss the formation of coal.
- Describe the several types of coal.
- Discuss coal mining, processing, and transportation.
- Discuss the environmental problems caused by coal mining and coal burning.
- Explain the concepts of coal liquification and coal gasification.

Coal is known as a fossil fuel. Its formation on the earth is the result of plants and animals that lived and died between one and 440 million years ago. When ignited it will burn and release heat energy. It is thought that the plants that now make up coal deposits grew in swampy areas of the earth. They died and accumulated over a period of time. The decaying plants first hardened into a material called peat. As time passed, sand and minerals formed into sedimentary rocks covering the peat and the increasing weight caused the peat to turn into coal.

Coal types vary according to how long they have been underground and how much pressure has been exerted on them. The harder the coal the higher the carbon content and the more efficient the burn. Lignite is the first stage of coal development; it is dark brown in color. The texture of wood is sometimes still distinct on its surface. Subbituminous coal is next in line followed by still

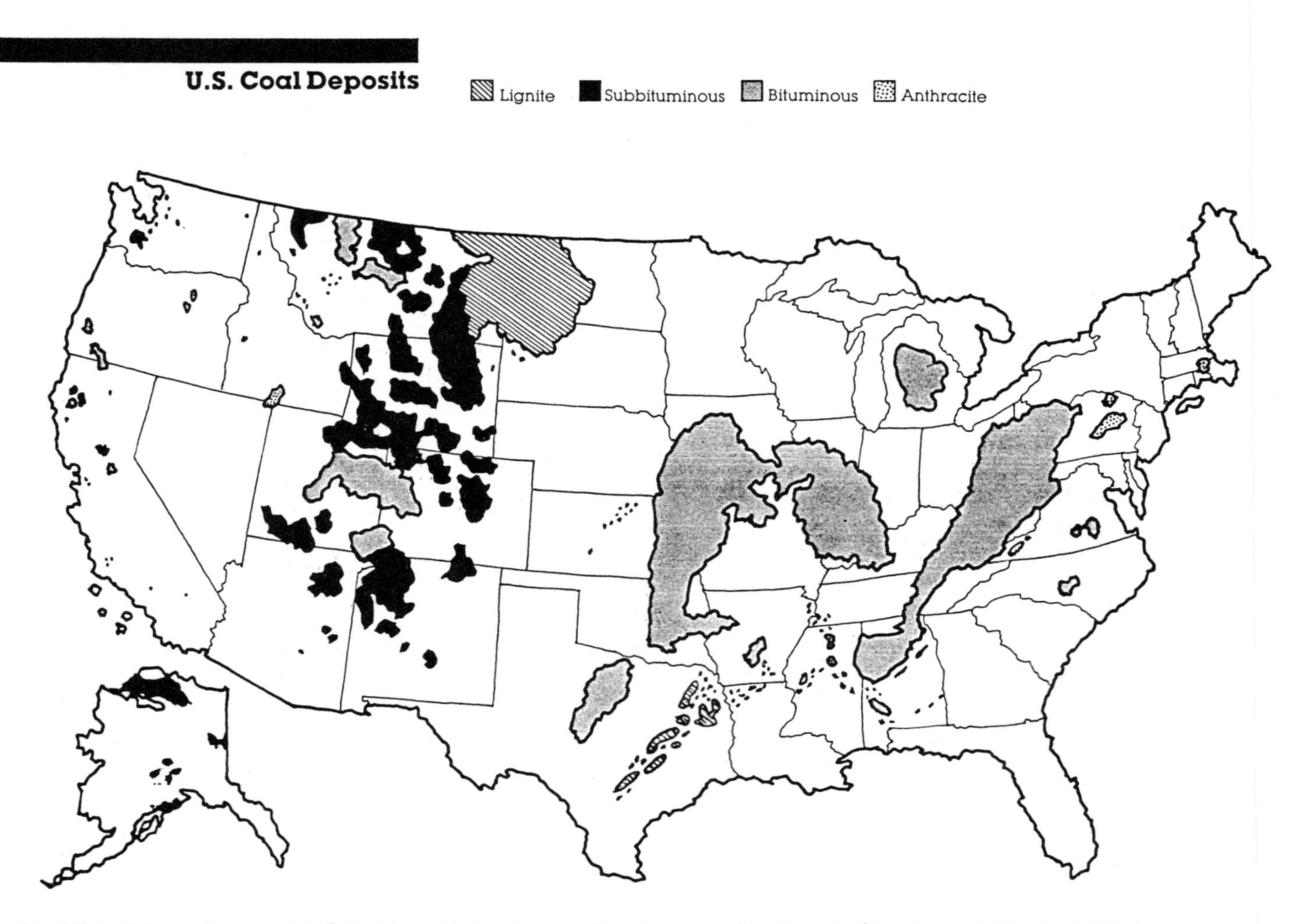

Fig. 25-1 U.S. coal deposits of lignite, subbituminous, bituminous, and anthracite (Courtesy of Standard Oil Company of Indiana)

harder bituminous coal. These two last are often called soft coal because they crumble easily. The hardest and oldest of coals is anthracite, figure 25-1.

In terms of their heating value measured in Btus (British thermal units—the amount of heat required to raise one pound of water one degree Fahrenheit), anthracite coal is the best, 13,000 Btus per pound. Bituminous coal has 10-13,000 Btus per pound; followed by subbituminous, 8,000 Btus per pound; and lignite, 6,000 Btus per pound. Bituminous coal is the most plentiful type of coal that is economically recoverable.

It is not known when man began to burn coal but it is recorded that the Chinese had a coal industry as early as 300 AD. In Europe, coal mining was underway in the 1200s. For most uses wood was the preferred fuel but by the 1600s wood was becoming so scarce that coal was the only alternative. England was the world's leading producer. Coal was often con-

Fig. 25-2 Machines at work in an underground mine (Courtesy of National Coal Association)

verted into coke for many industrial processes such as smelting metals. Coal was the fuel for the industrial revolution. It served as the fuel for steam engines that powered factories and the heat for industrial processes. Soon coal was also used in steam ships and railroad locomotives. Coal was king in industry, transportation and home heating for years, until the development of the petroleum industry.

MINING COAL

Coal layers are called seams or veins. They may be close to the earth's surface or buried very deep. Some seams are thin, perhaps an inch thick while others have been known to be 400 feet thick. Movements in the earth's crust have shifted some coal seams up so the layers can be seen in hillsides as outcroppings.

Fig. 25-3 Coal mining is a dangerous profession and the men and women who work the mines must be constantly alert. (Courtesy of National Coal Association)

How deep the coal is located will determine how it is to be mined. Underground mines are for coal that is more than 120 feet in the ground. These mines are the most expensive and the most dangerous. Even though the pick and shovel days are over and machines now do the backbreaking work that men used to, underground mining is still rugged work and a dangerous occupation, figure 25-3. Safety regulations are strict but falling rock and explosions are an ever-present hazard. Underground shaft mines generally have at least two shafts, one to provide access for the miners and the other for hauling out the coal. Both shafts provide ventilation to the mine. If the shafts go straight down (the deepest mines) it is called a shaft mine. Slope mines enter the earth at an angle, often following a rock seam until they meet the coal seam. A drift mine will enter parallel to a coal seam that is usually outcropping on a hillside.

Strip mining (surface mining) is done when the coal is relatively near the surface of the earth—no deeper than 120 feet, figure 25-4. The overburden, composed of earth and rock

Fig. 25-4 Surface mining of coal (Courtesy of Ashland Oil Inc., by American Petroleum Institute)

Fig. 25-5 Bucket wheel excavator used in surface mining (Courtesy of Environmental Protection Agency by American Petroleum Institute)

that is directly over the coal seam, must be removed. The land is cleared, then explosives are used to loosen the overburden. Power shovels as high as a twenty-story building remove the overburden down to the coal seam. A bucket wheel excavator is especially effective in mining to minimize the amount of reclamation necessary, figure 25–5. Smaller power shovels remove the coal seam, loading the coal into huge trucks which take the coal to processing plants for cleaning, washing, and grading.

In past years strip mining left acres of land as spoil banks—scarring the earth and rendering it virtually useless. Today, surface mining is done in such a way that the land can be reclaimed to make it attractive and useful, much as it was before the coal was removed, figure 25–6.

The upper layer of the earth (top soil) is removed and stored. After mining is completed

Fig. 25–6 Farming operations returned to reclaimed land(Courtesy of National Coal Association)

the overburden is graded and returned to its original contours and then the life-giving top soil is returned and the land is fertilized and reseeded. Reclamation and mining are done at virtually the same time. Today, 60% of the coal mined is done by surface mining.

TRANSPORTATION OF COAL

Coal is shipped within the country by barge, rail, or truck. Barges are the most economical method of transportation but this shipping method is limited to delivery at coastal or river ports, figure 25-7. Rail transportation accounts for about two-thirds of coal shipments in the United States. Most coal trains are called *unit* trains in that every rail car in the train carries coal. For example, the entire shipment goes from the mine processing plant to an electrical utility several hundred miles away. In figure 25-8, an automated coal-handling system offloads coal from railroad

Fig. 25-7 Coal barges on the river (Courtesy of American Petroleum Institute)

Fig. 25-8 Automated coal handling system at a power generating station can unload and store coal at the rate of 3,000 tons per hour (Courtesy of Wisconsin Power and Light Company)

cars at a generating station. This system includes a rotary railroad car dumper, a 1200-foot conveyor belt and a stacker-reclaimer. The stacker-reclaimer delivers the unloaded coal to storage areas on both sides of the conveyor at the rate of 3,000 tons per hour. Truck transportation is only economical for small shipments and short distances.

COAL USAGE AND RESERVES

There are several important reasons why coal continues to be important as an energy source. In the first place, it is our most abundant fossil fuel. Studies have shown that if consumption continues at its present rate there are enough coal reserves to last for 300 to 400 years; there are 700 billion tons of known reserves. Secondly, world reserves of petroleum and natural gas are being depleted much more rapidly than coal; therefore, it's understandable why coal remains an important source of energy. Coal supplies 25% of the world's energy today and this percentage is expected to rise in future years.

Only about one percent of the coal produced is sold for retail uses such as home heating. What about the rest? Seventy-eight

Fig. 25-9 Coal-burning power plants produce electricity. (Courtesy of National Coal Association)

percent of the coal goes to electric utilities where it is crushed and burned to heat water making steam to drive the steam turbines that power the electrical generators, figures 25-9 and 25-10. Eleven and one-half percent of coal production is converted into coke for smelting, while nine and one-half percent is consumed in general industrial use.

Coal is very competitive costwise with petroleum and natural gas but its impurities do present environmental problems. Sulfur is usually present in coal and is very undesirable. When sulfur is burned, a poisonous gas, sulfur dioxide, is produced. Special scrubbers in the utilities remove much of the sulfur dioxide before it reaches the outside air. Fly ash particles coming out of a smoke stack represent a hazard; these are controlled by filters, scrubbers, or electrostatic precipitators. Fly ash particles contain traces of lead, cadmium, and arsenic, and also block out sunlight. Acid rain and the greenhouse effect are to some extent caused by burning coal. Keeping the environment clean when using coal is expensive, however. It has been estimated that about forty percent of the cost of a new coal fired electric power plant goes to equipment and technology required to meet environmental protection requirements.

Fig. 25-10 Coal on a conveyor on the way to the crusher before it is burned at the power plant (Courtesy of National Coal Association)

Fig. 25-11 Research model of a coal gasification unit (Courtesy of Gulf Oil Corporation by American Petroleum Institute)

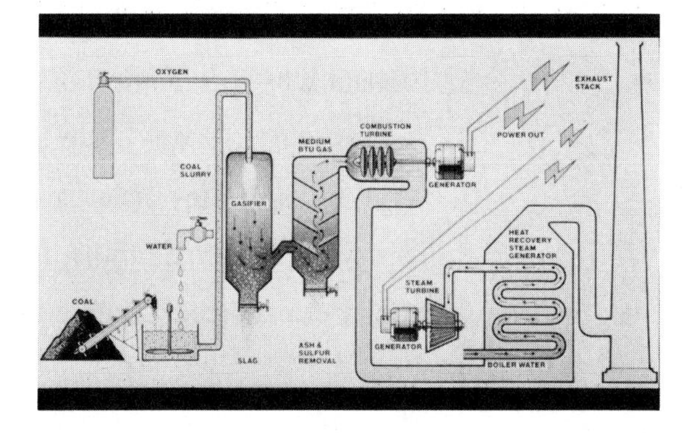

Fig. 25-12 Diagram of an electric power generator using coal gasification (Courtesy of Texaco, Inc., by American Petroleum Institute)

COAL GASIFICATION AND COAL LIQUIFICATION

Coal can be a versatile fuel–it does not have to be burned as a solid chunk or a powder. It is possible to change coal into a liquid that can be burned much the same as petroleum products or changed into a gas similar to natural gas.

Coal liquification is done by first crushing, drying, and grinding the coal. The coal is then combined with liquid to form a slurry that is heated and pressurized. When hydrogen is added the coal dissolves. Mineral matter and organic solids are removed as are gases. The result is a low-sulfur liquid fuel that can be burned in power plants designed to burn oil.

Coal gasification is designed to burn a small portion of the carbon in the coal, figure 25–11. Usually all of the carbon is burned or oxidized in normal burning. Coal gasification relies on carefully controlling the amount of oxygen admitted to the burning process and also controlling the temperature and time of burning, figure 25–12. The controlled oxidization of the coal results in the formation of hydrogen and carbon monoxide gases. In the presence of a catalyst, methane gas is then produced which is combustible.

Both processes are under intense development by the coal industry. The concepts are not new but the technologies to liquify and gasify coal are still developing. However, plants are being built and run successfully with the products of the two methods.

REVIEW QUESTIONS

1. What type of coal is the oldest and hardest?
2. How many Btus of energy does a pound of bituminous coal contain?
3. Explain how coal fueled the industrial revolution.

4. Briefly describe the two principal methods of mining coal.
5. Explain why reclamation of land is important in strip mining.
6. Discuss why coal usage is predicted to rise in years to come.
7. What industry uses the greatest amount of coal?
8. Briefly explain coal liquification.
9. Briefly explain coal gasification.

Unit 26 Petroleum

OBJECTIVES After completing this unit, the student should be able to:

- Discuss the importance of oil as an energy source.
- Explain how petroleum was formed.
- Briefly discuss oil exploration and drilling.
- Outline the refinery process.
- Explain the importance of petrochemicals.
- Discuss the potential of oil shale.

Oil is the main energy source for our modern industrial society. Some 43% of the energy the United States uses each year comes from oil (petroleum). This can be broken down as:

- The 120 million automobiles registered in the country consume 54% of the oil products used.
- Residential and commercial use is 12%.
- Industry uses 25%.
- Electric utilities use 9% of the oil.

Our country uses about 17 million barrels (42 gallons per barrel) of oil each day; of this we import about 6.8 million barrels. We are very aware that oil is vital to our nation's well-being but, at the same time, we cannot fail to know that the earth's oil supply is limited, figure 26-1. This is why, as we enjoy the benefits of oil, we must be conservative and wise in its use. At present, the United States represents only 6% of the world's population yet we consume one third of the world's energy. Each person in our country consumes energy equal to 60 barrels of oil each year, figure 26-2.

Petroleum (crude oil) that comes from the earth is made of many types of hydrocarbon (hydrogen and carbon combinations) molecules jumbled together. It began forming millions of years ago. The most accepted theory of its formation is that oil is the result of organic material formed by prehistoric remains of plants and animals that decayed in the ancient seas of the earth. Layer upon layer of

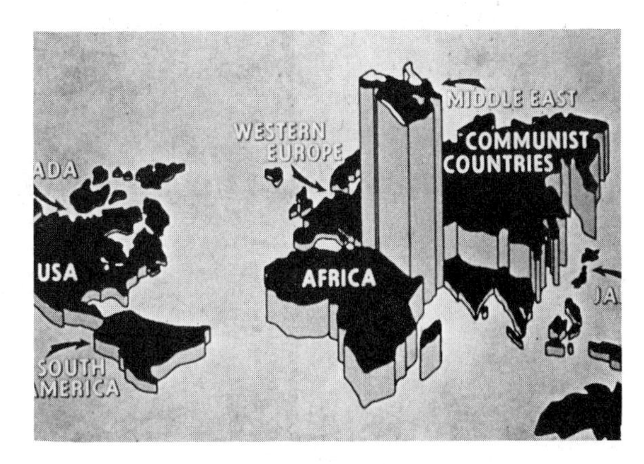

Fig. 26-1 Oil reserves of the world (Courtesy LA Offshore Oil Port, Inc., by American Petroleum Institute)

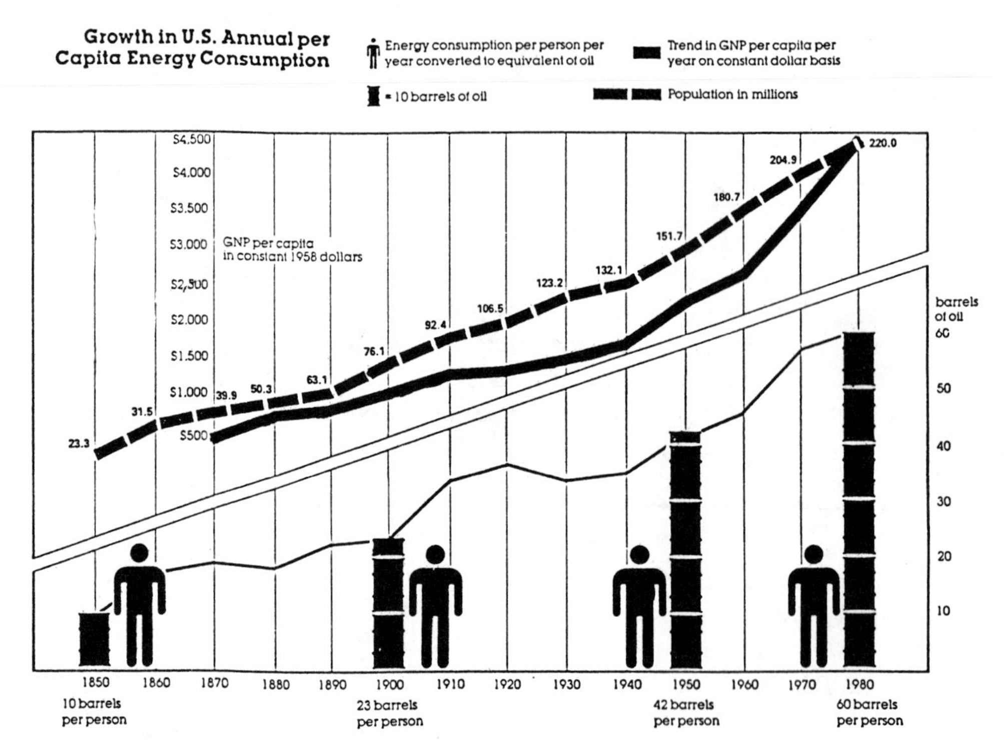

Fig. 26-2 Growth of energy consumption in the United States (Courtesy of Standard Oil Company of Indiana)

plant and animal matter accumulated along with mud and silt and pressure built up on the lower layers. With the increasing pressure came heat and the chemical, bacterial, and radioactive action which reformed the accumulated matter into the compounds of hydrogen and carbon that we refer to as petroleum.

Shifting layers of the earth's crust enabled some oil to drift to the surface where it could be easily collected, so it is very likely that petroleum has been known for centuries. As a matter of fact, it is known that oil was used in ancient Egypt and Greece. The Chinese even *drilled* successfully for oil in 300 BC. These early, limited uses of oil were for lighting, grease, and caulking ships.

In colonial times candles were the main source of illumination along with whale oil. Since whaling ships were well into the process of depleting the ocean's supply of whales, the search was soon on for other sources of illumination. It was already known that petroleum could serve this purpose but gathering oil from the few spots where it naturally broke through the earth's surface was limited, slow, and inefficient.

Edwin L. Drake, financed by a group of businessmen, began to drill for oil near an old oil spring in Titusville, Pennsylvania, figure 26-3. Drake wanted to recover petroleum in

Fig. 26-3 Titusville, Pennsylvania—site of the first commercial oil well (Courtesy of American Petroleum Institute)

large quantities so it could be refined into kerosene for illumination. In 1859, a dark green liquid rose to the top of the 69½ foot hole; Drake had struck oil! The oil rush was on; thousands of oil wells were being drilled by the emerging petroleum industry. Since the automobile was not yet on the scene the early refiners concentrated on kerosene for lighting. Oil fields were soon discovered and developed in Colorado, Texas, West Virginia, California, Ohio, Illinois, Oklahoma, Louisiana, and Kansas. Today, oil and/or natural gas are produced in 32 states, figure 26-4.

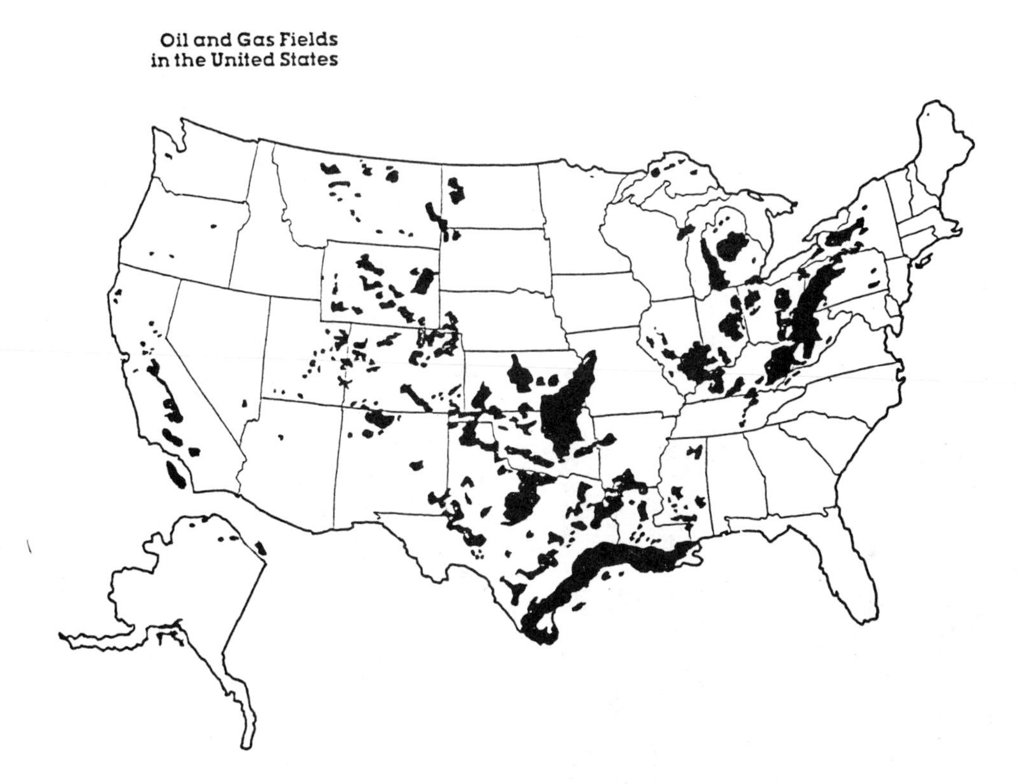

Fig. 26-4 Oil and gas fields in the United States (Courtesy of Standard Oil Company of Indiana)

OIL EXPLORATION

Most of the sedimentary basins of the earth where oil occurs are well known, but even this is no guarantee that every well that is drilled will have oil at the other end. A geophysicist continues the search aided by a magnetometer, gravimeter, and seismograph that more thoroughly map the underground structure of the earth. If the reports indicate a good chance for oil, test drilling is done and the rock formations are examined by geologists. Then, actual drilling of a well is done because it is the only way to be certain that oil is present. Even when a site looks very promising, only one in ten wells will strike oil and only one in fifty of these will strike oil that flows in quantities that produce enough petroleum to be commercially practical.

Alaska is the site of our latest petroleum and natural gas discoveries. Exploration, drilling, and transporting oil from the hostile arctic areas pose many costly problems but the quantities of oil resources are enormous, 42 billion barrels of oil and 181 trillion cubic feet of natural gas. The construction of the Trans-Alaska oil pipeline, figure 26-5, brings oil from the northern-most production areas

Fig. 26-5 The trans-Alaska pipeline (Courtesy American Petroleum Institute)

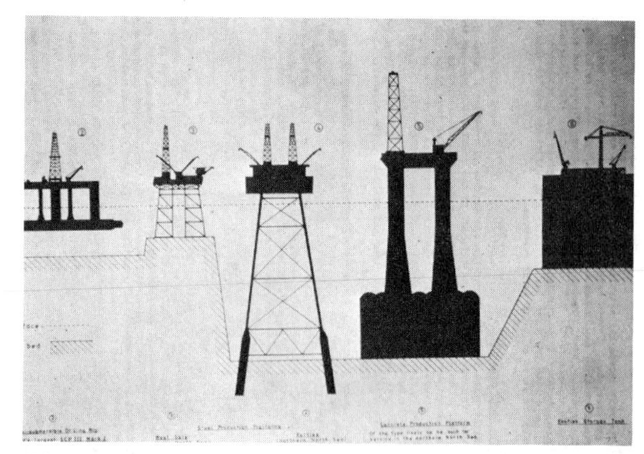

Fig. 26-6 Diagram and maps of offshore drilling rigs (Courtesy of American Petroleum Institute)

to shipping ports in southern Alaska where it is loaded onto tankers.

OIL DRILLING

Drilling for oil is a financially risky operation involving a drilling crew, a drilling rig, expensive rotary drill bits, and thousands of feet of drill pipe. The average depth of drilling is 5,000 feet; the average cost of a well is $380,000. As the rotary drill grinds away the rock and goes deeper into the earth, pipe must be added—length by length. Water, clay, and additives form a mud that is pumped into the drill hole to cool and clean the rotary bit. In recent years much drilling is done on drilling platforms that are located at sea on the continental shelf, figure 26-6. The average depth of offshore wells is 10,000 feet; the average cost of an offshore well is $2,545,000. Part of this cost is because drilling is carefully watched and monitored and many safeguards are installed to prevent accidents and oil spills.

TRANSPORTATION OF OIL

Transporting petroleum to the refineries is done by pipelines, barges, highway tank trucks, and railroad tank cars. Most domestic

Fig. 26-7 Ocean oil tanker (Courtesy of LA Offshore Oil Port, Inc., by American Petroleum Institute)

oil is delivered to the refinery by pipeline. More than 218,000 miles of petroleum pipelines deliver petroleum and petroleum products throughout the United States. When oil is imported from overseas it arrives by ocean tanker, figure 26-7.

REFINING OIL

There are 300 oil refineries in the country today and each is a huge complex of pipes, columns, and steel structures. Nothing seems to be moving, but the refinery is operating 24 hours a day to convert the hydrocarbons that make up the crude oil into many products. Not all crude oil is heavy and black, some is amber, green, or brown in color; some oil may be almost like water while some is thick as molasses, figure 26-8.

Fig. 26-8 Crude oil: samples of several types (Courtesy of Standard Oil of California by American Petroleum Institute)

In most cases, fractional distillation is the first process in refining, figure 26-9. Crude oil is heated to 725°F and then released into a vertical tower. Many of the various hydrocarbons vaporize under this temperature and begin to rise from the bottom of the tower toward the top. At different levels in the tower the hydrocarbon molecules cool and condense and are collected. As expected, the heavy molecules condense near the bottom and the lighter molecules condense near the top. From the bottom up the crude is separated as follows—asphalt and heavy fuel, lubricating oil, fuel oils, kerosene, and gasoline. Other lighter gases that do not condense are carried off for further processing.

Fractional distillation by itself does not produce enough gasoline to meet the demand so other processes such as thermal cracking,

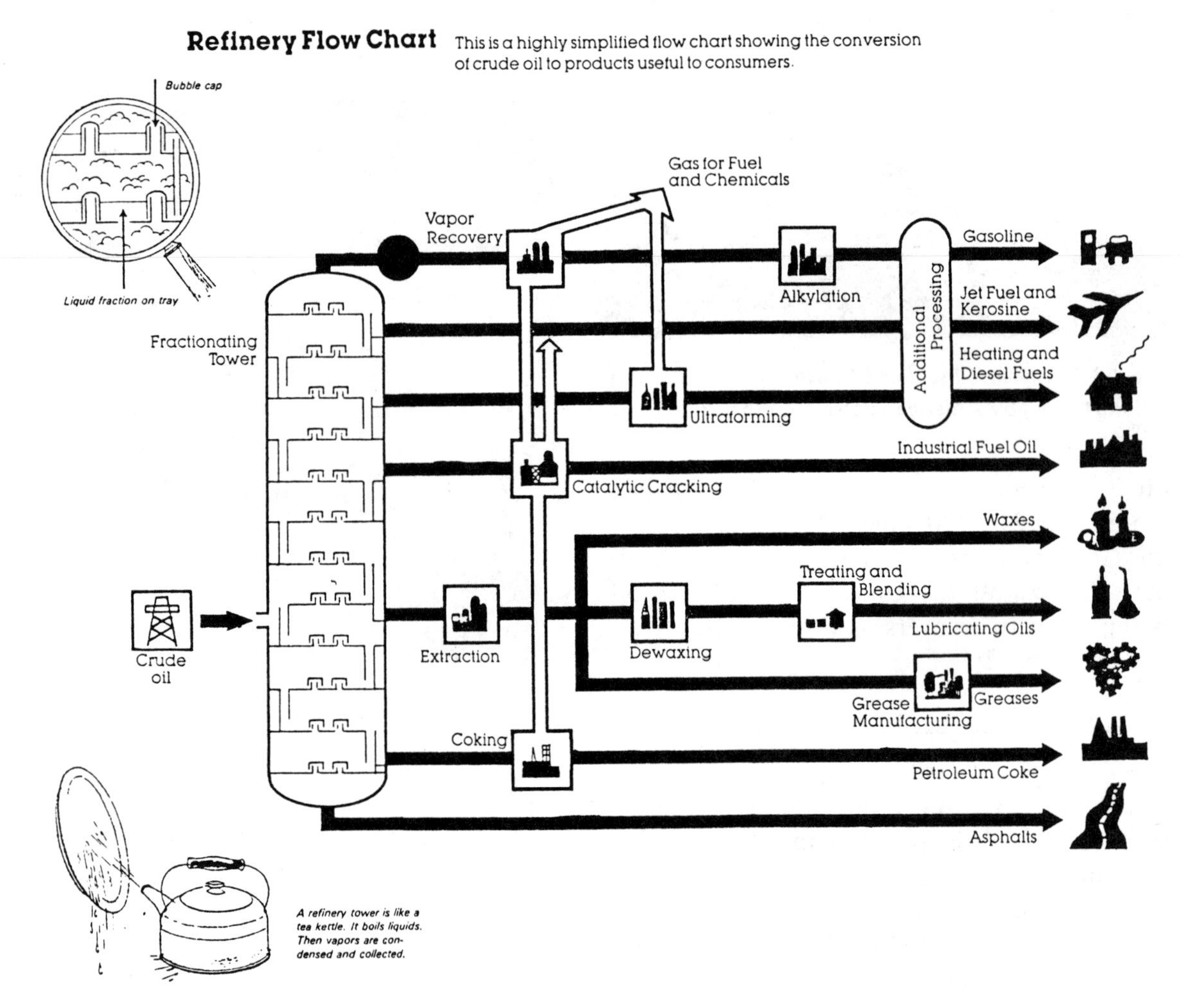

Fig. 26-9 Refinery flow chart (Courtesy of Standard Oil Company of Indiana)

catalytic cracking, and polymerization are used to reform the hydrocarbon molecules into gasoline molecules. Thermal cracking uses higher temperatures to break the bonds of the larger complex molecules of less volatile heating oils into gasoline. Catalytic cracking uses a catalyst to speed up the chemical changes in breaking up the heavier hydrocarbons into gasoline. Polymerization uses a catalyst to reform lighter molecules into liquids called polymers which are essential in high octane fuel. Using these processes refiners now obtain 19 gallons of gasoline from each 42 gallon barrel of crude oil.

PETROCHEMICALS

Petrochemical manufacturing has expanded the uses of petroleum beyond the wildest dreams of early oil pioneers, figure 26–10.

Fig. 26-10 Products from crude oil (Courtesy of the British Petroleum Co., Ltd., by American Petroleum Institute)

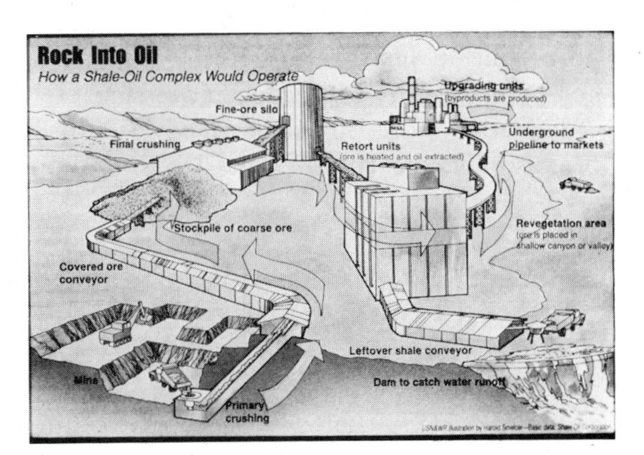

Fig. 26-11 Oil shale—how a shale oil complex would operate (Courtesy of U.S. News and World Report by American Petroleum Institute)

Fig. 26-12 This amount of oil shale rock can fill twelve 55-gallon drums with oil when it is processed. (Courtesy of Standard Oil Company of California by American Petroleum Institute)

About 3,000 different chemicals are produced from petroleum and natural gas in some 600 petrochemical plants. Many of the chemicals produced are seen in consumer products such as polyethylene plastic. Petrochemicals also play a part in the creation of synthetic fibers, films, antifreeze, construction materials, plastics, synthetic rubber, pharmaceuticals, cleaning compounds, plastic bottles, detergents, hydraulic fluids, explosives, dyes, photographic chemicals, lacquer, saccharin, food preservatives, insulation, paints, foam cushions, fertilizers, and insecticides.

OIL SHALE

Oil shale represents great potential for expanding oil production. The hard shale rock is light grey to charcoal in color and contains kerogen, a hydrocarbon. Kerogen will break down to produce oil, gases, and residual carbon when it is heated. Processes are under development to economically recover petroleum products from oil shale, figure 26–11.

One process involves first mining the oil shale and then crushing it. The crushed oil shale is then heated in huge vessels or retorts where the kerogen is released and is further broken down into hydrocarbon gases and carbon residue. To produce 25–35 gallons of oil about 2,000 pounds of rock must be processed, figure 26–12.

In another process, most of the action is done underground. After about one-third of the oil shale of a given area is removed the remaining two-thirds is loosened by explosives and then burned *in situ,* (in place). The oil shale burns at about 900°F and this heat causes the hydrocarbons to leave the burn zone in vapor form which is condensed and pumped away for further processing.

Processes for oil shale recovery show promise for the future. The deposits of oil shale are tremendous, especially in Colorado, Utah, and Wyoming. These states represent an oil source equal to 90% of the proven oil reserves of the world.

REVIEW QUESTIONS

1. How much energy is consumed by each person in a year?
2. What is Edwin Drake credited with?
3. List several oil production states.
4. What are the chances that drilling will produce a commercially practical oil well?
5. What is a hydrocarbon?
6. What products come from fractional distillation?
7. Explain the purposes of thermal cracking, catalytic cracking, and polymerization.
8. List several products produced by the petrochemical industry.
9. Briefly describe what is involved in extracting oil from oil shale.

Unit 27 Natural Gas

OBJECTIVES After completing this unit, the student should be able to:

- Discuss the advantages of natural gas.
- Explain how natural gas is distributed throughout the country.
- Discuss sources that can supplement our natural gas supply.

Natural gas is also a fossil fuel. Similar to coal and oil, it was formed millions of years ago as organic matter decayed and was subjected to heat and pressure as layers of material built up. Usually it is found in underground faults and traps in porous rock. Exploration for natural gas is done in much the same way as it is for oil.

Natural gas is a hydrocarbon like petroleum. In fact it is found with crude oil about 30% of the time, figure 27-1. The main hydrocarbon in natural gas is methane which will burn readily to release heat energy. The advantages of natural gas are quickly apparent: (1) it can be transported easily through pipelines, (2) when it burns there is no ash or smoke, and (3) the flow can be stopped and started by valves. It is the ideal fuel for home heating and cooking. In the 1940s natural gas replaced coal for much of our home heating needs. It is also ideal for industrial uses.

Fig. 27-1 Natural gas processing plant located in a gas field (Courtesy Standard Oil Company of California by American Petroleum Institute)

Natural gas provides about 25% of the nation's energy; only oil provides a larger percentage of our energy than gas. Like oil, large amounts of capital are required for the production of natural gas. Even after all these expenditures, only one of ten wells drilled for gas is successful.

The states of Alaska, Kansas, Louisiana, New Mexico, Oklahoma, and Texas account for 90% of our gas production. A vast network of pipelines gathers the gas and transports it to points of use. Underground pipes—approximately 68,000 miles of pipe—gather the gas to bring it to wellheads. Before transmission the gas goes to processing plants where water vapor, impurities, and hydrocarbon liquids are removed.

Large-diameter transmission pipelines—almost 278,000 miles of them—carry the gas to almost every major area of the nation. Gas in these transmission lines is moved along by

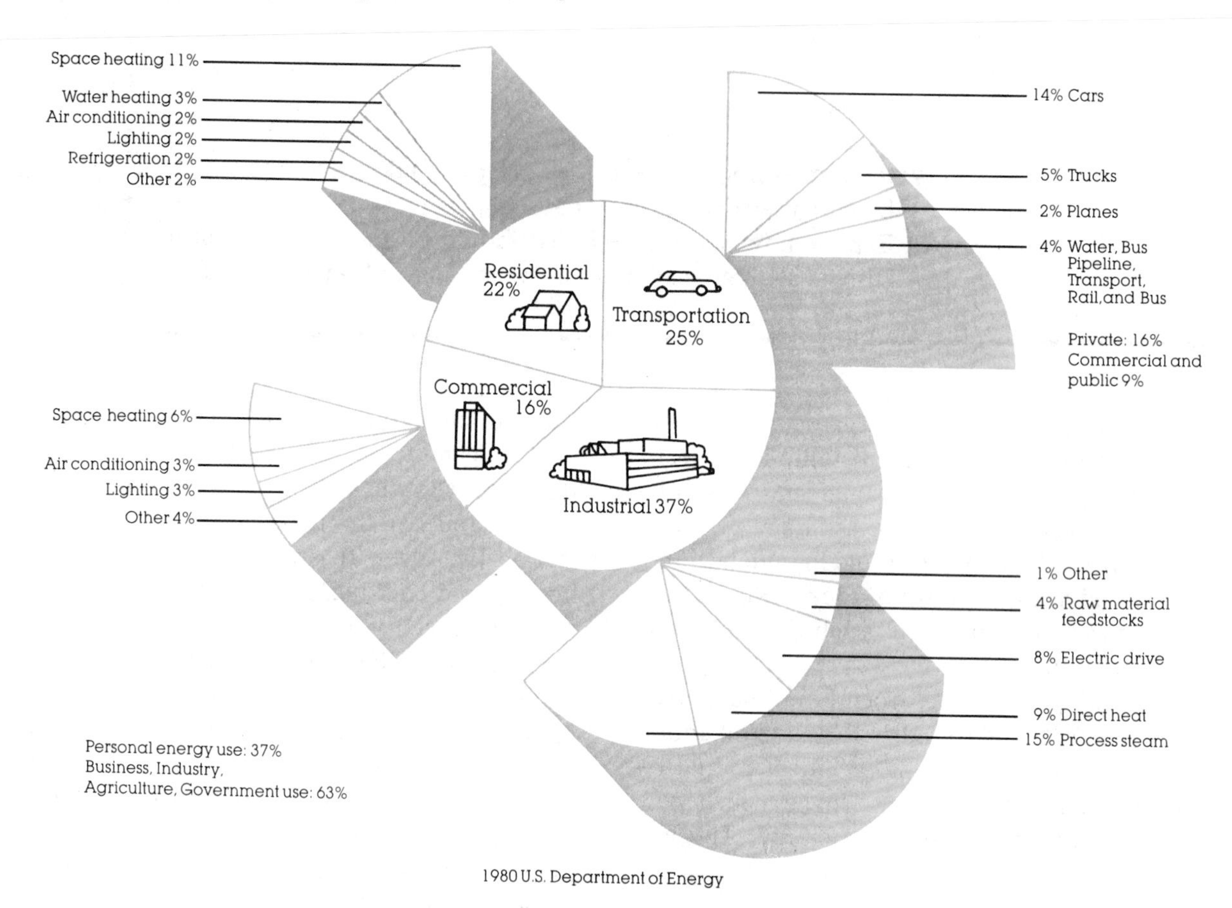

Fig. 27-2 United States energy usage (Courtesy of Standard Oil Company of Indiana)

pumping stations about 50 to 150 miles apart. In the transmission pipeline the gas is under a pressure of 1,000 pounds per square inch and moves along at 15 miles per hour. In the distribution lines, your local gas company will reduce the pressure to about 20 pounds per square inch; these gas mains are under the streets. When gas is brought into a home or other building, the pipe is called a service. These service lines carry only one-half pound of pressure.

Gas flows from a well constantly so distribution companies have huge underground storage reservoirs to store extra gas for cold winter usage when demand increases. Certain underground rock formations can also be used to store the natural gas.

Natural gas will continue to be important in the nation's overall energy supply, but present estimates are that gas can supply 25% of our energy needs for only the next forty years. These facts indicate that continued exploration and drilling will be required, conservation must be practiced, and new sources of substitute gases must be developed.

Sources that can supplement gas are coal gasification and biomass gases. Natural gas is also imported in increasing amounts by means of large ocean-going tankers which carry liquified natural gas. To liquify natural gas it is cooled to minus 260°F; the volume of 615 cubic feet of gas can be reduced to 1 cubic foot by this process. The insulated ships needed to carry the liquified gas increase the cost and complexity of obtaining gas for our use.

SUMMARY

It is important to remember that all forms of fossil fuels which took millions of years to form are on the verge of being used up in a matter of a few hundred years. Ultimately the answer to supplying our energy needs will be in new energy sources; in the meantime conservation of energy is essential. Examine the chart on energy usage, figure 27-2. There are many places where every person who uses energy can help to conserve it. We use energy in our homes, in the stores in which we shop, in the autos we drive, and in the products we buy.

REVIEW QUESTIONS

1. What is the main hydrocarbon in natural gas?
2. What percentage of the nation's energy is provided by natural gas?
3. What states produce most of our natural gas?
4. At what pressure does natural gas enter the home?
5. What is the purpose of liquifying natural gas?

SECTION 9 POWER TECHNOLOGY AND THE FUTURE

Unit 28 Occupational Opportunities in Power

OBJECTIVES

After completing this unit, the student should be able to:

- **Discuss the types of jobs related to power technology.**
- **Describe the training required for various job classifications.**

The scope of *Power Technology* is so wide that it touches the lives of everyone, figure 28-1. Every person is a consumer of power; it is part of our way of life. The automobile, electrical household appliances and machines, power lawn mowers, and recreational equipment are all woven very thoroughly into the American way of life. Everyone has a future in power, as a consumer if not as a vocation.

Understanding the concept of power technology can help to make a person a more intelligent consumer. It's easy, for instance, to understand that you heat your home with gas. It's quite another to know that at our present rate of consumption–in only 40 years–there will be *no more gas* unless certain things are done, and quickly. This understanding can increase everyone's appreciation of our resources too. Concepts of power and its relationship to the community can also help a person to enjoy life more.

OCCUPATIONS

But what about *occupations* in power technology? Is there a future for large numbers of people? What kind of skills, aptitudes, and talents are needed? These questions go back to the definition of power technology, energy

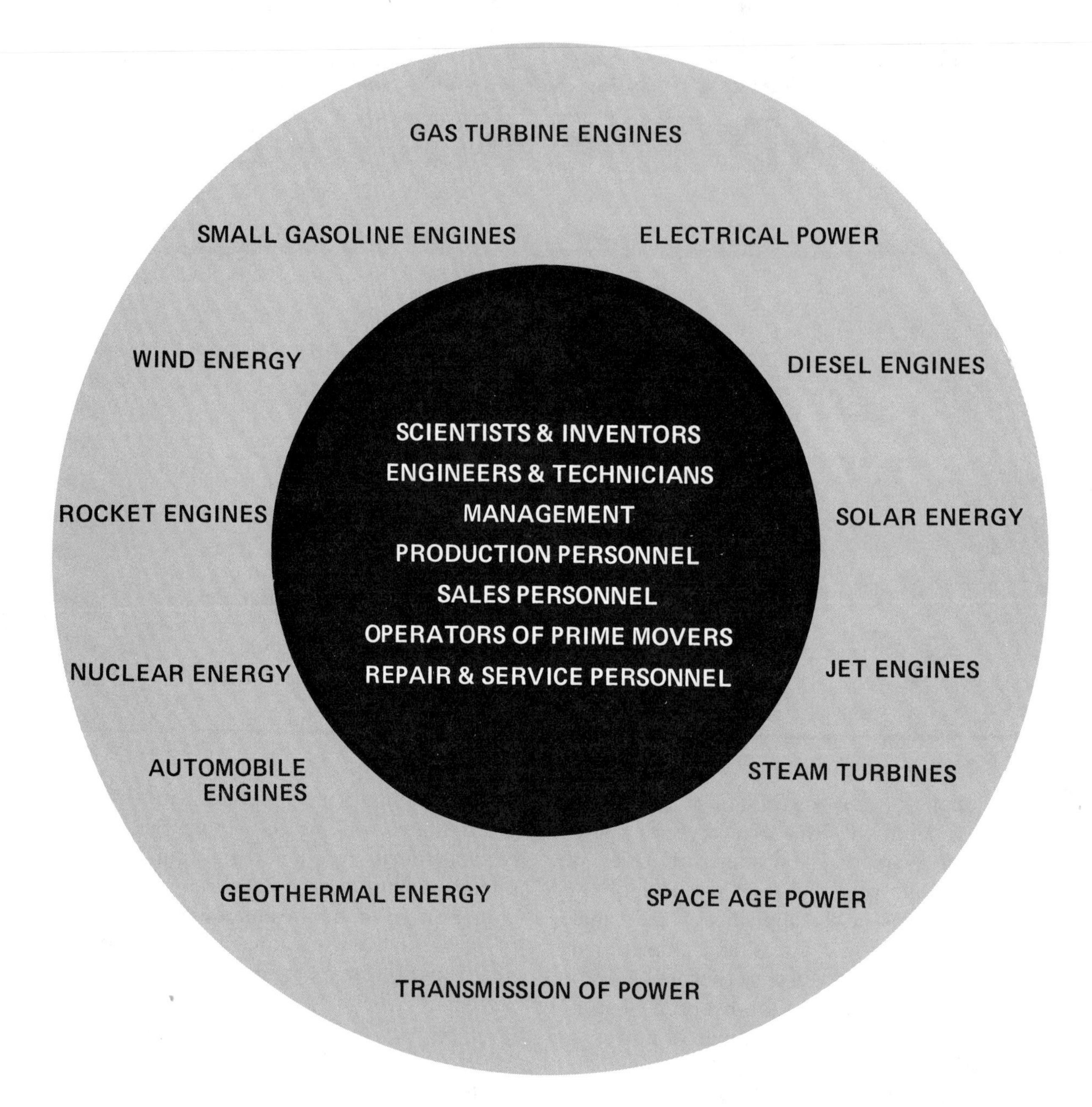

Fig. 28-1 A wide range of skills, talents, and interests lie within the field of power technology.

sources, and the machines that convert energy into useful work. The definition, in turn, leads to the units that lie within this book:

- Small gasoline engines
- Automobile engines
- Diesel engines
- Jet engines
- Rocket engines
- Gas turbine engines
- Electrical power
- Mechanical transmission of power
- Fluid transmission of power
- Steam turbines
- Alternative sources of energy
- Petroleum, coal, and natural gas energy
- Nuclear energy

The vastness of the field and the number of opportunities stagger the imagination. Millions of trained men and women are needed to meet the demands of our power-consuming society.

To the inexperienced, a career in power technology can have as narrow a meaning as a career as an auto technician. This is one phase of power technology that employs over 700,000 people. However, in its broad sense, the field of power technology also encompasses the electrical engineer and the nuclear physicist.

In every community, power fields employ many people, some in a very direct way, as service station managers or the employees of the public utility company. Others are employed in power in a more indirect way, operating trucks, trains, buses, or aircraft in the community's transportation network.

Every spoke on the power technology wheel requires persons with special talents, interests, and skills:

- Scientists, researchers, and inventors conceive ideas and principles, figure 28-2.
- Engineers and technicians develop scientific principles into practical applications, figure 28-3.
- Management organizes large-scale endeavors in power, figure 28-4.
- Production personnel manufacture the ever-lengthening list of prime movers, figure 28-5.
- Sales personnel understand power and fill the needs of business, industry, and individual consumers.
- Operators of prime movers include the professional airplane pilot, the railroad engineer, the truck driver, and the stationary engineer of the power plant, figure 28-6.
- Repair and service personnel have the skill needed to keep prime movers operating efficiently and safely.

The entire list of industries producing fuel for prime movers should also be listed since they represent the sources of energy consumed by prime movers: coal, natural gas, petroleum, nuclear fuel, and exotic fuels for space age power.

Fig. 28-2 Radio frequency plasma research done by scientists

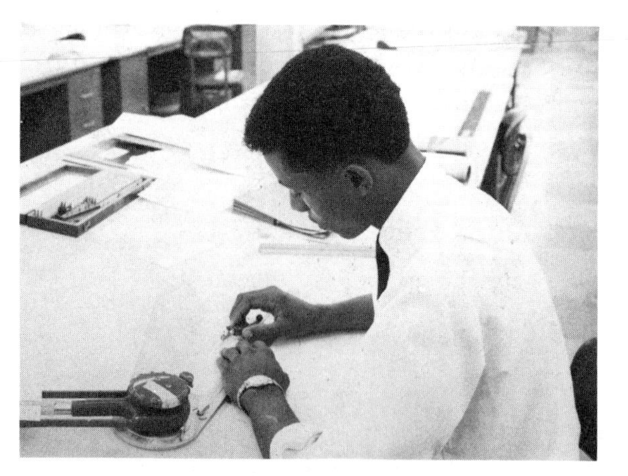

Fig. 28-3 Engineer at work

Fig. 28-4 Men and women are needed to manage industries related to power technology.

Fig. 28-5 Workers in a factory that produces jet engines

Fig. 28-6 Construction workers and farmers both use power equipment regularly. (Courtesy of Ford Tractor Operations, Ford Motor Company)

It is not the purpose of this unit to discuss in detail the thousands of occupations directly related or allied with power technology, but it might stimulate interest for further investigation. Many occupational books and pamphlets deal directly with the more common occupations that are available. These are usually on file in school and public libraries.

Training

Another way of viewing a future in power technology is to look at the training that is required for various job levels. This is one method that is commonly used for classifying occupations.

Professional. College education required, four or more years of formal education beyond the high school level. Scientists, engineers, researchers, industrial arts and vocational teachers, high level managers, and exectives are among these.

Technical. Some college or technical training required, but not necessarily a college degree. Draftsmen, engineering assistants, experimental mechanics, and technicians are among these, figure 28-7.

Skilled Workers. Lengthy preparation as a result of on-the-job training or through apprenticeship programs is required. Aircraft, diesel, and auto mechanics, machinists, and tool and die makers are among these occupations, figures 28-8, 28-9, 28-10, and 28-11.

Semiskilled Workers. Considerable on-the-job training is required. Generally, these persons are machine operators and are employed in every major industry, figures 28-12 and 28-13.

Fig. 28-7 Technician at work

Fig. 28-8 Work progresses in spacecraft assembly within a class 6 clean room.

Unskilled Workers. Little on-the-job training is required. Helpers, laborers, and gas station attendants fall into this occupational category.

EARNINGS

In terms of earnings, the occupations which require the most training, the most creative effort, the most attention to detail, and those that carry the greatest responsibility provide the greatest financial rewards. The financial aspect of an occupation is important, but a person should ask if the occupation is suited to his or her mental and physical abilities. Can it provide a lifetime of satisfying work? Opportunities are everywhere. Consider these facts:

- One person in seven earns a living from the manufacture, sale, service, and use of motor vehicles.

Fig. 28-9 A high degree of skill is necessary in the final assembly of aircraft.

Fig. 28-10 Workmen assembling a large gas turbine engine.

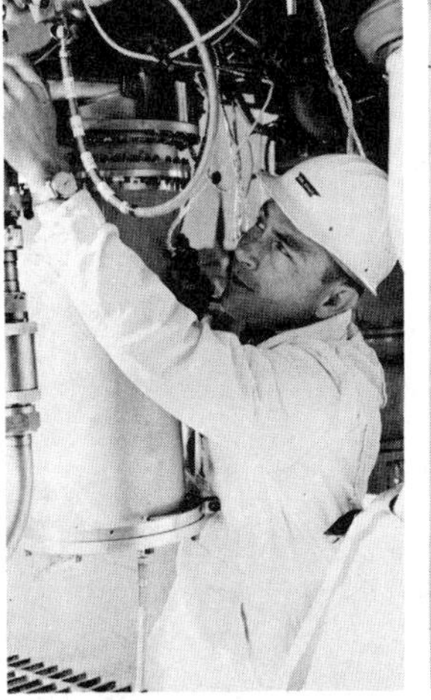

Fig. 28-11 Aero jet personnel at work on Titan 2 rocket engine components.

- One business out of every six is automotive in nature.
- More than 2,000,000 persons earn a living operating devices driven by prime movers, such as trucks, buses, trains, and taxis.
- Industrial production workers have at their disposal an average of 1249 horsepower per worker in installed electrical motors and prime movers.
- More than 1,400,000 persons are employed in aircraft, missile, and spacecraft manufacturing.
- The rapidly developing field of nuclear energy already employs more than 200,000 persons.

Fig. 28-12 A highly-skilled employee operates this numerically-controlled milling machine.

Fig. 28-13 Thousands of machine and equipment operators are employed in every state.

REVIEW QUESTIONS

1. Explain how a salesman might be involved in power technology.
2. Prepare a list of several occupational groups that are needed to bring a prime mover from the planning stage into actual use.
3. Briefly explain the educational requirements of the following job classifications:
   - Professional
   - Technical
   - Skilled workers
   - Semiskilled workers
   - Unskilled workers

Unit 29 Computers and Power Technology

OBJECTIVES

After completing this unit, the student should be able to:

- Define a computer in terms of what it does.
- List the major developments in the evolution of the computer.
- List the generations of computers and list the major distinguishing feature(s) of computers in each generation.
- Explain the function of each of the five major parts of a microcomputer.
- Define RAM and ROM.
- Describe several applications of computers in manufacturing.
- Describe how computers are used in modern automobiles.

In our society power technology has touched every person to varying degrees. None of us can escape the effects of power and its applications to automobiles, the electricity we use in our homes, at work and for recreation, space exploration, and all of the other uses of energy by means of numerous machines and devices that bring power into all aspects of our lives.

Computers are also a part of our everyday life and they play a part in all of power technology. It is not unrealistic to say that we are all involved with computers today. We may not realize that we are using computers, but many of our late model automobiles, home appliances, and office equipment contain computers.

The purpose of this unit is to help the reader develop a sense of computer awareness. We must all learn to be at ease with the computer world—it will not go away, figure 29-1. We must realize that computers are here to help us, not just to test our intellectual capacity; computers are merely tools to aid the mind.

Simply stated, a computer is an electronic device or machine used to solve problems and to organize, process, and store information (data) at a very high speed. A computer does its work by following a program or prescribed set of instructions.

Fig. 29-1 Young people are usually more comfortable with computers than the older generation.(Courtesy Apple Computer, Inc.)

A computer does not work by itself; it cannot be instructed what to do until a person has carefully thought out what needs to be done. Understanding, knowledge, and logic are the characteristics of the human mind that are needed to program a computer to accomplish a given task. Since computers can manipulate vast amounts of information at a high speed, a direct connection can be made to the steam engine of the industrial revolution. The steam engine multiplied physical capacity and the computer multiplies mental capacity.

The steam engine was the key to the industrial revolution, but other machines were needed to take advantage of its power. These machines included the locomotive, steamship, spinning jenny, and lathe. The digital computer is the heart of the computer revolution, but it is also necessary to devise ways to apply the computer to various jobs. Programs must therefore be written for every application.

All levels of skills and talent are needed in our computer society. Scientists and engineers are needed to design the computers, other people are required to develop the programs that will allow the computer to do a given job. Knowledgeable and highly-trained people are needed to maintain and repair computer systems, and millions of people must be trained to operate computers in offices and factories. If you are going to be one of the millions who operate a computer, figure 29–2, you should be aware that you do not have to know how to construct or program a computer to operate one. You do not have to be an automotive engineer to operate an auto, but the more you know about how it operates, the more enjoyment and better performance you will be able to obtain from it. Computers can be approached in this same way.

Fig. 29–2 At school, in the home, and in the world of work, the computer is revolutionizing the way work is done. (Courtesy Apple Computer, Inc.)

COUNTING AND CALCULATING MACHINES IN HISTORY

The average person has not had many opportunities until recently to learn anything with regard to computer technology even though we have been increasingly a part of a computerized society. That is because although most important inventions have taken 50 to 100 years to become integrated into our society, this is not true with computers. The growth of computers has been phenomenal—forty years ago there were none. Thirty years ago there were 250, twenty years ago there were 24,000, and today there are millions; tomorrow, there will be tens of millions.

Modern computing may be new, but the roots of computing extend far back into history. Many historians consider the abacus to be our first attempt at computing. The abacus was a counting board developed by the Babylonians about 5,000 years ago. A version developed by the Chinese consisted of wires stretched in a framework with wooden counters that could slide on the wires. This simple counting device is still used in many parts of the world.

In 1642 Blaise Pascal, an accountant who wanted to be liberated from the tedious calculations that had to be performed by hand, created a mechanical device that was capable of addition and subtraction. In his machine the counting was done by means of a gear-driven mechanism.

Another pioneer was Joseph-Marie Jacquard who used a punch card system to control the automatic weaving of complicated textile patterns. His system was put into use in 1801.

Charles Babbage was a man ahead of his time. He devised the *difference engine*—a device that could calculate scientific and astronomic tables. Between 1812 and 1822 he built a working model of his machine, but manufacturing techniques had not developed the precision needed to make his machine practical. In 1834, he began work on an *analytical engine* which was designed to be a universal automatic calculator. His design contained the essentials of modern digital computers—input, memory, control, processing, and output. Unfortunately, Babbage died in 1871 before completing the machine.

In 1890 Herman Hollerith (who later founded IBM) developed his famous punch card machine. The machine was developed specifically to tabulate the 1890 census. It was an electromechanical system that *sorted* punch cards. All types of statistics could be entered on the cards by punching small holes at specific locations. The cards were then sorted by the machine into groups depending upon the information desired; a tabulator then totalled the cards. The system is considered to be the first data processor because the machine counted, collated, and analyzed information.

DEVELOPMENT OF THE MODERN COMPUTER

Modern computers are often grouped into generations. Each generation is identified by a major change from a previous generation by increased speed and accuracy.

First Generation Computers

The computers in the period from 1944 to 1951 could perform thousands of calculations per second. The first truly electronic computer was the ENIAC (Electronic, Numerator, Integrator, Analyzer, and Computer). The ENIAC was a highly classified military project for the United States Navy. It was designed to calculate firing tables for the large guns on naval ships. The computer was huge, consisting of 17,468 vacuum tubes, 70,000 resistors, 10,000 capacitors, 1,500 relays, and 6,000 manual switches. To contain these parts, forty panels were arranged in a horseshoe pattern. The unit weighed 30 tons and was 8 feet high and 80 feet long. To program the computer, thousands of switches had to be set by hand, and hundreds of cable connections had to be made. The tremendous heat of the vacuum tubes had to be kept under control with large air blowers.

UNIVAC was the first of the large computers to be made commercially available. The first unit was placed in operation at the Bureau of the Census.

The IBM 650 computer was the most popular of these first generation computers of the 1940s and early 1950s. Generally, these computers used vacuum tubes, were large, and required extensive air conditioning.

Second Generation Computers

These computers used transistors in place of vacuum tubes. Transistors performed the same function as tubes but were much smaller, required far less electric power, and were more reliable. Second generation computers were produced from 1959 to 1965, and were used in scientific and commercial applications.

Third Generation Computers

Computers continued to become smaller and faster. The component that was responsible for these changes was the integrated circuit (IC). An integrated circuit is a group of transistors, capacitors, resistors, and other parts that are made into a solid, inseparable piece which performs the functions of the individual components. Third generation computers using integrated circuits can perform a million calculations in a second, more than a thousand times faster than first generation computers.

Fourth Generation Computers

These computers reflect the current state of computer technology with commercial applications. In these computers *chips* contain the integrated circuits. A process called microminiaturization enables a thousand circuits to be created on one tiny silicon chip about the size of a fingernail. This technology has resulted in reduction in the size of computers and in an increase in speed. Another result is a major decrease in computer prices.

Within forty years the computer has become an affordable consumer item. It is for this reason that one of the most significant developments in fourth generation computers is the *personal computer*. Stores that sell computers are not only located in almost every city of any size, large computer manufacturers, the Apple Computer Company—for instance—has an automated plant that is capable of manufacturing one million of its Macintosh personal computers a year, figure 29-3.

Fig. 29-3 Apple Computer's automated manufacturing plant for the Macintosh computer (Courtesy Apple Computer, Inc.)

THE MICROCOMPUTER

A brief explanation of a typical personal computer—also know as a microcomputer—will be helpful in understanding what a computer can do and will introduce some common computer terminology.

Most microcomputers have five major parts: video screen, central processing unit, keyboard, disk drive, and printer. All of these items make up a group of devices commonly known as computer *hardware*.

The *video screen (monitor)* is a very familiar item to all. The computer monitor is similar to an ordinary television set. The monitor, however, cannot be removed from the computer for use as the family TV set. Most monitors do not have a volume control and they do not have a tuner for receiving TV signals. Monitors have more lines of resolution than televisions to produce a crisper image. The monitor presents the information generated by the computer in visual form.

The *central processing unit* is the heart of the computer. All calculations and information processing take place in the central processing unit. In the modern microcomputer, the microprocessing chip is no larger than your thumbnail. A chip is a very small sliver of pure silicon. Many transistors and circuits are etched on the silicon using photographic processes. The end result is called an integrated circuit. The IC is so small that you can see the circuitry only through a microscope.

The *keyboard* is basically like an ordinary typewriter keyboard. The keyboard is used to put information into the computer instead of on a piece of paper. In many instances the information typed in will also be duplicated on the monitor.

The *disk drive* is about the size of a box of tissues. It is used to store and retrieve information by means of disks. The disks or diskettes, commonly called "floppy disks," are round, flexible circles of plastic with a coating of magnetic material similar to recording tape. The most common size for disks is 5¼ inches in diameter. These disks are referred to as *software*. The diskettes generally contain sets of instructions for the computer. These instructions are called programs. The program causes the computer to accomplish a certain job or task.

The *printer* can be thought of as a high technology typewriter. In many instances, it is necessary to have a copy of what is seen on the monitor. The image usually can be duplicated by using the printer. In computer language, a typed page is commonly referred to as "hard copy."

Information Capability of the Microcomputers

The key factor in classifying the size of a personal computer is the amount of information that it is capable of storing in its memory. There are two basic types of memory—RAM and ROM. RAM (Random Access Memory) chips make up the main memory of the computer. They store information while the computer is in use. Unless the data is transferred to a diskette, the information in RAM will be lost when the computer is turned off. On the other hand, ROM (Read Only Memory) is basically used to store the computer's internal instructions. These instructions are permanent and are not lost when the computer is turned off.

Computer storage capacity is usually expressed in RAM storage such as 16K, 48K or 128K (where K represents 1,000). What does this really mean? As an example, 16K means 16,000 bytes. A byte is roughly equivalent to one keystroke on the keyboard, whether this keystroke is a character or a space. A computer with 16K of memory is capable of holding approximately 3,200 words in its memory at any given time.

When a person operates a computer three things will typically happen. Information typed into a computer is considered to be INPUT. The computer acts on this information in a step called PROCESS. Finally, the result of input and process is usually seen on the monitor or is printed as hard copy and is referred to as OUTPUT.

COMPUTER APPLICATIONS

Wherever we look—at home, in business, in government, and in industry—we can see applications of computers, figure 29-4. In addition, new ideas for computer applications are

Fig. 29-4 A microcomputer in the home permits the entire family to enjoy and learn from the numerous programs that are available. (Courtesy of Apple Computer, Inc.)

being developed every day. Increasingly, we live in an information based society. Many of the most common computer applications include the accurate and rapid processing and storage of information. For example, a travel agent can quickly type a few commands into the computer and it will display airline flight information. A few more commands will allow the travel agent to reserve seats on your specified flight. The entire process takes only minutes and does not require letters or phone calls.

In some cases the speed and accuracy of computers is absolutely essential. For example, the calculations needed to control and direct the flight of a space craft can be done in microseconds. Time limitations would not permit such calculations to be done by hand or by slower machines.

In every application that now utilizes a computer, the computer is now doing the job because it is faster, more efficient, or more reliable than the old methods. Organizations that manage huge amounts of information have always been the largest users of computers. Government agencies keep track of mountainous amounts of information, insurance companies manage insurance policy records for millions of policy holders, utility companies use computers for billing their customers, financial institutions maintain records on all of their business and personal accounts, and large companies of all kinds use computers throughout their organizations. Figure 29-5 shows computers at the New York Stock Exchange where millions of shares are traded every day.

Fig. 29-5 Computers on the New York Stock Exchange have permitted trading to keep pace with the demands of stockholders. (Courtesy of the New York Stock Exchange)

Routine Applications of Computers

Computers can take care of the routine tasks of handling data, figure 29-6, but they also help business people analyze facts, figure 29-7. With a computer as a tool, the business people have more than experience, knowledge, and hunches to aid in decision making.

A short list of computer applications will show how widespread computer usage has become:

- Doctors–diagnosing diseases
- Police–checking vehicle registrations
- Stockbrokers–following the market
- Legislators–following the progress of legislation
- Lawyers–retrieving information on court decisions
- Railroads–keeping track of railroad cars
- Students–tutoring themselves
- Business–taking inventory
- Credit Bureaus–screening loan applications
- Realtors–locating homes to suit client needs
- Schools–keeping student records
- Weather forecasters–tracking storms
- Designers and engineers–developing new products

Computers are used to control and monitor many industrial processes and operations. A modern oil refinery is a good example of the use of computers to keep the various refining processes on track. The principle of feedback is used. Feedback allows one process to be adjusted to another process. For example, a thermostat turns on the furnace to maintain a given room temperature and then turns the furnace off again when the desired temperature is reached.

Feedback can also be seen in a business where information is fed through the organization as people and departments need to know the status of various situations if they are to accomplish a given task. As an example, when a retailer orders an item, the order af-

Fig. 29-6 Computers are a valuable tool for many students in high school and college. (Courtesy of Apple Computer, Inc.)

Fig. 29-7 The computer analyzes large amounts of data to help business people make decisions affecting the business. (Courtesy of AT & T Information Systems)

fects an entire organization, including all departments in the factory, and the companies that supply the factory. Because computers can supply information quickly, the results of an order can be communicated to all concerned without delay. Inventories at the distributor and the factory can be checked and factory suppliers can be notified immediately to ship materials. With the aid of computers, the chain of events that may otherwise take weeks is done in moments. The result is faster service and a reduction of expensive inventory stockpiles.

CAD, CAE, AND CAM

Increasingly, today's manufacturing industries are using computers. Three areas are commonly recognized–Computer Aided Design (CAD), Computed Aided Engineering (CAE), and Computer Aided Manufacturing (CAM).

In Computer Aided Design, figure 29-8, the drafting board and T square are replaced by a computer video screen. Using the computer, the designer can present parts on the screen where they can be studied. The parts

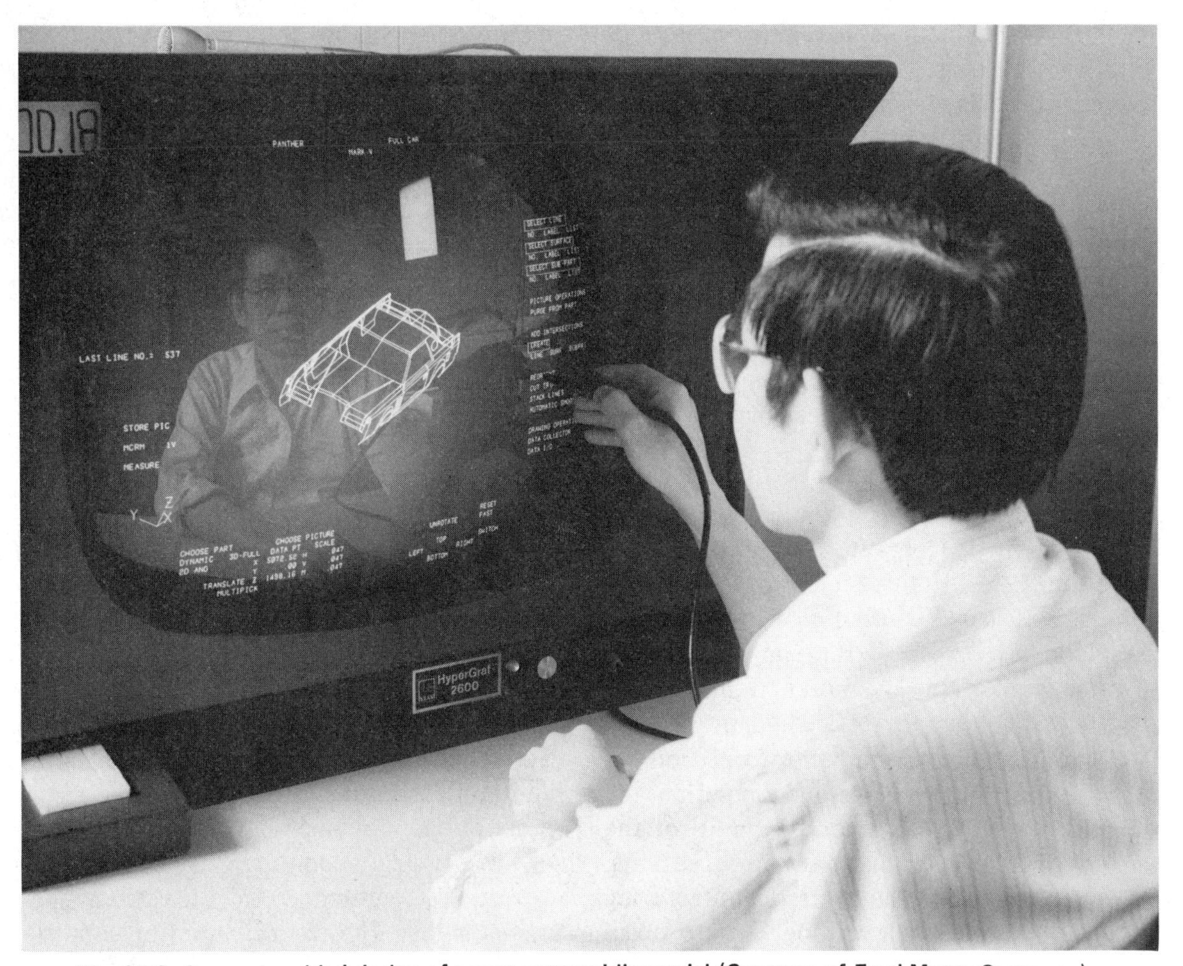

Fig. 29-8 Computer aided design of a new automobile model (Courtesy of Ford Motor Company)

are represented in computer graphics and they can be rotated, analyzed, and enlarged for better viewing. Alternative designs can also be considered. The designer is using computer graphics to help him see and understand the geometry of the part and its surfaces better than he could by presenting the part on a drafting board. Using CAD, designs or design changes can be made on the spot.

Computer Aided Engineering (CAE) is used to evaluate a part before prototypes are made or production is started. The computer helps to evaluate the function and performance of the part. When it is verified that the part meets the prescribed standards the design can be transferred to manufacturing.

Computer Aided Manufacturing (CAM) refers to the use of computers to run the machines needed for production. Computers are used for testing and monitoring quality. Robotics and numerically controlled machines are also a part of CAM.

The functions of CAD, CAE, and CAM are not separate, but rather overlap and interact. Experts in the field are finding new ways to use computers in industry every day.

CIM and Robotics

Manufacturing plants are moving rapidly in the direction of Computer Integrated Manufacturing (CIM). In this type of operation, the center is a large computer which contains the data base or all of the company's basic information. Many computer terminals are placed throughout the office and the factory itself. Design engineers use information on parts specifications that are located in the data base. Computer aided design allows the part to be designed quickly to specifications. The design now moves to the manufacturing engineering section where programs are developed for the numerically controlled machining centers. In the center of a cluster of machines is a robot that loads the part to be machined and then removes the completed parts from the machining center. Most parts move by conveyors or in computer controlled carts. Fewer workers are found in such a factory, but at almost every level of employment some knowledge of computers is required.

The science of robots is called robotics. Robotics promises to play an ever increasing role in manufacturing as industry strives to improve quality, cut production costs, and meet foreign competition. This is why manufacturers continue to use more and more robots in their operations. Often they perform jobs that are considered to be dangerous or uncomfortable for humans, such as welding, forging, or spray painting. Robots also are used to move parts from one place to another and, increasingly, they are used to perform exacting operations that are beyond human ability. The motion of the robots is not only directed by a computer program, but many robots can be reprogrammed easily to accomplish different jobs.

COMPUTERS IN AUTOMOBILES

Computer technology is increasingly a part of all modern automobiles. In many ways the use of computers in automobiles has been aided by both emission control requirements and the desire for increased gasoline mileage. Most automobiles manufactured in the United States now use microcomputers to monitor and control engine performance. The result is more efficient engine operation.

The typical computer is connected to sensors that are mounted on various engine components. The sensors monitor conditions and send information to the computer which then examines the data and sends signals to other engine components to correct or maintain proper operation. The primary job of the system is to provide precise control of the mixture of fuel and air through the carburetor or fuel injection system. For maximum efficiency of the catalytic converter and for best fuel economy, the air-fuel ration must be between 14.4:1 and 14.8:1. Several of the sensors associated with the system are: the coolant sensor mounted on the engine block, the catalytic converter temperature sensor mounted in the catalytic converter, the manifold pressure sensor, the throttle position sensor, and the oxygen sensor located in the exhaust manifold. Data fed to the computer from the sensors is analyzed. If the fuel mixture is too rich or too lean, signals are sent to a solenoid that controls the amount of gasoline mixing with the air.

The use of the computer, oxygen sensor, and low cost electronics has made it possible for manufacturers to use more electronic fuel injection systems to replace carburetors. The computer receives information about engine temperature, speed, load, and the oxygen content of the exhaust gases. This information is used to calculate the exact amount of gasoline needed for best combustion. A pulse of current is sent to the fuel injector to trigger the solenoid coil. As a result, the correct amount of fuel is delivered by the injector. The longer the pulse sent by the computer, the more fuel is delivered.

Two electronic fuel injection systems are commonly used. The multipoint or port system has a separate injector for each cylinder. The throttle body system has one or two injectors in a casting that bolts to the intake manifold. Its appearance is similar to the traditional carburetor. Both systems are very successful, but the multipoint system generally provides slightly better performance. The throttle body system is more economical.

In addition to the critical role of improving engine efficiency and controlling emission, computer and electronic systems perform many other functions. One of these functions is to monitor critical items such as engine temperature, fluid levels, and brake light operation. If a malfunction occurs appropriate lights go on or, in some cases, voice warnings are given.

Trip computers are found on several luxury automobiles. They can provide information on how much fuel has been used, how far you can travel on the remaining amount of fuel, estimated time of arrival, average speed traveled, fuel economy, and inside and outside temperatures.

Volkswagen has an upshift indicator light that goes on when the driver should shift to the next higher gear. Indications of engine speed and intake vacuum are sent to a computer. When the computer determines that the same power could be delivered more effectively at a higher gear the light goes on to indicate that it is time to shift.

Antiskid braking is a significant new safety feature. The danger of locking the brakes and loosing control while trying to stop on slippery roads has been a major safety concern. Within a few years this feature will be common on automobiles manufactured in the United States. Sensors at each of the four wheels relay information to the computer concerning the deceleration rate of each wheel during braking. If the computer determines that the rotation of the wheel is slowing too fast for the speed of the automobile, or if the wheel has locked, a signal is sent to decrease the hydraulic pressure at the brake. As a result, the wheel begins to rotate faster. While the driver's foot is on the brake pedal, the computer will control the slowing of all four wheels–applying pressure and releasing it rapidly to prevent the wheels from locking and putting the auto through a skid.

Electronic transmission control may become common before many years have passed. In addition to regulating the transmission by hydraulic pressure, these systems will use information from sensors on speeds within the transmission, the engine load, and the type of driving being done. A computer will determine the best gear for the given situation.

Computers are also in common use now in automotive repair shops to diagnose problems with various systems within the automobile, figure 29-9. Such computer analysis can help reduce troubleshooting time and repair costs. Computers are also used to monitor parts usage and inventories.

Fig. 29-9 Computer controlled diagnosis of automotive systems can benefit the consumer with reduced repair time and costs. (Courtesy of Apple Computer, Inc.)

Fig. 29-10 Aero 2000, the automobile of the future, contains many high technology features. (Courtesy of General Motors Corporation)

Designers are including computers to a greater extent in future automobile models. For example, the General Motors Aero 2000 experimental subcompact auto, figure 29-10, has a sophisticated computer-controlled electronic system. Power windows and the heating/air-conditioning system are programmed to respond to spoken commands. Other features include a satellite navigation system, a video camera that provides a 180° view of the rear to replace mirrors, radar to assist in timely braking, and a complete electronic diagnostic system to monitor engine performance.

COMPUTER PROGRAM AREAS

Applications for microcomputers are unlimited today for business, educational and personal uses. Thousands of written programs or software are available. Several common program areas are listed with examples of each.

- ACCOUNTING PROGRAMS: accomplish business/accounting functions such as a general ledger, accounts payable and receivable, and payroll.
- BUSINESS GRAPHICS PROGRAMS: make it possible to produce bar graphs, line graphs, and technical illustrations.
- COMMUNICATIONS PROGRAMS: allow the operator to exchange data with other companies and to access on-line information data bases.
- DATA BASE PROGRAMS: help the user to store, find, and keep track of information.
- OFFICE MANAGEMENT PROGRAMS: keep track of appointments, phone numbers, various office forms, and other common office functions.
- SPECIFIC APPLICATION PROGRAMS: designed for special users such as architects, engineers, electronic designers, medical offices, and law offices.
- SPREADSHEET PROGRAMS: designed to help answer the "what if" questions; they project results in response to changes in variables.
- TRAINING PROGRAMS: designed to teach users how to operate computers, how to type, how to program, or how to do almost anything.
- WORD PROCESSING PROGRAMS: designed to make writing easy; they are used for correspondence, reports, memorandums, and for writing books.
- EDUCATIONAL PROGRAMS: help a person learn math, reading, writing skills and many other subjects.
- PERSONAL AND HOME FINANCE PROGRAMS: help an individual manage money matters such as checkbooks, financial and investment planning, and various home accounting needs.

THE COMPUTER INDUSTRY

Computer companies are part of the high technology industry. These companies deal with process systems, telecommunications, computer assisted design, computer aided manufacturing, robotics, fiber optics, information processing, electronic mail, and many other products and services.

High technology companies tend to be relatively small and able to react quickly to market demands. These companies generally locate where highly trained personnel are readily available, such as near universities which promote research and also provide training programs that are in step with advancing technology.

Many leaders in high technology industries are relatively young—under 30. Changes are taking place rapidly and there is room for those with the necessary skills.

To be sure, not everyone will work in the computer and high technology industries. However, everyone is now and will continue to be affected by these industries. Completing this unit on computers should not be the end of your computer studies, but only the beginning.

REVIEW QUESTIONS

1. Write a simple definition of a computer.
2. Explain the various occupational levels needed in the computer society.
3. Does a computer operator need to know how to program a compu-puter? Explain.
4. Trace the rapid growth in the numbers of computers in the country.
5. Discuss the abacus and its history.
6. List several early pioneers who worked with computer principles and machines.
7. State a significant characteristic of each of the four generations of computers.
8. What is an integrated circuit? A chip?
9. List the five major parts of a microcomputer and write a statement about each.
10. Explain the term "software."
11. Explain why the government has always been a major user of computers.
12. List five occupations and how they might use computers.
13. Explain the term "feedback."
14. Briefly explain CAD, CAE, and CAM.
15. Explain why computers are used in various automobile systems. List several systems which now include computers.

Unit 30 Power Technology and the Environment

OBJECTIVES

After completing this unit, the student should be able to:

- Explain how the use of power contributes to pollution.
- List the major sources of pollution.
- Discuss the role of government in controlling pollution.
- Explain the need for recycling.

Throughout the centuries civilizations have regarded this planet earth as a challenge. They met the challenge almost as if the earth was the enemy. They have been thorough and sometimes ruthless in the way they have cut the trees, mined the minerals, and killed the wildlife. And remember, all this was done in the name of *progress.*

The earth has been considered something limitless, figure 30-1. It was assumed that people could go on taking forever, using what they wanted and discarding what did not suit them as waste. These *civilized* people have proved themselves to be nature's most untidy and careless animal, even though they possess as intelligence superior to any other creature.

Even in today's enlightened scientific age, most persons share to some degree these careless, shortsighted views about the earth. Our environment and our world are threatened by a huge appetite for manufactured goods. Can our environmental problems be solved? Solutions require sacrifice and a change in our life style.

It is important that students of power technology understand the problem and do their part in the overall solution, figures 30-2 and 30-3. Power has much to do with our present high degree of civilization. However, energy is being used on an ever increasing scale. In 1945 there was a total of 31 million automobiles in the United States. Today there

Fig. 30-1 Spaceship earth. Its resources are limited.

Fig. 30-2 Solar energy is a clean solution to heating. (Courtesy of Electric Power Research Institute)

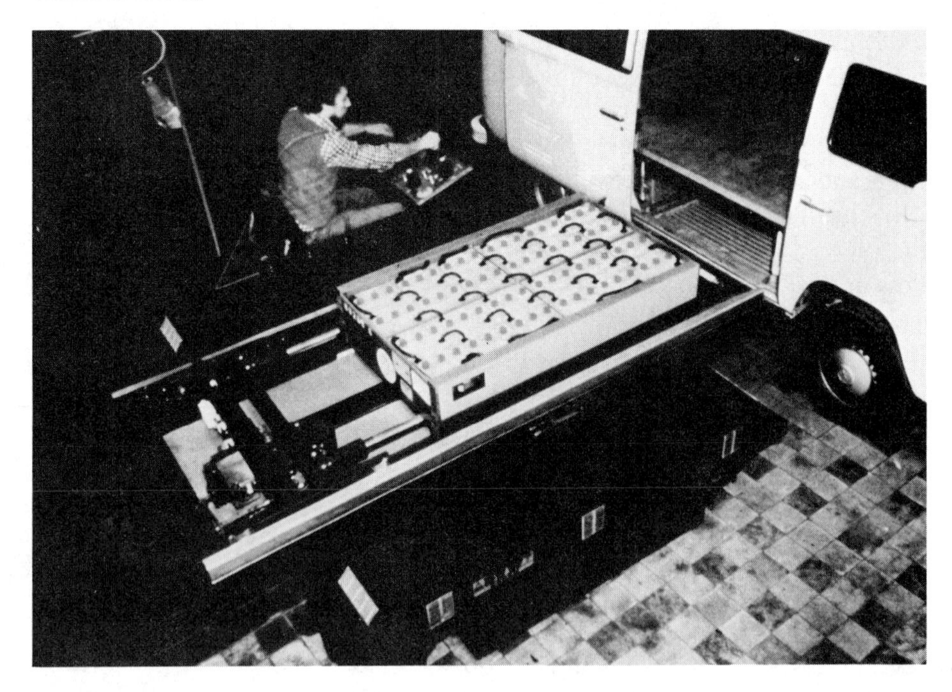

Fig. 30-3 Electric-powered vehicles would help reduce pollution. (Courtesy of Electric Power Research Institute)

are more than 100 million automobiles on our highways. Our electrical consumption has doubled every decade for the last 50 years. Only in recent years has it become obvious that a huge price is being paid for all this power: mineral depletion, scarred landscape, and polluted air.

The scope of the problem is a vast one with many interrelated parts. It is important to define three important terms.

- **Environment** is the sum total of the things, conditions, and influences that affect the development of life.
- **Pollution** is the waste that makes the environment unclean or foul. Pollution can affect health, safety, and even technology itself. Pollution also offends moral and aesthetic sensibilities, decreasing the quality of life.
- **Ecology** is a branch of biology that concerns itself with the relationship of organisms and their environment. Ecology is the study of the *interdependence* of all forms of life.

THE GOVERNMENT AND POLLUTION

Conservationists have tried through the years to inform the citizens of a rapidly approaching environmental crises. They experienced moderate success in arousing the public and in securing the passage of legislation aimed at environmental problems. The late Rachel Carson's book, *Silent Spring,* which was published in 1962, did much to stimulate widespread interest in the environment. Later in the decade, oil spills fouled beaches and destroyed marine life, and smog endangered the health of people living in urban areas. These events finally awakened the public to the fact that the environment was indeed being threatened.

The federal government expressed concern with the passage of the Air Pollution Control Act in 1955 which funded air pollution research. The Clean Air Act of 1963 poured millions of dollars into solving pollution problems. It shifted the emphasis to developing, establishing, and improving pollution prevention and control programs.

In 1965 the Motor Vehicle Air Pollution Control Act provided for establishment of emission standards for new motor vehicles. In 1968 the auto industry began to meet federal emission standards, which have become increasingly stringent in recent years.

Federal involvement in all forms of pollution continued to grow. In 1970 a master agency called the Environmental Protection Agency (EPA) was formed. The EPA has authority over municipalities, states, and industries in pollution control of air, water, radiation hazards, solid wastes, and pesticides.

The Council on Environmental Quality, CEQ, and the National Oceanic and Atmospheric Agency, (NOAA), were established in 1970. The CEQ is assigned the task of supervising the government itself. This agency checks out all projects, plans, and proposals that may affect the environment. The NOAA is designed to be a superior scientific agency. It determines how badly air and water are already polluted and suggests what can be done about it. In addition, the NOAA determines existing resources in the sea and air. It also explores ways to shape the weather.

AIR POLLUTION

The major sources of air pollution are transportation, in which the main culpret is the automobile; industrial and residential heating plants; refuse burning; pollution from industrial processes; and utilities, especially coal-burning electrical power generation stations. These sources of pollution are highly concentrated in urban areas where 70% of the population lives.

People are quick to single out the billowing industrial stacks—the other guy, so to speak—as the prime polluter, but the fact is that the individual should turn around and look at the effects of his own pollution. These include furnaces, incinerators (Note: It is best to mention here that the majority of states have outlawed backyard incinerators. If your state has not, they should be considered a major source of pollution by an individual.), power mowers, and automobiles, which are the largest single source of air pollution.

Several types of air pollutants have been identified. Carbon monoxide is the colorless, odorless, and tasteless killer in auto exhaust. Other pollutants include sulfur oxides, carbon dioxide, lead, nitrogen oxides, assorted hydrocarbons, ash, mists of oil, aerosols, metallic oxides, and various soots. At this point researchers are just beginning their list of harmful air pollutants.

Property damage caused by pollution has a price tag of ten billion dollars per year. Photochemical smog is partially produced by the reactions of nitrogen oxides with oxygen in the sunlight. It is agreed that smog produces damage in the human body, but specific research results are lacking. It can be said, however, that pollutants do aggravate all types of respiratory conditions. Millions of people have experienced the effects of smog: stinging eyes, irritated nasal passages, and breathing difficulties. Breathing polluted air over a lifetime may well be costing far more than is now realized.

SOLUTIONS TO AIR POLLUTION

Air pollution is being reduced, figure 30-4. The big stacks of industry are being equipped with electrostatic precipitators, centrifugal

Fig. 30-4 Scientists collecting samples of rain water to be tested for pollution. (Courtesy of Electric Power Research Institute)

collectors, filters, baffles, and water scrubbers to reduce pollution hazards. These devices are expensive, but industries are moving rapidly to control the emissions from their stacks. Nuclear-powered electrical plants are the cleanest power source, but even nuclear reactors pose environmental threats from radiation and thermal pollution.

The individual can help reduce air pollution. The cumulative power demands of the public are reflected at the power station as it consumes mountainous stockpiles of coal. Instead of loading the home with labor-saving devices to the point where it takes on the appearance of a miniature factory, the family should do more under their own power. Burning a cleaner fuel for residential heating also reduces pollution.

Choosing a Residential Heating Fuel

Natural gas is quite clean, but it is also in short supply. The most plentiful fuel is coal, but the cheapest coal is high in sulfur content. Taking the sulfur out of coal before burning it–though possible–is an expensive process. Much of the fuel oil that is priced competitively with coal is also high in sulfur content. Home heating oil has most of the sulfur removed, but it costs about twice as much as the heavier oils.

POLLUTION FROM AUTOMOBILES

Many environmentalists favor deemphasizing the auto and its place in today's life-style. In other words, they advocate a return to mass transit systems. They even want to prohibit automobiles from the most congested area of a city. The millions of autos in a metropolis pollute the air, take acres of parking space, require fantastic highway complexes, and are generally inefficient in moving people. Ranked in order of efficiency of people moving, the bus, train, bicycle, and even walking are all ahead of the automobile.

No doubt there is a renewed interest in mass transit, but there is little evidence that people are about to willingly give up the family car. It is so woven into our economy and our lives that some people feel that life without the car is as unpalatable as life with polluted air. Hence, the automobile must be cleaned up. Spurred on by the pollution reduction standards of the Environmental Protection Agency, automobile manufacturers are doing just that.

Automobile pollution comes from three principal sources: vapors from the crankcase, fuel vaporization from the gas tank and carburetor, and combustion by-products from the tailpipe. The solutions to automobile pollution are discussed in detail in Unit 9.

RECYCLING

Society cannot afford to continually discard the by-products of living. This practice not only depletes resources but, literally buries us in garbage. Recycling of the earth's resources must someday become the rule rather than the exception, figure 30-5. While recycling is common with things such as scrap iron, copper, brass, aluminum, and paper, the field of recycling is still new. Generally, industrial wastes are located, collected, separated, and controlled or recycled, but the costs are high for industry. The collective refuse from a community is not nearly so easy to separate and control or recycle. Up to now it has been dumped in the ocean or on the land, buried in landfills, or burned in incinerators. A totally closed system must be developed where the wastes that cannot be recycled are controlled, yet do not pollute the environment or damage the ecological balance of the earth.

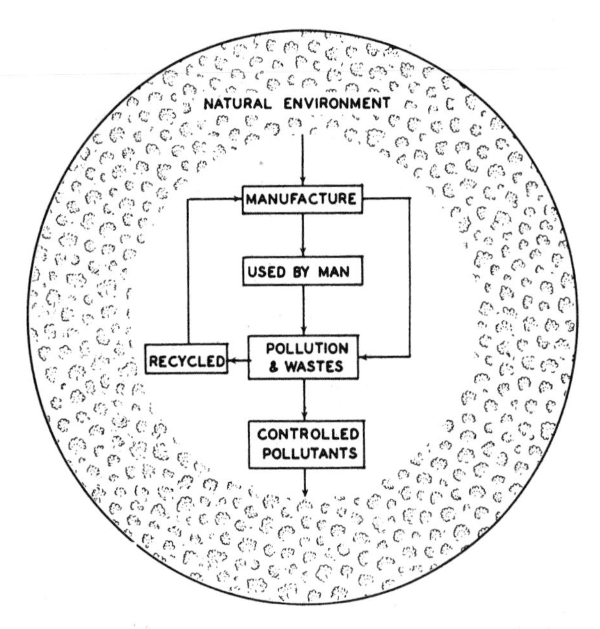

Fig. 30-5 Our natural environment can be saved with a system that recycles resources and controls pollutants.

SUMMARY

Most air pollutants come from transportation; engines and turbines on autos, trucks, trains, ships, and aircraft; and the production of electric power at coal-burning power stations. But identifying and controlling emissions from these sources is not enough. What about the scars on the earth left by the coal strip mines, the millions of wrecked and abandoned automobiles, the chemical pollution in lakes and waterways?

The noise pollution of the city and industrial site has a relationship to power usage. Power has played a part in the production of thousands of nonbiodegradable plastics that will litter the landscape for years to come. The psychology of our prepackaged, throwaway society creates mountains of refuse. Pollution in all forms has a relationship to power usage.

Attitudes have a lot to do with success or failure. The attitudes of individuals, communities, and nations are important. The old attitudes of conquering nature and taking all types of resources must be replaced with attitudes of conserving, obeying the laws of nature, and learning to live on the earth without destroying it, figure 30-6.

Fig. 30-6 Industry can live in harmony with agriculture and nature. (Courtesy American Petroleum Institute)

REVIEW QUESTIONS

1. Explain how the use of power has contributed to pollution.
2. Define the following terms:
   environment pollution ecology
3. What is the function of the Environmental Protection Agency?
4. Name five major sources of air pollution.
5. List five air pollutants.
6. Explain why rapid transit systems pollute less than automobiles.
7. What are the three sources of pollution from the automobile?
8. List three features of auto design that reduce pollution.

Appendixes

Appendix A Troubleshooting, Tune-up, and Reconditioning

Troubleshooting is the intelligent, step-by-step process of locating engine trouble. The troubleshooter examines and/or tests the engine to determine the cause of its disorder. To engage in this process, a person must thoroughly understand the how and why of engine operation. Troubleshooting is more a mental activity than a physical one.

The troubleshooter first establishes the engine's symptoms. The most common causes for these symptoms are then checked. If the solution is not found, another possible cause for the engine's problem is explored. The possible cause for engine failure must quickly be narrowed down because there are many afflictions that can creep into an ailing engine.

An experienced mechanic troubleshoots quickly and with little apparent effort. Years of experience and a thorough understanding of engine operation allow the mechanic to work with such competence.

Engine owners can do their own troubleshooting when they understand engine operating principles. Many owner's manuals contain a troubleshooting chart that is of great assistance. With the aid of an engine troubleshooting chart, the engine owner or operator can locate and correct many engine difficulties without calling in a professional mechanic.

Troubleshooting requires a knowledge of engine operation as well as some common sense. For example, if the engine does not start, there are two possible reasons: either (1) fuel is not getting into the combustion chamber, or (2) fuel is being provided but it is not being ignited. Therefore, the trouble is either in the fuel system or the ignition system. Lubrication or cooling systems have little influence on the ability of the engine to start so these systems are not inspected.

The troubleshooter starts with the simplest and most frequent causes of trouble. In inspecting the fuel system, the first step is to check for fuel in the tank. If this is not the cause of the trouble, the fuel shutoff valve is then checked. The inspection continues in a logical manner until the trouble is found. It would be a mistake to completely disassemble the carburetor or magneto as the first step in troubleshooting.

Look at the troubleshooting charts (figures A-1 to A-3) and discover how many of the common engine troubles can be corrected by an owner who has the tools on hand.

ENGINE TUNE-UP

Engine tune-up does not involve major engine repair work. Rather, it is a process of cleaning and adjusting the engine so that it gives top performance. A tune-up can be done by an experienced engine owner or it can be done by a mechanic. Before beginning tune-up THINK SAFETY. Study the following safety rules.

GENERAL SAFETY RULES FOR TUNE-UP AND RECONDITIONING

- Do not run an engine in a closed building without proper ventilation.

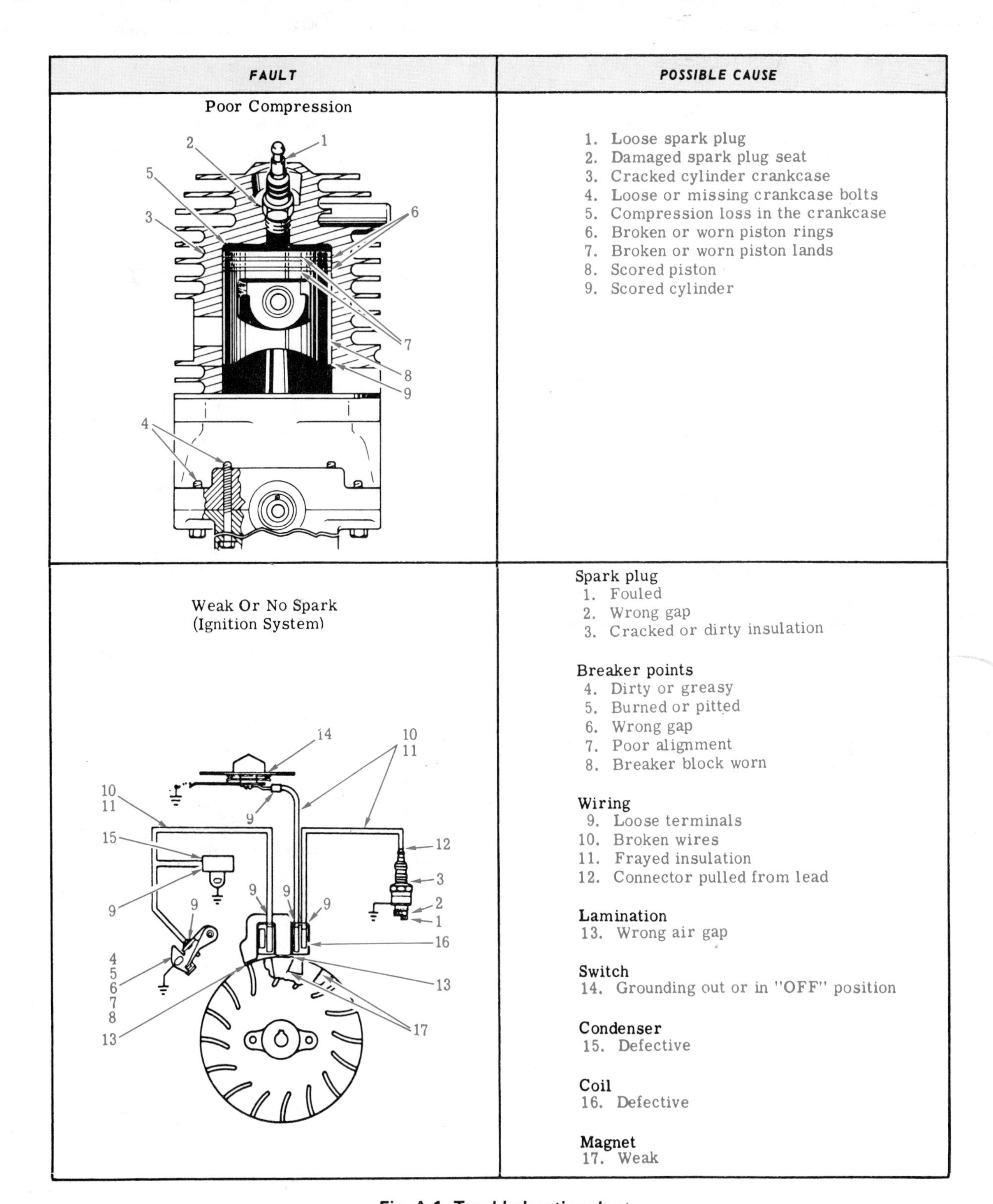

FAULT	POSSIBLE CAUSE
Poor Compression	1. Loose spark plug 2. Damaged spark plug seat 3. Cracked cylinder crankcase 4. Loose or missing crankcase bolts 5. Compression loss in the crankcase 6. Broken or worn piston rings 7. Broken or worn piston lands 8. Scored piston 9. Scored cylinder
Weak Or No Spark (Ignition System)	Spark plug 1. Fouled 2. Wrong gap 3. Cracked or dirty insulation Breaker points 4. Dirty or greasy 5. Burned or pitted 6. Wrong gap 7. Poor alignment 8. Breaker block worn Wiring 9. Loose terminals 10. Broken wires 11. Frayed insulation 12. Connector pulled from lead Lamination 13. Wrong air gap Switch 14. Grounding out or in "OFF" position Condenser 15. Defective Coil 16. Defective Magnet 17. Weak

Fig. A-1 Troubleshooting chart

- Exhaust fumes contain carbon monoxide which is poisonous and can be fatal.
- Do not touch the cylinder head, cylinder block, or exhaust system on an engine that is running or one that has just been stopped. The parts may be hot enough to cause a severe burn.
- Do not add fuel to a hot engine or an engine that is running. Gasoline spilled on the engine could burn.
- Do not operate an engine without all safety guards in place. Take care that your clothing, hands, feet, and hair do not get caught in any moving parts.
- Do not touch any electrical wires or electric components on a running engine. A faulty part or unguarded terminal represents an electrical shock hazard that could cause injury or death.
- If the engine has a battery system that must be disconnected, disconnect the negative or ground cable first. When reconnecting the battery, reconnect power (positive) cable first. Be careful not to spill any battery acid as it can cause serious burns. However, if a spill occurs, it can be neutralized using baking soda.

A typical tune-up procedure includes the following steps. The mechanic's repair manual for the particular engine should be at hand.

1. Inspect the air cleaner then clean and reassemble it. Is the cleaner damaged in any way? Is the filter element too clogged for cleaning? If so, replace the cleaner and/or element. Clean the unit according to the manufacturer's instructions.
2. Clean the gas tank, fuel lines, and any fuel filters or screens.

Problem	No Fuel	Improper Fuel	Dirt In Fuel Line	Dirty Air Screen	Incorrect Oil Level	Engine Over-Loaded	Dirty Filter Element	Faulty Spark Plug
Will not start	X		X			X	X	X
Hard starting	X	X	X	X		X		X
Stops suddenly	X		X	X	X	X	X	
Lacks power		X	X	X		X	X	X
Operates erratically		X	X	X		X	X	X
Knocks or pings		X		X		X		X
Skips or misfires		X	X	X			X	X
Backfires			X			X	X	X
Overheats			X	X	X	X	X	
High fuel consumption							X	X

Fig. A-2 Troubleshooting chart for the engine owner

Engine Will Not Crank

Hydrostatic drive not in neutral.
Battery discharged or defective.
PTO drive engaged.
Defective safety switch.
Defective starter.
Defective solenoid.
Loose electrical connections.
Defective key switch.

Engine Cranks But Will Not Start

Empty fuel tank.
Restricted fuel tank vent.
Fuel shut-off valve closed.
Clogged or restricted fuel line.
Breaker points worn or pitted.
Defective spark plugs.
Battery not fully charged.
Loose electrical connections.
Faulty condenser.
Defective ignition coil.
Dirt in fuel system.

Engine Starts Hard

Defective spark plugs.
Breaker points worn or pitted.
Loose wires or connections.
Restricted fuel tank vent.
Clogged or restricted fuel line.
Broken choke or throttle cable.
Dirt or water in fuel system.
Carburetor not properly adjusted.
Wrong valve clearance.
Head gasket leaking.
Low compression.

Engine Starts But Fails To Keep Running

Restricted fuel tank vent.
Carburetor not properly adjusted.
Broken choke cable.
Dirt or water in fuel system.
Breaker points not properly adjusted.
Loose wires or connections.
Defective head gasket.
Faulty condenser.

Engine Runs But Misses

Breaker points not properly adjusted.
Defective spark plugs.
Loose wires or connections.
Carburetor float leaking or not properly adjusted.
Dirt or water in fuel system.
Wrong valve clearance.
Faulty coil.
Air intake or shrouding plugged.

Engine Misses Under Load

Defective spark plugs.
Carburetor not properly adjusted.
Incorrect spark plugs.
Breaker points not properly adjusted.
Ignition out of time.
Dirt or water in fuel system.

Engine Will Not Idle

Idle speed too slow.
Idle mixture needle not properly adjusted.
Dirt or water in fuel system.
Restricted fuel tank filler cap.
Defective spark plugs.
Wrong valve clearance.
Low engine compression.

Engine Loses Power

Crankcase low on oil.
Air intake or shrouding plugged.
Excessive engine load.
Restricted air filter.
Dirt or water in fuel system.
Carburetor not properly adjusted.
Defective spark plugs.
Too much oil in crankcase.
Low engine compression.
Worn cylinder bore.

Engine Overheats

Air intake or shrouding plugged.
Carburetor not properly adjusted.
Too much oil in crankcase.
Crankcase low on oil.
Excessive engine load.

Engine Knocks

Engine out of time.
Excessive engine load.
Crankcase low on oil.

Engine Uses Excessive Amount of Oil

Clogged breather assembly.
Breather not assembled properly.
Worn or broken piston rings.
Worn cylinder bores.
Wrong size piston rings.
Worn valve stems and/or valve guides.
Incorrect oil viscosity.

Fig. A-3 Troubleshooting chart for the mechanic

3. Check the compression. This can be done by slowly turning the engine over by hand. As the piston nears top dead center, considerable resistance should be felt. As top dead center is passed, the piston should snap back down the cylinder. A more thorough test can be made with a compression gauge, figure A-4. The manufacturer's repair manual should be consulted for normal and minimum acceptable compression pressure.

   NOTE: Many engines have an automatic compression release mechanism which holds the intake valve slightly open for cranking the engine during starting. This allows the operator to start the engine with less pull. However, the automatic compression release makes checking the compression difficult.

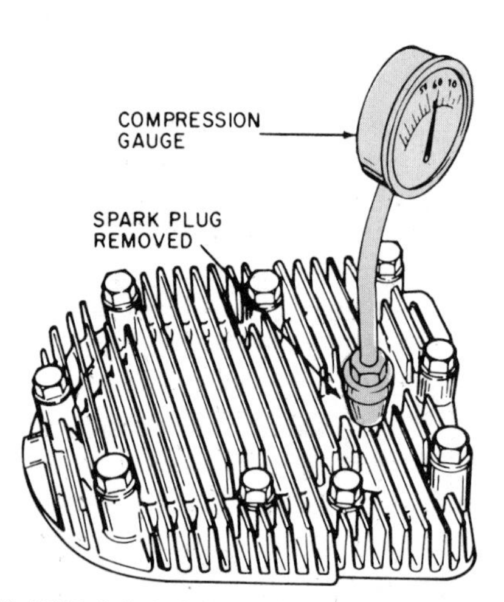

Fig. A-4 Checking compression with a gauge

4. Check the spark plug: clean, regap, or replace it. Remove the spark plug and attach a grounded spark plug tester to the spark plug, figure A-5. Reattach the spark plug cable and clamp the tester to the engine block. Turn the engine over. If a good spark jumps at the plug electrodes the plug is good, figure A-6. This is not an absolute test but is a good indication. If the plug is questionable, replace it.

5. Check the operation of the governor, figure A-7. Be certain the governor linkages do not bind at any point.

6. Check the magneto. On most engines, the flywheel must be removed for access to the magneto parts. Use a flywheel or gear puller if available, figure A-8. Otherwise, pop the flywheel loose by removing the flywheel nut and delivering a sharp hammer blow to a lead block held against the end of the crankshaft.

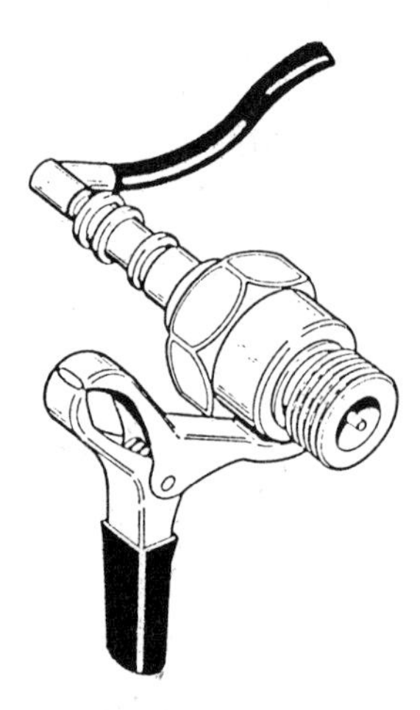

Fig. A-5 Checking for spark (Courtesy of Tecumseh Products Company)

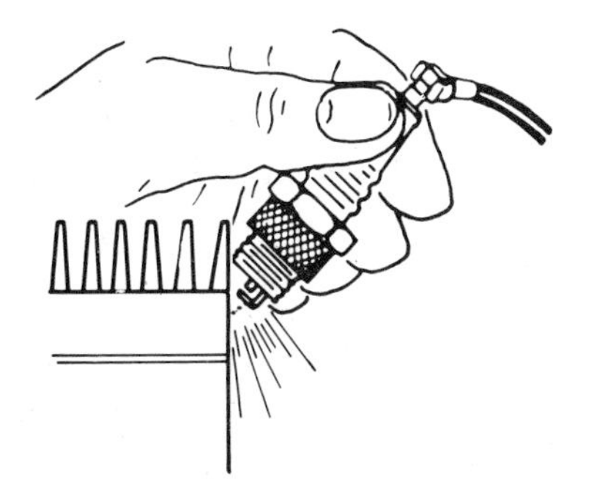

Fig. A-6 Checking spark plug

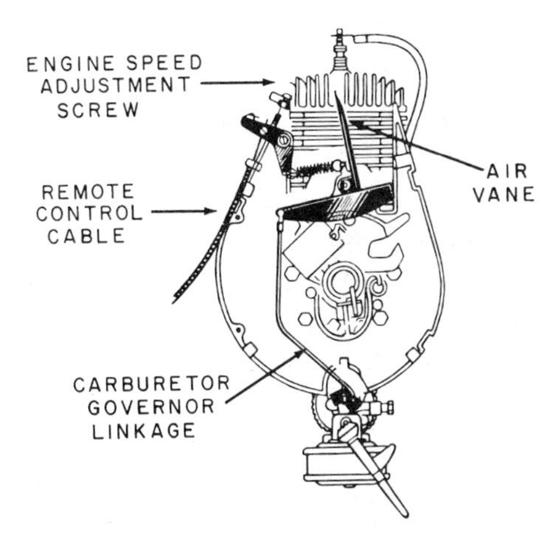

Fig. A-7 Air vane governor parts

Another method is to back off the flywheel nut until it is about 1/3 off the crankshaft and then strike it sharply. Do not hammer on the end of the crankshaft as this can damage the threads, figure A-9.

Adjust the breaker point gap and check the condenser and breaker point terminals for tightness, figure A-10. Breaker points are set for 0.020 inch in most cases although the correct setting may vary according to the manufacturer. Turn the engine over until the breaker points reach their maximum opening, then check with a flat feeler gauge or wire gauge. Adjust the points if the setting is incorrect. Breaker points must line up, figure A-11. Points can be cleaned if necessary. If the points are pitted or do not line up, they should be replaced, figure A-12.

7. Fill the crankcase with clean oil of the correct type. (four-stroke cycle engines)

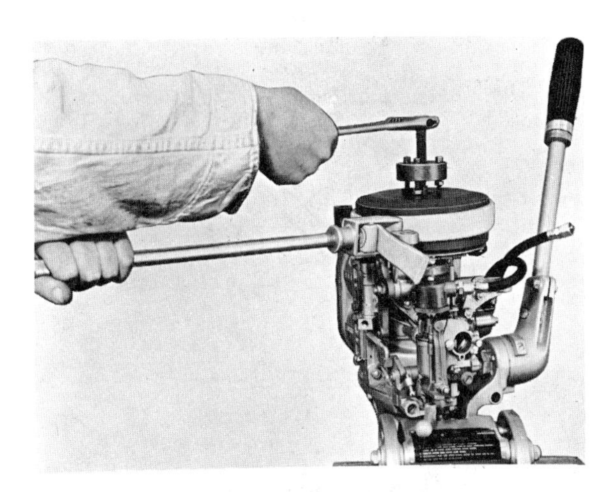

Fig. A-8 Removing a flywheel using a special puller

Fig. A-9 Removing the flywheel with a soft hammer

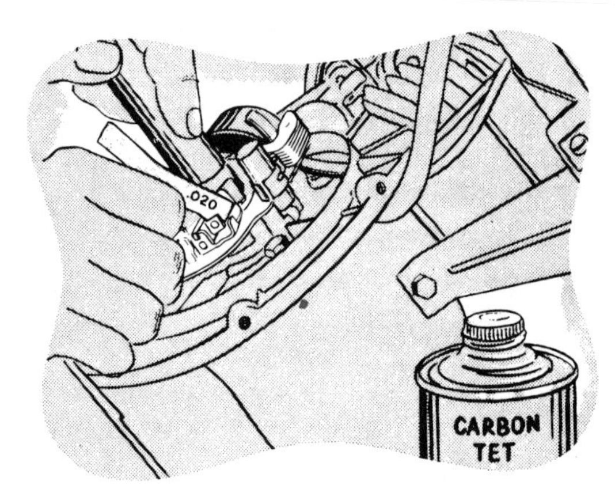

Fig. A-10 Clean and check breaker points.

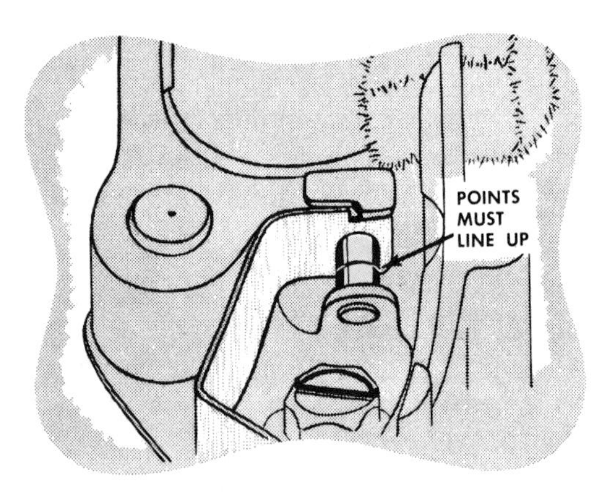

Fig. A-11 Breaker points must line up.

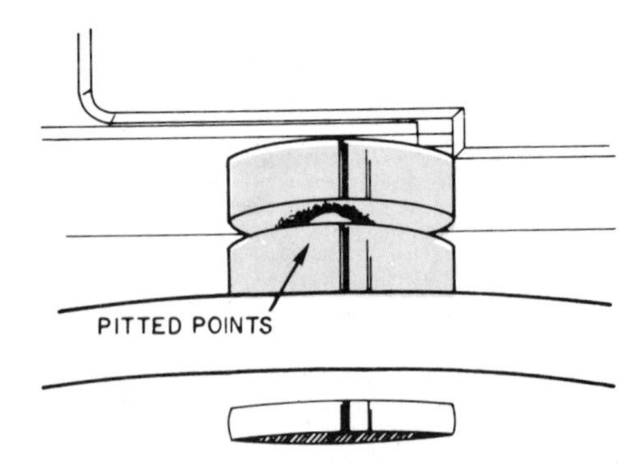

Fig. A-12 Pitted breaker points should be replaced.

8. Fill the gasoline tank with regular gasoline. (four-stroke cycle engines)

   Fill the gasoline tank with the correct mixture of gasoline and oil. (two-stroke cycle engines)
9. Start the engine.
10. Adjust the carburetor for peak performance. See carburetor parts shown in figure A-13.

RECONDITIONING

Reconditioning or overhauling an engine is generally the job for the mechanic – one who has the tools and know-how to do the job correctly. However, much reconditioning and overhauling can be done by the amateur who has the basic tools and the correct approach. Care and precision are very important. Every part must be in place correctly, with none missing or left over.

Before beginning the disassembly of an engine, provide a spot to put the parts as they are removed. One good method is to lay out a large sheet of paper. As the parts are placed

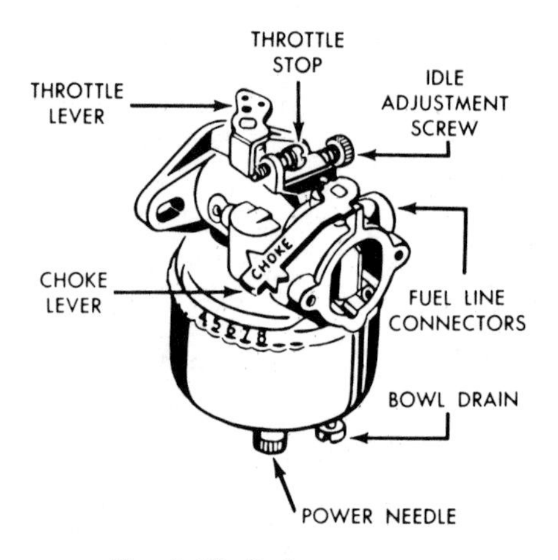

Fig. A-13 Carburetor parts

on the paper, label them so they will not be lost. It is often difficult to remember just where a part came from when forty or fifty pieces are laid out; also, it may be several days before the engine is reassembled.

A mechanic's handbook or service manual is essential for top-quality overhaul work. These books can sometimes be obtained directly from the engine manufacturer or borrowed from a mechanic. In only a few instances is this type of book supplied with the new engine. The service manual is quite detailed, fully explaining each step of overhaul or reconditioning. Allowances, clearances, torque data and other specifications are also given, figure A-14. Without the use of the service manual there is too much guesswork involved. The amateur mechanic should have the service manual for the engine before any repair work is started.

Tools and Equipment

Unless the engine owner is a skilled mechanic and has an adequately equipped repair shop, the owner cannot perform all types of engine repair and overhaul. Most owners

CLINTON ENGINES TORQUE DATA – INCH POUNDS
"Red Horse"

		1600 A1600 A1690	1800 1890	2500 A2500	B2500 B2590 2790
Bearing Plate P.T.O.	Min.	160	160	16C	160
	Max.	180	180	180	180
Back Plate to Block	Min.	70	70	70	70
	Max.	80	80	80	80

CLINTON ENGINES SERVICE CLEARANCES
"Red Horse"

		1600	A1600	A1690	1800	1890	2500	B2500	B2590	A2500	2790
Piston Skirt Clearance	Min.	.007	.007	.007	.007	.007	.0065	.0065	.0065	.005	.005
	Max.	.009	.009	.009	.009	.009	.0085	.0085	.0085	.007	.007
	Rework	.010	.010	.010	.010	.010	.010	.010	.010	.0085	.0085
Ring End Gap	Min.	.007	.007	.007	.007	.007	.010	.010	.010	.010	.010
	Max.	.017	.017	.017	.017	.017	.020	.020	.020	.020	.020
	Rework	.025	.025	.025	.025	.025	.028	.028	.028	.028	.028

CLINTON ENGINES TOLERANCES AND SPECIFICATIONS
"Red Horse"

	1600	A1600	A1690	1800	1890	2500	B2500	B2590	A2500	2790
Cylinder - Bore	2.8125 2.8135	2.8125 2.8135	2.8125 2.8135	2.9995 3.0005	2.9995 3.0005	3.1245 3.1255	3.1245 3.1255	3.1245 3.1255	3.1245 3.1255	3.1245 3.1255
Skirt - Diameter	2.8045 2.8055	2.8045 2.8055	2.8045 2.8055	2.9915 2.9925	2.9915 2.9925	3.117 3.118	3.117 3.118	3.117 3.118	3.1185 3.1195	3.1185 3.1195

Fig. A-14 Torque data, clearances, tolerances, and specifications as excerpted from a service manual

TORQUE SETTINGS – GENERAL

Tightening Torque Into Cast Iron or Steel

Size	Grade 2	Grade 5	Grade 8
8-32	20 in-lb	25 in-lb	
10-24	32 in-lb	40 in-lb	
10-32	32 in-lb	32 in-lb	
1/4-20	70 in-lb	115 in-lb	165 in-lb
1/4-28	85 in-lb	140 in-lb	200 in-lb
5/16-18	150 in-lb	250 in-lb	350 in-lb
5/16-24	165 in-lb	270 in-lb	30 ft-lb
3/8-16	260 in-lb	35 ft-lb	50 ft-lb
3/8-24	300 in-lb	40 ft-lb	60 ft-lb
7/16-14	35 ft-lb	55 ft-lb	80 ft-lb
7/16-20	45 ft-lb	75 ft-lb	105 ft-lb
1/2-13	50 ft-lb	80 ft-lb	115 ft-lb
1/2-20	70 ft-lb	105 ft-lb	165 ft-lb
9/16-12	75 ft-lb	125 ft-lb	175 ft-lb
9/16-18	100 ft-lb	165 ft-lb	230 ft-lb
5/8-11	110 ft-lb	180 ft-lb	260 ft-lb
5/8-18	140 ft-lb	230 ft-lb	330 ft-lb
3/4-10	150 ft-lb	245 ft-lb	350 ft-lb
3/4-16	200 ft-lb	325 ft-lb	470 ft-lb

Tightening Torque Into Aluminum

Size	Grade 2	Grade 5
8-32	20 in-lb	20 in-lb
10-24	32 in-lb	32 in-lb
1/4-20	70 in-lb	70 in-lb
5/16-18	150 in-lb	150 in-lb

CONVERSIONS

in-lb X 0.083 - ft-lb
ft-lb X 12 - in-lb
ft-lb X 0.1383 - kgm
ft-lb X 1.3558 - N m

Fig. A-15 General torque settings for bolts

would not find it practical to purchase the equipment necessary to do all types of overhaul and repair. Professional repairers may have invested from five hundred to several thousand dollars in tools and equipment. Equipment may consist of: magneto testers, valve seat resurfacers, air compressors, steam cleaners, special sharpeners, grinders, lapping stands, special factory tools, general repair tools, repair parts, and others.

A small investment in tools allows a person to perform many repair jobs. Most home workshops have some of the basic tools that are necessary. Several specialized tools can be added, figure A-16. A tentative list of necessary tools include the following:

- Screwdriver (various sizes)
- Combination wrenches (set)
- Adjustable wrench
- Socket set
- Torque wrench
- Deep well spark plug socket
- Needle nose pliers
- Feeler gauge/wire gauge
- Spark gap gauge
- Piston ring compressor
- Piston ring expander
- Valve spring compressor

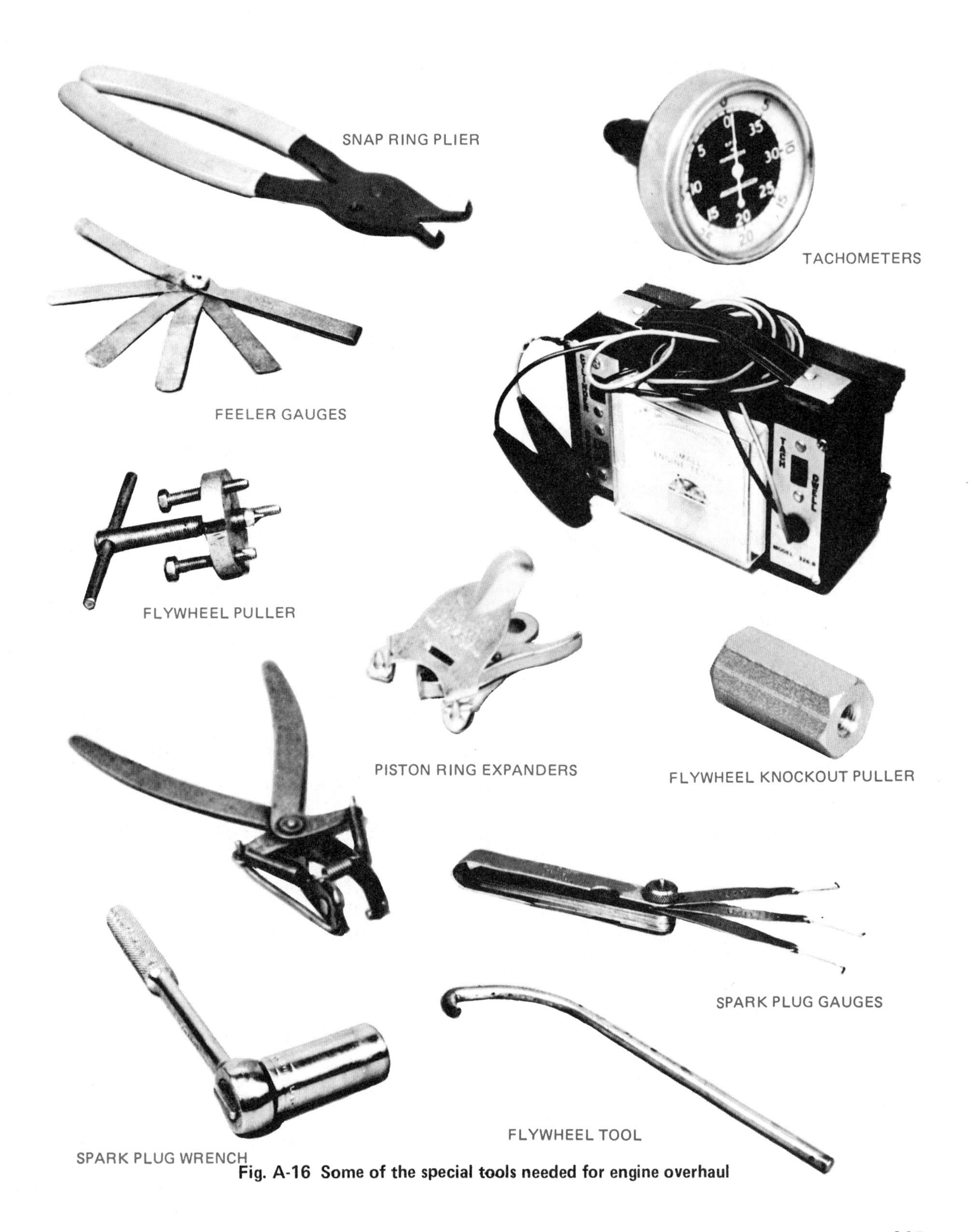

Fig. A-16 Some of the special tools needed for engine overhaul

Additional tools:

- Flywheel puller
- Compression gauge
- Tachometer
- Valve grinder – hand operated
- Arbor press

Basic Steps

Reconditioning or overhaul of an engine involves four basic steps: disassembly, inspection of parts, repair or replacement of worn or broken parts, and reassembly. In actual practice, a service manual would be followed for these steps. However, the following disassembly procedure is a general guide for a four-stroke cycle engine. Before disassembly, however, check for compression, spark, and crankshaft end-play. These items cannot be checked once the engine is disassembled.

1. Disconnect the spark plug lead and remove the spark plug.
2. Drain the fuel system – tank, lines, carburetor.
3. Drain the oil from the crankcase.
4. Remove the air cleaner.
5. Remove the carburetor.
6. Remove the metal air shrouding, gas tank, and recoil starter.
7. Remove the flywheel.
8. Remove the breaker assembly and push rod.
9. Remove the magneto plate assembly.
10. Remove the breather plate assembly (valve spring cover). Check the valve clearance.
11. Remove the cylinder head. Check for warpage.
12. Remove the valves.
13. Remove the base.
14. Remove the piston assembly.
15. Remove the crankshaft.
16. Remove the camshaft and tappets.
17. Remove the mechanical governor.

As general information, the disassembly, inspection, repair, and reassembly for each engine part are discussed in this unit.

FLYWHEEL

Removal of the flywheel is necessary to gain access to the ignition system parts. If a special flywheel puller designed for the specific engine is available, use it. Otherwise use a flywheel puller. Be sure the flywheel nut or any threaded starter mechanism is removed before trying to pull the flywheel. Opposing pressure against the flywheel and the end of the crankshaft will break the contact of the tapered hole and the tapered shaft. Care should be taken with aluminum flywheels since excessive force can cause them to break.

If a flywheel puller is not available, the flywheel can be removed by first removing the flywheel nut and then striking the end of the crankshaft with a plastic, lead, or other soft-face hammer. A sharp blow on the shaft while exerting some outward pull on the flywheel will separate the flywheel from the shaft. Take care not to damage the end of the crankshaft, especially the threads. Engines that rotate clockwise have right-handed threads on the crankshaft and engines that

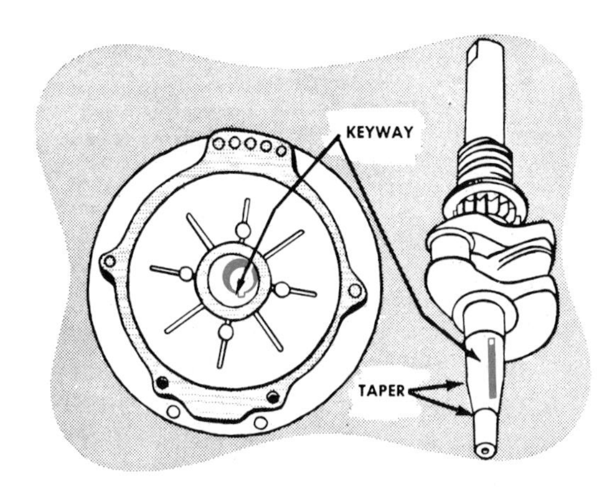

Fig. A-17 Keyway on tapered end of the crankshaft and the flywheel

rotate counterclockwise have left-handed threads.

The magnets in the flywheel can be damaged if the flywheel is dropped or if the magnets are hammered. Excessive heat causes the magnets to lose magnetism.

The key that aligns the crankshaft and flywheel is small; care should be taken not to lose it. If the key is damaged or worn, it should be replaced since poor alignment can affect engine timing. When reassembling the flywheel onto the crankshaft, carefully fit the key into the keyway, figure A-17. Also, be sure that no washers are forgotten when the flywheel is reinstalled.

MAGNETO PARTS

The magneto parts (high-voltage coil, breaker points, condenser) may need to be removed for checking or replacement. The coil may be visually inspected for cracks and gouges in insulation, evidence of overheating, and the condition of the leads where they go into the coil. If an ignition coil tester is available, the coil can be checked for firing, leakage, secondary continuity, and primary continuity. This checking should be done using the procedure recommended by the manufacturers.

Breaker points that are corroded, pitted or dirty, provide a poor electrical contact. This lowers the magneto's ability to produce the proper voltage in the primary windings. Also, when the breaker points open, less voltage is induced because a clean, quick break is not produced. Breaker points that are in poor condition can seriously reduce the voltage available to the spark plug.

The setting of the breaker points (usually 0.020 inch) affects the engine timing. If the points are set for too small an opening, the spark is retarded. The spark is late because the cam must rotate further before the points open. Too wide an opening causes the spark to be advanced. Both conditions cause an engine to lose power. The two points should meet each other flat with their surfaces in full contact with each other. Points have a chrome or silvery appearance when new, and become gray as they are used.

The condenser can be visually inspected for dents, terminal lead damage and broken mounting clips. A condenser tester is used to check the condenser for capacity, leakage and series resistance, figure A-18. The test procedure suggested by the test equipment manufacturer should be followed.

Manufacturers often recommend replacing the condenser when the points are replaced even though the condenser is still reliable and in good condition. If one of the breaker points is a part of the condenser, the entire condenser must be replaced.

When the magneto is reassembled, the laminated iron core must be as close to the flywheel and its magnets as possible without

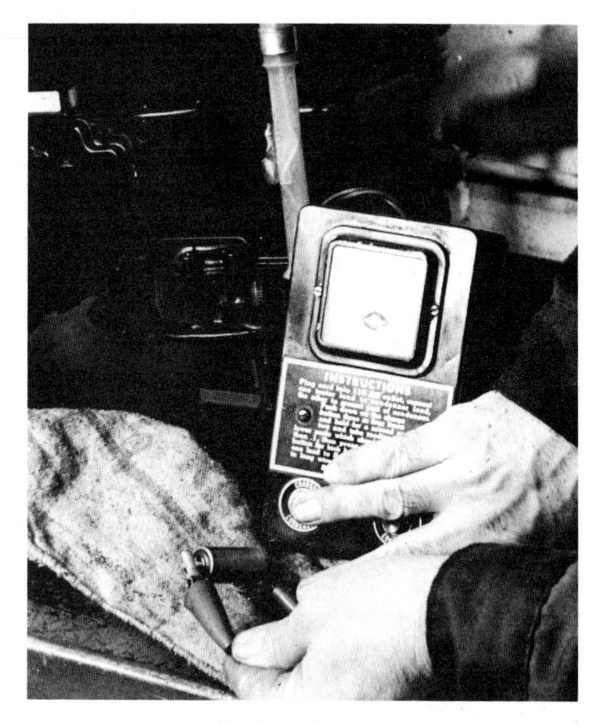

Fig. A-18 Checking a condenser

touching or rubbing as the flywheel revolves. The closer the magnet is to the core (or pole shoes) the stronger the magnetic influence of the flywheel magnets. A space of about 0.007 inch to 0.012 inch is usually suggested, about the thickness of a postal card.

ENGINE TIMING

Engine timing refers to the magneto timing to the piston – the position of the piston just as the breaker points start to open. Timing is set at the factory but it is possible for timing to cause engine trouble. If the breaker points open too late in the cycle, power is lost. If the breaker points open too early in the cycle, detonation can result. Improper engine timing can be caused by the

- crankshaft and camshaft gears being installed one tooth off
- spark advance mechanism being stuck
- breaker points incorrectly set
- magneto assembly plate loose or slipped
- rotor incorrectly positioned.

These causes of trouble depend on the engine and the type of magneto.

The position of the piston for timing a magneto varies from engine to engine. The piston can be at top dead center (TDC) or slightly before the top dead center (BTDC). Check engine specifications for this information.

One common method of checking timing is to locate TDC. This is the point where the piston does not seem to move as the flywheel is rotated. A dial indicator can be used for greater accuracy. With TDC located, place a reference mark on the flywheel and the magneto assembly plate, figure A-19. Now remove the flywheel. Rotate the crankshaft until the breaker points barely open (0.001 inch). Carefully replace the

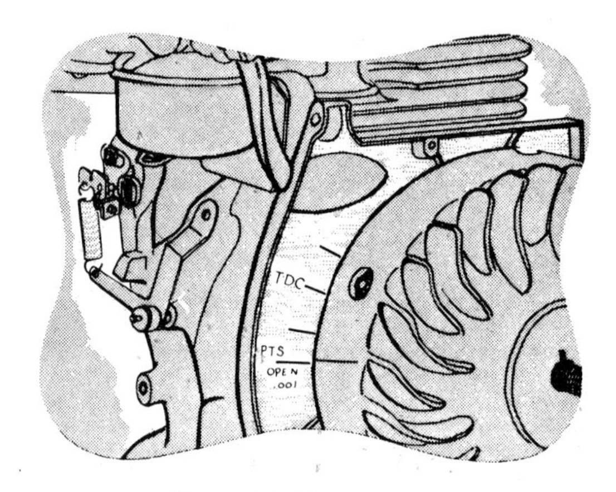

Fig. A-19 Timing marks

flywheel. Put another mark on the magneto plate aligned with the flywheel reference mark. The difference between the marks on the magneto plate is the magneto timing to the piston. This can be figured in degrees by dividing the number of flywheel vanes into 360 degrees. (Twenty vanes; each vane 18 degrees.) The correct number of degrees of firing before top dead center is found in the engine specifications.

If the engine has a magnetic rotor magneto, the rotor and armature must be timed to the piston, figure A-20. Timing is correctly set when the engine leaves the factory. However, if the armature has been removed or the crankshaft or cam gear replaced, it is necessary to retime the rotor. Breaker points are first set correctly (0.020 inch). The rotor is on the shaft correctly and tightened. The armature is mounted but mounting screws are not tight. Turn the crankshaft until the breaker points just start to open. (Place a piece of tissue paper between the points to detect when the points let go.) Now turn the armature slightly until the timing marks on the rotor and the armature line up.

Provisions are made on many engines to check the timing of the engine when the engine is running. This timing is done with a timing light, figure A-21. Follow the instructions given for the particular timing light and engine. The timing light flashes each time the breaker points open. Reference marks on the engine must be located – one on a stationary part of the engine and one on the flywheel. The timing light is directed at the flywheel. If the two marks line up, timing is correct. If the marks do not line up, the breaker plate must be loosened and shifted to a position where the marks do line up.

Two-cycle engine timing is quite similar,

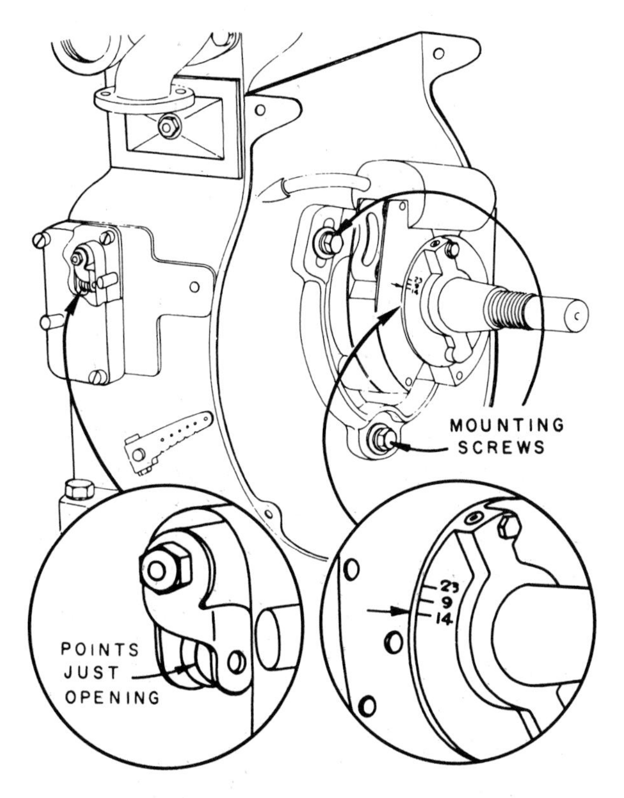

Fig. A-20 Timing the rotor and armature to the piston. The numbers represent model numbers.

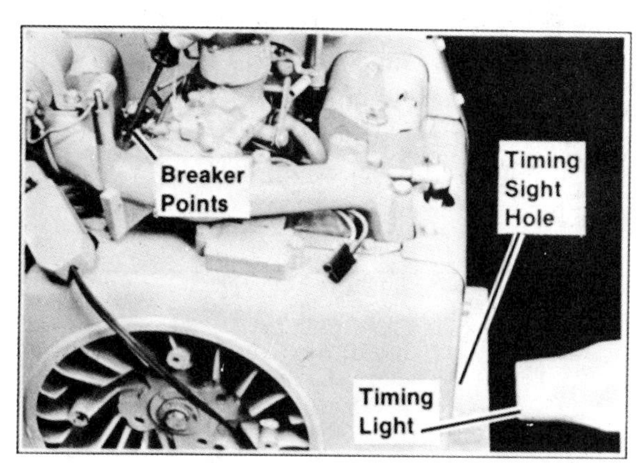

Fig. A-21 Timing light method

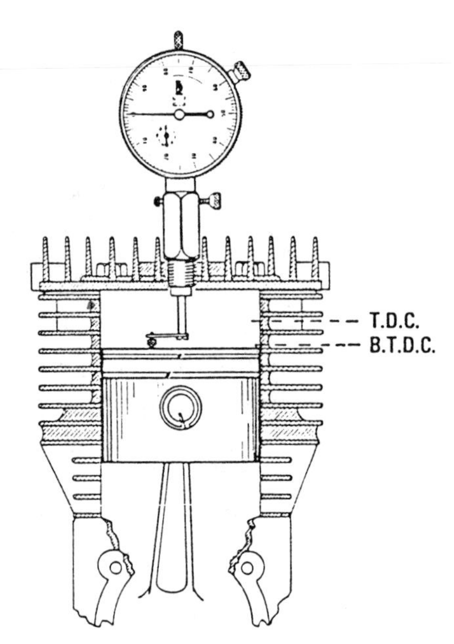

Fig. A-22 Two-cycle engine timing (Courtesy of Tecumseh Products Company)

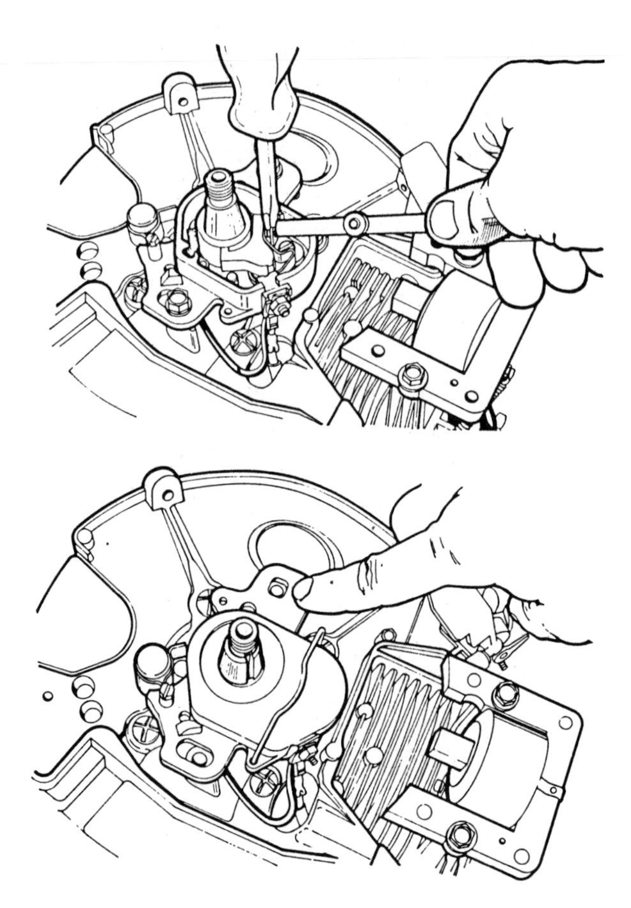

Fig. A-23 Two-cycle timing (Courtesy of Tecumseh Products Company)

figure A-22. One common procedure is to remove the spark plug and locate top dead center using a dial indicator. With TDC located, back the piston down the cylinder the correct distance (check manufacturer's specifications for the measurement before TDC). Now the breaker points should just begin to open. If timing is incorrect, loosen the stator plate setscrew and rotate the stator plate slightly until the points just begin to open. Retighten the stator plate setscrew. See figure A-23 for an example of two-cycle timing.

CYLINDER HEAD

The cylinder head must be removed if work is to be done on the piston, piston rings, connecting rod, or valves. It must also be removed if the combustion chamber is to be cleaned. The head bolts or nuts should be removed and set aside. The head gasket may be stuck to the head, cylinder, or both. In many cases the gasket is badly damaged while the head is being removed. If the gasket is damaged, it should be discarded.

Carbon deposits should be carefully scraped from the cylinder head and all old gasket material scraped away from the contact surfaces, figure A-24. The cylinder head can be checked for warpage by laying it on a *surface plate* (a large block of metal with a perfectly flat surface), figure A-25. If the cylinder head is warped, it should be replaced.

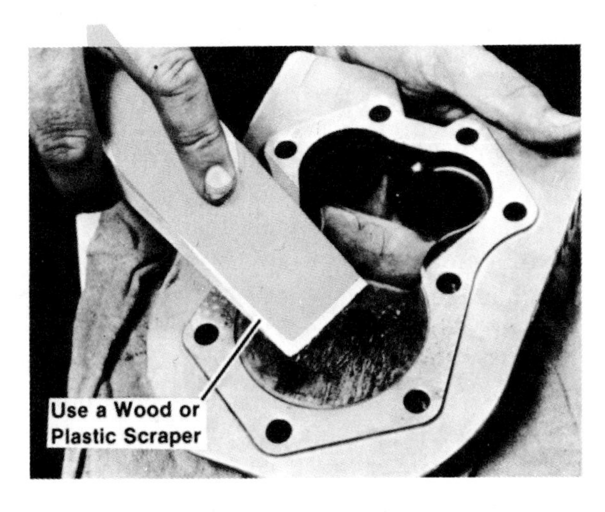

Fig. A-24 Cleaning cylinder heads

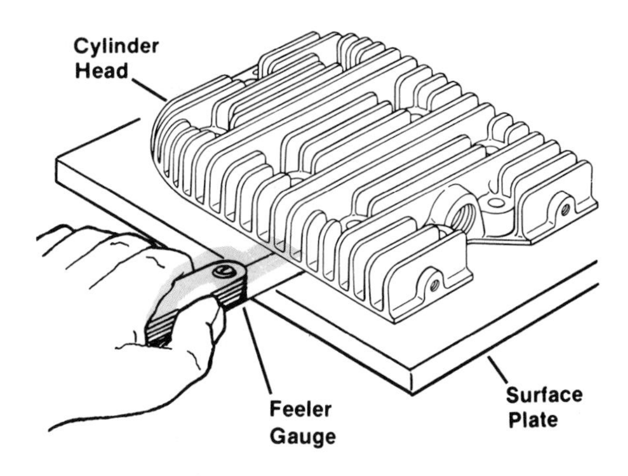

Fig. A-25 Checking cylinder head flatness

A new head gasket should be used upon reassembly. The cylinder head bolts or nuts should be tightened in the correct sequence. To be certain of the correct degree of tightness, a torque wrench should be used (generally 14 to 18 ft-lb), figure A-26.

VALVES

Valve problems are fairly common and can be quite serious. Faulty valves affect engine power and cause hard starting. A poorly seated valve is one of the possible causes for poor compression. If it is reasonably certain that the piston rings are in good shape and well-oiled and the head gasket is air tight and a compression test indicates leakage, the leakage is probably caused by a valve problem. A quick test is to remove the spark plug and hold a finger over the spark plug hole while turning over the engine. Compression should blow the finger away from the hole with noticeable force.

With the cylinder head removed the valves can be visually inspected. Look for problems

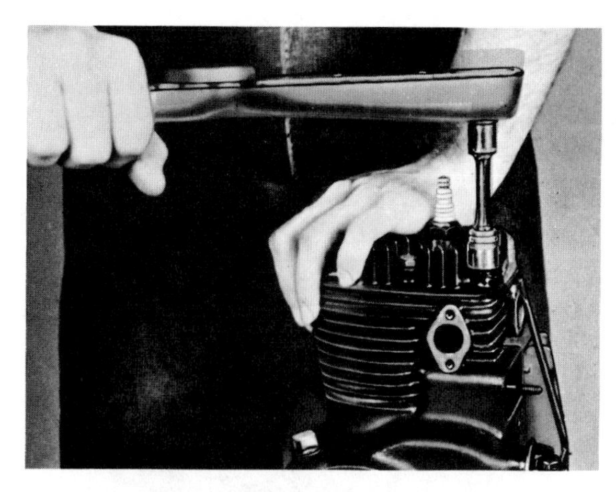

Fig. A-26 Tightening the cylinder head bolts with a torque wrench

such as carbon fouling, burned valves, valve corrosion, or valve sticking.

Carbon Fouling. Carbon deposits may have built up to the point where the valves do not fully seat. Also, carbon particles that have broken loose may be under the valve seat.

Burned Valves. Burned valves are often caused by improper tappet clearance, figure

A-27. The valve may seat nicely when the engine is cold. However, when the parts heat up they expand and the valve cannot fully seat. Being constantly exposed to the hot gases of combustion, the valve may burn and the valve seating becomes progressively worse.

Valve Corrosion. Valve corrosion can be seen when the valve is removed, figure A-28. The top of the valve stem (neck) is worn away or thinned out. The problem is usually caused by having too lean a fuel mixture or by condensation. Condensation can form during storage or when the engine is frequently stopped before it reaches its normal operating temperature.

Fig. A-27 Burned valve

Valve Sticking. A stuck valve means that the valve stem and valve guide have built up deposits of carbon and gum, figure A-29. The deposits build up more rapidly in hot weather and under heavy engine loads. An engine that runs hot due to clogged radiating fins or a clogged air inlet screen builds up valve stem deposits faster than normal. Old gasoline often causes a rapid buildup of gum deposits. When the engine is cold, the valves may not stick; as the engine warms up the deposits expand and the valves stick. To correct the problem, the valve guides must be reamed out and the valves cleaned or replaced. Exhaust valves take more punishment than intake valves and are more subject to valve problems.

Engine owners may be able to do some valve work themselves. However, a large amount of work or a complete renewal of the valve system should be done by a mechanic who has the tools and necessary experience.

Fig. A-28 Valve corrosion

Fig. A-29 Gum deposits cause valve sticking

Fig. A-30 Checking tappet clearance

A complete valve job may include:

- Installing new valves
- Installing new valve guides
- Installing new valve seats
- Installing new valve springs
- Grinding valve seats
- Lapping valves
- Setting valve clearances

One of the first check points is the valve *clearance* (space between tappet and end of valve). This clearance is checked with a feeler gauge, figure A-30. On most small engines the clearance can be enlarged by grinding a small amount from the end of the valve. If the clearance is already too large, the valve must be replaced. Some small engines have adjustable tappet clearances.

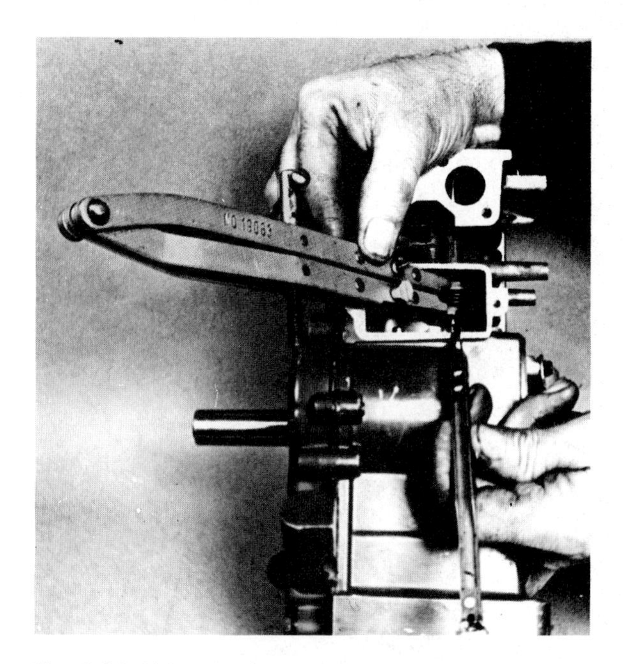

Fig. A-31 Using a valve spring compressor to remove the valve springs

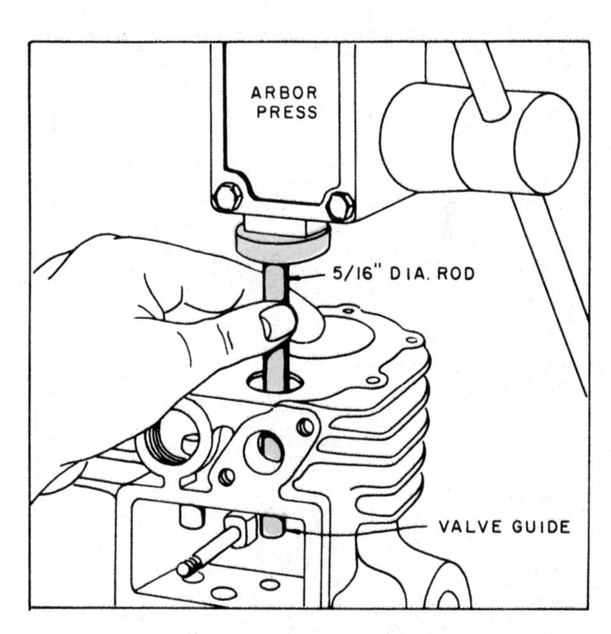

Fig. A-32 Removing a worn valve guide

To remove the valves, first compress the valve spring; then slip off or slip out the valve spring retainers, sometimes called keepers. Pull the valve out of the engine. The valve spring and associated parts will also come out, figure A-31.

With the valve out, the valve stem, face, and head can be closely inspected and cleaned. The valve guide and valve seat can also be inspected.

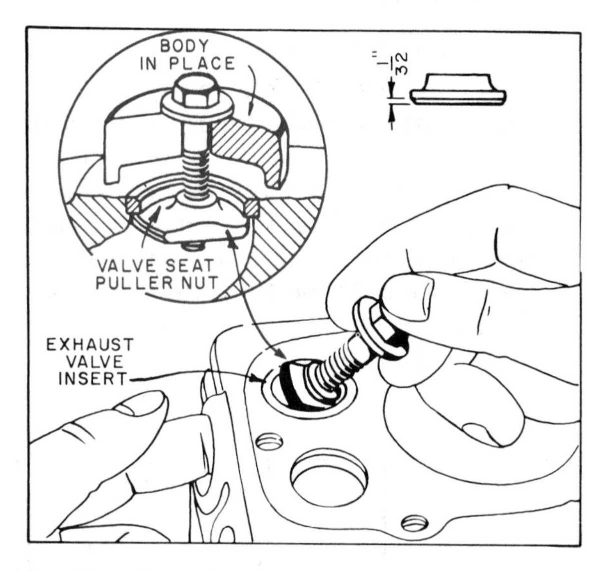

Fig. A-33 Removing the exhaust valve seat with a special puller

Many engines have replaceable valve guides. If these are worn or other wise damaged, they must be pressed out using an arbor press or carefully driven out with a special punch, figure A-32. Also, the exhaust valve seat is removable on many engines. If the seat is beyond regrinding, remove it and reinstall a new valve seat. A special valve seat extracting tool is used for this job, figure A-33.

If the valve and/or valve seats need regrinding, this can be done with special valve grinding equipment, figure A-34. However, in some cases hand valve grinders can be used, figure A-35.

If either or both of these parts have been replaced or reground, the parts must be lapped to provide the perfect seal necessary for valve operation, figure A-36. When lapping, use a small amount of lapping compound. Rotate the valve against the seat a few times until the compound produces a dull finish on the valve face. Do not lap the valves too heavily. When the valves are replaced, oil the stems and be sure the exhaust valve goes in the exhaust side and that the intake valve goes in the intake side.

Fig. A-34 Valve grinder

Fig. A-35 Reconditioning a valve seat

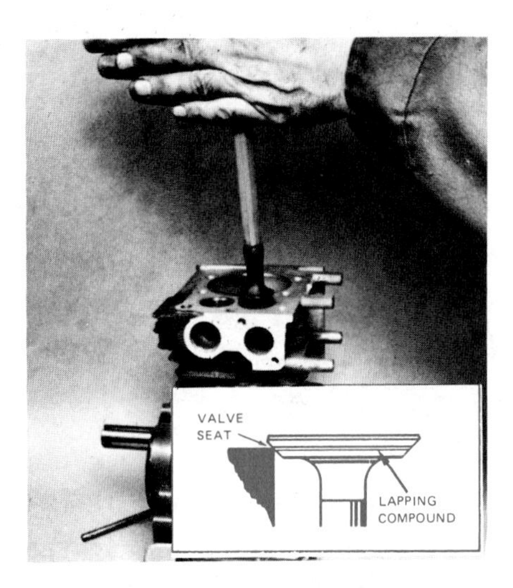

Fig. A-36 Lapping valves

PISTON AND ROD ASSEMBLY

In order to gain access to the piston and connecting rod assembly, the connecting rod bolts must be reached. On many engines the mechanic must first remove the engine from the base or sump, figure A-37. This is done by loosening the bolts and breaking the seal. In most cases, a new gasket should be installed upon reassembly to prevent oil leakage. On engines that do not have a separate base, the bearing plate on the power takeoff side will need to be removed.

To remove the piston rod assembly, remove the connecting rod cap from around the crankshaft. Look for the boss marks that indicate how the end of the connecting rod and the rod cap go together. They must be reassembled in the same position. Push up on the bottom of the connecting rod and push the piston rod assembly up and out of the cylinder, figure A-38. If there has been carbon buildup in the cylinder above the point of ring travel, the carbon should be scraped clean before the piston is removed.

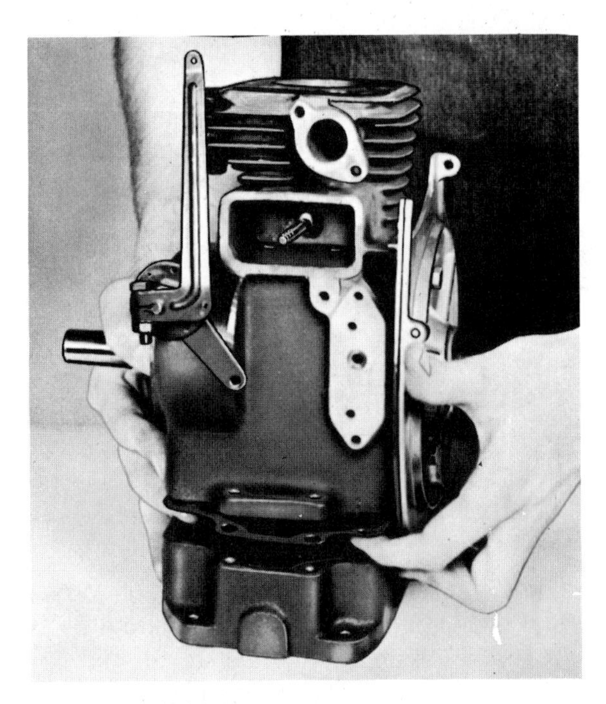

Fig. A-37 Engine base, gasket, engine

Fig. A-38 Connecting rod and piston removal

Mark the piston so it can be reinstalled the same way it came out.

CYLINDER

The cylinder should be checked for score marks. Scoring or wear scratches in the area of ring travel cause increased oil consumption and reduced engine power. Also the cylinder

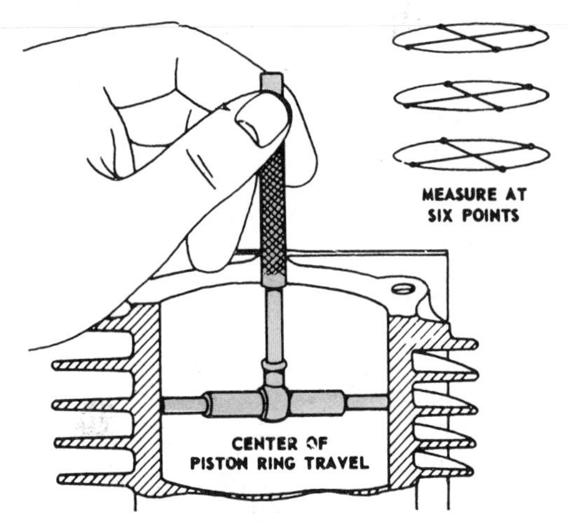

Fig. A-39 Check cylinder bore

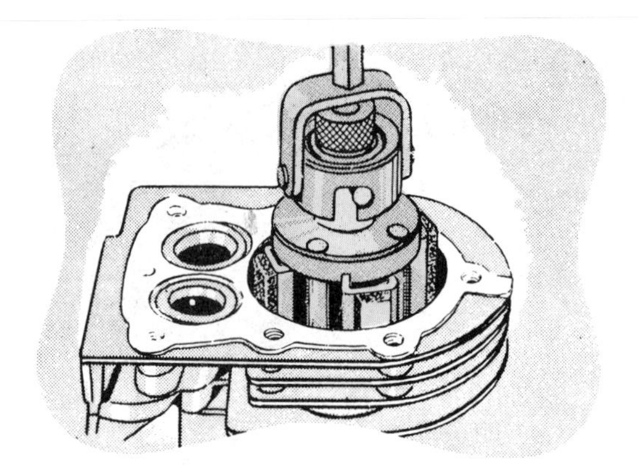

Fig. A-40 Deglazing a cylinder with a finish hone

should be checked for distortion or warpage and size with a cylinder gauge, figure A-39.

If the cylinder appears to be in good condition, it can be deglazed with a finish hone to prepare the cylinder for new rings, figure A-40. However, a cylinder in poor condition may have to be rebored to a larger size which will also require a new and oversize piston for the cylinder. Some manufacturers do not recommend honing or deglazing.

PISTON

The piston should be checked for scuffing, scoring and scratches. Scoring marks are relatively deep marks that are caused by overheating and the metal fusing against the cylinder and wiping away metal. Lack of lubrication is usually the cause. Scuffing marks are not as deep. Scratches may be caused by abrasive materials entering the engine through the oil supply or air cleaner, figure A-41.

The size of the piston itself can be checked with a micrometer. Generally it is 0.005 inch to 0.010 inch smaller than the cylinder. If the piston is too small, the piston skirts may slap the cylinder as it moves up and down.

PISTON PIN

The piston pin can be removed by dislodging the retainer rings from their grooves. Needle nose pliers work quite well for this operation. The piston pin is usually quite tight in the piston so it must be driven out with a soft punch and hammer. Reassemble the piston pin and connecting rod in the same positions.

PISTON RINGS

If the piston size and condition have checked out all right, the piston rings should be checked next. Check the side groove clearance with a feeler gauge, with the rings still on the piston, figure A-42. The correct side groove clearance is found in the engine overhaul specifications. Remove the piston rings from the piston with a piston ring expander, figure A-43. Once the rings are removed, start at the top and look at the end of each ring to see how it was installed. Many rings will have a bevel or notch that must face up or down. Carefully put the ring in the cylinder, square it with the piston, and check the end gap with a feeler gauge, figure A-44. Too little end gap can cause the ring to freeze

Fig. A-41 Piston damage

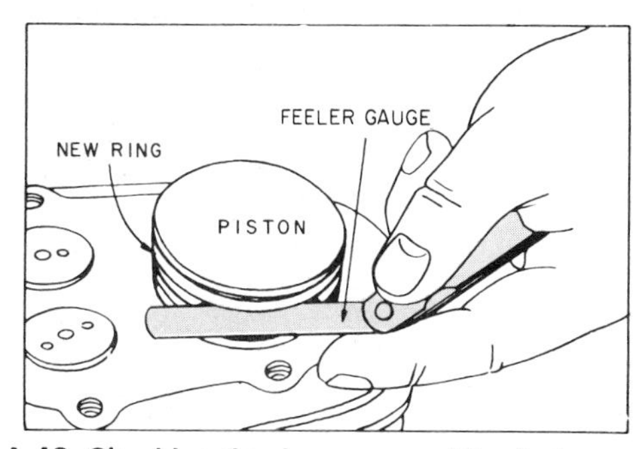

Fig. A-42 Checking the ring groove with a feeler gauge

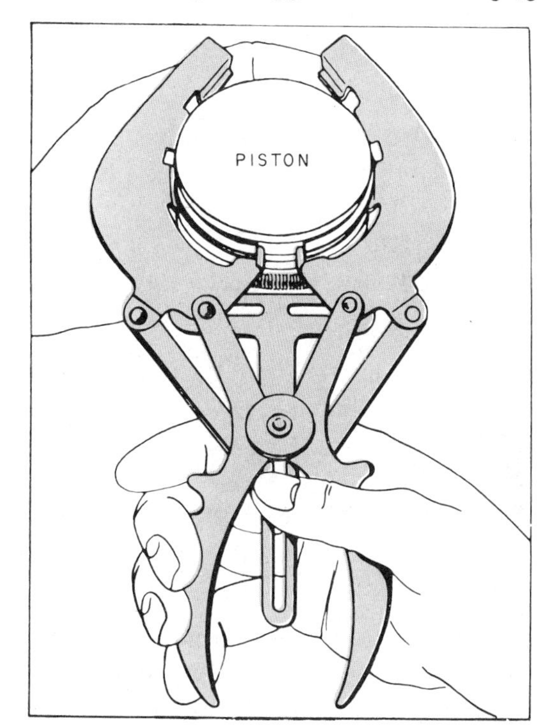

Fig. A-43 Piston ring tools

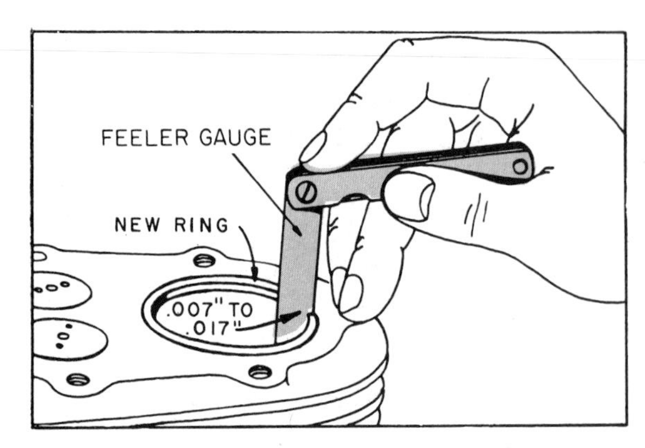

Fig. A-44 Checking the ring gap with a feeler gauge (end gap)

Fig. A-45 Piston ring compressor

or snap when it becomes hot and expands. Too much end gap may allow blow-by and a resulting loss of power. Piston ring grooves should be cleaned to remove any carbon accumulations. New piston rings are usually installed during engine overhaul.

Use a piston ring compressor to reinstall the piston assembly, figure A-45. Remember to place the piston in the cylinder the same way it came out. Also, put the connecting rod cap on the same way it came off, finding the match marks, figure A-46. A torque wrench

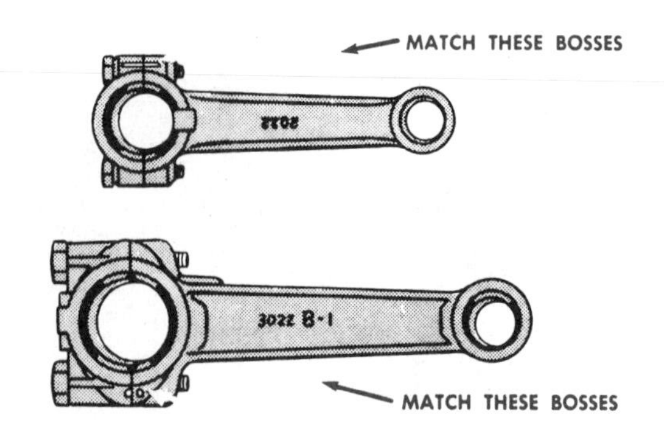

Fig. A-46 Match boss marks when reassembling the connecting rod and connecting rod cap.

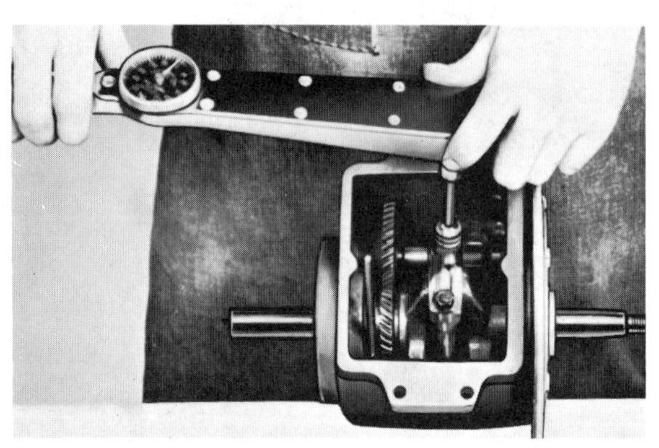

Fig. A-47 Torque wrench used to correctly tighten connecting rod cap

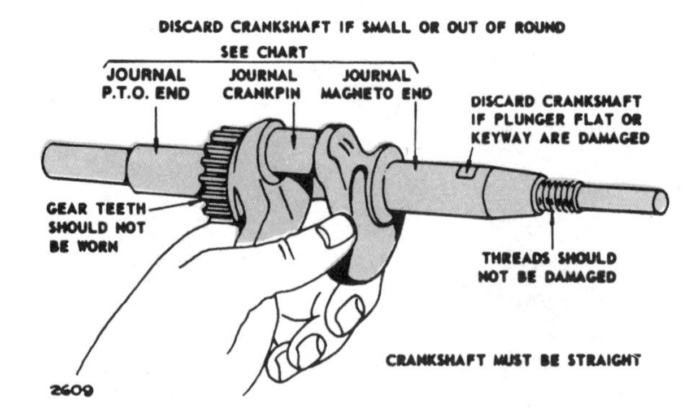

Fig. A-48 Crankshaft check points

Fig. A-49 Pressing in the main ball bearing

should be used to tighten the connecting rod cap bolts according to the manufacturer's specifications, figure A-47.

CRANKSHAFT

The crankshaft should be removed and checked for scoring and any metallic pickup. The journals and crankpin should be checked with a micrometer for roundness. The gear and keyway should be checked for wear. Also, check the crankshaft for straightness, figure A-48. In some cases, the main ball bearings may come out with the crankshaft. Upon reinstallation, the bearings may have to be pressed into place. An arbor press is good for this job, figure A-49.

CAMSHAFT AND LIFTERS

The camshaft and valve lifters can also be removed for inspection and repair. The camshaft pin should be driven out with a drift punch from the power takeoff side of the engine. Upon reassembly of the camshaft and crankshaft be certain to line up the timing marks, figures A-50 and A-51.

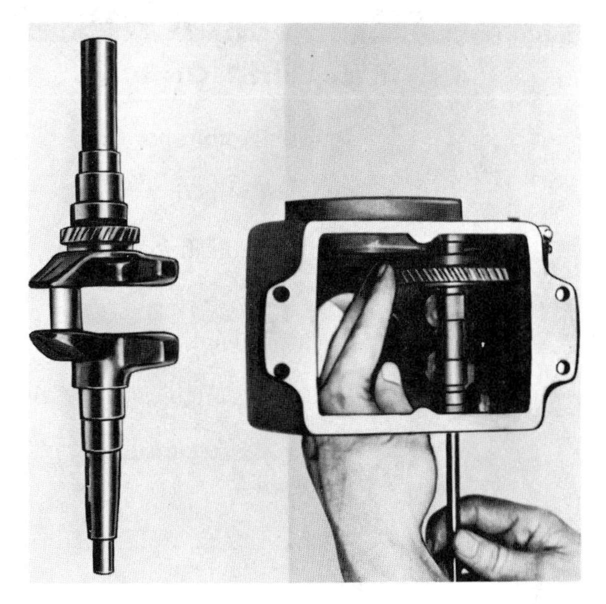

Fig. A-50 Reinstalling the camshaft

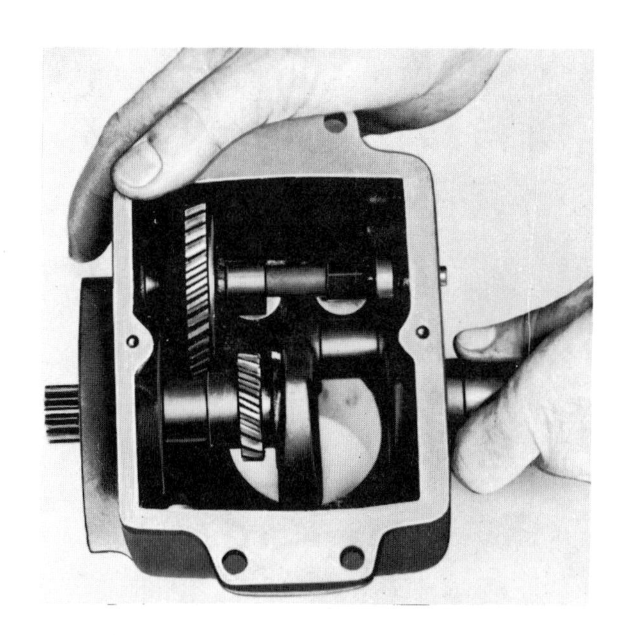

Fig. A-51 Reinstalling the crankshaft

REVIEW QUESTIONS

1. Define troubleshooting.
2. Of what value is a troubleshooting chart?
3. Can troubleshooting be done by an individual engine owner or operator?
4. Define engine tune-up.
5. What does engine reconditioning or overhaul involve?
6. Can any reconditioning or overhaul be accomplished by the amateur mechanic?
7. Why is the careful layout of parts important during disassembly?
8. Why is a mechanic's handbook or service manual essential for the reconditioning of an engine?
9. Explain engine timing.
10. What might a complete valve job include?

Appendix B Horsepower and Buying Considerations

Horsepower (hp) is the yardstick of the engine's power, its capacity to do work. James Watt, the inventor of the steam engine, devised the unit. He assumed that the average horse could raise 33,000 pounds one foot in one minute. An engine that can lift 33,000 pounds one foot in one minute is a one-horsepower engine. The engine can lift 8250 pounds four feet in one minute, and still be delivering 1 horsepower.

As discussed in unit 1, the formula for horsepower is:

$$\text{Horsepower} = \frac{\text{Work}}{\text{Time (in minutes)} \times 33{,}000}$$

Remember that work is the energy required to move a weight through a distance. For example, lifting one pound, one foot, is one foot-pound of work. Lifting 50 pounds, two feet, is 100 foot-pounds of work. The element of time is not a factor.

In horsepower the time factor is added. For example, a boat with an outboard motor might cross a river in ten minutes. An identical boat with a larger motor might make the crossing in five minutes. The same amount of work has been accomplished by each engine but the larger horsepower engine did the work faster.

There are a variety of terms related to horsepower, such as brake horsepower, frictional horsepower, indicated horsepower, SAE or taxable horsepower, rated horsepower, developed horsepower, maximum horsepower, continuous horsepower, corrected horsepower, observed horsepower, and others. Several of these are discussed in the following paragraphs.

Brake horsepower is usually used by manufacturers to advertise their engine's power. Brake horsepower is measured either with a prony brake or with a dynamometer. The *prony brake* consists basically of a flywheel pully, adjustable brake band, lever, and scale measuring device. With the engine operating, the brake band is tightened on the flywheel and the pressure or torque is transmitted to the scale. The readings on the scale and other data are used to calculate the horsepower.

The *dynamometer* is a device that measures power produced by an internal combustion engine. The electric dynamometer contains a dynamo. As the engine drives the dynamo the current output can be carefully recorded. The more powerful the engine, the more current is produced. This current is correlated to horsepower.

Frictional horsepower is the power that is used to overcome the friction in the engine itself. The parts themselves absorb a certain amount of power. Losses included are mainly due to the friction of the piston moving in the cylinder but also included are losses due to oil and coolant pumps, valves, fans, ignition parts, and other accessories. These losses can amount to 10 percent of the brake horsepower.

Indicated horsepower is the power that is actually produced by the burning gases within the engine. It does not take into account the power that is absorbed by or used to move the engine parts. It is the sum of the brake horsepower plus the power used to drive the engine (frictional horsepower).

SAE or taxable horsepower is the horsepower used to compute the license fee for automobiles in some states.

$$\text{SAE hp} = \frac{D^2 \times \text{No. of Cylinders}}{2.5}$$

D = Diameter of bore in inches

$$\text{Example: } \frac{(3.5 \times 3.5) \times 6}{2.5} = 29.4 \text{ hp}$$

FACTORS AFFECTING HORSEPOWER

Many factors affect the horsepower of an engine. Some of these factors are discussed in the following paragraphs.

Engine torque is a factor that relates to horsepower. *Torque* is the twisting force of the engine's crankshaft. Torque can be compared to the force a person uses to tighten a nut. At first a small amount of torque is used, but more and more torque is applied as the nut tightens. Even when the nut stops turning, the person still may be applying torque. Motion is not necessary to have torque. Torque is measured in foot-pounds or inch-pounds. On an engine, the maximum torque is developed at speeds below the maximum engine speed. At top speeds, frictional horsepower is greater and volumetric efficiency is less.

Volumetric efficiency relates to horsepower also. It refers to the engine's ability to breathe or take in a full charge of fuel mixture in the short time allowed for intake. Engine design largely determines the volumetric efficiency of the engine. At high speeds the volumetric efficiency drops off. It can be increased with superchargers and turbochargers that force or blow air into the intake manifold. Also, volumetric efficiency can be increased by using multiple-barrel carburetors (two-barrel and four-barrel) with the extra barrels opening up at high speed when the demand for air is greater.

Compression ratio is another factor that affects horsepower. This is the relationship of the volume of the cylinder when the piston is at the bottom of its stroke compared to the volume of the cylinder (and combustion chamber) when the piston is at top dead center. For small gasoline engines it may be 6:1; for automobile engines, the compression ratio may be 8:1 or higher. The higher the compression ratio, the greater the horsepower delivered when the fuel mixture is ignited.

Piston displacement is the volume of air the pistons displace from the bottom of their stroke to top dead center. Piston displacement also relates to horsepower. Generally, the greater the piston displacement the greater the horsepower. Most engines deliver 1/2 to 7/8 horsepower per cubic inch displacement. High performance engines develop about 1 horsepower per cubic inch displacement. Supercharged engines can deliver much more than 1 horsepower per cubic inch displacement. Displacement equals *area of bore* × *stroke* × *number of cylinders,* and is expressed in cubic inches.

Estimating Horsepower

The following formula can be used to estimate the maximum horsepower of a four-stroke cycle engine.

$$\text{Horsepower} = \frac{D^2 \times N \times S \times \text{rpm}}{11{,}000}$$

D^2 (Bore in Inches)

N (Number of Cylinders)

S (Stroke in Inches)

11,000 (Experimental Constant)

For a two-cycle engine, the formula can be used with 9000 as the experimental constant.

BUYING CONSIDERATIONS

The person who is shopping for a gasoline engine will have no trouble finding an engine for the job. There are many manufacturers of small gasoline engines who produce engines in a variety of horsepower that are designed and engineered to satisfy every need. It is not uncommon to find a manufacturer's basic model with variations to adapt the engine for many different jobs. Manufacturers of high-quality engines conduct many tests on their engines to assure quality performance, figures B-1 to B-3. Several points to consider in buying an engine are: cost, reputation of manufacturer, availability of repair parts, power requirements, engine warranties.

Cost. Beware of bargains. A good bargain can be found now and then but the general rule, "You get what you pay for," is a good one. This does not mean that the most expensive engine is the best buy. The engine must be the right one for the intended usage. For cutting the grass once a week, an economical, dependable engine should be enough. However, if the engine will be exposed to continuous heavy usage it would be wise to buy a heavy-duty engine, one made to operate for a long period of time without failure. The additional cost is worthwhile in this case.

Fig. B-1 Manufacturers of high-quality engines maintain elaborate testing laboratories.

Reputation of the Manufacturer. Be certain the manufacturer's past record in the field justifies faith in the new engines. Talk with engine dealers and individual owners of engines. Their opinions and experience can help in making the best selection.

Availability of Repair Parts. If an engine breaks down and requires a new part, will it be readily available? Can the part be found at a local dealer or will it be necessary to send to

Fig. B-2 Cold room freezing test insures dependable operation in cold weather.

a factory that may be many miles away? Delays caused by breakdowns can be costly as well as annoying. The availability of service and repair parts is an important consideration.

Power Requirements. Be certain the engine is big enough for the job. Constant overloading or constant operating at full throttle shortens engine life. A dealer can assist in finding the correct power plant for the job.

Manufacturers today produce engines that operate on either the four-cycle or the two-cycle principle; both have their place and both are entirely successful. Some engines are water cooled, some are air cooled; some have simple splash lubrication, some have pump lubrication; some have fuel pumps, and some do not. In fact, no two engines look exactly alike or operate exactly alike.

Engine Warranties. Most engine manufacturers give a warranty to the purchaser of

Fig. B-3 Endurance tests of engines insure their dependability.

a new engine. A common warranty is good for ninety days. If during the ninety days after purchase any parts fail due to defective material or workmanship, the manufacturer will repair or replace the defective part.

In most cases the part or engine must be returned to the factory or to an authorized distributor. Most manufacturers require the owner to pay shipping charges to the factory. In some cases, the owner must also pay any labor costs involved in the repair.

If the engine has been damaged through misuse, negligence, or accident, most warranties are void. A few manufacturers require that the engine be registered shortly after purchase; if the engine is not registered, the warranty is not valid. Generally, engine components such as magnetos, carburetors, starters, etc. are only covered by the terms of their individual manufacturer.

REVIEW QUESTIONS

1. If an engine can lift 150,000 pounds a distance of one foot in three minutes, what is its horsepower?
2. In general, can larger horsepower engines do work faster than smaller horsepower engines?
3. What type of horsepower rating do most manufacturers use in advertising their engines?
4. What is a dynamometer?
5. In what way is SAE horsepower used?
6. Estimate the horsepower of a two-stroke cycle having a bore of 2 inches, one cylinder, a stroke of 1 1/2 inches, and operating at 3600 revolutions per minute.
7. What is engine torque?
8. What is volumetric efficiency?
9. What is the relationship between compression ratio and horsepower?
10. What is the relationship between piston displacement and horsepower?
11. List four important considerations for a person to think about before purchasing an engine.

Appendix C

1. **Spark plug gap: .030 All Models**
2. **Condenser capacity: .18 to .24 MFD. All Models**
3. **Contact point gap: .020 All Models**

	BASIC MODEL SERIES	IDLE SPEED	ARMATURE TWO LEG AIR GAP	ARMATURE THREE LEG AIR GAP	VALVE CLEARANCE INTAKE	VALVE CLEARANCE EXHAUST	VALVE GUIDE REJECT GAGE	TORQUE SPECIFICATIONS FLYWHEEL NUT FT. LBS.	TORQUE SPECIFICATIONS CYLINDER HEAD IN. LBS.	TORQUE SPECIFICATIONS CONN. ROD IN. LBS.
ALUMINUM	6B, 60000	1750	.006 .010	.012 .016	.005 .007	.009 .011	19122	55	140	100
	8B, 80000, 81000, 82000	1750	.006 .010	.012 .016	.005 .007	.009 .011	19122	55	140	100
	92000, 94000	1750	.006 .010		.005 .007	.009 .011	19122	55	140	100
	100000	1750	.010 .014	.012 .016	.005 .007	.009 .011	19122	60	140	100
	110000	1750	.006 .010		.005 .007	.009 .011	19122	55	140	100
	130000	1750	.010 .014		.005 .007	.009 .011	19122	60	140	100
	140000	1750	.010 .014	.016 .019	.005 .007	.009 .011	19151	65	165	165
	170000, 171700•	1750 **	.010 .014		.005 .007	.009 .011	19151	65	165	165
	190000, 191700•	1750 **	.010 .014		.005 .007	.009 .011	19151	65	165	165
	220000, 250000	1750 **	.010 .014		.005 .007	.009 .011	19151	65	165	190
CAST IRON	5, 6, N	1750		.012 .016	.007 .009	.014 .016	19122	55	140	100
	8	1750		.012 .016	.007 .009	.014 .016	19122	55	140	100
	9	1200			.007 .009	.014 .016	19151	60	140	140
	14	1200			.007 .009	.014 .016	19151	65	165	190
	19, 190000, 200000•	1200 **	.010 .014	.022 .026	.007 .009	.014 .016	19151	115	190	190
	23, 230000	1200 **	.010 .014	.022 .026	.007 .009	.017 .019	19151	145	190	190
	243000	1200 **	.010 .014		.007 .009	.017 .019	19151	145	190	190
	300000	1200 **	.010 .014		.007 .009	.017 .019	19151	145	190	190
	320000	1200 **	.010 .014		.007 .009	.017 .019	19151	145	190	190

COMMONLY USED TOOLS FOR SERVICING

	19051	Spark Tester, all models
	19061	Carburetor jet screwdriver, all models
	19062	Carburetor jet screwdriver, all models
	19063	Valve spring compressor, all models
†	19161	Starter clutch wrench, use with ½" drive torque wrench
	19203	Flywheel Puller, 170000 thru 320000 Aluminum Models & Cast Iron Models

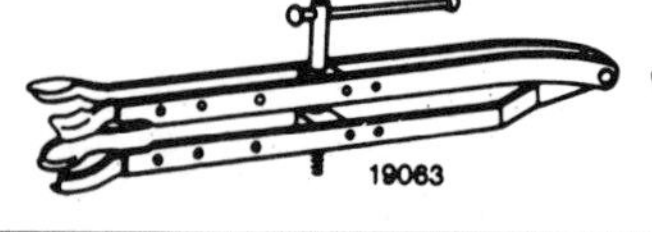

4. Top governed speed: See Briggs & Stratton Service Bulletin No. 467 or Engine Replacement Data

5. Crankshaft End Play: .002-.008 All Models

CRANKSHAFT REJECT SIZE			MAIN BEARING REJECT GAGE	CYLINDER BORE STD. ▲
MAG. JOURNAL	CRANKPIN	P.T.O. JOURNAL		
.8726	.8697	.8726	19166	2.375* 2.374
.8726	.9963	.8726	19166	2.375 2.374
.8726	.9963	.8726	19166	2.5625 2.5615
.8726	.9963	.9976	19166 Mag. 19178 PTO	2.500 2.499
.8726	.9963	.8726	19166	2.7812 2.7802
.8726	.9963	.9976	19166 Mag. 19178 PTO	2.5625 2.5615
.9975	1.090	1.1790	19178	2.750 2.749
.9975 1.1790•	1.090	1.1790	19178	3.000 2.999
.9975 1.1790•	1.122	1.1790	19178	3.000 2.999
1.3760	1.2470	1.3760		3.4375 3.4365
.8726	.7433	.8726	19166	2.000 1.999
.8726	.7433	.8726	19166	2.250 2.249
.9832	.8726	.9832	19117	2.250 2.249
1.1790	.9964	1.1790	19117	2.625 2.624
1.1800	.9964 1.1219•	1.1790	19117	3.000 2.999
1.3769	1.1844	1.3759	19117	3.000 2.999
Ball	1.3094	Ball	Ball	3.0625 3.0615
Ball	1.3094	Ball	Ball	3.4375 3.4365
Ball	1.3094	Ball	Ball	3.5625 3.5615

CARBURETOR TYPE	INITIAL CARBURETOR ADJUSTMENT ALL MODELS TURNS OPEN FROM SEAT NEEDLE VALVE	IDLE VALVE
Pulsa-Jet Vacu-Jet	1½	
Two Piece Flo-Jet	1½	1
One Piece Flo-Jet	2½	1½
One Piece Flo-Jet (6, 7, 8, 10 & 11 H.P Vertical Crankshaft)	1½	1¼
One Piece Flo-Jet (11 H.P. Horizontal Crankshaft)	1½	1

CYLINDER RESIZING

▲ Resize if .003 or more wear or .0015 out of round on C.I. cylinder engines, .0025 out of round on aluminum alloy engines.

Resize to .010, .020 or .030 over Standard.

*Model 6B series and early Models 60000 and 61000 series engines have cylinder bore of 2.3125 - 2.3115.

RING GAP REJECT SIZES

MODEL	COMP. RINGS	OIL RING
Alum. Cylinder Models	.035"	.045"
C.I. Cylinder Models	.030"	.035"

****GOVERNED IDLE**

For Adjustment Procedures, see Service & Repair Instructions 270962, Section 5, for Single Cylinder Models and Repair Instructions MS-7000 or 271172, Section 5, for Twin Cylinder Models.

■ With Valve Springs Installed.

• Synchro-Balance.

BRIGGS & STRATTON ENGINES

19114

19165

19230

19167

Flywheel puller, all models thru 130000	19069
Piston ring compressor, all models	19230
Starter clutch wrench, all rewind starter models	19114
Flywheel Puller, 140000, 170000, 190000 & 250000 Models	19165
Flywheel holder, all models 6B thru 130000	19167

See Section 13 for Complete List of Tools

Fig. C-1 Engine specifications for all popular Briggs and Stratton engine models. A good service and repair manual contains specifications such as these for their engines.

GENERAL INFORMATION
Check-up

CHECK - UP

Most complaints concerning engine operation can be classified as one or a combination of the following:

1. Will not start
2. Hard starting
3. Kicks back when starting
4. Lack of power
5. Vibration
6. Erratic operation
7. Overheating
8. High oil consumption

When the cause of malfunction is not readily apparent, perform a check of the Compression, Ignition and Carburetion Systems. This check-up, performed in a systematic manner, can usually be done in a matter of minutes. It is the quickest and surest method of determining the cause of failure. This check-up will point up possible cause of future failures, which can be corrected at the time. The basic check-up procedure is the same for all engine models, while any variation, by model will be shown under the subject heading.

NOTE: What appears to be an engine malfunction may be a fault of the powered equipment rather than the engine. If equipment is suspect, see Equipment, affecting engine operation.

Check Compression

Spin flywheel in reverse rotation (counterclockwise) to obtain accurate compression check. The flywheel should rebound sharply, indicating satisfactory compression.

If compression is poor, look for –

1. Loose spark plug
2. Loose cylinder head bolts
3. Blown head gasket
4. Burnt valves and/or seats
5. Insufficient tappet clearance
6. Warped cylinder head
7. Warped valve stems
8. Worn bore and/or rings
9. Broken connecting rod

Check Ignition

Remove the spark plug. Spin the flywheel rapidly with one end of the ignition cable clipped to the 19051 tester and with the other end of the tester grounded on the cylinder head. If spark jumps the .166'' tester gap, you may assume the ignition system is functioning satisfactorily. Try a new spark plug.

If spark does not occur look for –

1. Incorrect armature air gap
2. Worn bearings and/or shaft on flywheel side
3. Sheared flywheel key
4. Incorrect breaker point gap
5. Dirty or burned breaker points
6. Breaker plunger stuck or worn
7. Shorted ground wire (when so equipped)
8. Shorted stop switch (when so equipped)
9. Condenser failure
10. Armature failure
11. Improperly operating interlock system

NOTE: If engine runs but misses during operation, a quick check to determine if ignition is or is not at fault can be made by inserting the 19051 tester between the ignition cable and the spark plug. A spark miss will be readily apparent. While conducting this test on Magna-Matic equipped engines, Models 9, 14, 19 and 23, set the tester gap at .060''.

Fig. C-2 General check-up procedures (troubleshooting) suggested for Briggs and Stratton engines

GENERAL INFORMATION
Check-up

Check Carburetion

Before making a carburetion check, be sure the fuel tank has an ample supply of fresh, clean gasoline. On gravity feed (Flo-Jet) models, see that the shut-off valve is open and fuel flows freely through the fuel line. On all models, inspect and adjust the needle valves. Check to see that the choke closes completely. If engine will not start, remove and inspect the spark plug. If plug is wet, look for –

1. Overchoking
2. Excessively rich fuel mixture
3. Water in fuel
4. Inlet valve stuck open (Flo-Jet carburetor)

If plug is dry, look for –

1. Leaking carburetor mounting gaskets
2. Gummy or dirty screen or check valve (Pulsa-Jet and Vacu-Jet carburetors)
3. Inlet valve stuck shut (Flo-Jet carburetors)
4. Inoperative pump (Pulsa-Jet carburetors)

A simple check to determine if the fuel is getting to the combustion chamber through the carburetor is to remove the spark plug and pour a small quantity of gasoline through the spark plug hole. Replace the plug. If the engine fires a few times and then quits, look for the same condition as for a dry plug.

Equipment - Effecting Engine Operation

Frequently, what appears to be a problem with engine operations, such as hard starting, vibration, etc., may be the fault of the equipment powered rather than the engine itself. Since many varied types of equipment are powered by Briggs and Stratton engines, it is not possible to list all of the various conditions that may exist. Listed are the most common effects of equipment problems, and what to look for as the most common cause.

Hard Starting, Kickback, or Will Not Start

1. Loose blade – Blade must be tight to shaft or adaptor.
2. Loose belt – a loose belt like a loose blade can cause a back-lash effect, which will counteract engine cranking effort.
3. Starting under load – See if the unit is disengaged when engine is started; or if engaged, does not have a heavy starting load.
4. Check remote Choke-A-Matic control assembly for proper adjustment.
5. Check interlock system for shorted wires, loose or corroded connections, or defective modules or switches.

Vibration

1. Cutter blade bent or out of balance – Remove and balance
2. Crankshaft bent – Replace
3. Worn blade coupling – Replace if coupling allows blade to shift, causing unbalance.
4. Mounting bolts loose – Tighten
5. Mounting deck or plate cracked – Repair or replace.

Power Loss

1. Bind or drag in unit – If possible, disengage engine and operate unit manually to feel for any binding action.
2. Grass cuttings build-up under deck.
3. No lubrication in transmission or gear box.
4. Excessive drive belt tension may cause seizure.

Noise

1. Cutter blade coupling or pulley – an oversize or worn coupling can result in knocking, usually under acceleration. Check for fit, or tightness.
2. No lubricant in transmission or gear box.

Specifications/model **K91**

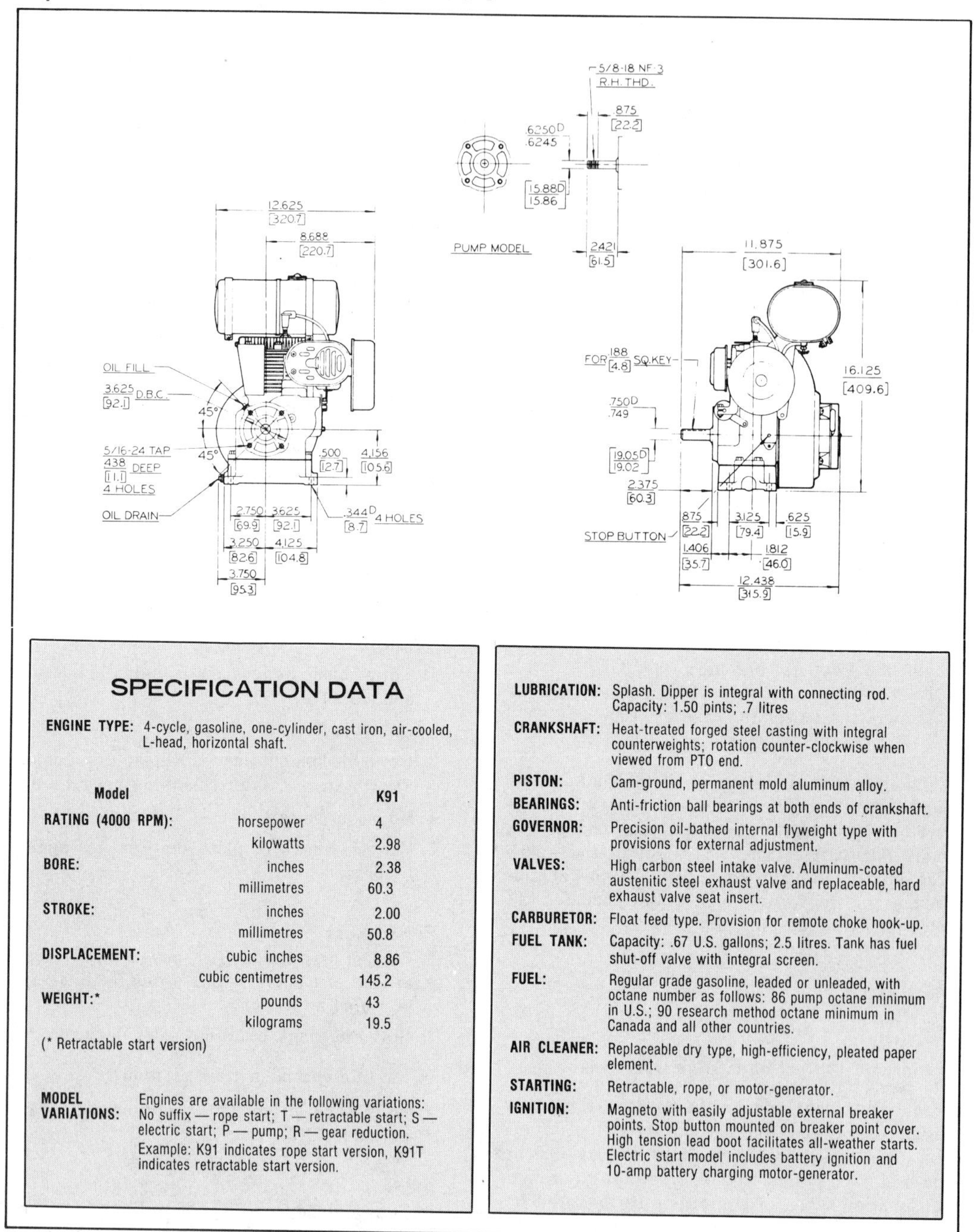

SPECIFICATION DATA

ENGINE TYPE: 4-cycle, gasoline, one-cylinder, cast iron, air-cooled, L-head, horizontal shaft.

Model		K91
RATING (4000 RPM):	horsepower	4
	kilowatts	2.98
BORE:	inches	2.38
	millimetres	60.3
STROKE:	inches	2.00
	millimetres	50.8
DISPLACEMENT:	cubic inches	8.86
	cubic centimetres	145.2
WEIGHT:*	pounds	43
	kilograms	19.5

(* Retractable start version)

MODEL VARIATIONS: Engines are available in the following variations: No suffix — rope start; T — retractable start; S — electric start; P — pump; R — gear reduction. Example: K91 indicates rope start version, K91T indicates retractable start version.

LUBRICATION: Splash. Dipper is integral with connecting rod. Capacity: 1.50 pints; .7 litres

CRANKSHAFT: Heat-treated forged steel casting with integral counterweights; rotation counter-clockwise when viewed from PTO end.

PISTON: Cam-ground, permanent mold aluminum alloy.

BEARINGS: Anti-friction ball bearings at both ends of crankshaft.

GOVERNOR: Precision oil-bathed internal flyweight type with provisions for external adjustment.

VALVES: High carbon steel intake valve. Aluminum-coated austenitic steel exhaust valve and replaceable, hard exhaust valve seat insert.

CARBURETOR: Float feed type. Provision for remote choke hook-up.

FUEL TANK: Capacity: .67 U.S. gallons; 2.5 litres. Tank has fuel shut-off valve with integral screen.

FUEL: Regular grade gasoline, leaded or unleaded, with octane number as follows: 86 pump octane minimum in U.S.; 90 research method octane minimum in Canada and all other countries.

AIR CLEANER: Replaceable dry type, high-efficiency, pleated paper element.

STARTING: Retractable, rope, or motor-generator.

IGNITION: Magneto with easily adjustable external breaker points. Stop button mounted on breaker point cover. High tension lead boot facilitates all-weather starts. Electric start model includes battery ignition and 10-amp battery charging motor-generator.

Fig. C-3 Specification sheet for Kohler Model K91. Specification sheets like these help the engineer, designer, and buyer select the type of engine suitable for a particular application or job

DISASSEMBLY and REASSEMBLY PROCEDURES

With the disassembly operations, instructions on reassembly are also given, as often it will not be necessary to disassemble the entire engine. If it is desired to disassemble the entire engine, the reassembly instructions can be looked up under the headings of the various parts.

Tighten the capscrews of the cylinder head, gear cover, connecting rod, flywheel and spark plug to the specified torque readings indicated in the related paragraphs of reassembly. Use caution in tightening all other screws and nuts so as not to strip threads or break flanges.

RECOIL STARTER

1. Remove four mounting screws and take off recoil starter assembly.
2. ***In reassembly;*** check condition of rope for possible replacement. Repair per the *Repair Instructions* beginning on *Page 20.*

FLYWHEEL SHROUD

1. Disconnect coil primary wire from stop button wire.
2. Take off flywheel shroud from cylinder-case and cylinder head.
3. Remove baffle from cylinder block.

Fig. C-4 Disassembly and reassembly procedures for a Wisconsin ROBIN engine Model W1-145. This is an excellent example of the type of detailed information that a good repair manual will provide the mechanic

FUEL TANK

1. Disconnect fuel line at carburetor. ***In reassembly;*** it is advisable to use a new fuel line.
2. Remove fuel tank from cylinder head.
3. Remove cover from cylinder head.

AIR CLEANER

1. Remove air cleaner cover, element and retainer.
2. Disconnect breather tube from inspection cover.
3. Take out two lock nuts and remove air cleaner body and gasket from carburetor studs.
4. ***In reassembly;*** wash element per *Air Cleaner Instructions* on *Page 4*. Mount new gasket and assemble back cover, elements and cover per *Fig. 3*. Connect breather line.

GOVERNOR LEVER and CARBURETOR

1. Disconnect governor spring from lever to speed control assembly.
2. Remove governor lever from shaft, and disconnect rod and spring from lever to carburetor.
3. Remove carburetor, insulating plate and gaskets from studs in cylinder-case.
4. If necessary, the speed control assembly can be removed from cylinder head by taking off flange bolt.
5. ***In reassembly;*** *see 'Governor Adjustment', Page 6.*

MUFFLER

Take off the two hex nuts and remove muffler and gasket from cylinder-case.

STARTING PULLEY and FLYWHEEL (magneto)

1. Place a ***19mm*** socket wrench on to the flywheel nut and give the wrench a sharp blow with a soft hammer. Remove nut, spring washer and pulley.
2. As illustrated in *Fig. 10,* attach puller to flywheel — turn center bolt clockwise until flywheel becomes loose enough to be removed.

   If puller is not available: Screw flywheel nut flush with end of crankshaft to protect shaft threads from being damaged. Place the end of a large screwdriver between the crankcase and flywheel in line with the keyway. Then, strike the end of the flywheel nut with a babbitt hammer and at the same time wedge outward with the screwdriver.
3. The ignition coil along with attached high tension cable can then be removed by taking out two mounting screws.
4. Take off point cover and remove contact breaker and condenser.
5. ***In reassembly;*** refer to *'Breaker Point Adjustment'* and *'Timing'* instructions.
6. Securely tighten flywheel nut after timing is finalized, but first be sure woodruff key is in position on shaft. ***Do not*** drive flywheel on to taper of crankshaft and ***do not*** overtighten flywheel nut. Simply turn nut until lock-washer collapses. Then, tighten by placing wrench on nut and giving handle of wrench 1 or 2 sharp blows with a soft hammer. If torque wrench is available, tighten ***44*** to ***47 ft. lbs.*** (6 to 6.5 kgm).

CYLINDER HEAD and SPARK PLUG

1. Remove spark plug from cylinder head.
2. Loosen mounting screws and remove cylinder head along with gasket.
3. Clean carbon from combustion chamber and dirt from between the cooling fins. Check the cylinder head mounting face for distortion. If ***warpage*** is evident, replace head.
4. ***In reassembly;*** use new cylinder head gasket and spark plug. ***Torque*** head screws ***14*** to ***17 ft. lbs.*** (1.9 to 2.3 kgm).

   Leave spark plug out temporarily, for ease in turning engine over for remainder of assembly and for timing adjustments. When mounting spark plug, tighten ***8.7*** to ***10.8 ft. lbs.*** (1.2 to 1.5 kgm).

INTAKE and EXHAUST VALVES, Fig's. 11, 12, 13

1. Remove the valve inspection cover, breather plate and gaskets from cylinder-case.
2. Position valve spring retainer so that ***notch*** on outer diameter is toward the outside of the valve compartment.

Fig. 10

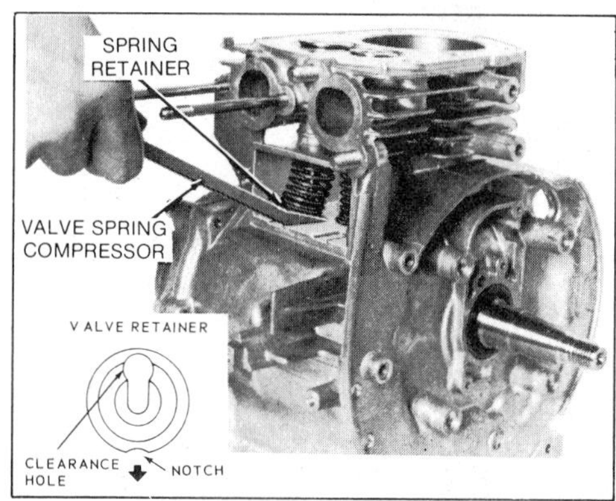

Fig. 11

Place ***compressor tool*** EY2079500307 under spring retainer and compress valve spring. At the same time, draw retainer outward toward notch so that ***clearance hole*** in retainer allows for valve to be pulled out from top of cylinder block.

Caution: ***Do not damage*** gasket surface of tappet chamber with the compressor tool.

3. Clean carbon and gum deposits from the valves, seats, ports and guides.
4. ***In reassembly;*** replace valves that are badly burned, pitted or warped. With retainer notch facing outward, place retainer under valve spring with clearance hole over valve stem. While compressing valve spring with compressor tool, force retainer inward so that it locks in place on valve stem. *See Fig. 11.*

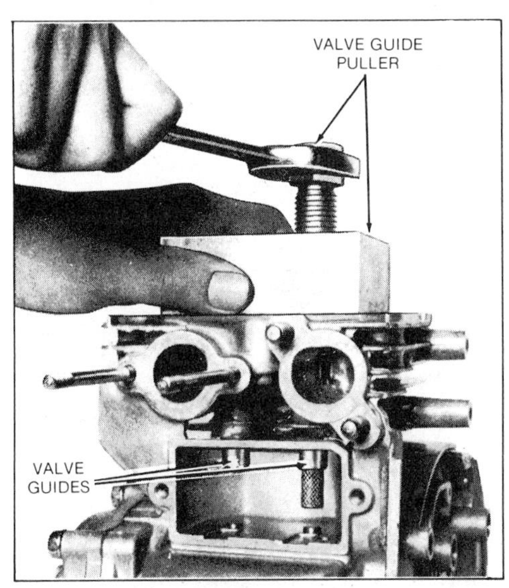

Fig. 12

5. ***Valve guides*** should be replaced when valve stem clearance becomes excessive. Use valve guide puller tool ***EY2279500107*** as illustrated in *Fig. 12.* Draw valve guides out and press new guides in, using the same puller tool. Refer to *Fig. 13* for clearance specifications and proper assembly.
6. After replacing valves and valve guides, lap valves in place until a uniform ring will show entirely around the face of the valve. Clean valves, and wash block thoroughly with a hot solution of soap and water. Wipe cylinder walls with clean lint free rags and light engine oil. ***Do not assemble valve springs*** until tappet clearance has been checked. *See 'Tappet Adjustment'.*

TAPPET ADJUSTMENT, Fig. 14

With tappet in its lowest position, hold valve down and insert feeler gauge between valve and tappet stem. The clearance for both intake and exhaust, with engine ***cold: 0.003*** to ***0.005 inch*** (0.08 to 0.12mm).

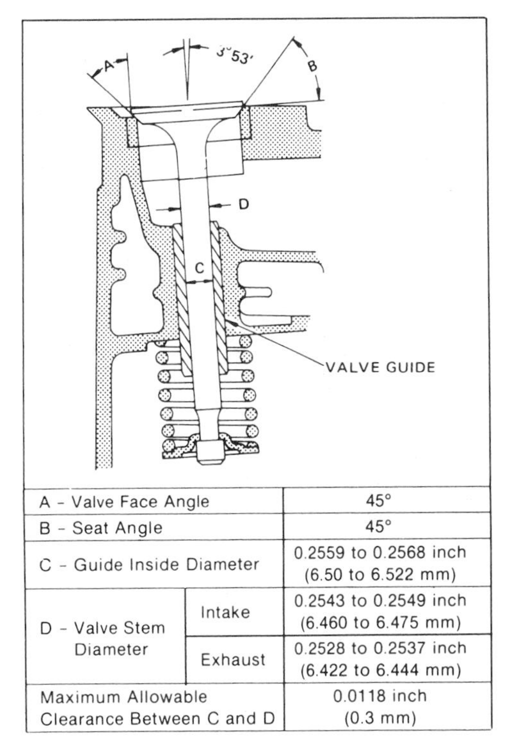

A - Valve Face Angle		45°
B - Seat Angle		45°
C - Guide Inside Diameter		0.2559 to 0.2568 inch (6.50 to 6.522 mm)
D - Valve Stem Diameter	Intake	0.2543 to 0.2549 inch (6.460 to 6.475 mm)
	Exhaust	0.2528 to 0.2537 inch (6.422 to 6.444 mm)
Maximum Allowable Clearance Between C and D		0.0118 inch (0.3 mm)

Fig. 13

If the clearance is less than it should be, grind the end of valve stem a very little at a time and remeasure. Stems must be ground square and flat.

If clearance is too large, replace the valve and, or tappet.

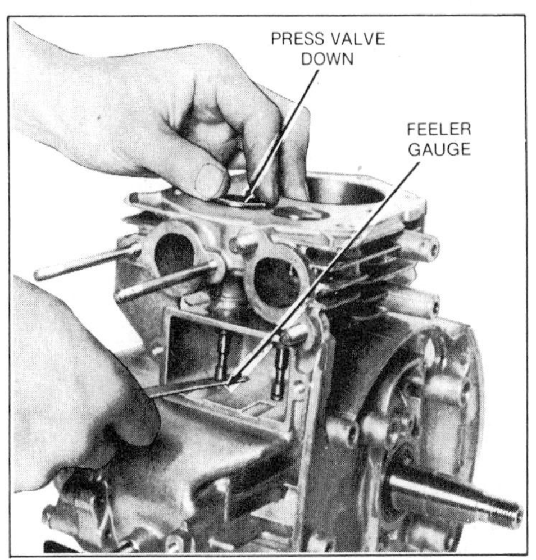

Fig. 14

After obtaining correct clearance, assemble valve springs and retainers. Check operation of valves by turning crankshaft over — remeasure tappet clearance.

GEAR COVER, Fig's. 15, 16 and 17

Place a rag under engine to absorb excess oil, and remove gear cover mounting screws. Use a soft hammer and evenly tap around outer surface of gear cover until it breaks free from crankcase face. Carefully remove cover so as not to damage oil seal. Dowel pins will remain in crankcase flange.

Inspect adjusting shim, oil seal, governor cross shaft, gear and thrust sleeve. If any of these parts are damaged or worn — they should be replaced.

Governor Gear-Flyweight Assembly; can be removed by straddling the handles of a pliers under the gear hub and across the gear cover face. Then, press down on the jaw end of the pliers and gear-thrust sleeve will loosen from shaft. ***In Reassembly;*** place spacing washer on shaft. Insert sleeve between flyweights, mount complete assembly to shaft and press down until unit ***snaps*** in place. ***NOTE: Thrust sleeve cannot be assembled or removed when gear assembly is mounted to shaft.***

Crankshaft End Play is regulated by a selected thickness of adjusting shim. The ***end play*** should be ***0*** to ***0.008 inch*** (0 to 0.2mm) ***engine cold.***

The ***adjusting shim*** is located between the crankshaft gear and main bearing, and replacement is seldom necessary unless the crankshaft or gear cover are replaced.

To determine what thickness ***adjusting shim*** to use for regulating the end play, refer to *Fig. 15* and the following instructions:

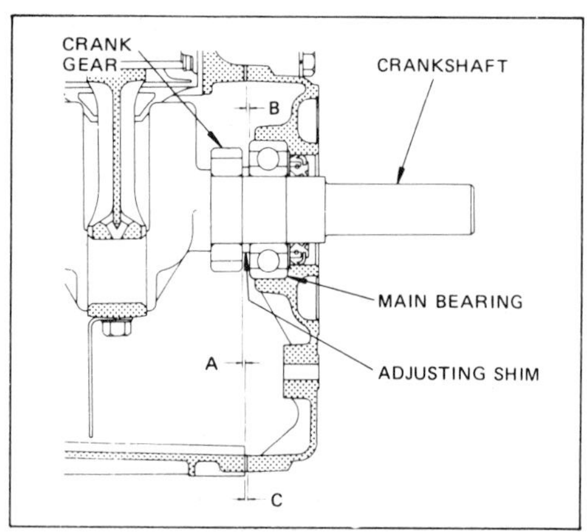

Fig. 15

1. With the gear cover removed, tap end of crankshaft slightly to insure that shaft is shouldered against main bearing at flywheel end.
2. Use a Depth Micrometer and measure the distance between the machined surface of crankcase face and end of crank gear (dim. ***A***).
3. Measure the distance between the machined face of gear cover and outer face of main bearing (dim. ***B***).
4. The compressed thickness of gear cover gasket is ***0.0086 inch*** (0.22mm) dim. ***C***.
5. From the following chart, select an ***adjusting shim*** that is ***0*** to ***0.008 inch*** (0 to 0.2mm) ***less*** that the total length, ***A + B + C***.

Adjusting Shim Thickness	Part Number
.0215 to .0257 inch (.546 to .654 mm)	EY0230250110
.0287 to .0343 inch (.728 to .872 mm)	EY0230250120
.0362 to .0425 inch (.920 to 1.080 mm)	EY0230250130

In reassembly of gear cover, apply oil to bearing surfaces, gear train and tappets. Also lubricate lips of oil seal and add a light film of oil on gear cover face to hold gasket in place.

Mount adjusting shim to crankshaft at crank gear face.

When assembling gear cover, as illustrated in *Fig. 16,* be sure that governor gear-flyweight assembly is in place, and that thrust sleeve aligns with governor cross shaft. If available, use an oil seal sleeve mounted on to the crankshaft, to prevent damage to the oil seal lips.

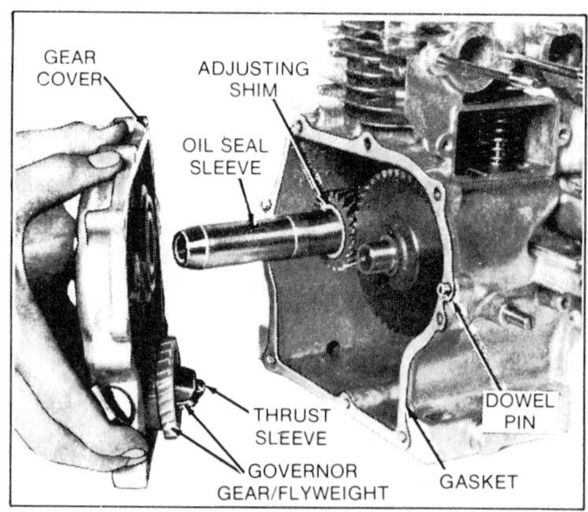

Fig. 16

Caution: ***Be sure timing marks on crankshaft and camshaft gear, Fig. 18, remain correctly mated when end of camshaft is fitted into bearing hole of gear cover, and that governor gear engages camshaft and gear.***

Tap gear cover in place with a soft hammer, remove oil seal sleeve and tighten gear cover capscrews to ***5.8*** to ***7.2 ft. lbs. torque*** (0.8 to 1.0 Kgm).

Check End Play after gear cover and flywheel are in place. If the proper thickness of adjusting shim was selected as explained in *'Crankshaft End Play'* paragraphs and *Fig. 15,* then a satisfactory end play of ***0*** to ***0.008 inch*** (0 to 0.2mm) should have been obtained. However, on instal-

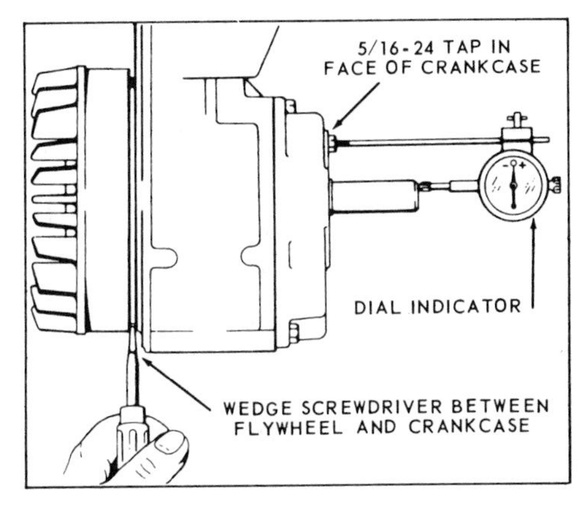

Fig. 17

lations where end play is critical, and to confirm the actual amount of end play, adhere to the following instructions and *Fig. 17*.

1. Tap end of crankshaft with a soft hammer so that crankshaft will shoulder against main bearing at flywheel end.
2. Tap crankshaft in the opposite direction (from flywheel end) to seat adjusting shim against main bearing at take-off end.
3. Attach a Dial Indicator to one of the 5/16-24 tapped holes on the face of the crankcase, and set the dial at 0.
4. Wedge a screwdriver between the flywheel and crankcase. The movement of the flywheel away from the engine block will register as end play on the Indicator dial.

If ***end play*** is ***more*** than the limits of ***0*** to ***0.008 inch*** (0 to 0.2mm) use the reading from the Indicator dial as a guide in selecting what thickness of adjusting shim to use. If there is ***no end play,*** revert to *Gear Cover* paragraphs on Page 12 and follow procedures for adjusting ***crankshaft end play*** and determine exactly what adjusting shim thickness is required.

CAMSHAFT, TAPPETS and TIMING MARKS, Fig. 18

1. To prevent tappets from falling out and becoming damaged when camshaft is removed, turn crankcase over on its side as shown in *Fig. 18.* Push tappets inward to clear cam lobes, and remove camshaft.
2. Withdraw tappets and mark them for identification with the hole that they were removed from.
3. ***In reassembly,*** put tappets back in their corresponding guide hole. This will eliminate unnecessary valve stem grinding for obtaining correct tappet clearance.
4. Lubricate bearing surfaces of camshaft, push valve tappets to their uppermost position and mount camshaft assembly to crankcase.
5. ***Timing marks*** on camshaft gear and crankshaft gear must match up. Mount camshaft so that the ***marked tooth on crankshaft gear*** is between the ***two marked teeth*** of the ***camshaft gear,*** *Fig. 18.* If valve timing is off, engine will not function properly or may not run at all.

CONNECTING ROD and PISTON, Fig's. 19 thru 24

1. Straighten out the bent tabs of lock plate and remove bolts from connecting rod.
2. Take off lock plate, oil dipper and connecting rod cap.
3. Scrape off all carbon deposits that might interfere with removal of piston from upper end of cylinder.
4. Turn crankshaft until piston is at top, then push connecting rod and piston assembly upward and out thru top of cylinder.
5. Remove piston from connecting rod by taking out one of the snap rings and then removing the piston pin. ***A new snap ring should be used in reassembly.***
6. ***Reassembly***

   a. ***Piston Rings***

   As illustrated in *Fig. 19,* use a Ring Expander tool to prevent ring from becoming distorted or broken when installing onto piston.

   If an expander tool is not available, install rings by placing the open end of the ring on first land of

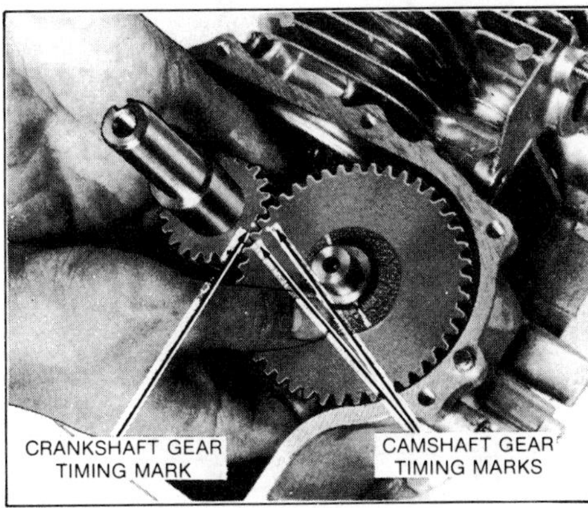

Fig. 18

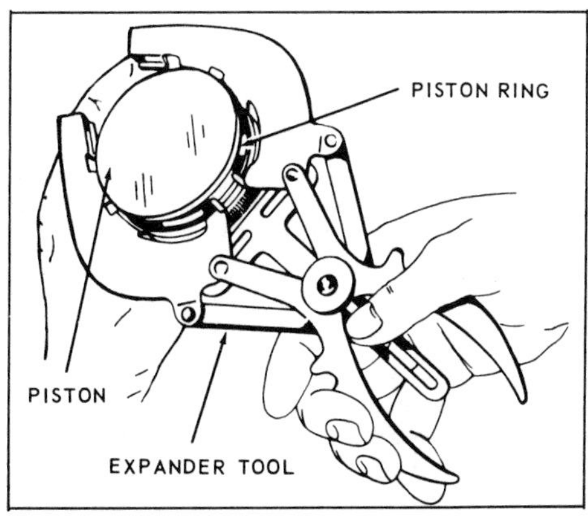

Fig. 19

piston. Spread ring only far enough to slip over piston and into correct groove, being careful not to distort ring.

With or without expander tool, assemble bottom ring first and work upward, installing top ring last.

Mount scraper ring with scraper edge down, otherwise oil pumping and excessive oil consumption will result. Refer to *Fig. 20* for correct placement of rings.

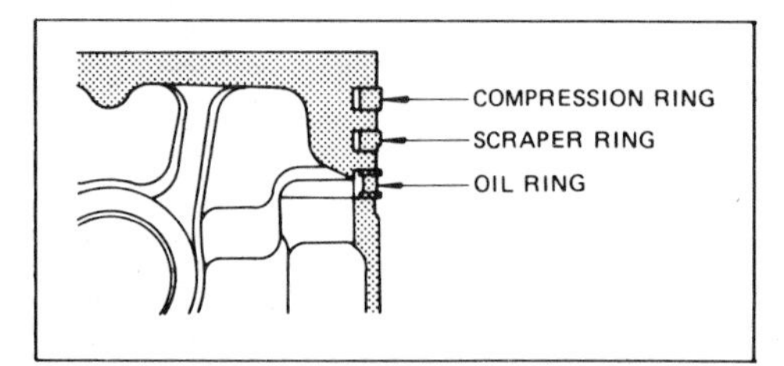

Fig. 20

b. ***Piston, Cylinder and Connecting Rod***

Measure the diameter of the piston in the center of the thrust faces at the bottom of the piston skirt, as illustrated in *Fig. 21*.

Measure the cylinder bore and inspect for out-of-round and taper. If cylinder is scored or worn more than ***.005"*** (0.125mm) over standard size, it should be rebored and fitted with an oversize piston and rings.

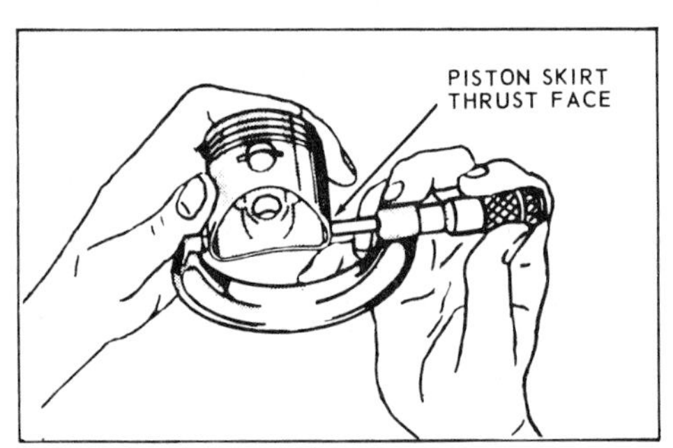

Fig. 21

Refer to chart, *Fig. 22,* for clearance between piston and cylinder. Size, clearance and wear limits are given in more detail on *Page 17*.

When installing the piston in the cylinder, oil the piston, rings, wrist pin, rod bearings and cylinder wall before assembly. Stagger the piston ring gaps 90° apart around the piston. Use a piston ring compressor as illustrated in *Fig. 23*.

Turn crankshaft to bottom of stroke and tap piston down until rod contacts crank pin. Mount connecting rod cap so that the ***cast rib*** between face of rod and bolt boss matches up with the cast rib on the connecting rod. *See Fig. 24*. Assemble ***oil dipper*** to cap. The dipper should be ***toward the flywheel*** end of the connecting rod cap. If engine is operated on a ***tilt toward the take-off end,*** reverse dipper mounting.

		Inches	Millimeters
Piston to Cylinder at Piston Skirt Thrust Face		0.0008 0.0023	0.020 0.059
Connecting Rod to Crank Pin	Dia.	0.0015 0.0025	0.037 0.063
	Side	0.0039 0.0118	0.1 0.3
Piston Pin to Connecting Rod		0.0004 0.0012	0.010 0.029
D - Crankshaft Pin Diameter		0.9429 0.9434	23.950 23.963
W - Crankshaft Pin Width		0.9055 0.9094	23.0 23.1
Piston Ring Gap		0.008 0.016	0.2 0.4
Piston Rings - Side Clearance In Grooves	Top	0.0035 0.0053	0.090 0.135
	2nd	0.0024 0.0041	0.060 0.105
	Oil	0.0004 0.0025	0.010 0.065
Piston Pin to Piston		0.00035 tight to 0.00039 loose	0.009 0.010

STANDARD CRANK PIN DIMENSIONS

D

W

Fig. 22 PISTON, RING AND ROD CLEARANCE CHART

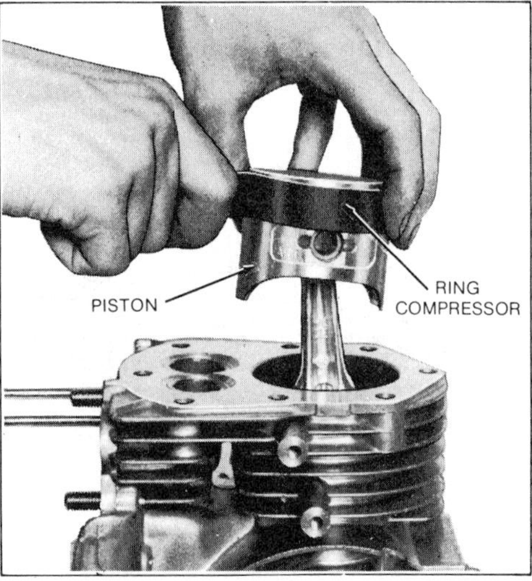

Fig. 23

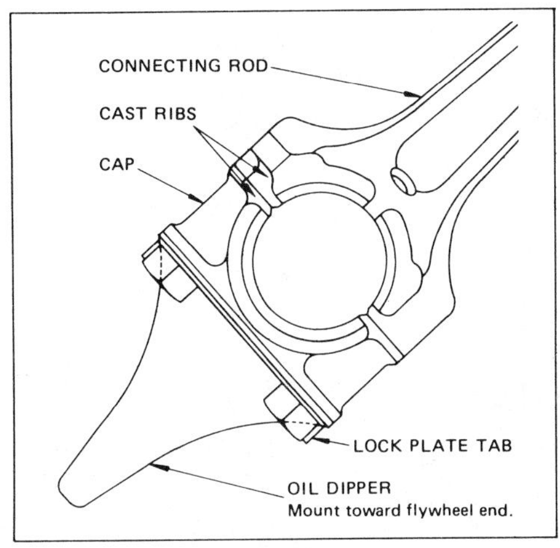

Fig. 24

Install a new rod bolt lock plate. Mount connecting rod bolts and tighten to the following torque specifications — ***6.5*** to ***8.3 ft. lbs.*** (0.90 to 1.15 Kgm).

Check for free movement of connecting rod by turning crankshaft over slowly. If satisfactory, bend tabs on lock plate against hex head flat of connecting rod bolts.

CRANKSHAFT, Fig. 25

1. Remove flywheel woodruff key.
2. Pull crankshaft out from open end of crankcase and take care not to damage the oil seal. If necessary, loosen shaft by tapping lightly at flywheel end with a soft hammer.
3. ***In reassembly,*** inspect crankcase oil seal and main bearing for possible replacement. Mount crankshaft with extreme care so as not to damage lips of oil seal. Use an oil seal sleeve if available.
4. ***End Play*** is regulated by means of an ***adjusting shim*** at the gear end of the crankshaft. This should be set immediately before mounting gear cover as explained in *'Crankshaft End Play'* paragraphs on *Page 12.*

Fig. 25

ENGINE REASSEMBLY

Rebuild the engine in reverse order, and by following the reassembly instructions which are included with the disassembly procedures.

After complete reassembly, turn the engine over by means of the starting pulley to check for any abnormal conditions and loose fitting parts. Then, review and proceed with the following:

1. Crankshaft end play — *Page 12.*
2. Check ignition spark — *Page 4.*
3. Timing — *Page 5.*
4. Governor adjustment — *Page 6.*
5. Carburetor adjustments — *Page 20.*
6. Fill crankcase with correct grade of oil — *Page 4.*
7. Fill fuel tank with a well known brand of *'Regular Grade'* gasoline.
8. Test engine per the following instructions.

TESTING REBUILT ENGINE

An engine that has been completely overhauled by having the cylinder rebored and fitted with a new piston, rings, valves and connecting rod should be thoroughly ***'RUN-IN'*** before being put back into service. Good bearing surfaces and running clearances between the various parts can only be established by operating the engine under reduced speed and loads for a short period of time.

Load		Speed	Time
No Load		2500 r.p.m.	10 minutes
No Load		3000 r.p.m.	10 minutes
No Load		3600 r.p.m.	10 minutes
1.35 hp.	1.01 kw.	3600 r.p.m.	30 minutes
2.7 hp.	2.01 kw.	3600 r.p.m.	60 minutes

While engine is being tested — check for oil leaks. Make final carburetor adjustments and regulate the engine operating speed.

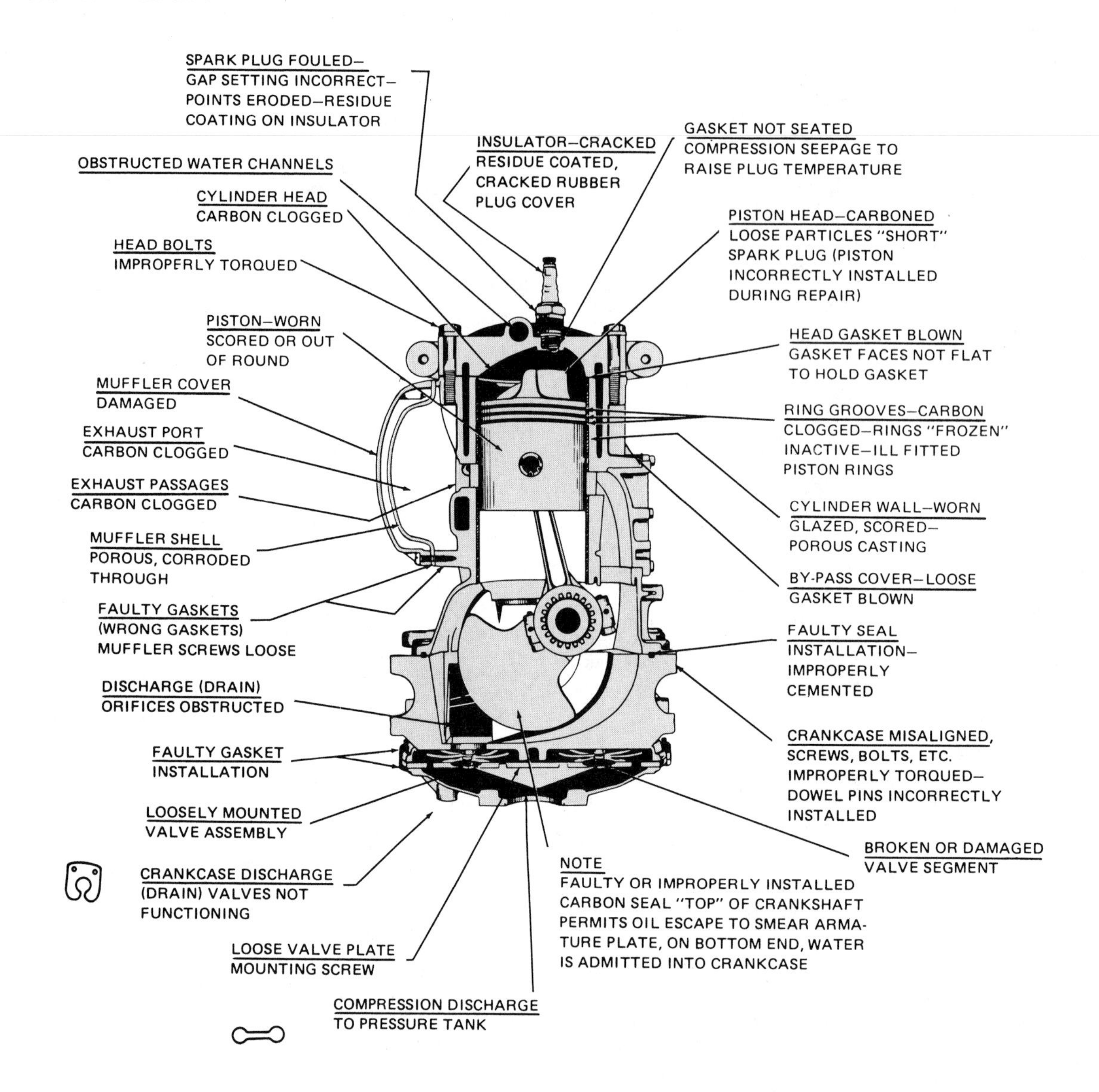

Fig. C-5 Power head diagnosis chart

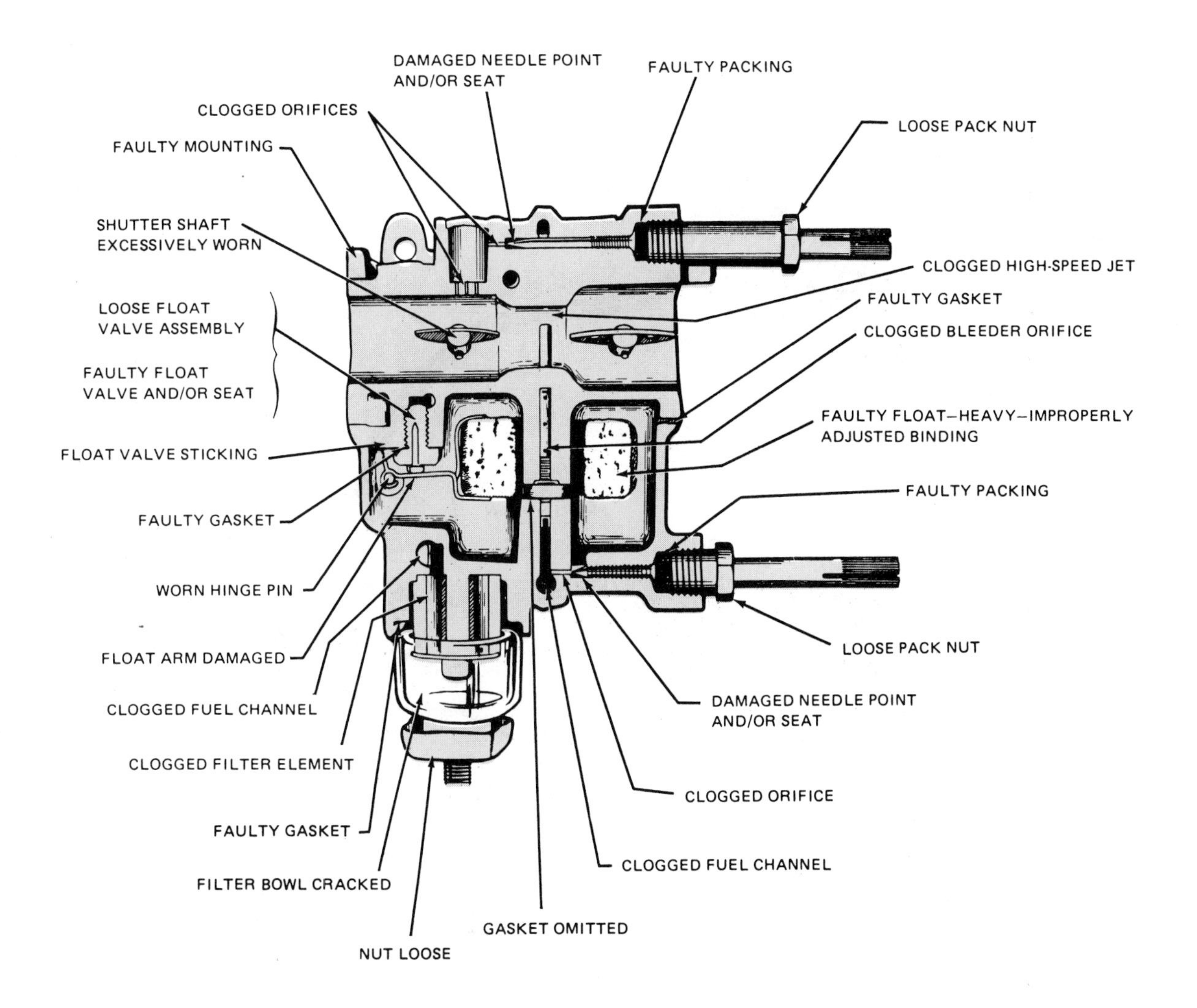

Fig. C-6 Carburetion diagnosis chart

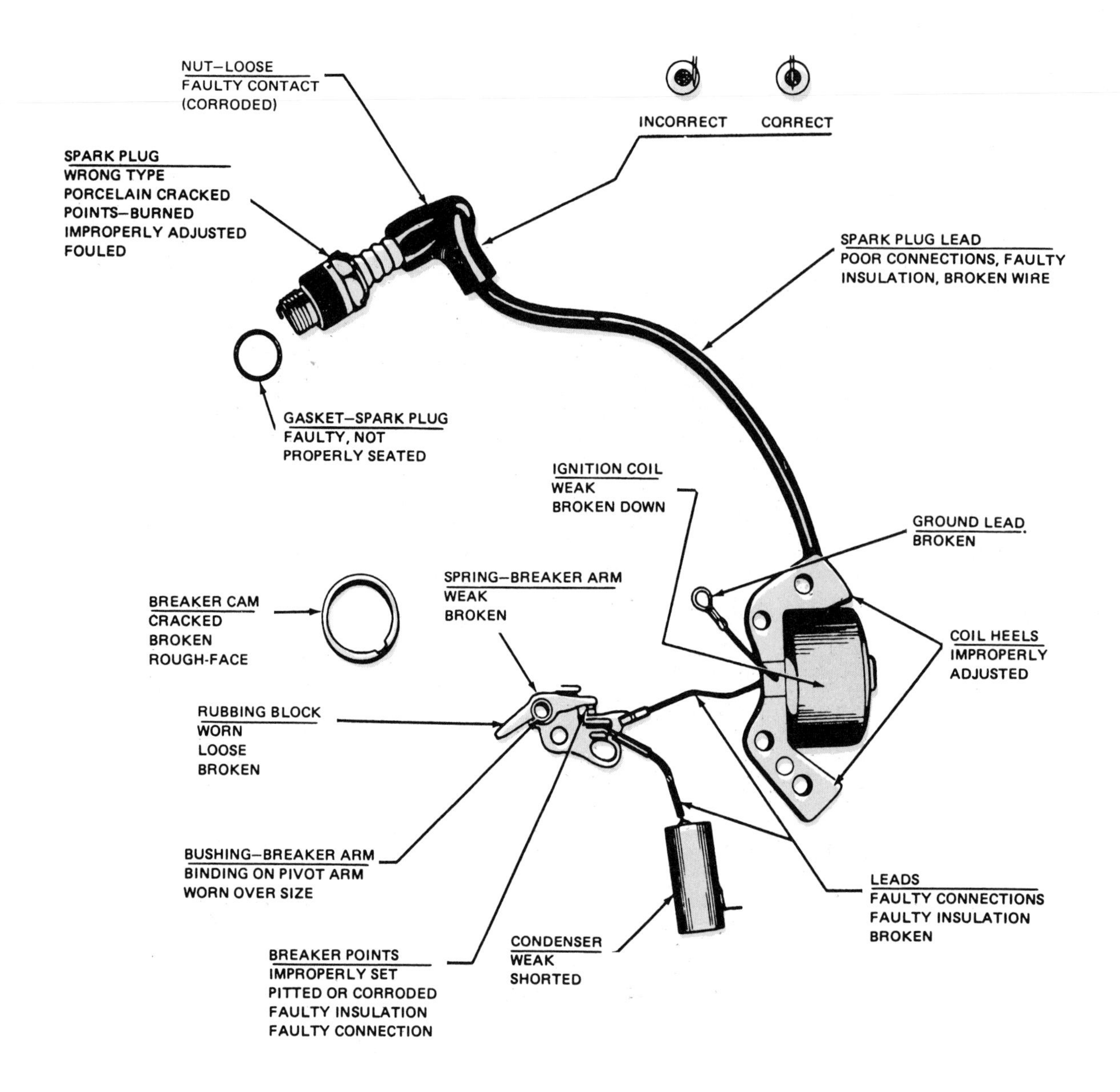

Fig. C-7 Magneto diagnosis chart

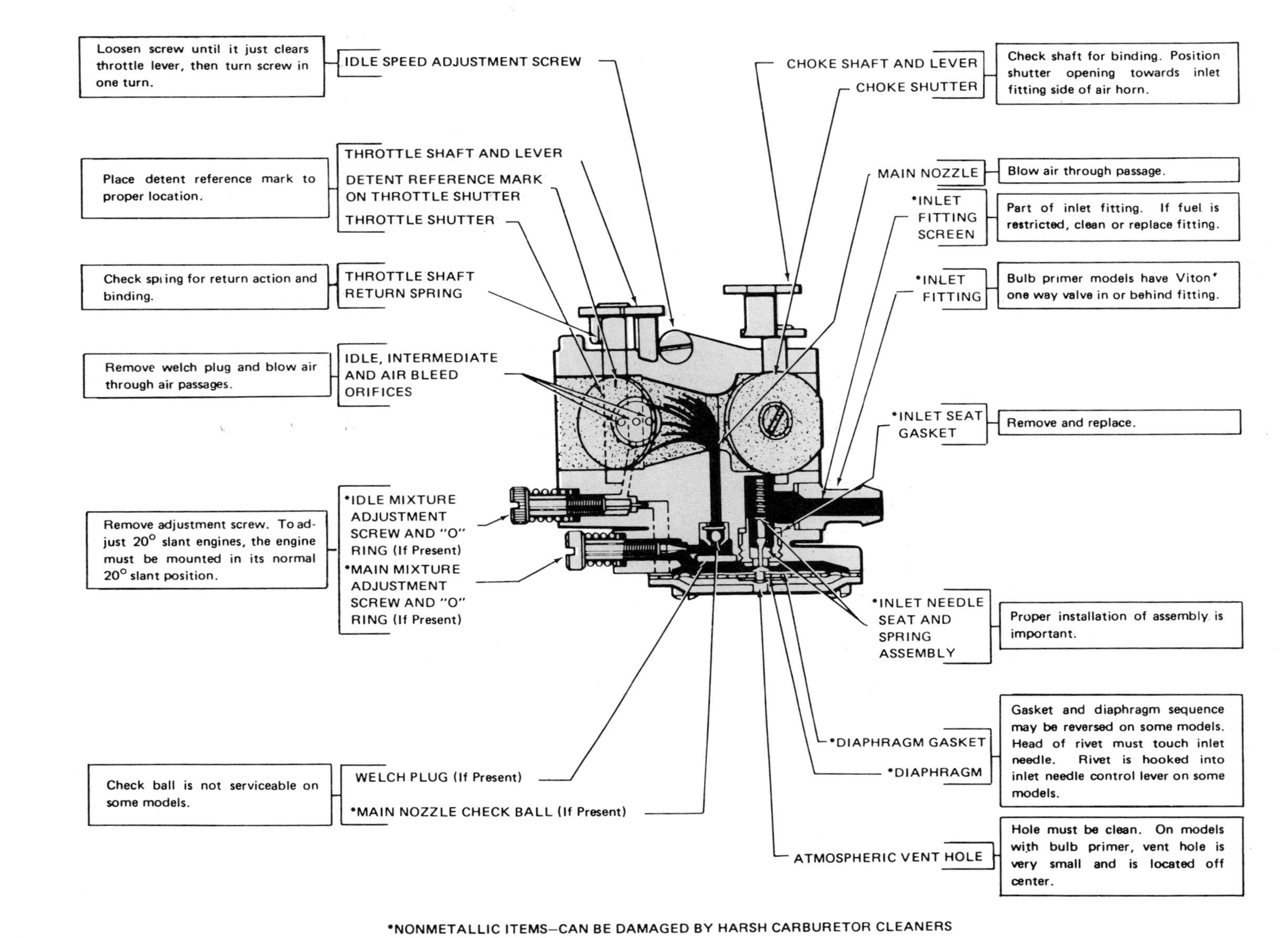

Fig. C-8 Service hints for diaphragm carburetors

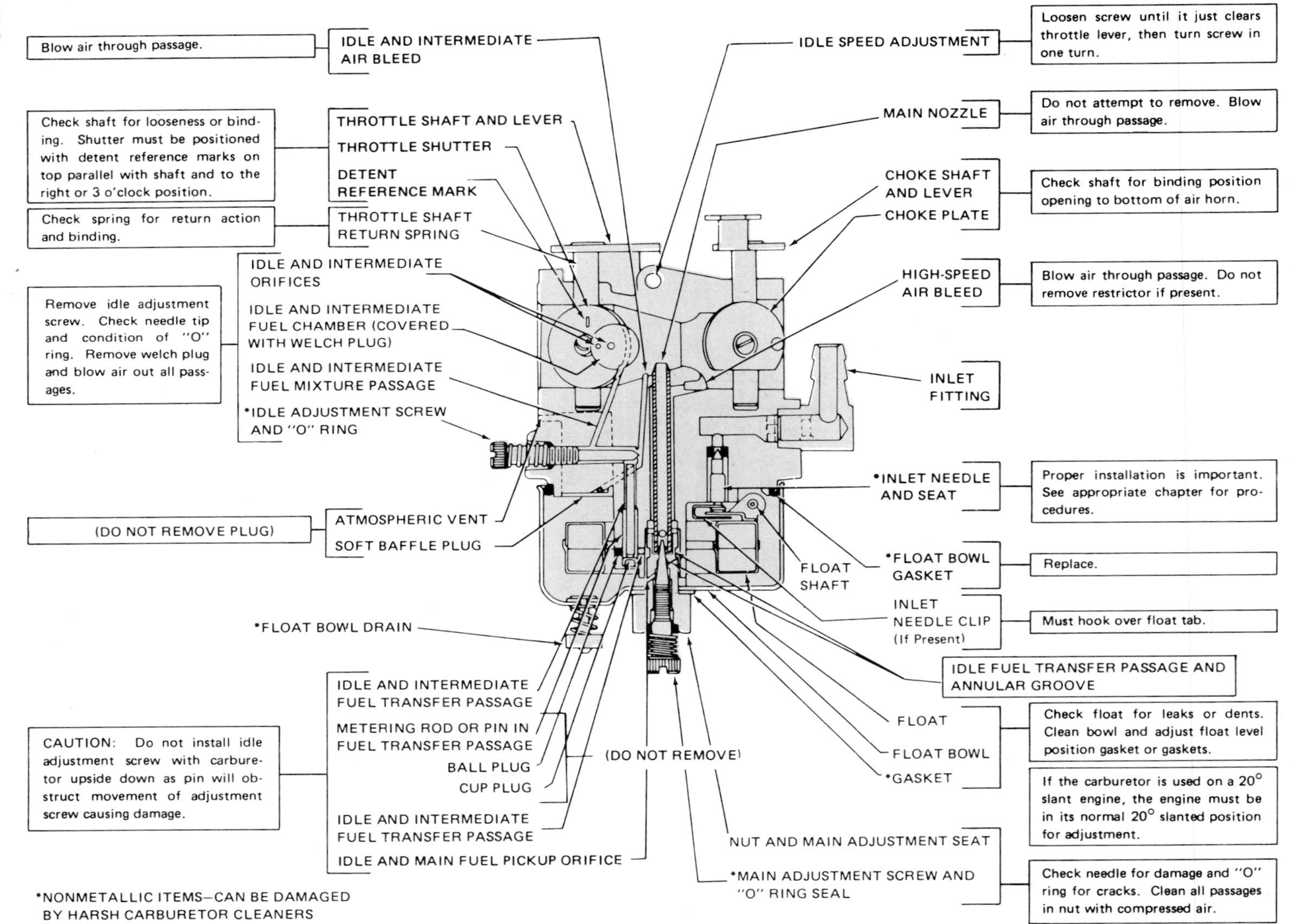

Fig. C-9 Service hints for float-feed carburetors

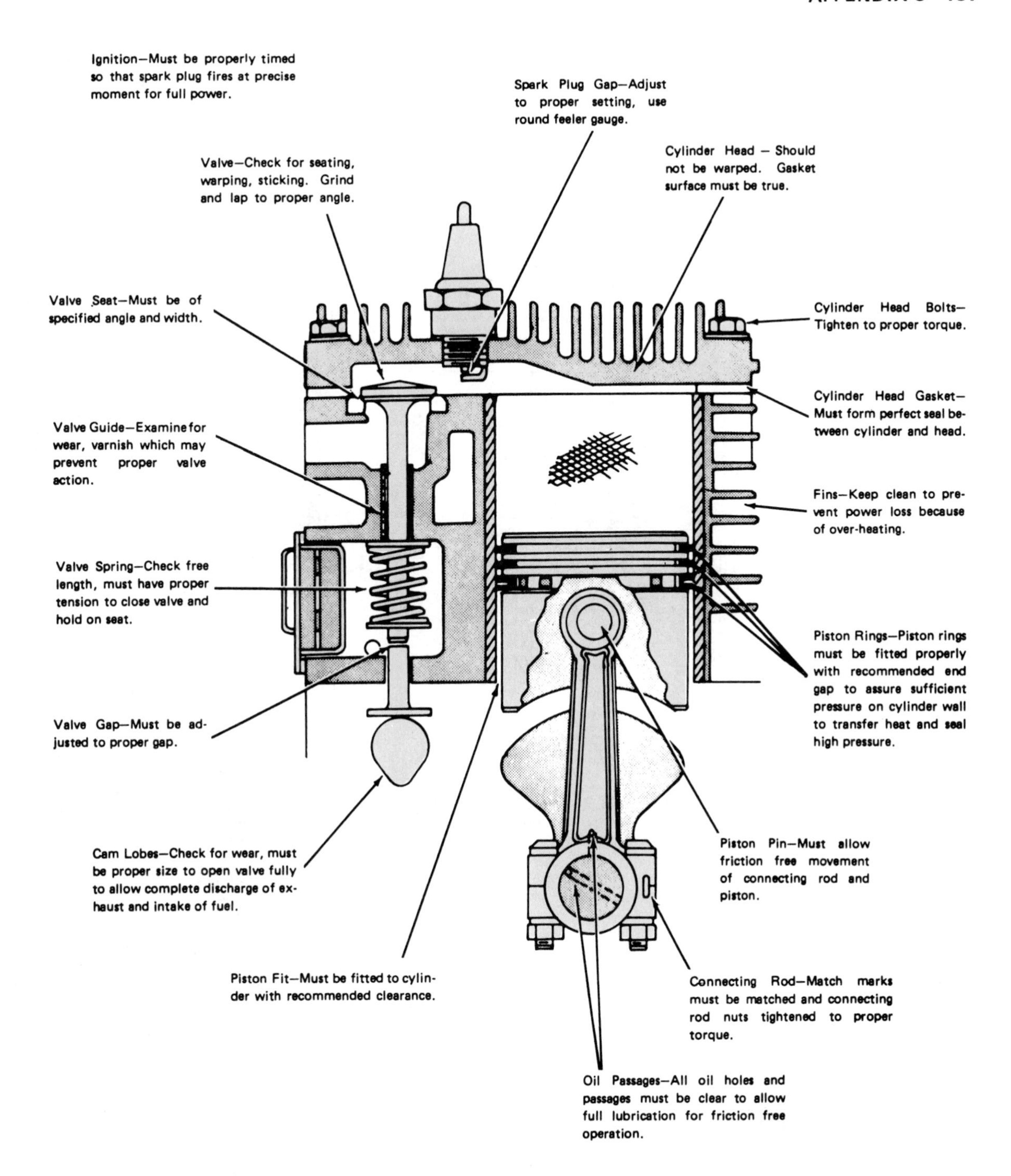

Fig. C-10 Points to check for engine power

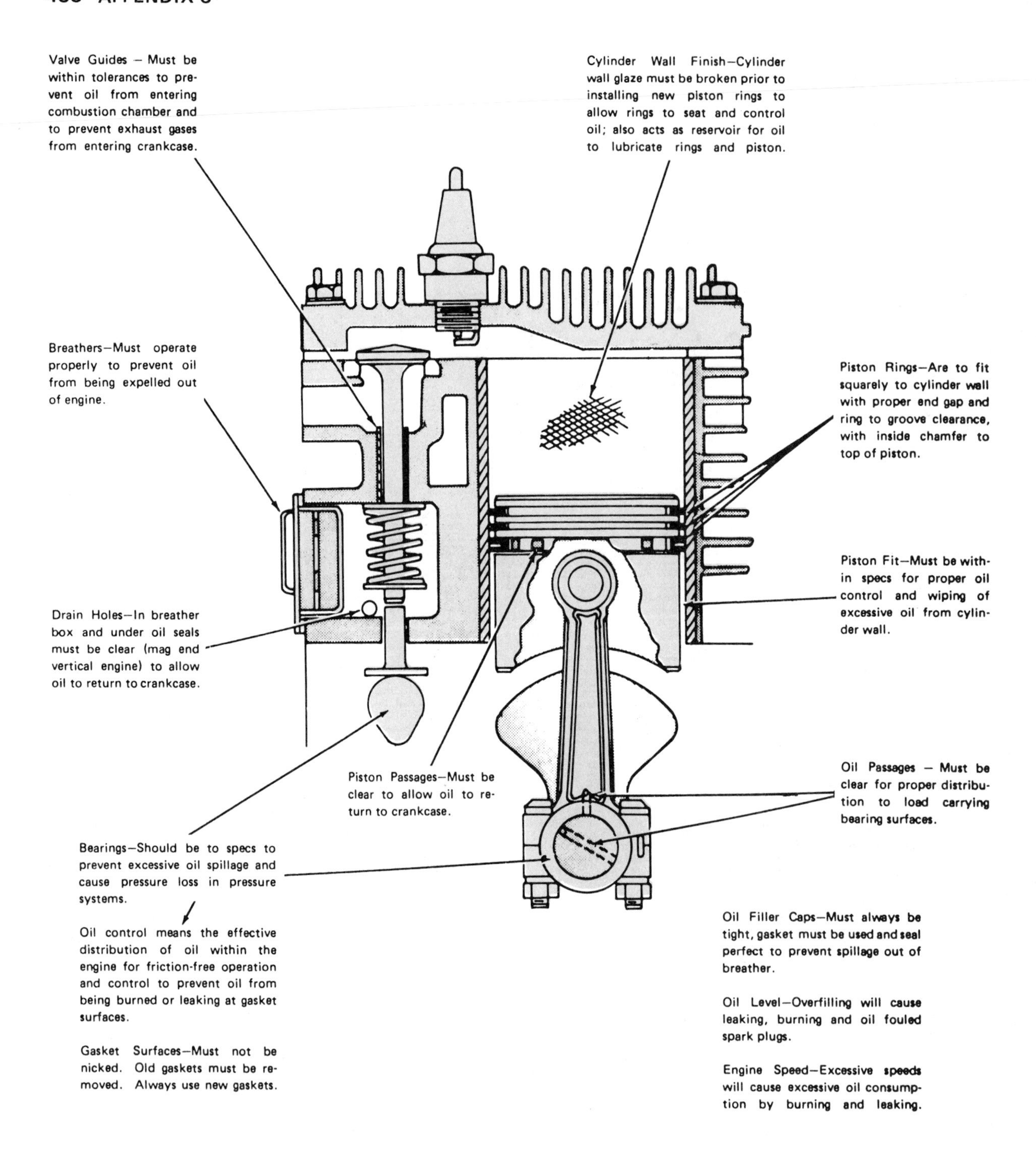

Fig. C-11 Points to check for engine oil control

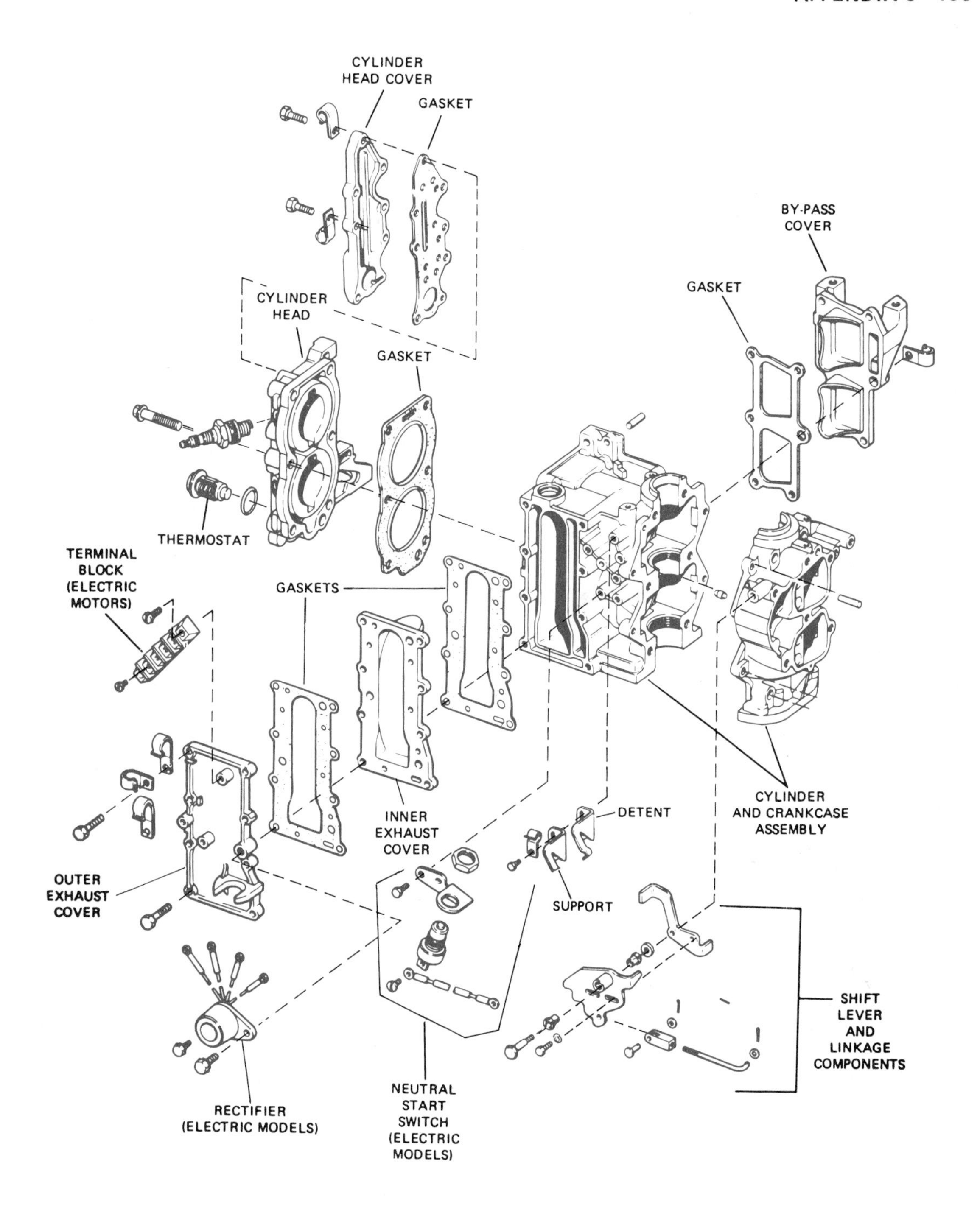

Fig. C-12 Power head – exploded view

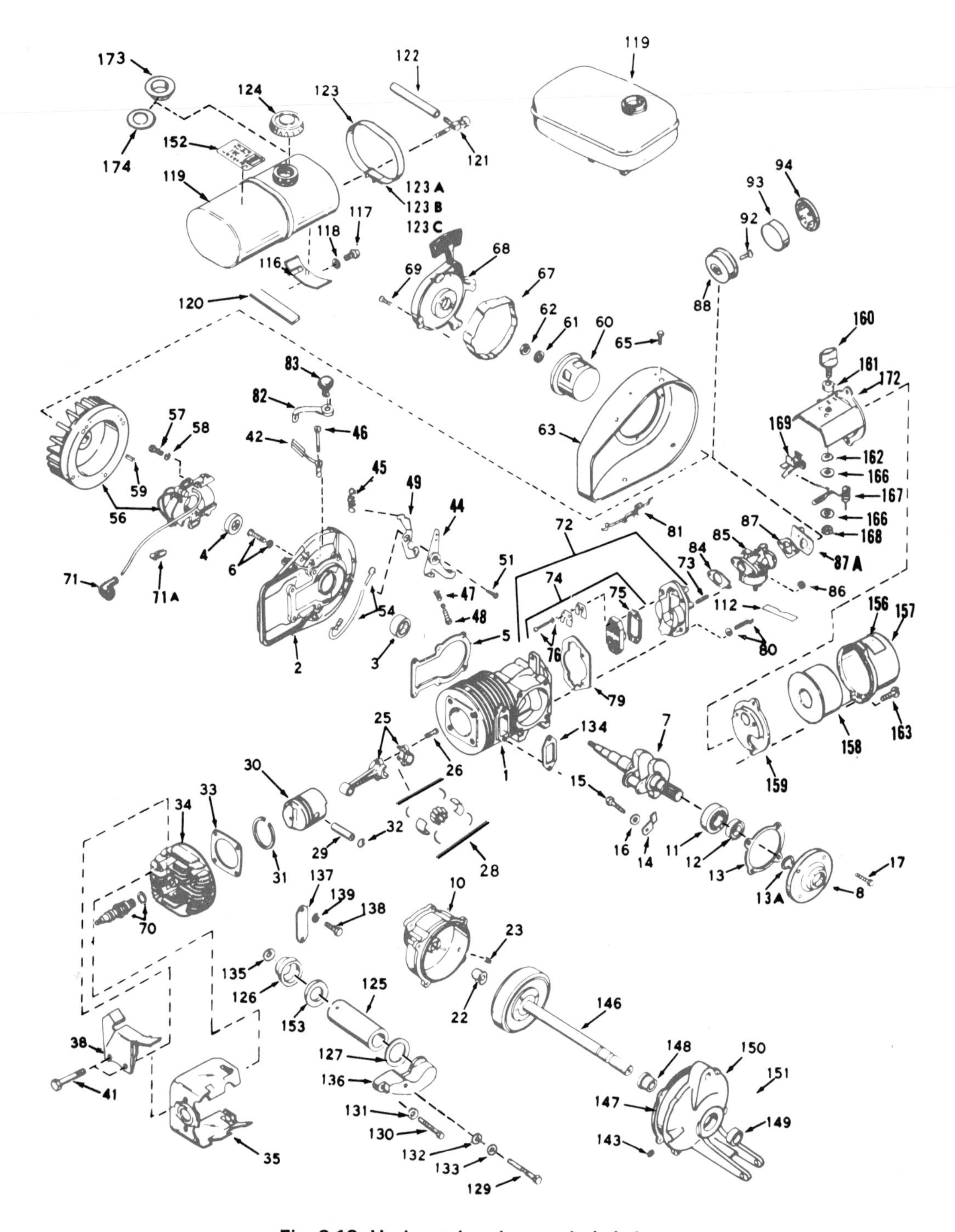

Fig. C-13 Horizontal engine, exploded view

1. Cylinder and Crankcase
2. Backplate with Bearing and Seal
3. Backplate Bearing
4. Backplate Oil Seal
5. Backplate Gasket
6. Screw and Lockwasher
7. Crankshaft
8. Crankcase Head
10. Crankcase Head
11. Bearing
12. Oil Seal
13. Crankcase Head Gasket
13A. Ring, Snap
14. Bearing Retainer Clip
15. Screw
16. Lockwasher
17. Screw
22. Gearcase Bushing
23. Plug
25. Connecting Rod Assembly
26. Screw
28. Bearing Assembly, 28 Rollers, 2 Liners, 4 Guides
29. Piston Pin
30. Piston
31. Piston Ring
32. Piston Pin Retaining Ring
33. Cylinder Head Gasket
34. Cylinder Head
35. Air Deflector
38. Fuel Tank Mounting Bracket
41. Screw
42. Governor Vane Assembly
44. Control Lever Assembly
45. Governor Vane Spring
46. Screw, Governor Vane
47. Spring, Control
48. Screw, Speed Control
49. Lever, Governor Spring
51. Screw, Control Lever Mounting
54. Stop Switch Wire
56. Magneto Assembly
57. Screw and Lockwasher
58. Washer
59. Flywheel Key
60. Rewind Starter Hub
61. Washer, Belleville
62. Nut
63. Fan Housing
65. Screw
67. Starter Screen
68. Rewind Starter Assembly
69. Screw and Lockwasher
70. Spark Plug
71. Spark Plug Cap
71A. Terminal, Spark Plug
72. Carburetor Adapter and Reed Plate Assembly
73. Stud
74. Reed Plate
75. Reed Plate Gasket
76. Screw, Reed Plate Mounting
79. Adapter Mounting Gasket
80. Screw and Lockwasher
81. Choke Link Assembly
82. Governor Link
83. Push Rivet
84. Carburetor Mounting Gasket
85. Carburetor Assembly
86. Nut
87. Plate Mounting Gasket
87A. Bracket, Throttle Wire
88. Air Filter Housing
92. Screw
93. Air Filter
94. Air Filter Cover
112. Decal, Choke, and Shutoff
116. Tank Mounting Bracket
117. Screw
118. Lockwasher (Washer)
119. Fuel Tank Assembly
120. Pad, Fuel Tank Mounting
121. Shutoff Valve Assembly
122. Fuel Line
123. Tank Mounting Strap
123A. Screw
123B. Lockwasher
123C. Nut
124. Fuel Tank Cap Assembly
125. Muffler
126. Muffler Cap
127. Muffler Mounting Gasket
128. Bolt
129. Screw
130. Screw
131. Lockwasher
132. Muffler Head Gasket
133. Washer
134. Exhaust Flange Gasket
135. Lockwasher
136. Muffler Head
137. Exhaust Flange Cover
138. Screw
139. Lockwasher
143. Plug
146. Gear and Shaft Assembly
147. Gear Reducer Cover Gasket
148. Gear Reducer Cover Bearing
149. Gear Reducer Cover Seal
150. Gear Reducer Cover
151. Screw
152. Decal, Name, Mix
153. Muffler Gasket
156. Decal, Air Cleaner
157. Air Filter Body
158. Air Filter
159. Bracket
160. Knob
161. Spacer
162. Washer
163. Screw
166. Washer
167. Spring
168. Nut
169. Stop Switch Assembly
172. Bracket Assembly
173. Baffle Assembly
174. Gasket, Fuel Tank Cap

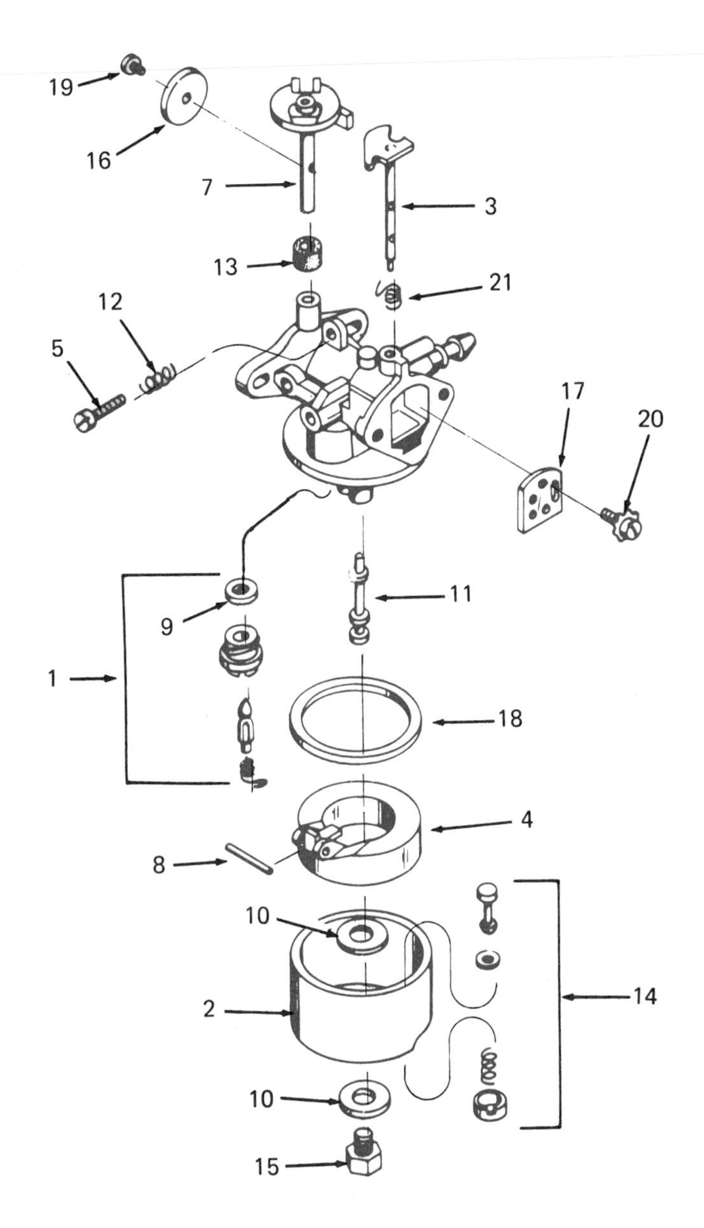

1. Matched, Float Valve Seat, Spring and Gasket Assembly
2. Bowl Assembly—Float Bowl
3. Shaft Assembly—Choke
4. Float Assembly
5. Screw—Throttle Adjustment
7. Shaft Assembly—Throttle
8. Shaft—Float
9. Gasket—Float Valve Seat
10. Gasket—Nut to Bowl
11. Main Metering Nozzle
12. Spring Throttle Adjustment
13. Seal—Throttle Shaft
14. Bowl Drain Assembly
15. Retainer Screw
16. Throttle Plate
17. Choke Plate
18. Gasket—Bowl to Body
19. Screw
20. Screw
21. Spring—Choke Return

Fig. C-14 Fixed-jet carburetor, exploded view

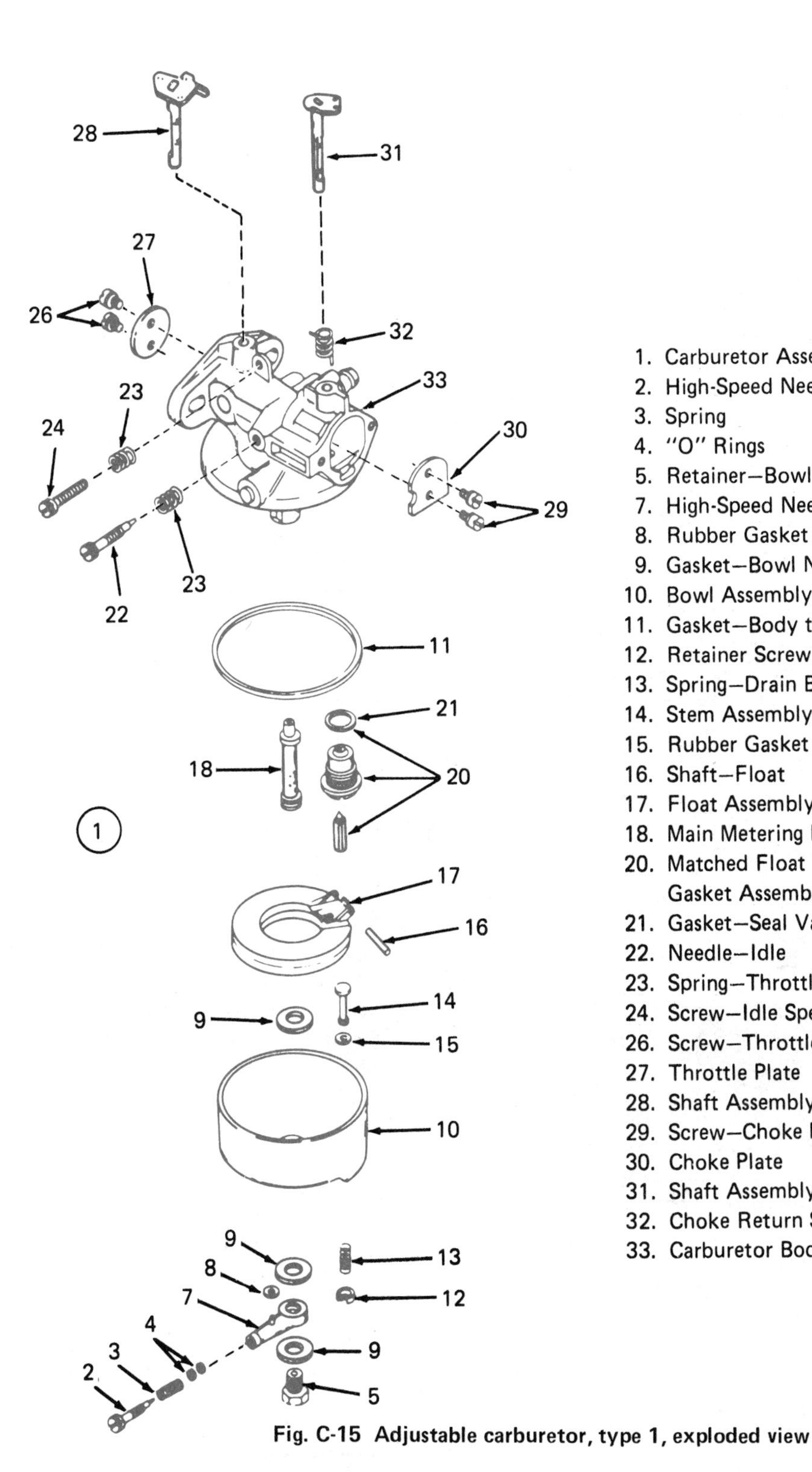

1. Carburetor Assembly
2. High-Speed Needle
3. Spring
4. "O" Rings
5. Retainer–Bowl
7. High-Speed Needle Housing
8. Rubber Gasket
9. Gasket–Bowl Nut to Bowl
10. Bowl Assembly–Float Bowl
11. Gasket–Body to Bowl
12. Retainer Screw
13. Spring–Drain Bowl
14. Stem Assembly–Drain Bowl
15. Rubber Gasket
16. Shaft–Float
17. Float Assembly
18. Main Metering Nozzle
20. Matched Float Valve, Seat, Spring and Gasket Assembly
21. Gasket–Seal Valve Float
22. Needle–Idle
23. Spring–Throttle Adjustment Screw
24. Screw–Idle Speed
26. Screw–Throttle Plate Mounting
27. Throttle Plate
28. Shaft Assembly–Throttle
29. Screw–Choke Plate Mounting
30. Choke Plate
31. Shaft Assembly–Choke
32. Choke Return Spring
33. Carburetor Body

Fig. C-15 Adjustable carburetor, type 1, exploded view

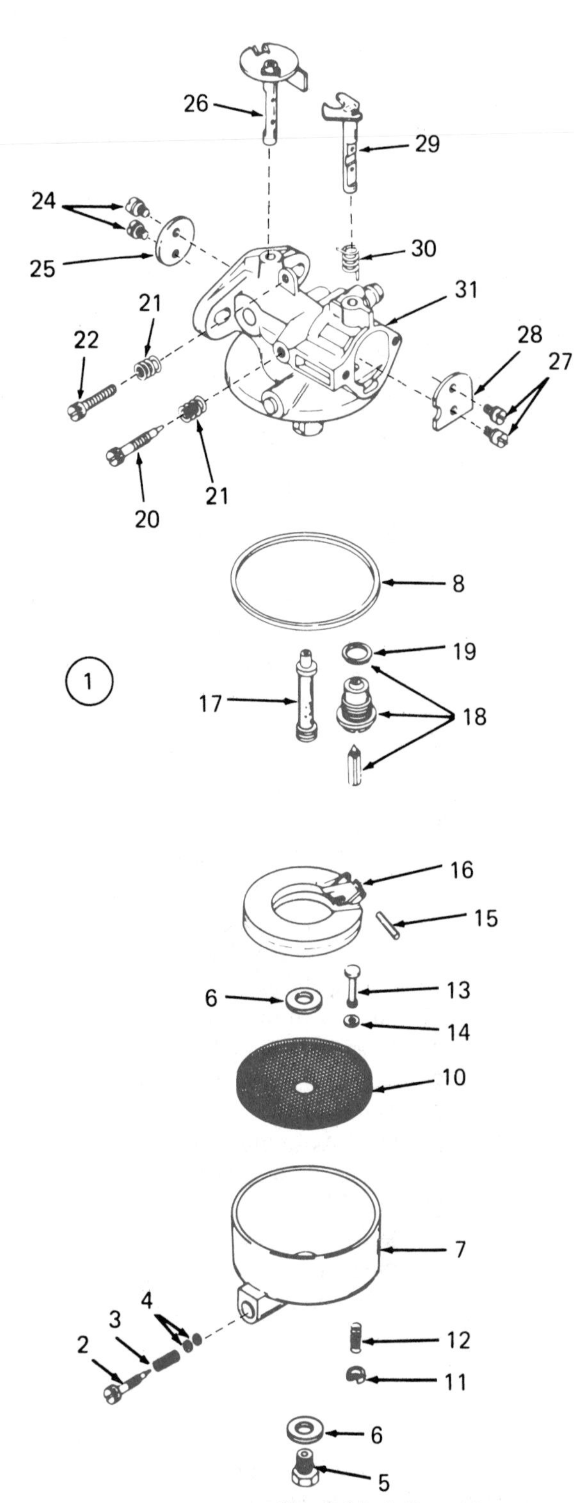

1. Carburetor Assembly
2. High-Speed Needle
3. Spring
4. "O" Rings
5. Retainer Screw
6. Gasket—Bowl Nut to Bowl
7. Bowl Assembly—Float Bowl
8. Gasket—Body to Bowl
10. Screen
11. Retainer Screw
12. Spring—Drain Bowl
13. Stem Assembly—Drain Bowl
14. Rubber Gasket
15. Shaft—Float
16. Float Assembly
17. Main Metering Nozzle
18. Matched Float Valve Seat, Spring and Gasket Assembly
19. Gasket—Seal Valve Float
20. Needle, Idle
21. Spring
22. Screw, Throttle Adjustment
24. Screw—Throttle Plate Mounting
25. Throttle Plate
26. Shaft Assembly Throttle
27. Screw—Choke Plate Mounting
28. Choke Plate
29. Shaft Assembly—Choke
30. Spring Choke Return
31. Carburetor Body

Fig. C-16 Adjustable carburetor, type 2, exploded view

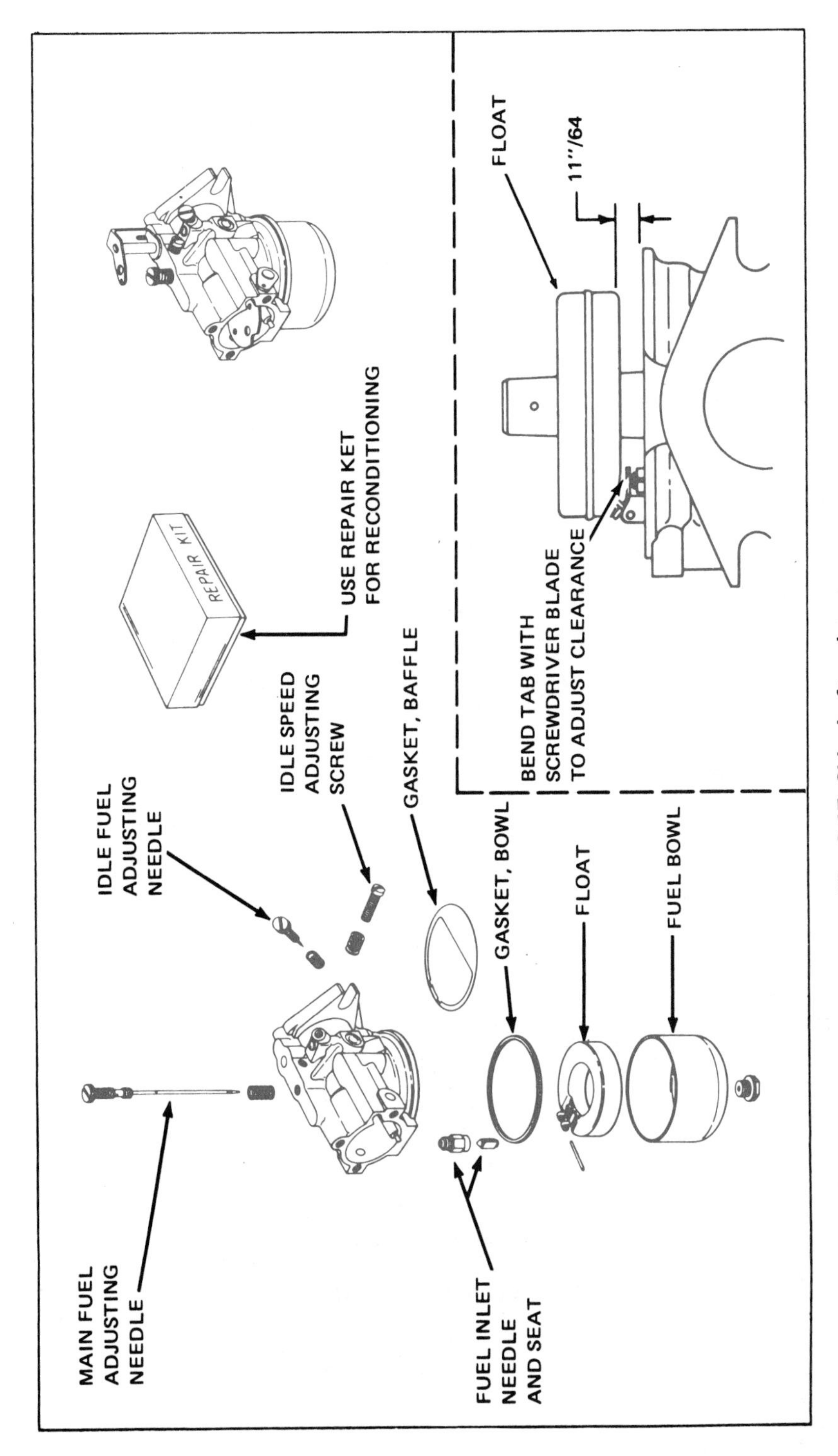

Fig. C-17 Side-draft carburetor

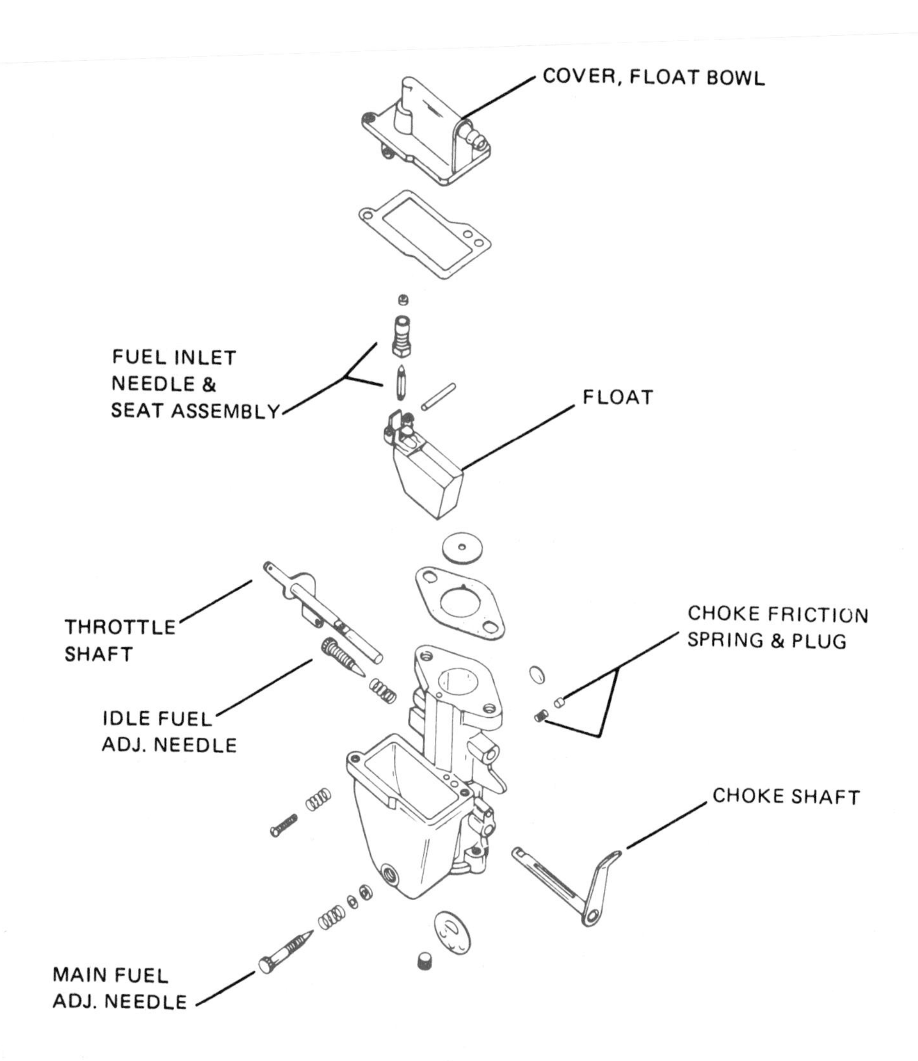

Fig. C-18 Up-draft carburetor

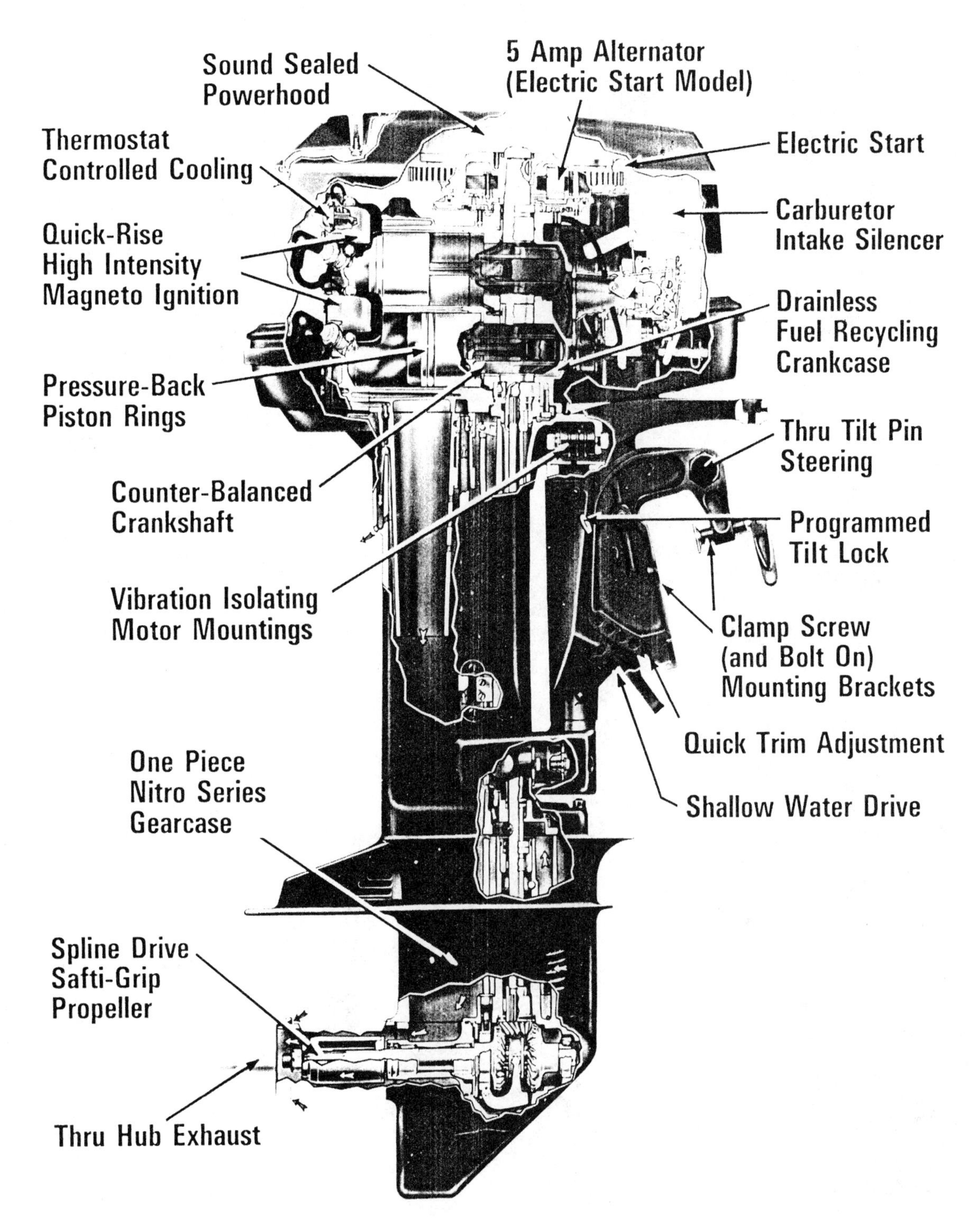

Fig. C-19 Two-cylinder outboard engine

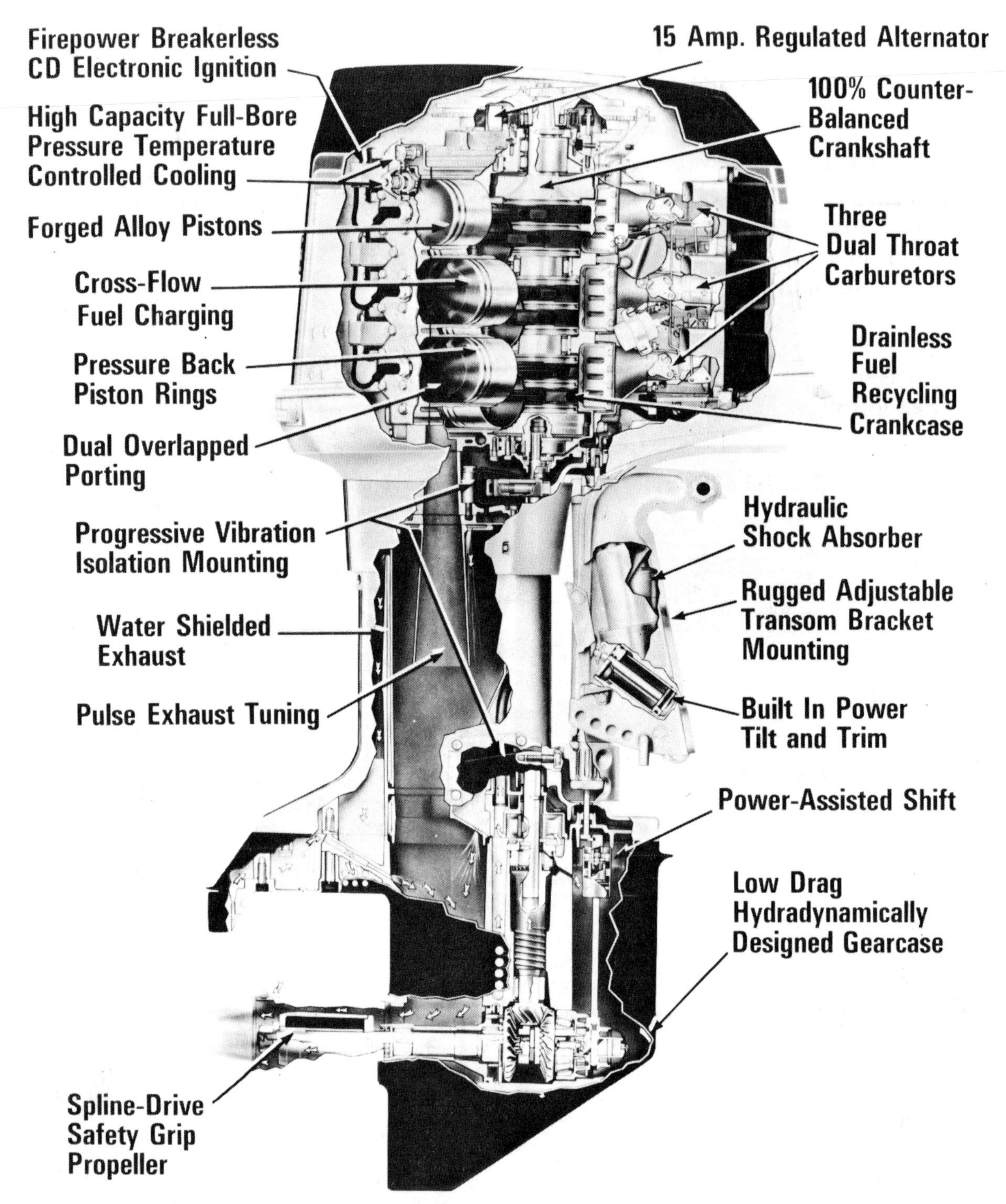

Fig. C-20 Six-cycle outboard engine

Glossary

Additives. Substances added to oils to improve their ability to function properly.

Alternating current. Electrical current that changes its direction back and forth regularly in a cycle (AC).

Alternator. An electrical device that produces AC electricity with stationary coils of wire and a rotating magnet.

Ammeter. An electrical instrument for measuring current flow in a circuit.

Ampere. The electrical measurement of the rate of current flow in a conductor.

Anthracite. The hardest type of coal which also contains the most heat energy.

Antifriction bearing. A bearing that substitutes rolling friction for sliding friction such as a roller bearing or ball bearing.

API. American Petroleum Institute

Atmospheric pressure. The weight of the atmosphere above the earth: 14.7 psi at sea level.

Atom. The smallest particle of an element that still retains the properties of that element.

Backfire. Unwanted ignition of fuel mixture in the intake manifold.

Battery. A number of primary cells grouped together to store chemical electrical energy.

BDC. Bottom dead center or the lowest point of the piston's travel.

Bearing. A supportive part in which a shaft revolves or moves.

Biomass. All organic matter which can be used as a fuel directly or be converted into a fuel.

Bituminous coal. The most plentiful type of coal that is economically recoverable. It is softer than anthracite but harder than subbituminous coal.

Blow-by. Refers to combustion gases that pass between the cylinder wall and the piston and rings and enters the crankcase.

Bore. The diameter of a cylinder.

Brake horsepower. The measurement of the engine's horsepower taken at the flywheel.

Breaker points. An electrical switch used to interrupt the current flow in the primary coil of an ignition coil in an ignition system. The opening of the points triggers the spark.

Breeder reactor. A nuclear reactor that is capable of producing more fissionable material than it consumes.

British thermal unit (BTU). The unit of measurement of heat energy. The amount of heat required to raise the temperature of one pound of water 1°F.

Calorie. The metric unit of measurement for heat. The amount of heat required to raise the temperature of one gram of water 1°C.

Camshaft. A shaft with lobes that convert rotary motion into linear motion for opening valves.

Carbon monoxide. A poisonous gas produced by the burning of carbon-type fuels.

Carbonaceous. Referring to all hydrocarbon-type fuels.

Carburetor. A device for mixing fuel and air in the correct proportions for burning in an engine.

Carburetor icing. The formation of frost on the throttle plate under certain humidity and temperature conditions.

Catalyst. A substance that speeds up a chemical reaction without entering into the reaction.

Catalytic converter. An emission control device in the engine's exhaust system that removes hydrocarbons and carbon monoxide.

Centrifugal force. The force that tends to throw objects away from the center of rotation.

Choke. A device for controlling the amount of air that can enter the carburetor. It is used mainly to provide a rich mixture for starting and warmup.

Clutch. A device for engaging and disengaging the engine from the transmission.

Coal gasification. A method of producing methane gas from coal by burning it with a controlled amount of oxygen.

Coal liquification. The process of converting coal into a liquid fuel that can be burned much the same as heating oil.

Combustion chamber. The small volume formed by the cylinder head, the top of the cylinder and the top of the piston, where combustion of fuel takes place.

Compression ignition. The principle of operation of a diesel engine.

Compression ratio. The ratio of maximum volume to minimum volume in the cylinder caused by the piston's motion.

Compression stroke. The stroke of the piston in which the trapped fuel mixture is squeezed into a small space just before combustion.

Condenser. An electrical device for storing electrical energy. It prevents arcing at the breaker points. Also, a device for converting a gas into a liquid.

Connecting rod. The part that connects the crankshaft with the piston pin and piston.

Coolant. The liquid circulated through the engine's cooling system.

Crankcase. The main body of the engine which holds the crankshaft and many other engine parts.

Crankcase breather. A part that vents the crankcase to the atmosphere to prevent pressure buildup.

Crankshaft. The main shaft of the engine to which the connecting rods are attached. The crankshaft converts reciprocating motion into rotary motion.

Critical. When the process of fission is proceeding at a controlled rate.

Crude oil. Oil in its unrefined state—just as it comes from the ground.

Cylinder. A round hole bored into the engine block to accept the piston.

Cylinder block. Commonly called the block; the largest single part of the engine which contains the cylinders.

Cylinder head. The top of the cylinder, usually bolted down tightly but sometimes cast as a part of the cylinder.

Detergent. Used in reference to oils, these are additives that keep the dirt, soot, and other minute particles in suspension.

Detonation. Sometimes called knock; occurs when two flame fronts in the combustion chamber meet, exploding the fuel mixture between them.

Diesel engine. An engine that uses the heat of compression to ignite the fuel. Diesel fuel is used; no spark plug is necessary. It is named after its inventor Rudolph Diesel.

Differential. An arrangement of gears with the rear axle which permits one wheel to turn faster than the other when corners are turned.

Direct current. Electrical current that flows in only one direction, such as current from a storage battery (DC).

Dynamometer. A device used to determine the horsepower developed by an engine.

Ecology. That branch of biology that concerns itself with the relationship of organisms to their environment.

Electron. A very small part of an atom. It is negatively charged and orbits around the nucleus.

Emissions. Term commonly applied to the harmful portions of the exhaust gases of an engine.

Energy. The ability to do work. Heat energy is measured in BTUs (British Thermal Units) or joules (metric).

Environment. The sum total of the things, conditions and influences that affect the development of life.

EPA. Environmental Protection Agency.

ERDA. Energy Research and Development Administration.

Ethanol. Made by the distillation of grain or plants to produce ethyl alcohol (grain alcohol) a combustible liquid.

Exhaust stroke. The stroke of the piston in which the exhaust gases are removed from the cylinder.

External combustion engine. An engine that burns fuel in a fire box under a boiler outside the engine, to boil water and produce steam to drive the engine.

Fluid power. The use of a fluid–either liquid or gas–to operate a mechanism: hydraulics and pneumatics.

Flywheel. A relatively large wheel that absorbs energy and stores it by momentum.

Fossil fuels. Fuels that have been formed in the earth over million of years–petroleum, coal and natural gas. Originally deposited as dead plants and organisms, they were changed into hydrocarbon molecules by the action of heat, pressure, and chemical changes.

Fouling. Term used to denote damage done to spark plugs.

Four-stroke cycle. An engine that completes its operating cycle in four strokes of the piston and two revolutions of the crankshaft.

Fractional distillation. A basic refinery process in which many of the hydrocarbons that make up crude oil vaporize when heated and then condense at various levels in a cooling tower.

Friction. The resistance to motion between two parts in contact with each other.

Fuel cell. A device that converts chemical energy directly into electricity. A reaction between hydrogen and oxygen is used.

Fuel injection. Refers to engines, either gasoline or diesel, which inject the fuel into the cylinder after the air is compressed.

Fuel pump. A device that is operated by the motion of a diaphragm to deliver fuel from the tank to the carburetor.

Fuel system. The entire contents of an internal combustion engine. Includes fuel tank, fuel line, fuel pump, and carburetor.

Gas turbine. An internal-combustion engine in which hot gases are directed against vanes mounted on a rotary shaft.

Gasket. A thin, relatively soft material made of rubber, fiber, metal, or cork, used to form an air-tight seal.

Gasohol. A blended fuel containing 9 parts gasoline to 1 part ethanol (ethyl alcohol).

Generator. A device for converting mechanical energy into electrical energy.

Geothermal energy. Energy released from the magma (molten core of the earth). This energy is usually seen in certain parts of the earth where the heat is close to the surface and underground water comes in contact with the hot rocks to produce steam.

Governor. A device used to control or regulate engine speed.

Ground. An electrical term for a large conductive body that acts electrically neutral to accept a flow of electrons.

Heat engine. An engine that converts the heat of a burning fuel into mechanical energy.

High voltage coil. A coil containing a primary and a secondary coil. Its transformer action steps up low voltages to high voltages which can jump the spark gap in the spark plug. Sometimes it is called the ignition coil.

Horsepower. Unit of measurement for the rate of doing work. Lifting 33,000 pounds 1 foot in 1 minute is 1 horsepower.

Hydraulics. A fluid transmission system that uses a liquid within a dynamic system as the transmission medium.

Hydrocarbon. A compound consisting of molecules of hydrogen and carbon. Petroleum is a hydrocarbon.

Hydroelectric. Electricity produced by water-driven

turbines which in turn drive a generator. Water for driving the turbines is collected behind dams.

Industrial revolution. The period in history beginning in the mid 1800s when the rapid growth of industry and transportation took place using the steam engine as its main power source.

Intake stroke. The stroke of the piston in which the fuel mixture is brought into the cylinder.

Internal combustion engine. An engine that burns fuel inside the engine itself.

Jet propulsion. An engine that uses Newton's principle; for every action there is an equal but opposite reaction. Rapidly escaping gases from one end of the engine produce thrust at the other end of the engine.

Journal. The part of an axle or shaft that makes contact with the bearing.

Key. A device for locking two parts together, such as the crankshaft and the flywheel. Also called a shear pin.

Kinetic energy. The energy of motion.

Lignite. A type of soft coal with a relatively low heat value.

Magnet. A hard steel alloy in which a magnetic field is permanently maintained due to its molecular arrangement.

Magneto. An electrical device that converts magnetism into high voltage for the spark plug of the ignition system. It is self-contained and uses only the energy of a moving magnetic field.

Manifold. A pipe that has several openings to connect cylinders with a single inlet or outlet—intake manifold or exhaust manifold.

Methane. A combustible gas produced by the decomposition of organic matter.

Methanol. Made by the destructive distillation of wood or plants to produce methyl alcohol (wood alcohol) a combustible liquid.

Metric. The decimal system of weights and measures with units such as meters, liters, and grams.

Molecule. The smallest portion of a chemical compound that still has the properties of the compound.

NASA. National Aeronautics and Space Administration.

Natural gas. A gaseous hydrocarbon fuel found in underground formations. Its principle ingredient is methane, a combustible gas.

Needle valve. A tapered end of a threaded shaft that fits into a tapered hole. Used to regulate the amount of gasoline that can flow out of the hole.

Nuclear battery. An electrical device that produces electricity from radioactive materials.

Nuclear energy. Energy released when the nucleus of an atom is split (nuclear fission). Uranium is the only natural element that can be split.

Nuclear fission. Splitting the nucleus of an atom to release energy.

Nuclear fusion. The coming together of the nuclei of light elements. When they collide and merge energy is released.

Nuclear reactor. A device in which a chain reaction of nuclear fission can be sustained and controlled. Heat produced from the reactor is used to produce electricity.

Octane number. A measurement of a fuel's ability to resist detonation or knocking; its anti-knock quality.

Ohm. The electrical measurement for the resistance to electrical current flow in a conductor.

Oil shale. A shale rock that contains the hydrocarbon kerogen.

Orbit. The path of a satellite around the earth during its periodic revolution.

Overdrive. An arrangement of gears in the transmission in which the output shaft is going faster than the engine (input shaft).

Parallel circuit. An electrical circuit in which there is more than one path for electrical current to flow.

Petrochemicals. A wide variety of chemical products made from petroleum.

Petroleum. A fossil fuel formed in the earth. It is made up of many types of hydrocarbon molecules. Commonly called crude oil.

Photosynthesis. The process by which plants grow—converting the energy of the sun, carbon dioxide, and water into plant tissue.

Piston. A cylindrical part that is fastened to the connecting rod and moves up and down in the cylinder.

Piston displacement. The volume that the piston displaces between TDC and BDC. Displacement is measured in cubic inches, cubic centimeters, or liters.

Piston pin or wrist pin. A sturdy, finely machined steel pin used to connect the piston and the connecting rod.

Piston rings. Cast iron bands which fit into grooves machined into the piston to provide a tight seal between the piston and cylinder walls.

Pneumatics. That branch of fluid power that deals with the use of gases, usually compressed air or nitrogen, to transmit power from one place to another.

Pollution. The waste that makes the environment unclean or foul.

Ports. Openings or holes usually in the cylinder wall.

Potential energy. The energy a body has due to its position, condition, or chemical state.

Power. The rate of doing work or the rate of energy conversion, usually measured in horsepower or watts.

Power stroke. The stroke of the piston in which the pressures of combustion force the piston down the cylinder.

Power technology. A study of energy sources and the machines and devices that are used to convert and apply energy.

Propellant. The fuel and oxidizer that are carried on board a rocket. They may be blended together—a monopropellant; or they may be contained in separate tanks—a bipropellant.

Reciprocation. Motion that is back and forth, such as the motion of a piston in a cylinder.

Reconditioning. The process of repairing or replacing worn parts of an engine to return it to peak performance.

Reed valves. Thin, springy sheet of metal that can be opened by air pressure to admit a fuel mixture. Sometimes called leaf valves.

Rocket engine. An engine that carries its own fuel supply and oxidizer. The operating principle is based on the reaction principle, as is the jet engine.

SAE. Society of Automotive Engineers.

Satellite. A manufactured object orbiting the earth.

Scavenging. To clean out, such as to clean the exhaust out of the cylinder. Usually, air is used.

Semiconductors. Materials that do not have the clear properties of insulators or conductors, but become electrically conductive when they are heated, exposed to light energy, or doped with an additive.

Series circuit. An electrical circuit in which there is only one path for electric current to flow.

Solar cell. A device that converts the energy of the sun directly into electrical energy.

Solar energy. Energy from the sun. Almost all energy sources trace their origin to the sun.

Solar heating. The use of solar energy to heat interior spaces. The system may be active, using solar panels, pumps and/or fans, and heat storage media. The system may be passive by simply storing the heat energy of the day in large collectors such as masonry construction or water tanks. The heat energy is then released naturally at night.

Solid-state ignition. An ignition system that has no moving parts. Its parts are transistors, diodes, and a silicon-controlled rectifier.

Space shuttle. A space craft that is capable of orbiting the earth in space and then returning safely to earth where it is prepared for successive flights.

Spark advance. A device used to cause the spark to occur earlier in the engine's operating cycle. Spark advance is needed most at high speeds.

Spark plug. A part that fits into the cylinder head and extends into the combustion chamber. Its two electrodes are separated by an air gap. When voltage arcs across the gap, the fuel mixture is ignited.

Steam engine. An external combustion engine in which the fuel burned heats water. The water is converted into steam which moves the piston.

Steam turbine. An external combustion engine in

which the fuel burned heats water. The water is converted into steam which is directed against the turbine blades, rotating them.

Stroke. Movement of the piston in a cylinder from top to bottom or from bottom to top.

Supercharger. A blower device that forces air into the cylinders. It permits more fuel to be burned and increases engine power.

Surface mining. A mining technique used for coal and other minerals that is used when the ore is relatively near the earth's surface. Also called strip mining.

Tachometer. A device for measuring the revolutions per minute of a shaft such as a crankshaft.

Tappets. Located between the valve stem and the cams. They transmit the motion of the cam to the valve. They are often referred to as valve lifters.

TDC. Top dead center or the upper limit of the piston's travel.

Throttle. The part of the carburetor that controls the amount of fuel mixture that can enter the engine, thereby controlling the engine's speed.

Thrust. Driving force which pushes a jet or rocket engine forward.

Torque. Effort applied toward twisting or turning.

Transaxle. A transmission device that combines the transmission and axle in one housing.

Transformer. An electrical device with coils of wire that will change voltage and amperage from one circuit to another by induction.

Transmission. A system of gears that can be placed in various combinations that can change the output speed. When speed is increased, torque is reduced.

Troubleshooting. The intelligent, step-by-step process for locating engine troubles. The process is based on knowledge of engine operations and uses observation and tests.

Tuneup. The process of cleaning, adjusting, and testing an engine so it will achieve peak performance.

Turbocharging. A device used to force or blow air into the cylinder of a gasoline engine. It is driven by the engine's exhaust gases.

Turbofan. A jet engine, similar to a turboprop engine, in which large ducted fans replace the propeller.

Turbojet engine. A jet engine that uses the principle of jet propulsion.

Turboprop engine. Essentially a turbojet engine in which the turbine drives the propeller to produce most of the engine's thrust.

Two-stroke cycle. An engine that completes its operating cycle in two strokes of the piston and one revolution of the crankshaft.

Valves. Devices for opening and sealing off a hole, port, or aperture.

Vapor lock. A condition in which the fuel vaporizes in the fuel line restricting its flow to the carburetor. Vapor lock is most likely to occur at high altitudes on hot days.

Venturi. The constricted part of the carburetor where the incoming air speeds up and air pressures are reduced.

Viscosity. The tendency of oil or any liquid to resist flowing. The rate at which a liquid flows.

Volatility. The ability of a liquid to vaporize.

Volt. The electrical measurement for electromotive force which will cause current to flow within a conductor.

Volumetric efficiency. The ability of an engine to take in a full charge of fuel mixture and to discharge the exhaust gases efficiently.

Warranty. A written agreement covering the responsibility of the manufacturer and the owner for the engine's performance during a certain length of time.

Wind turbine. A device that harnesses wind energy—similar to a wind mill—and converts it into electrical energy by driving a generator.

Work. Applying a force to cause motion. Measured in foot-pounds or joules.

Acknowledgments

AC Spark Plug Division, General Motors Corporation: Figures 7-28, 7-29
Aero Jet General Corporation: Figures 28-3, 28-11
Agriculture Canada: Figure 21-2
Allison: Figure 13-18
American Motors Corporation: Figure 9-4
American Petroleum Institute: Figures 18-11, 19-9, 19-10, 21-1, 23-15, 23-16, 24-12, 25-4, 25-5, 25-7, 25-11, 25-12, 26-1, 26-3, 26-5, 26-6, 26-7, 26-8, 26-10, 26-11, 26-12, 27-1, 30-6
Amoco Oil: Figures 4-26, 4-27
Apple Computers Incorporated: Figures 29-1, 29-2, 29-3, 29-4, 29-6, 29-9
Ashland Oil Incorporated: Figure 25-4
Atomic Energy Commission: Figures 23-1, 23-2, 23-3, 23-4, 23-5, 23-6, 23-7, 23-8, 23-9, 23-10, 23-11, 23-12, 24-1, 24-2, 24-3, 24-22, 24-23, 24-25, 24-26, 24-32
AT&T Information Systems: Figure 29-7
AVCO Everett Research Laboratories: Figures 24-34, 24-35
AVCO Lycoming Division: Figures 11-15, 13-19
Basic Designs Incorporated: Figure 18-8
Beaird Poulan/Weed Eater Division, Emerson Electric Company: Figure 4-20
Boston Edison Company: Figure 23-13
Briggs and Stratton Corporation: Figures 3-5, 3-19, 3-20, 3-29, 4-1, 4-4, 4-6, 4-13, 4-14, 4-23, 4-24, 4-34, 4-37, 4-39, 5-9, 5-15, 7-9, 7-19, 7-26, 8-12, A-20, A-32, A-33, A-39, A-42, A-44, A-48, C-1, C-2
The British Petroleum Company: Figure 26-10
Caterpillar Tractor Company: Figures 10-5, 10-6, 10-8, 10-11, 10-13, 10-14, 10-15, 28-7
Champion Spark Plug Company: Figures 4-30, 4-31, 4-32, 4-33, 7-27, 8-3, 8-4, 8-5, 8-6, 8-7
Chevrolet Division, General Motors Corporation: Figures 9-10, 9-17, 9-54, 9-58, 9-59, 13-11, 13-12, 13-13, 13-14, 13-15, 13-16, 17-7
Chrysler Corporation: Figures 9-2, 9-5, 9-9, 9-12, 9-20, 9-38, 9-39, 9-40, 9-49, 9-57, 13-5, 13-6, 13-7, 13-8, 13-9, 13-10
Clinton Engines Corporation: Figures 3-6, 3-18, 3-33, 4-2, 5-4, 8-16, A-7, A-10, A-11, A-12, A-13, A-14, A-17, A-19, A-40, A-46
Deere and Company: Figure 3-30B

Delco-Remy Division, General Motors Corporation: Figures 9-35, 9-36, 9-37, 9-41
Detroit Diesel Engine Division, General Motors Corporation: Figures 10-4, 10-9, 10-12, 10-17
Dominion Engineering Works Limited: Figures 20-1, 20-2, 20-3, 20-4, 20-5, 20-6, 20-7, 20-8
Electric Power Research Institute (EPRI): Figures 11-1, 23-19, 30-2, 30-3, 30-4
Electronics Industry Magazine: Figures 24-14, 24-15
Electronics Magazine: Figures 24-12, 24-17, 24-18, 24-19, 24-20, 24-21, 24-28, 24-29, 24-30, 24-31, 24-33
Environmental Protection Agency (EPA): Figure 25-5
Esso Research and Engineering Company: Figure 24-16
Ethyl Corporation: Figures 3-15, 4-28
Evinrude Motors Division, Outboard Marine Corporation: Figures 8-9, C-19, C-20
Exxon Corporation: Figures 3-23, 3-24, 3-25, 3-26, 5-7, 7-1
Fairbanks-Morse Engine Accessories Operation: Figures 7-8, 7-10, 7-11, 7-13, 7-14, 7-21, 7-23
J. B. Foote Foundry Company: Figure 16-24
Ford Motor Company: Figures 9-7, 9-8, 9-11, 9-13, 9-21, 9-22, 9-23, 9-24, 9-25, 9-26, 9-27, 9-28, 9-34, 9-42, 9-50, 9-51, 9-52, 10-7, 10-20, 13-17, 29-8, 29-10
Ford Tractor Operation, Ford Motor Company: Figure 28-6
General Electric Company: Figures 11-4, 14-16
General Motors Corporation: Figures 4-11, 14-1, 14-2, 14-3, 28-4, 29-9, 29-10, 29-11, 29-12, 29-14
Gulf Oil Corporation: Figures 23-16, 25-11
Institute of Energy Conversion: Figure 18-1
International Harvester Company: Figures 10-16, 10-19
International Telephone and Telegraph Corporation (ITT): Figure 12-1
Jacobsen Manufacturing: Figures 3-1, 8-15, C-13, C-14, C-15, C-16
Johnson Motors Company: Figures 1-4, 3-2, 3-12, 3-13, 4-7, 4-8, 4-9, 4-10, 4-18, 4-29, 5-17, 5-18, 5-19, 6-8, 6-9, 6-10, 6-11, 7-12, 7-22, B-1, B-2, B-3, C-5, C-6, C-7, C-12
Kawasaki Motor Corporation, USA: Figures 6-2, 8-9
Kohler Company: Figures 3-16, 3-17, 3-36, 6-4, 7-30, 7-31, 7-32, 7-33, 8-1, A-2, A-3, A-15, A-24, A-25, A-27, A-28, A-29, A-38, A-41, C-3, C-17, C-18
Lausen Power Products Department, Tecumseh Products Company: Figures 4-38, 5-13
Lincoln-Mercury Division, Ford Motor Company: Figures 3-4, 6-6, 6-7
Los Angeles Offshore Oil Port, Incorporated: Figures 26-1, 26-7
Lovejoy Incorporated: Figure 16-23
McCulloch Corporation: Figures 3-35, 4-15, 4-16, 4-17, 7-18, 8-2, A-1
McDonnell-Douglas: Figures 28-2, 28-8, 28-9, 28-12
Mack Trucks Incorporated: Figure 10-18
Massey-Ferguson: Figures 17-9, 28-13
Susan A. Menchel: Figure 3-37
Mercedes Benz of North America, Inc.: Figures 9-14, 9-16, 10-3, 10-21
Miller Fluid Power: Figures 17-21, 17-33
Motor Vehicle Manufacturers Association of the United States: Figure 9-45

National Aeronautics and Space Administration (NASA): Figures 12-2, 12-3, 12-4, 12-6, 12-7, 12-8, 12-9, 12-12, 12-13, 12-14, 12-15, 12-16, 12-17, 12-18, 12-19, 24-4, 30-1
NASA, Lewis Research Center: Figures 19-1, 19-4, 19-5, 19-6, 19-7
National Coal Association: Figures 25-2, 25-3, 25-6, 25-9, 25-10
National Research Council of Canada: Figure 19-9
New York Stock Exchange: Figure 29-5
Newport News Ship Building: Figure 23-15
Olympic Resource Management: Figure 18-11
OMC Lincoln: Figures 3-31, 4-19
Parker Pneumatic Division: Figures 17-18, 17-29, 17-30
Perfect Circle Valve Division, Dana Corporation: Figures 3-8, 3-9, 3-10, A-43
Philadelphia Electric Company: Figure 15-21
Phillips Petroleum Company: Figure 19-10
Pittsburg Plate Glass: Figures 18-3, 18-9
Pontiac Motor Division, General Motors Corporation: Figures 9-1, 9-3, 9-10, 9-15, 9-18, 9-29, 9-30, 9-31, 9-32, 9-33
Portland Stove Foundry, Incorporated: Figure 22-6
Pratt & Whitney Aircraft Group, United Technologies: Figures 11-3, 11-7, 11-9, 11-10, 11-12, 11-13, 11-14, 11-17, 11-18, 11-19, 11-20, 28-5
Reach-All Manufacturing and Engineering Company: Figure 17-8
Rife Hydraulic Engine Manufacturing Company: Figures 20-9, 20-10, 29-11
Riteway-Dominion Manufacturing, Inc.: Figures 22-7, 22-8
Rocketdyne Division, Rockwell International: Figures 12-5, 12-10, 12-11, 12-21
Senecbaugh Wind Electric: Figure 19-3
Standard Oil Company of California: Figures 23-21, 24-12, 26-8, 26-12, 27-1
Standard Oil Company of Indiana: Figures 22-9, 25-1, 26-2, 26-4, 26-9, 27-2
Sun Oil Company: Figures 17-10, 17-16, 17-20
Joseph Tardi Associates: Figures 2-3, 2-7, 2-9, 2-11, 3-7, 3-14, 4-3, 4-5, 5-16, 8-10, 8-11, A-16, A-18, A-31, A-34, A-35, A-36
Tecumseh Products Company: Figures 3-11, 3-27, 3-32, 3-34, 4-12, 4-22, 4-36, 5-6, 7-24, 8-13, 8-14, 16-19, 16-20, 16-21, 16-24, 16-25, 16-26, A-4, A-5, A-6, A-22, A-23, A-30, C-8, C-9, C-10, C-11
Teledyne Total Power, Figures 4-35, 5-10, 5-11, 6-5, A-26, A-37, A-45, A-47, A-49, A-50, A-51, C-4
Terry Corporation: Figures 14-7, 14-8, 14-9, 14-12, 14-13, 14-14, 14-15
Texaco Incorporated: Figure 25-12
3-M: Figures 24-24, 24-27
Toro Company: Figure 2-10
United States Air Force: Figures 11-4, 11-6, 11-8, 11-11, 11-16, 12-20, 12-22, 12-23, 12-24
U. S. Department of Energy, Office of Scientific and Technical Information: Figures 18-2, 18-6, 18-10, 19-8, 22-1, 22-2, 22-3, 22-4, 23-17, 23-18, 23-20
United States Navy: Figures 17-31, 17-32, 17-34, 17-35, 17-36

U. S. News and World Report: Figure 26-11
Vickers, Incorporated: Figures 17-1, 17-2, 17-3, 17-4, 17-5, 17-6, 17-11, 17-12, 17-13, 17-14, 17-15, 17-24
Waukesha Motor Company: Figure 16-13
Westinghouse Electric Corporation, Power Generation Group: Figures 13-20, 13-21, 13-22, 18-4, 18-5, 28-10
Wico Electric Company, Figures 7-16, 7-17, 7-20
J. Wiley & Sons, Incorporated: Figures 14-4, 14-5, 14-11
Wisconsin Power and Light Company: Figure 25-8
Womack Machine Supply Company: Figures 17-17, 17-19, 17-22, 17-23, 17-37, 17-38
Zenith Carburetor Division, Bendix Corporation: Figures 4-21, 4-25

INDEX